CASE STUDIES IN ECOHEALTH

CASE STUDIES IN ECOHEALTH

Examining the Interaction between Animals and their Environment

Edited by

Susan C. Cork

Douglas P. Whiteside

First published 2024

Published by
5M Books Ltd
Lings, Great Easton
Essex CM6 2HH, UK
Tel: +44 (0)330 1333 580
www.5mbooks.com

Follow us on
Twitter @5m_Books
Instagram 5m_books
Facebook @5mBooks
LinkedIn @5mbooks

A Catalogue record for this book is available from the British Library

ISBN 9781789182408
eISBN 9781789183313
DOI 10.52517/9781789183313

Book layout by Cheshire Typesetting Ltd, Cuddington, Cheshire
Printed by CPI Antony Rowe Ltd, UK
Photos and illustrations by the authors unless otherwise indicated

Contents

Contributors

Laura A. Adamovicz College of Veterinary Medicine, University of Illinois, USA

Matthew C. Allender College of Veterinary Medicine, University of Illinois, USA

Colleen Arnison Faculty of Environmental Design, University of Calgary, Alberta, Canada

Kathyrn E. Arnold York Graduate Research School, University of York, YO10 5DD, UK

Rachel Bolus Department of Biology, Southern Utah University, USA

Roisin Campbell-Palmer Beaver Trust, 61 Bridge Street, Kington, UK

Louise Caplan University of Calgary, Faculty of Veterinary Medicine, Calgary, Alberta, Canada

Aurelie Castinel Ministry for Primary Industries, Wellington, New Zealand

Sylvia Checkley University of Calgary, Faculty of Veterinary Medicine, Calgary, Alberta, Canada

Bruce Chessman Chessman Ecology, Pymble, NSW, Australia 2073

Susan C. Cork University of Calgary Faculty of Veterinary Medicine, Calgary, Alberta, Canada

Isabelle Couloigner University of Calgary Faculty of Veterinary Medicine, Calgary, Alberta, Canada; Department of Geography, University of Calgary

Pádraig Duignan The Marine Mammal Centre, Sausalito, California, USA

Cara Field The Marine Mammal Centre, Sausalito, California, USA

Melissa Giese Department of Planning and the Environment, Coffs Harbour, NSW, Australia 2450

Simon J. Girling Royal Zoological Society of Scotland, 134 Corstorphine Road, Edinburgh, Scotland, UK

Jacqualine Grant Department of Geosciences, Southern Utah University, USA

Ratna B. Gurung Dzongdag, Dzongkhag Administration, Bumthang, Bhutan

David C. Hall University of Calgary Faculty of Veterinary Medicine, Calgary, Alberta, Canada

Allison C. Hanes One Health Productions, Brooklyn, New York, NY, USA; Anthropology Centre for Conservation, Environment and Development, Department of Social Sciences, Oxford Brookes University, Oxford, UK

Ashlee Hardin Department of Biology, Southern Utah University, USA

Kimberley Hockings Centre for Ecology and Conservation, Faculty of Environment, Science and Economy, University of Exeter, Penryn, TR10 9FE, UK

Ian Hogg Canadian High Arctic Research Station, Polar Knowledge Canada, Cambridge Bay, Nunavut, Canada; School of Science, University of Waikato, Hamilton, New Zealand

Richard Jakob-Hoff IUCN-SSC Conservation Planning Specialist Group (Australasia), Auckland, New Zealand

Gladys Kalema-Zikusoka Conservation Through Public Health, Uganda

Regina M. Krohn University of Calgary Faculty of Veterinary Medicine, Calgary, Alberta, Canada

Karen Landman School of Environmental Design and Rural Development, University of Guelph, Guelph, Ontario, Canada

Henry Lane National Institute of Water and Atmospheric Research, New Zealand

Caroline Lees IUCN-SSC Conservation Planning Specialist Group, Apple Valley, MN, USA 55124

Zachary G. MacDonald University of California Los Angeles (UCLA), La Kretz Center for California Conservation Science, Institute of the Environment and Sustainability, Los Angeles, California, USA

Alessandro Massolo Ethology Unit, Department of Biological Sciences, University of Pisa, Pisa, Italy

Michael McFadden Taronga Conservation Society Australia, Sydney, Australia, 2088

Gerry McGilvray Department of Planning and the Environment, Coffs Harbour, NSW, Australia

Oana Catalina Moldoveanu Department of Biological Sciences, Università degli Studi di Firenze, Florence, Italy

V.R. Monger Highland Development Project, Bhutan, South Asia

Emiliano Mori Consiglio Nazionale delle Ricerche, Istituto di Ricerca sugli Ecosistemi Terrestri, Sesto Fiorentino (FI), Italy

Marco Musiani Dipartimento di Scienze Biologiche, Geologiche e Ambientali (BiGeA). Università di Bologna, Italy

Jamyang Namgyal University of Calgary, Faculty of Veterinary Medicine, Alberta, Canada

Matt Ogburn Department of Biology, Southern Utah University, USA

Jacob Olvera Department of Biology, Southern Utah University, USA

Luca Petroni Department of Biological Sciences, University of Pisa, Pisa, Italy

Kyle Plotsky University of Calgary, Faculty of Veterinary Medicine, Calgary, Alberta, Canada

Julie Pynn Department of Psychology, Southern Utah University, USA

Sangay Rinchen National Centre for Animal Health, Serbithang, Bhutan

Karrie Rose Australian Registry of Wildlife Health, Taronga Conservation Society Australia, Sydney, Australia, 2088

Shane Ruming Department of Planning and the Environment, Coffs Harbour, NSW, Australia

Joy Sammy Joy Sammy Consulting, Guelph, Ontario, Canada

Danna M. Schock Palustris Environmental, Athabasca, Alberta, Canada

H. Bradley Shaffer University of California Los Angeles (UCLA), Department of Ecology and Evolutionary Biology, and La Kretz Center for California Conservation Science, Institute of the Environment and Sustainability, Los Angeles, California, USA

Donna J. Sheppard Wilder Institute/Calgary Zoo, Calgary, Alberta, Canada; Rhino Ark Kenya Charitable Trust, Kenya

Adam Skidmore Taronga Conservation Society Australia, Sydney, Australia

Claire Smith Department of Biology, Southern Utah University, USA

Connor Smith Department of Biology, Southern Utah University, USA

Judit E.G. Smits University of Calgary, Faculty of Veterinary Medicine, Alberta, Canada

Issac Sorensen Department of Biology, Southern Utah University, USA

Ricky Spencer 1Million Turtles, School of Science, Western Sydney University, Penrith, NSW, Australia

Felix A. H. Sperling University of Alberta, Edmonton, Canada

Wlodek L. Stanislawek Ministry for Primary Industries, Animal Health Laboratory, Upper Hutt, New Zealand

Nancy J. Stevens Ohio University Distinguished Professor, Department of Biomedical Sciences Heritage College of Osteopathic Medicine, Ohio University, Athens, Ohio, USA and Museum of Natural History/Department of Anthropology, University of Colorado, USA

Tenzin Tenzin World Organization for Animal Health, Botswana

Matilde Tomaselli Canadian High Arctic Research Station, Polar Knowledge Canada, Cambridge Bay, Nunavut, Canada

Sareeha Vasanthakumar Water Engineering and Management, School of Engineering and Technology, Asian Institute of Technology, Pathum Thani, Thailand

Samuel Wells Department of Biology, Southern Utah University, USA

Douglas P. Whiteside University of Calgary Faculty of Veterinary Medicine, Calgary, Alberta, Canada; Wilder Institute/Calgary Zoo, Calgary, Alberta, Canada

Jessica Wu University of Calgary, Faculty of Veterinary Medicine, Alberta, Canada

Leeloo Yutac Department of Biology, Southern Utah University, USA

Foreword

This foreword is being written in the post-COVID-19 period. As the reader will likely know, sometime during 2019, the SARS-CoV2 virus, originally infecting bats, jumped host, and – probably via another animal – started to infect humans. A pandemic followed, and by the time it ran its course, an estimated three million people had died. There were lockdowns, vaccination campaigns, and an unprecedented degree of misinformation circulated. This is not the first nor the last such event, in fact such host switching to humans is becoming more and more frequent as we encroach into wildlife habitat. While we realised that various environmental factors underpin, modify or determine the outcome of disease, little has been done to amend our habits. Nonetheless, the realization that we need to consider our interaction with nature more is correct. Even the calls for re-thinking our relationship with nature no longer seem excessive or outlandish. Some guidance towards more acceptable practices can be found in traditional uses of natural resources, when the desire to lord over nature may have been present but our level of technology made it patently ridiculous. So, 'sustainable' practices were followed. In other words, nature set the rules, not us. There is wisdom in that attitude. The chapters in this book reinforce that general message with manyfold examples, from the New World Arctic to the Himalayas and the southern seas.

The terms mentioned include One Health, Ecosystem Health and EcoHealth. Except the first one, they are not yet familiar or commonly used by most scientists. Given that numerous pathogens, originally of animal origin, from bubonic plague to SARS-CoV2, have killed uncounted numbers of humans, especially since we started to domesticate wild animals, the realization that (mostly domestic) animal health and human health should not be considered separately is not really a 'difficult-to-understand' idea. An important and new factor in all this is globalization. Globalization relates to the intensive cross-continental transport of goods and this is without precedent in human history. So is the massive human mobility – so poignantly illustrated by the lightning-fast spread of emerging diseases world-wide. Yes, throughout history humans have always travelled and transported goods from one end of the world to another (mostly to Europe from elsewhere) but now we have an unprecedented increase in these activities, creating a qualitatively different landscape of conditions. Literally, a pathogen can be carried from one end of the world to another within 24 hours. This all may be best captured by Phil Hulme's recently suggested concept of 'One Biosecurity'.[1] Invasive organisms, pathogens of wild plants, of crops, of animals and of humans all use the multiple pathways we created for them. Yes, we created them, even if without intent or just being thoughtless, careless or inattentive. Once the undesirable consequences manifested themselves, we introduced various mechanisms and laws to lessen the harm. These means are not useless, but they are not perfect, either. No matter what we do, we will not be able to prevent unexpected, sudden spread and appearance of new species in new places. So, thinking about these as components of 'one biosecurity' is quite logical.

It also started to dawn on us that a 'healthy' ecosystem has inherent diversity. We may not yet precisely know what the criteria of a healthy ecosystem are – but as we continue chipping away at biodiversity, obvious signs of non-health started to appear. Humans have already exterminated much of the former megafauna on various continents, as well as on the mega-islands of Madagascar and New Zealand. Defaunation is still ongoing. Large predators are globally threatened with extinction. Arthropods started to decline. Pollinators are becoming scarce in places. Apart from utilitarian consequences that the ecosystem services concept tries to capture, the human experience of a full, satisfying life is what is in danger if this continues.

All efforts point in the same direction, even though the vision is not yet clear. It is like a dawn: we see the coming of light but are unsure of its details, and do not yet see clearly. Several of the chapters in this book encompass multiple fields and multiple locations, and may not initially seem to hang together. Nonetheless, they all instruct us to move in one direction: we need to take the health of our environment as well as those of other living beings more into account. We are not independent of nature, and as Richard Feynman so aptly reminded us: 'Nature cannot be fooled'.[2] It remains to be seen if we can move fast enough, and the necessary adjustments are substantial enough.

Gábor L. Lövei
Aarhus University, Denmark

Endnotes

1 Hulme, P. E. (2020). One biosecurity: A unified concept to integrate human, animal, plant and environmental health. *Emerging Topics in Life Sciences*, 4(5), 539–549. https://doi.org/10.1042/ETLS20200067

2 Report of the PRESIDENTIAL COMMISSION on the Space Shuttle Challenger Accident, Volume 2: Appendix F – Personal Observations on Reliability of Shuttle. https://history.nasa.gov/rogersrep/v2appf.htm.

Preface

The term 'EcoHealth' (also known as Eco-Health) has sometimes been used interchangeably with other terms such as 'Ecosystem Approaches to Health' (EAH) or 'One Health' although it is recognized that there are differences in how these approaches are developed and used. In this book we refer to EcoHealth as a field of research and practice that uses sustainable interdisciplinary and transdisciplinary approaches to promote the health and wellness of animals (including humans) and ecosystems. As a field of study, 'EcoHealth' draws on the natural sciences, conservation medicine, and social sciences and encourages the engagement of communities and policy makers to address complex problems at the interface of ecosystems, animals and society (see Chapter 1).

To illustrate the EcoHealth approach we have engaged a broad range of EcoHealth practitioners to prepare and present a number of case studies from around the world describing practical approaches to addressing current Ecohealth management challenges. All of these case studies illustrate the value of working across health disciplines while also drawing on other areas of technical expertise. We venture from the African jungles to the seas around New Zealand, to the high Himalayas, the North American Prairies and the Arctic. We consider the case of the disappearing pollinators in Utah, the impact of climate change on Arctic ecosystems and on the life of yak herders in Bhutan. We explore innovative responses to turtle conservation in Australia and sustainable forest management in Canada, as well as how to deal with disease risks to endangered gorilla populations in Uganda, and how to raise awareness about the impact of toxins in the environment. We have grouped the chapters into one of four themes reflecting the focus of the case studies: Conservation medicine; Climate change and land use; Wetlands, oceans and aquaculture; and Outreach, research and education. These four principal themes capture key characteristics of the EcoHealth approach and we have used them to provide a structure for grouping the case studies presented. At the same time, there are also several cross-cutting features that arise in several of the case studies outlined in the book, for example; the importance of implementing the principles of biosecurity, the value of knowledge translation and citizen science, the impact of research in policy formulation and how to engage local communities in wildlife conservation.

We developed this book to appeal to a broad audience, although the academic content is largely aimed at undergraduate and graduate university students with some knowledge of ecology and biological sciences. We hope that the broad scope of the case studies included, and the range of viewpoints presented, will also be of interest to academics, policymakers and 'EcoHealth' practitioners.

Susan C. Cork and Douglas P. Whiteside

Chapter 1

Introduction to EcoHealth, One Health and ecosystem health

Susan C. Cork and Douglas P. Whiteside

1.1 What is 'EcoHealth'?

The terminology surrounding 'EcoHealth' (sometimes Eco-Health) can be a bit confusing for anyone new to the subject. This is partially due to the fact that other terms such as 'One Health' and 'Ecosystem Approaches to Health' (EAH) are often used interchangeably with the term 'EcoHealth'. However, there are subtle differences between these approaches as the terms can imply a different scope of activities and may engage different stakeholders. These terms also have different origins. It is recognized though that not everyone agrees on a single definition for EcoHealth (Harrison et al., 2019; Lerner and Berg, 2017). In this introductory chapter we will examine ways in which the term EcoHealth is used and how it has evolved.

As a general rule, EcoHealth involves research and practice(s) to promote the sustainability of human populations, and animal and environmental biodiversity, by considering the complex interaction of ecosystems, socio-cultural, political and economic factors (Bergquist et al., 2017; Harrison et al., 2019). EcoHealth, and EAH, tend to identify 'sustainability' as the central focus of global health issues as compared with international 'sustainable development goals', which have a primary economic focus (Charron, 2012a, b). EcoHealth

approaches also consider disease dynamics in the broader context of sustainable agriculture and food security, socioeconomic development and environmental protection. In comparison, One Health approaches often, but not always, have a focus on human health with animal health and environmental concerns viewed from an anthropogenic perspective (Harrison et al., 2019; Rüegg et al., 2019).

EAH were first mentioned in public health literature in the 1990s (Webb et al., 2010) but there is evidence of the use of similar integrated approaches well before that time. For example, the Greek physician Hippocrates suggested that human health was intricately linked to the health of the environment around 2600 years ago. Webb et al. (2010), have also suggested that EAH have been strengthened through the emerging field of EcoHealth. In a similar context, Pham et al. (2018) describe EcoHealth as a transdisciplinary research approach that considers cultural, socioeconomic and environmental factors. They also suggest that there is a need to better understand how EcoHealth approaches can be integrated, and adapted in practice to support the sustainability of this broad applied approach. In general, both the EcoHealth approach and the One Health approach recognize the importance of transdisciplinary and interdisciplinary research and practice(s), stakeholder engagement and knowledge exchange within a wider ecological context (Harrison et al., 2019).

Susan C. Cork and Douglas P. Whiteside (eds) **Case Studies in EcoHealth**
DOI: 10.52517/9781789183368.001, © 5m Books Ltd, 2024

EcoHealth definition

EcoHealth (in this book) is a field of research and practice that uses sustainable interdisciplinary and transdisciplinary approaches to promote the health and wellness of animals (including humans) and ecosystems. As a field of study, EcoHealth draws on the natural sciences, conservation medicine and social sciences, and encourages the engagement of communities and policy makers to address complex problems at the interface of ecosystems, animals and society.

Figure 1.1 illustrates the varied roles provided by bats as ecosystem service providers. Although we know that they can be reservoirs for many potentially zoonotic diseases their presence is also important for the health of our shared environment.

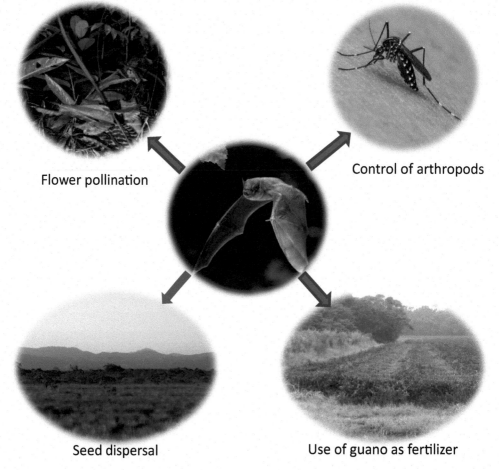

Flower pollination

Control of arthropods

Seed dispersal

Use of guano as fertilizer

Figure 1.1 Bats as ecosystem service providers. Adapted from Soliman and Emam (2022).

It is well recognized that transdisciplinarity is fundamental in both One Health and EcoHealth approaches to addressing complex societal issues. When done well, transdisciplinarity facilitates consideration of multiple perspectives, ensures a high level of community engagement and promotes enhanced opportunities for collaboration between diverse stakeholders (Harrison et al., 2019; Parkes et al., 2012; Saint-Charles et al., 2014). These transdisciplinary approaches can promote a genuinely collective endeavour providing a mechanism for mutual respect, knowledge integration and broad engagement (Parkes, 2012). Success, however, does require good leadership and a shared vision. Max-Neef (2005) differentiates between disciplinarity, multidisciplinarity, interdisciplinarity and transdisciplinarity. Essentially, he suggests that adopting a transdisciplinary approach facilitates an advanced perspective beyond 'linear logic' and a limited perspective of 'reality'.

There is recognition that integrated approaches to human, animal and environmental health can be adequately described using different terms and that all of these terms can be used when tackling complex health challenges (Assmuth et al., 2020). One Health approaches can address challenges at the interface between people, animals and their environments but human health often (but not always) remains the core priority. In contrast, EcoHealth approaches tend to take a broader perspective of health and well-being with more focus on natural resources and the environment in a socioeconomic, cultural and political context (Harrison et al., 2019). Both One Health and EcoHealth approaches require transdisciplinary collaboration among multiple stakeholders from different sectors. A key element of the One Health approach is the integration of knowledge across sectors, disciplines and stakeholders (Cork et al., 2015). These can focus on a local problem but can also be applied at a national or global scale where health innovation can be shared as a result of successful knowledge integration across regional and national boundaries.

1.2 So, what is the difference between EcoHealth and One Health?

From the literature it can be seen that the terms EcoHealth and One Health have frequently been used interchangeably and to some extent this depends on who is using the term. Roger et al. (2016) wrote an article with the title 'Same wine in different bottles' and in it they suggest that both the One Health and EcoHealth approaches have historically struggled to sufficiently distinguish between themselves and they acknowledge the apparent similarities regarding core objectives. However, from examples discussed in the current literature it is evident that health professionals and researchers tend to regard the two approaches differently, even though they have overlapping scopes of practice

(Harrison et al., 2019). For many academics, both approaches can be applied to develop inter- and transdisciplinary frameworks for research activities. This is especially true for research addressing zoonotic diseases. However, EcoHealth approaches are perhaps more widely applied to projects with a focus on environmental conservation, climate change adaptation and biodiversity.

Barrett and Bouley (2015) suggest that EcoHealth approaches have been progressively converging with the One Health approach and some commentators suggest that there may be a good case to combine the two approaches (Harrison et al., 2019; Nguyen-Viet et al., 2015). For health professionals, and some decision makers, both approaches have been used to promote official cooperation among different sectors and stakeholders to address public health issues in low- and middle-income countries, especially in Africa and Southeast Asia. However, it is also recognized that a better distinction between the two approaches is needed because different sets of stakeholders are needed for broader environmental issues than those more typical for issues focused on human health and disease.

Lapinski et al. (2015) emphasized the lack of certain key components in some examples of One Health approaches, for example, social sciences are often marginalized. The wildlife component, and associated ecological perspectives, are also frequently neglected in One Health examples (Barrett & Bouley, 2015; Rostal et al., 2013). Both approaches should engage multiple stakeholders, include core interdisciplinary concepts, comprise interconnected scientific elements and consider complex socio-political drivers.

In addition to the above, the concept of One Health has also been used to examine biomedical questions, mostly with an emphasis on zoonoses and the study of animals to understand human disease. In contrast, the EcoHealth concept has been predominantly considered

as an ecosystem approach to health, focusing on broader environmental and socioeconomic and cultural issues. It values traditional and non-expert knowledge, and emphasizes community empowerment and capacity-building (Charron, 2012b). It is significant that the EcoHealth approach was initially designed by disease ecologists working in the field of biodiversity conservation (Roger et al., 2016) and it evolved from the amalgamation of different fields including conservation medicine, disease ecology, and a framework developed by the Millennium Ecosystem Assessment (2005).

Although the field of One Health is not new (Hall and Cork, 2016) it has more recently evolved into common language and has been widely adopted by research funders and global institutions. In contrast, the field of EcoHealth continues to function at a more practical level with a focus on a shared environment at the local, regional, national or international level. Both approaches aim to deal with the complexity of imbalance (dis-ease) and health but the emphasis differs (Bunch et al., 2015). It is noted that the One Health approach is currently used and promoted by a number of international standards agencies (e.g. FAO, UNEP, WHO, WOAH) but that many One Health case studies have a human-centric focus. Some commentators remain concerned that many high profiled examples of the One Health approach do not take sufficient notice of environmental factors and argue that the EcoHealth approach generally takes a broader view of 'health' and links public health and public well-being to ecosystem approaches to natural resource management (Harrison et al., 2019; Lerner and Berg, 2017). They suggest that EcoHealth approaches generally include a more robust examination of the environmental aspects of disease ecology. It is suggested that developing effective and sustainable solutions for the prevention and control of zoonotic disease (e.g. rabies, tuberculosis, avian influenza, vector-borne diseases, etc.) and other health issues (e.g. food and water security, and natural resource management) requires an integrated interdisciplinary approach engaging a broad range of stakeholders working at the human–animal–environment interface (Rupprecht and Burgess, 2015) and that environmental and socioeconomic factors must be considered. It is also becoming evident that we should take account of a broader context with regard to climate change, land use and changing regional and global demographics. As a result of the above considerations we will use the term EcoHealth in this book.

> One Health is a collaborative, multisectoral, and transdisciplinary approach — working at the local, regional, national, and global levels — with the goal of achieving optimal health outcomes recognizing the interconnection between people, animals, plants, and their shared environment.
>
> https://www.cdc.gov/onehealth/basics/index.html (see also OIE-WHO-FAO-UNEP statement)

The above definition aligns well with what we mean in this book by EcoHealth. Although many of our examples do not focus on human health we take the view that 'health' includes 'wellness' and that healthy ecosystems are essential for 'wellness' in humans and other animals.

In this book we have engaged talented experts from around the globe with the aim of providing contemporary and historic examples of environmental and animal health issues that have been studied and addressed by transdisciplinary teams. We also explore how some of these teams have been able to translate science into policy.[1] The scope of the book includes examples of environmental and anthropogenic factors, including land use and climate change,

that impact ecosystems with subsequent effects on wild and domestic animal health. We also examine the direct impacts of aquaculture, forest management, and medical and agricultural practices on human, animal and environmental health. The book is aimed at undergraduate and graduate university students with some knowledge of ecology and biological sciences and is also intended to be inclusive and engaging for a more general audience. The broad scope of viewpoints provided will also be of interest to academics, policy makers and EcoHealth practitioners. The key themes explored in the book include:

1. Conservation medicine
2. Climate change and land use
3. Wetlands, oceans and aquaculture
4. Outreach, research and education.

The case studies presented within each theme have been selected to represent topics such as wildlife conservation, wild and domestic animal health, sustainable agriculture, aquaculture, climate change, forest management, globalization and biosecurity from a scientific and traditional knowledge approach.

1.3 Conclusion

In summary, EcoHealth approaches aim to develop sustainable and environmentally friendly health and conservation interventions to tackle a wide range of complex issues at the human–animal–environmental interface. Given the above points we wanted to present this book of case studies to illustrate how the EcoHealth approach can be applied to environmental and conservation challenges that may not always have human health as the core component.

Endnote

1 In some chapters the contributing authors have used the term One Health to explain their approach to address problems with a specific human health and wellness component.

References

Assmuth, T., Chen, X., Degeling, C., Haahtela, T., Irvine, K. N., Keune, H., Kock, R., Rantala, S., Rüegg, S., & Vikström, S. (2020). Integrative concepts and practices of health in transdisciplinary social ecology. *Socio-Ecological Practice Research*, 2(1), 71–90. https://doi.org/10.1007/s42532-019-00038-y

Barrett, M. A., & Bouley, T. A. (2015)_9. Need for enhanced environmental representation in the implementation of one health. *EcoHealth*, 12(2), 212–219. https://doi.org/10.1007/s10393-014-0964-5

Bergquist, R., Brattig, N. W., Chimbari, M. J., Zinsstag, J., & Utzinger, J. (2017). Ecohealth research in Africa: Where from—Where to? *Acta Tropica*, 175, 1–8. https://doi.org/10.1016/j.actatropica.2017.07.015

Bunch, M. J., & Waltner-Toews, D. (2015). Grappling with complexity: The context for One Health and the Eco-Health approach. In J. Zinsstag, E. Schelling, D. Waltner-Toews, M. Whittaker & M. Tanner (Eds.), *One Health: The theory and practice of integrated health approaches*. CABI Publishing.

Charron, D. F. (2012a). Eco-health: Origins and approaches. In D. F. Charron (Ed.), *Eco-health research in practice: Innovative applications of an ecosystem approach to health*. International Development Research Centre.

Charron, D. F. (Ed.). (2012b). Ecohealth research in practice. In *Eco-Health Research in Practice: Innovative Applications of an Ecosystem Approach to Health* (pp. 255–271). Springer.

Cork, S. C., Geale, D. W., & Hall, D. C. (2015). One health in policy development: An integrated approach to translating science into policy. In J. Zinstag, E. Schelling, D. Waltner-Toews, M. Whittaker & M. Tanner (Eds.), *One health: The theory and practice of integrated health approaches* (pp. 304–316). CABI International. https://doi.org/10.1079/9781780643410.0304

Hall, D. C., & Cork, S. C. (2016). One health, introduction. In S. C. Cork, D. C. Hall, & K. Liljebjelke (Eds.) *One health case studies: addressing complex problems in a changing world*. 5m Publishing.

Harrison, S., Kivuti-Bitok, L., Macmillan, A., & Priest, P. (2019). EcoHealth and One Health: A theory-focused review in response to calls for convergence. *Environment International*, 132, 105058. https://doi.org/10.1016/j.envint.2019.105058

Lapinski, M. K., Funk, J. A., & Moccia, L. T. (2015)_60. Recommendations for the role of social science research in one health. *Social Science and Medicine*, 129, 51–60. https://doi.org/10.1016/j.socscimed.2014.09.048

Lerner, H., & Berg, C. (2017). A comparison of three holistic approaches to health: One health, ecohealth, and planetary health. *Frontiers in Veterinary Science*, 4, 163. https://doi.org/10.3389/fvets.2017.00163

Max-Neef, M. A. (2005). Foundations of transdisciplinarity. *Ecological Economics*, 53(1), 5–16. https://doi.org/10.1016/j.ecolecon.2005.01.014

Millennium Ecosystem Assessment. (2005). *Ecosystems and human well-being: Synthesis report*. Island Press. https://www.millenniumassessment.org/documents/document.356.aspx.pdf.

Nguyen-Viet, H., Doria, S., Tung, D. X., Mallee, H., Wilcox, B. A., & Grace, D. (2015). Ecohealth research in Southeast Asia: Past, present and the way forward. *Infectious Diseases of Poverty*, 4, 5. https://doi.org/10.1186/2049-9957-4-5

Parkes, M. W., Charron, D. F., & Sanchez, A. (2012). Better together: Field-building networks at the frontiers of eco-health research. In D. F. Charron (Ed.) *Eco-health research in practice: Innovation applications of an ecosystem approach to health, insight and innovation in international development*. International Development Research Centre.

Parkes, M. (2012). Diversity, emergence, resilience: Guides for a new generation of Eco-Health research and practice. *Eco-Health*, 8, 137–139.

Pham, G., Lam, S., Dinh-Xuan, T., & Nguyen-Viet, H. (2018). Evaluation of an eco-health approach to public health intervention in Ha Nam, Vietnam. *Journal of Public Health Management and Practice,*:24 (Suppl. 2), S36–S43. https://doi.org/10.1097/PHH.0000000000000732

Roger, F., Caron, A., Morand, S., Pedrono, M., de Garine-Wichatitsky, M., Chevalier, V., Tran, A., Gaidet, N., Figuié, M., de Visscher, M. N., & Binot, A. (2016). One Health and Eco-Health: The same wine in different bottles? *Infection Ecology and Epidemiology, 6(1). Published on line, 30978. http://doi.org/10.3402/iee.v6.30978*

Rostal, M. K., Olival, K. J., Loh, E. H., & Karesh, W. B. (2013). Wildlife: The need to better understand the linkages. *Current Topics in Microbiology and Immunology, 365,* 101–125. https://doi.org/10.1007/82_2012_271

Rüegg, S. R., Buttigieg, S. C., Goutard, F. L., Binot, A., Morand, S., Thys, S., & Keune, H. (2019). Editorial: Concepts and experiences in framing, integration and evaluation of One Health and EcoHealth. *Frontiers in Veterinary Science, 6,* 155. https://doi.org/10.3389/fvets.2019.00155

Rupprecht, C. E., & Burgess, G. W. (2015). Viral and vector zoonotic exploitation of a homo-sociome memetic complex. *Clinical Microbiology and Infection, 21*(5), 394–403. https://doi.org/10.1016/j.cmi.2015.02.032

Saint-Charles, J., Webb, J., Sanchez, A., Mallee, H., van Wendel de Joode, B. W., & Nguyen-Viet, H. (2014). Eco-Health as a field: Looking forward. *EcoHealth, 11*(3), 300–307. https://doi.org/10.1007/s10393-014-0930-2

Soliman, K. M., & Emam, W. W. (2022). Bats and ecosystem management. In H. Mikkola (Ed.), *Bats, Disease-prone but beneficial.* IntechOpen. https://doi.org/10.5772/intechopen.101600

Webb, J. C., Mergler, D., Parkes, M. W., Saint-Charles, J., Spiegel, J., Waltner-Toews, D., Yassi, A., & Woollard, R. F. (2010). Tools for thoughtful action: The role of ecosystem approaches to health and enhancing public health. *Canadian Journal of Public Health, 101*(6), 439–441. https://doi.org/10.1007/BF03403959

World Health Organization (1986). Ottawa Charter for Health Promotion.

PART I

CONSERVATION MEDICINE

Chapter 2

EcoHealth, conservation medicine, and the role of the modern zoo

Douglas P. Whiteside

Abstract

The ongoing global loss of biodiversity requires transformative changes to the relationship between nature and people to address it. With more than 700 million visitors visiting over 10,000 zoos and aquariums each year, the ability of the modern zoological global network to engage with the public, and to inspire conservation actions beyond their institutional walls, is unparalleled. Modern zoos and aquariums play a vital role in the conservation of wild species and their habitats, by contributing significant resources to support in situ conservation, leading the preservation and recovery of species through ex situ conservation breeding programs, and raising awareness of call-to-action conservation and environmental issues with visitors to the zoos and the general public. They are committed to advancing animal health and welfare and also serve to enhance human wellbeing. Increasingly, modern zoos and aquariums are important collaborators in transdisciplinary disciplines of EcoHealth and conservation medicine. This chapter focuses on the important ways that modern zoos and aquariums contribute to EcoHealth and the conservation medicine movement through conservation initiatives, diverse research programs, comparative medicine, capacity building and education, and a commitment to urban environmental health.

With a burgeoning population that has surpassed 7.9 billion people, the ever-increasing negative impacts of human activity on global biodiversity, such as habitat degradation, destruction, and fragmentation, wildlife trafficking, and climate change, has resulted in the current epoch being labelled the Anthropocene. Over the past 50 years, there has been a 69% decline in studied species populations around the world (WWF, 2022). Free-ranging wildlife populations subsist on shrinking habitat 'islands', often restricted and intensively managed on often inadequate conservation budgets, in the context of human expansion and rapid ecological change (Bottrill et al., 2008; Sulzner et al., 2021; Walzer, 2017). This alarming loss of biodiversity calls for novel interdisciplinary approaches to develop effective strategies for sustainable

Susan C. Cork and Douglas P. Whiteside (eds) **Case Studies in EcoHealth**
DOI: 10.52517/9781789183368.002, © 5m Books Ltd, 2024

management of populations, species, and entire ecosystems (Corbyn, 2022). Transformative changes in conservation are needed to combat declining biodiversity, with a focus on human populations and nature co-existing in dynamic configurations in natural, rural, and urban spaces (Massarella et al., 2021; Romanelli et al., 2014). To accomplish this, there needs to be a shift in people's attitudes and behaviour by raising awareness of the importance of biodiversity and re-framing human–nature relationships through positive experiences with animals and nature (Deem, 2015; Massarella et al., 2021; Rose and Riley, 2022).

Each year, with more than 700 million people visiting over 10,000 zoos and aquariums, the ability of the modern zoological global network to engage with the public, and to inspire conservation actions beyond their institutional walls, is unparalleled (Barongi et al., 2015; Gussett and Dick, 2011; Rose and Riley, 2022; Zordan, 2021). Modern zoos and aquariums (hereafter 'zoos') are often well respected and influential sources of advice and knowledge within their community (see Box 2.1). These zoos have embraced a core philosophy of actively leading, and collaboratively contributing to, the conservation of biodiversity, continuously improving animal welfare or 'wellbeing', engaging in conservation focused research, and facilitating transgenerational environmental education. On an annual basis, accredited zoos and aquariums spend more than US$350 million on in situ conservation projects that focus on habitat protection and reducing threats to species (Barongi et al., 2015). In addition, many modern zoos also have integrated botanical gardens to conserve and educate the public about habitats and flora. By modelling environmentally responsible actions, such as using sustainable energy sources, conserving water, and considering their carbon footprint, zoos serve to promote planetary wellbeing and the associated positive impacts on wildlife, humans, and the environment. The

---

Box 2.1 Defining the modern zoo and aquarium

Modern zoos and aquariums are conservation-based organizations that are accredited through regional or national associations with robust accreditation standards that evaluate animal health and welfare, biodiversity conservation initiatives and funding, research and education, and green practices. Examples of these associations in North America are the Association of Zoos and Aquariums (AZA) and Canada's Accredited Zoos and Aquariums (CAZA), while other associations around the globe include the European Association of Zoos and Aquariums (EAZA), the Zoo and Aquarium Association Australasia (ZAA), and the World Association of Zoos and Aquariums (WAZA).

---

overarching aim of modern zoos globally is to increase conservation-mindedness and conservation related behaviours in society (Howell et al., 2019; Padda et al., 2018; Rose and Riley, 2022; Routman et al., 2022).

Recent decades have seen escalating challenges for biodiversity conservation and public health, and zoos have made significant changes to address these. One of the most significant shifts that has occurred in modern zoos is moving beyond ensuring good animal welfare and the conservation of the species under managed care to the adoption of a more holistic and integrative EcoHealth approach that encompasses animals, humans, and ecosystem health (Braverman, 2021; Padda et al., 2018; Robinette et al., 2017; Sulzner et al., 2021). This approach recognizes that the health and wellbeing of animals and humans is directly related to the health of their environment in a complex and multifaceted socio-ecological context (Harrison et al., 2019; Romanelli et al., 2014; Schmeller, 2022).

Box 2.2 What is conservation medicine?

Conservation medicine as a term first appeared in the 1990s, highlighting the important linkages between conservation biology, wildlife health, veterinary science, public health, and epidemiology, to better understand and manage conservation efforts and disease challenges. Although many definitions of conservation medicine exist, the most consistent theme among all of them is a transdisciplinary approach to study and promote the health of animals, humans, and ecosystems to ensure the conservation of all biodiversity (Deem et al., 2000, 2015; Jakob-Hoff and Warren, 2012; Sulzner et al., 2021; Vitali et al., 2011). Conservation medicine scientists from across disciplines work to understand the ecological context of health from all levels of biological organization, from cells to ecosystems, by building on existing knowledge frameworks (Braverman, 2021; Vitali et al., 2011).

With a diverse group of multidisciplinary professionals, zoos are well positioned to lead or collaboratively participate in transdisciplinary approaches to study and address EcoHealth challenges globally, in an ecologically driven and conservation focused manner, through the practice of conservation medicine (see Box 2.2) (Lerner and Berg, 2017; Robinette et al., 2017; Sulzner et al., 2021; Walzer, 2017).

There is an increasing movement in modern zoos to financially and culturally support conservation medicine as a core duty of conservation and EcoHealth initiatives. In a survey by Sulzner et al. (2021), 85% of responding zoos were actively engaged in conservation medicine and One Health/EcoHealth activities, with formal programs present at six institutions, and with several others in early development of formal programs. Most zoological institutions do not have the financial or staff resources, or the expertise, to independently undertake conservation medicine projects, thus building and maintaining strong transdisciplinary relationships is one of the most important steps. Zoos partner with other zoological institutions, government organizations, non-government organizations and agencies, universities with and without veterinary/medical schools, companies, independent consultants, and local peoples to create successful programs. It is imperative that the leadership of modern zoos recognize the need to develop and support conservation medicine as a core responsibility of a zoo's mandate, as irrespective of size, zoos can make meaningful collaborative contributions to conservation medicine and EcoHealth initiatives (Deem, 2015; Sulzner et al., 2021; Vitali et al., 2011).

With their strong and committed focus on local and global conservation, education, research, and wellbeing, modern zoos should strive to be active participants in collaborative EcoHealth and conservation medicine movements (Corbyn, 2022; Robinette et al., 2017). This is accomplished through the following six strategies (see Figure 2.1).

2.1 Ex situ management of zoological species to preserve biodiversity

One of the core components of the modern zoo's profile is the conservation management of species via ex situ breeding and support of field-based recovery programs to ensure long-term survival. Significant financial and time resources are invested in saving species for current or future potential reinforcement of populations or reintroductions of extirpated ('wiped out') species. In North America, more than one-half of

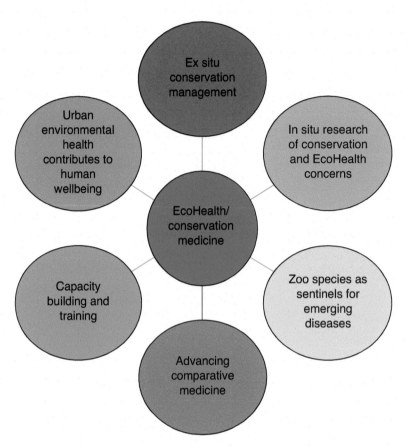

Figure 2.1 Strategies employed by modern zoos and aquariums to collaboratively contribute to EcoHealth and conservation medicine movements.

animal conservation translocations utilize ex situ breeding (Brichieri-Colombi et al., 2018; Conde et al., 2011; McCleery et al., 2014). Zoo species not only are ambassadors for their wild counterparts, but also serve to increase knowledge on diseases, natural behaviours, small population models, and reproduction. The recognition that there is a continuum between the health concerns of species under managed care and those that are free living is an important aspect of conservation medicine. Zoos invest heavily in creating global databases, archives, and serum and tissue banks to continuously learn from the species under their care and ensure the long-term preservation of genetic material (Bowkett, 2009; Zhang and Rawson 2007). This comparative knowledge of

animals and their needs has been advanced by bridging knowledge gaps between free-living animals and those under managed care, and this has resulted in constant improvement to facilities, husbandry and medical practices, and overall improvements in animal welfare. For example, the Wilder Institute/Calgary Zoo (WICZ) has worked collaboratively with other organizations to review and improve quarantine, biosecurity, husbandry, and pre-release medical programs for endangered conservation species that are part of zoo-based breeding for release and head starting programs, such as the Vancouver Island marmot (*Marmota vancouverensis*), whooping cranes (*Grus americana*), burrowing owls (*Athene cunicularia hypugaea*), greater sage grouse (*Centrocercus urophasianus*), and northern leopard frogs (*Lithobates pipiens*). This work has provided vital baseline information for interpreting the health of wild populations and guiding veterinary interventions, which contributes to strong conservation medicine outcomes (Braverman, 2021; Greenwell et al., 2023; Vitali et al., 2011; Walzer, 2017).

2.2 Researching in situ issues of conservation and EcoHealth concerns

Accredited zoological institutions are significantly invested in in situ global conservation efforts, with AZA institutions spending more than US$217 million on field conservation, while engaging with nearly 1000 conservation partners in over 117 countries (AZA, 2021). Studying issues of conservation or EcoHealth concerns is another way in which modern zoos participate in conservation medicine (see Box 2.3). Understanding wildlife health is crucial for wildlife conservation. Pathogen emergence is associated with biodiversity loss, habitat degradation, climate change, and increased contact between humans (directly or indirectly via their domestic animals) and wildlife. The assessment of population viability of conservation sensitive species requires an understanding of expected health parameters as well as information on infectious and non-infectious diseases that threaten their existence. Often research conducted on biodiversity conservation coincides with emerging diseases at the animal and human health interface (Deem, 2015; Schmeller et al., 2020; Schmeller, 2022; Vitali et al., 2011). Zoological health professionals lead research that highlights the epidemiological, pathological and clinical implications of diseases that threaten species and may have direct human health consequences. (Deem, 2015; Schmeller et al., 2020).

Box 2.3 Diseases of conservation and EcoHealth concern

Zoo health professionals undertake conservation medicine research that spans beyond the walls of their institutions that includes wildlife health monitoring, disease investigation and outbreak response, intra- and interspecies disease prevention, and biodiversity protection. Examples include chytridiomycosis and ranaviral infections in amphibians, canine distemper in carnivores, endotheliotropic herpesvirus in elephants, and Tasmanian devil facial tumour disease. This research also includes zoonotic disease emergence, that often stem from human activities or disturbed or degraded environments. The rise of rabies in Asia associated with steep declines of *Gyps* spp. vulture populations due to diclofenac toxicity from previously treated deceased cattle, Ebola virus outbreaks in human and non-human primates, and the recent SARS-COV-2 pandemic are examples of these, where zoos have been at the forefront in collaboratively researching these issues (Deem, 2015; Padda et al., 2018; Sulzner et al., 2021).

2.3 Recognition of diseases in zoo species as sentinels for emerging diseases of humans and other animals

Modern zoos function as mini-ecosystems by caring for animals from diverse geographical areas of the world with varying susceptibilities to disease and thus can function as early warning systems to new diseases that may pose a threat to wild populations or humans. The emergence of West Nile virus in North America in 1999 was discovered by veterinarians at the

Bronx Zoo in New York by recognizing that bird species that were being infected and dying with the virus were different than the species normally infected by known encephalitis viruses in the Americas. Working with public health officials who were noting encephalitic infections in humans at the same time, the virus was identified as a novel pathogen in North America (Ludwig et al., 2002). As it continued to spread throughout the continent, the virus had a significant negative impact on wild bird populations (LaDeau et al., 2007). As equids also are affected by the virus, a commercial equine vaccine was developed, and zoos evaluated the efficacy and safety of this vaccine and other experimental vaccines to protect numerous bird species under their care, many of which are endangered (Jiménez de Oya et al., 2019; Okeson et al., 2007; Vaasjo et al., 2021).

A more recent example, the first reported transmission of SARS-COV-2 from humans to wild species occurred in an accredited zoological facility in April 2020, most likely from an asymptomatic or pre-symptomatic animal care professional. As the pandemic progressed and the virus mutated, the varying susceptibility of wild mammal species was uncovered, ranging from mild disease to death (Gutierrez et al., 2021). Zoos responded by enhancing biosecurity measures for the species in their care and evaluated the safety and efficacy of an experimental COVID-19 vaccine (Murphy and Ly, 2021).

Finally, zoos are often an urban oasis for numerous transient or resident wildlife species and may be involved in early triage or rehabilitation of injured and diseased animals. Modern zoos have robust preventive medicine and pathology programs that includes surveillance for disease in local wildlife that may impact the health of zoological species or humans (Deem, 2015; Jakob-Hoff and Pas, 2023).

2.4 Advancing comparative medicine to enhance conservation medicine efforts

Modern zoos are living classrooms and collaborative research institutions which serve to advance comparative medicine (see Box 2.4). This can include providing biological samples and data to further comparative understanding of basic sciences (e.g. anatomy, physiology,

---

Box 2.4   Case example: Zoos as comparative and conservation medicine education centres

The Wilder Institute/Calgary Zoo (WICZ) has a long-standing collaboration with the University of Calgary Faculty of Veterinary Medicine (UCVM) to increase veterinary student exposure to comparative medicine and introduce conservation medicine and EcoHealth concepts. This has included avian health and handling clinical skill sessions at WICZ that are paired with annual examinations and West Nile virus vaccination, with discussions on the conservation and One Health implications of the disease (Figure 2.2). Comparative and conservation medicine lectures are infused into the curriculum. Annually, over 20% of the graduating veterinary class elect to participate in senior year zoological and conservation medicine externship rotations at WICZ. Collaborative conservation medicine graduate student projects that occur at WICZ have been developed. Finally, anatomical specimens collected at necropsy at WICZ are donated to UCVM to be used for anatomy and other laboratories. This is one example of the successful collaborations that can be developed between modern zoos and aquariums and academic institutions.

England offers a four-week clinical elective rotation to human medical students in their senior year that is built on comparative and translational medicine with broader discussions on the importance of conserving biodiversity and ecosystem health for the protection of human health. Similar collaborations exist between other modern zoos and universities to provide clinical experiences, interactive lectures, and opportunities for graduate work. All of these opportunities promote the transdisciplinary approach that is needed to enhance conservation medicine (Baitchman and Deem, 2023).

2.5 Capacity building and training

Biodiversity loss associated with climate change, habitat loss and degradation, and increasing human–wildlife contact is most concentrated in the subtropical and tropic regions of the world, while most accredited zoos are concentrated in temperate regions. This mismatch coupled with negative cumulative effects of persisting poverty and a growing human population poses challenges for implementation of effective conservation actions (Conde et al., 2011; Schmeller et al., 2020). However, active engagement in local conservation and EcoHealth initiatives requires participation, equity, and knowledge-to-action championed by community empowerment and capacity building which allows the development of solutions that reflect local priorities (Kock and Kock, 2003; Vansteenkiste, 2014; Harrison et al., 2019). Modern zoos can aid in facilitating collaborative transdisciplinary, ecologically driven, and conservation focused capacity building and education to train local people to lead conservation programs and promote the relevance of conservation for their communities. This can be accomplished either through well-established ex situ training programs, such as the Durrell Training Academy in Jersey, UK, which has trained hundreds of international

Figure 2.2 Zoo veterinarians at the Wilder Institute/Calgary Zoo instruct first year veterinary medicine students in an avian health laboratory at the zoo with discussions on the EcoHealth/One Health implications of West Nile virus.

and microbiology) and clinical medicine and pathology, to active public and student engagement in conservation medicine practices. Collaborative comparative medicine projects such as the Great Ape Heart Project (https://greatapeheartproject.org/) or the Exotic Species Cancer Research Alliance (https://escra.cvm.ncsu.edu/) are excellent examples of transdisciplinary collaborations. Increasingly, many new modern zoo animal hospitals have in person or remote viewing opportunities for the public to demonstrate the medical care of the species under human care and educate on EcoHealth and conservation medicine. Some modern zoos have become education centres for human health professionals to introduce them to EcoHealth, One Health and conservation medicine concepts. For example, Zoo New conservationists, or by modern zoos providing staff expertise and technical support to train local people to lead local conservation and EcoHealth based action (Greenwell et al., 2023; Koch and Koch, 2003; Walzer, 2017; Wolf, 2015; Wyatt and Jennings, 2023).

As significant financers of thousands of collaborative conservation projects across hundreds of different countries, modern zoos can contribute to significant and beneficial societal impacts far beyond their primary role of species conservation (Breuer et al., 2018). For example, the WICZ has heavily financed staffing needs and provided technical support for the Wechiau Community Hippo Sanctuary (WCHS), Ghana, for over two decades. This support has precipitated true empowerment leading to measurable conservation and EcoHealth benefits (stable hippopotamus populations with increased connectivity to northern populations, increased biodiversity within the WCHS compared to areas outside the sanctuary), socioeconomic benefits associated with increased tourism and the development of shea butter exportation which contributes to human wellbeing, and increased ecological awareness and support (Sheppard et al., 2010). This is but one example of modern zoos engaging local stakeholders in conservation efforts which is a critical factor in the successful recovery of threatened species' populations (Greenwell et al., 2023).

Box 2.5   Case example: Sister zoo relationships

One of the strong ways for modern zoos to participate in capacity building in countries with lower socioeconomic status is for them to form 'sister zoo' relationships with a zoo found in those countries. The ultimate goal of these partnerships is to build the capacity of the sister zoo to create their own opportunities in a way that works best for them. This is accomplished through providing assistance and access to resources to improve their zoo and the health and living conditions of the animals in their care, and to assist with their conservation and education programming. For almost two decades, a few North American zoos including the WICZ and the Jacksonville Zoo, have had a sister zoo relationship with the Guyana Zoo in Guyana, South America, developing strong foundations for the Guyana Zoo in conservation, education, and animal welfare. By providing resources and experiential learning for the Guyana Zoo staff, and by extending out continuing education opportunities for local veterinarians, the local animal humane society, and agriculture students from the University of Guyana, this has led to improvements in animal and human wellbeing in Guyana. Training has included education on zoonotic, anthroponotic, and interspecies transfer of disease, veterinary and zoological medicine, animal welfare, nutrition, behavioural husbandry, and pathology (see Figure 2.3). In addition, Guyana Zoo professionals were trained to monitor wildlife use rates and trends and create their own programs to educate the public on species conservation and promote in situ conservation management. Continuity of communication is achieved through video communication platforms. It is important to highlight that this is not the simple transfer of technologies and practices from the modern zoo, as this is often inappropriate, impractical, and unsustainable within the context of the sister zoo. Rather, the partnership facilitates the zoo to create their own opportunities and engage the local population to become more involved in the zoo's conservation efforts and see the relevance of conservation in their own lives.

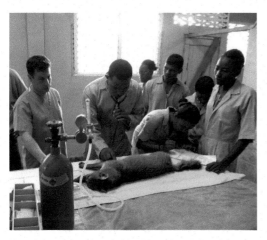

Figure 2.3 Training of Guyanese university students enrolled in an agricultural health program on optimal veterinary anaesthesia practices at the Guyana zoo. This training builds capacity within the country for optimizing animal health and welfare utilizing available anaesthetics and equipment.

2.6 Commitment to local urban environmental health is beneficial to human wellbeing

For a large majority of the population, zoos are the only place where they can visit and physically and emotionally connect with non-native animal species. This engagement opportunity with the natural world affords zoos the opportunity to engage the public with the zoo's conservation, research and education objectives to ensure biodiversity conservation and promote 'planetary friendly' behavioural changes that support the United Nations Sustainability Development Goals (Howell et al., 2019; Maynard et al., 2020; Rose and Riley, 2022; United Nations, 2021). In addition to being strongly grounded in conservation and animal welfare, modern zoos can play a pivotal role in advancing how society thinks and cares about the natural world, especially the anthropogenic effects on the viability of biodiversity and overall planetary health. This starts at the local level with the recognition that

nature is beneficial to human health (Greenwell et al., 2023). Many zoos occupy a significant footprint in urban spaces, and serve as green spaces that are heavily planted with trees, flowers, and other vegetation. They are often leaders in sustainability initiatives. While serving as an urban oasis for many indigenous invertebrate and vertebrate species, zoos allow humans of all ages exposure to the outdoors and biodiversity in a safe environment. Green spaces are known to promote positive affective states in humans, and by interweaving the green environment of a zoo within and around the animal habitats, this is potentially even more enhancing for positive human wellbeing than visiting other forms of green spaces (Akiyama et al., 2021). For the vast majority of the human population, which is predominately urbanized and often sedentary, zoos connect people with nature and encourage physical activity (Dopko et al., 2019; Akiyama et al., 2021). This connection with nature aids in promoting and restoring positive physical and mental states, and has been shown to lower blood pressure, decrease anxiety and the risk of depression, promote feelings of happiness and energy, and improve symptoms associated with attention deficit and autism spectrum disorders (Coolman et al., 2020; Corbyn, 2022; Dopko et al., 2019; Mitchell et al., 2016; Sahrmann et al., 2016; Sakagami and Ohta, 2010).

2.7 Conclusion

The extinction crisis is compelling zoos to transform their commitment to the global conservation of biodiversity. This requires a broad transdisciplinary approach that bridges the biological, socioeconomic, and political variables that influence animal and human wellbeing in in situ and ex situ environments. Zoos are in a unique influential position to address current and future conservation and public health issues by connecting people to nature and inspiring action locally, nationally, and internationally. The important contributions of the modern zoo to conservation and wildlife research, the understanding of wildlife health, and the promotion of human wellbeing have served to strengthen the global EcoHealth and conservation medicine movement. (Robinette et al., 2017; Sulzner et al., 2021).

References

Akiyama, J., Sakagami, T., Uchiyama, H., & Ohta, M. (2021). The health benefits of visiting a zoo, park, and aquarium for older Japanese. *Anthrozoös*, *34*(3), 463–473. https://doi.org/10.1080/089279 36.2021.1898211

Association of Zoos and Aquariums (2021). Statistics (AZA). *Report on conservation and science*. https://assets.speakcdn.com/assets/2332/aza_arcshighlights_2021_final_web.pdf

Bainbridge, D. R. J., & Jabbour, H. N. (1998). Potential of assisted breeding techniques for the conservation of endangered mammalian species in captivity: A review. *Veterinary Record*, *143*(6), 159–168. https://doi.org/10.1136/vr.143.6.159

Baitchman, E., & Deem, S. L. (2023). Zoos as one health education centers for students in the human health professions. In R. E. Miller, P. P. Calle & N. Lamberski (Eds.), *Fowler's zoo and wild animal medicine current therapy*, 10 (pp. 79–83). Elsevier/Saunders.

Barongi, R., Fisken, F. A., Parker, M., & Gusset, M. (Eds.). (2015). *Committing to conservation: The world zoo and aquarium conservation strategy*. WAZA Executive Office.

Bottrill, M. C., Joseph, L. N., Carwardine, J., Bode, M., Cook, C., Game, E. T., Grantham, H., Kark, S., Linke, S., McDonald-Madden, E., Pressey, R. L., Walker, S., Wilson, K. A., & Possingham, H. P. (2008). Is conservation triage just smart decision making? *Trends in Ecology and Evolution*, *23*(12), 649–654. https://doi.org/10.1016/j.tree.2008.07.007

Bowkett, A. E. (2009). Recent captive-breeding proposals and the return of the ark concept to global species conservation. *Conservation Biology*, *23*(3), 773–776. https://doi.org/10.1111/j.1523-1739.2008.01157.x

Braverman, I. (2021). *Zoo veterinarians: Governing care on a diseased planet*. Routledge.

Breuer, T., Manguette, M., & Groenenberg, M. (2018). Gorilla gorilla spp conservation–from zoos to the field and back: Examples from the Mbeli Bai Study. *International Zoo Yearbook*, *52*(1), 137–149. https://doi.org/10.1111/izy.12181

Brichieri-Colombi, T. A., Lloyd, N. A., McPherson, J. M., & Moehrenschlager, A. (2018). Limited contributions of released animals from zoos to North American conservation translocations. *Conservation Biology*, *0*, 1.e7. https://doi.org/10.1111/cobi.13160

Butchart, S. H. M., Stattersfield, A. J., & Collar, N. J. (2006). How many bird extinctions have we prevented? *Oryx*, *40*(3), 266–278. https://doi.org/10.1017/S0030605306000950

Carr, N., & Cohen, S. (2011). The public face of zoos: Images of entertainment, education and conservation. *Anthrozoös*, *24*(2), 175–189. https://doi.org/10.2752/175303711X12998632257620

Clayton, S., & Le Nguyen, K. D. (2018). People in the zoo: A social context for conservation. In *The ark and beyond* (pp. 204–211). University of Chicago Press.

Conde, D. A., Flesness, N., Colchero, F., Jones, O. R., & Scheuerlein, A. (2011). Conservation. An emerging role of zoos to conserve biodiversity. *Science*, *331*(6023), 1390–1391. https://doi.org/10.1126/science.1200674

Coolman, A. A., Niedbalski, A., Powell, D. M., Kozlowski, C. P., Franklin, A. D., & Deem, S. L. (2020). Changes in human health parameters associated with an immersive exhibit experience at a zoological institution. *PLOS ONE*, *15*(4), e0231383. https://doi.org/10.1371/journal.pone.0231383

Corbyn, A. (2022). An initiative to improve health at a zoological park. *Health Economics and Outcome Research*, *8*(10), 4.

Daszak, P., Tabor, G. M., Kilpatrick, A. M., Epstein, J. O. N., & Plowright, R. (2004). Conservation medicine and a new agenda for emerging diseases. *Annals of the New York Academy of Sciences*, *1026*(1), 1–11. https://doi.org/10.1196/annals.1307.001

Deem, S. L. (2015). Conservation medicine to one health: The role of zoologic veterinarians. In R. E. Miller & M. E. Fowler (Eds.), *Fowler's zoo and wild animal medicine* (pp. 698–703). Saunders Elsevier.

Deem, S. L. (2016). *Conservation medicine: A solution-based approach for saving nonhuman primates. Ethnoprimatology: Primate Conservation in the 21st Century* (pp. 63–76).

Deem, S. L., Kilbourn, A. M., Wolfe, N. D., Cook, R. A., & Karesh, W. B. (2000). Conservation medicine. *Annals of the New York Academy of Sciences, 916*(1), 370–377. https://doi.org/10.1111/j.1749-6632.2000.tb05315.x

Dopko, R. L., Capaldi, C. A., & Zelenski, J. M. (2019). The psychological and social benefits of a nature experience for children: A preliminary investigation. *Journal of Environmental Psychology, 63*, 134–138. https://doi.org/10.1016/j.jenvp.2019.05.002

Feilen, K. L., Guillen, R. R., Vega, J., & Savage, A. (2018). Developing successful conservation education programs as a means to engage local communities in protecting cotton-top tamarins (*Saguinus oedipus*) in Colombia. *Journal for Nature Conservation, 41*, 44–50. https://doi.org/10.1016/j.jnc.2017.10.003

Fuller, R. A., Irvine, K. N., Devine-Wright, P., Warren, P. H., & Gaston, K. J. (2007). Psychological benefits of greenspace increase with biodiversity. *Biology Letters, 3*(4), 390–394. https://doi.org/10.1098/rsbl.2007.0149

Greenwell, P. J., Riley, L. M., Lemos de Figueiredo, R., Brereton, J. E., Mooney, A., & Rose, P. E. (2023). The societal value of the modern zoo: A commentary on how zoos can positively impact on human populations locally and globally. *Journal of Zoological and Botanical Gardens, 4*(1), 53–69. https://doi.org/10.3390/jzbg4010006

Gusset, M., & Dick, G. (2011). The global reach of zoos and aquariums in visitor numbers and conservation expenditures. *Zoo Biology, 30*(5), 566–569. https://doi.org/10.1002/zoo.20369

Gutierrez, S., Canington, S. L., Eller, A. R., Herrelko, E. S., & Sholts, S. B. (2023). The intertwined history of non-human primate health and human medicine at the Smithsonian's National Zoo and Conservation Biology Institute. *Notes and Records, 77*(1), 73–96. https://doi.org/10.1098/rsnr.2021.0009

Harrison, S., Kivuti-Bitok, L., Macmillan, A., & Priest, P. (2019). EcoHealth and One Health: A theory-focused review in response to calls for convergence. *Environment International, 132*, 105058. https://doi.org/10.1016/j.envint.2019.105058

Howell, T. J., McLeod, E. M., & Coleman, G. J. (2019). When zoo visitors "connect" with a zoo animal, what does that mean? *Zoo Biology, 38*(6), 461–470. https://doi.org/10.1002/zoo.21509

Jakob-Hoff, R., & Pas, A. (2023). Auckland Zoo: Applying one health in New Zealand. In R. E. Miller, P. P. Calle & N. Lamberski (Eds.), *Fowler's zoo and wild animal medicine current therapy,* 10 (pp. 89–94). Elsevier/Saunders.

Jakob-Hoff, R., & Warren, K. (2012). Zoo veterinarians and conservation medicine. In R. E. Miller & M. E. Fowler (Eds.), *Fowler's zoo and wild animal medicine current therapy,* 7 (pp. 15–23). Elsevier/Saunders.

Jiménez de Oya, N., Escribano-Romero, E., Blázquez, A. B., Martín-Acebes, M. A., & Saiz, J. C. (2019). Current progress of avian vaccines against West Nile virus. *Vaccines, 7*(4), 126. https://doi.org/10.3390/vaccines7040126

Kock, M. D., & Kock, R. A. (2003). Softly, softly: Veterinarians and conservation practioners working in the developing world. *Journal of Zoo and Wildlife Medicine, 34*(1), 1–2. https://doi.org/10.1638/1042-7260(2003)34[0001:SSVACP]2.0.CO;2

LaDeau, S. L., Kilpatrick, A. M., & Marra, P. P. (2007). West Nile virus emergence and large-scale declines of North American bird populations. *Nature, 447*(7145), 710–713. https://doi.org/10.1038/nature05829

Lerner, H., & Berg, C. A. (2017). A Comparison of three holistic approaches to health: One Health, ecohealth, and planetary health. *Frontiers in Veterinary Science, 4*, 163. https://doi.org/10.3389/fvets.2017.00163

Ludwig, G. V., Calle, P. P., Mangiafico, J. A., Raphael, B. L., Danner, D. K., Hile, J. A., Clippinger, T. L., Smith, J. F., Cook, R. A., & McNamara, T. (2002). An outbreak of West Nile virus in a New York City captive wildlife population. *American Journal of Tropical Medicine and Hygiene, 67*(1), 67–75. https://doi.org/10.4269/ajtmh.2002.67.67

Massarella, K., Nygren, A., Fletcher, R., Büscher, B., Kiwango, W. A., Komi, S., Krauss, J. E., Mabele, M. B., McInturff, A., Sandroni, L. T., Alagona, P. S., Brockington, D., Coates, R., Duffy, R., Ferraz, K. M. P. M. B., Koot, S., Marchini, S., & Percequillo, A. R. (2021). Transformation beyond conservation: How critical social science can contribute to a radical new agenda in biodiversity conservation. *Current Opinion in Environmental Sustainability, 49*, 79–87. https://doi.org/10.1016/j.cosust.2021.03.005

Maynard, L., Jacobson, S. K., Monroe, M. C., & Savage, A. (2020). Mission impossible or mission accomplished: Do zoo organizational missions influence conservation practices? *Zoo Biology, 39*(5), 304–314. https://doi.org/10.1002/zoo.21557

McCleery, R., Hostetler, J. A., & Oli, M. K. (2014). Better off in the wild? Evaluating a captive breeding and release program for the recovery of an endangered rodent. *Biological Conservation, 169*, 198–205. https://doi.org/10.1016/j.biocon.2013.11.026

Miller, B., Conway, W., Reading, R. P., Wemmer, C., Wildt, D., Kleiman, D., Monfort, S., Rabinowitz, A., Armstrong, B., & Hutchins, M. (2004). Evaluating the conservation mission of zoos, aquariums, botanical gardens, and natural history museums. *Conservation Biology, 18*(1), 86–93. https://doi.org/10.1111/j.1523-1739.2004.00181.x

Mitchell, D., Tippins, D. J., Kim, Y. A., Perkins, G. D., & Rudolph, H. A. (2016). Last child in the woods: An analysis of nature, child, and time through a lens of eco-mindfulness. In *Mindfulness and educating citizens for everyday life* (pp. 135–158). Brill.

Murphy, H. L., & Ly, H. (2021). Understanding the prevalence of SARS-CoV-2 (COVID-19) exposure in companion, captive, wild, and farmed animals. *Virulence, 12*(1), 2777–2786. https://doi.org/10.1080/21505594.2021.1996519

Okeson, D. M., Llizo, S. Y., Miller, C. L., & Glaser, A. L. (2007). Antibody response of five bird species after vaccination with a killed West Nile virus vaccine. *Journal of Zoo and Wildlife Medicine, 38*(2), 240–244. https://doi.org/10.1638/1042-7260(2007)038[0240:AROFBS]2.0.CO;2

Padda, H., Niedbalski, A., Tate, E., & Deem, S. L. (2018). Member perceptions of the One Health Initiative at a zoological institution. *Frontiers in Veterinary Science, 5*, 22. https://doi.org/10.3389/fvets.2018.00022

Robinette, C., Saffran, L., Ruple, A., & Deem, S. L. (2017). Zoos and public health: A partnership on the One Health frontier. *One Health, 3*, 1–4. https://doi.org/10.1016/j.onehlt.2016.11.003

Robovský, J., Melichar, L., & Gippoliti, S. (2020). *Zoos and conservation in the Anthropocene: Opportunities and problems. Problematic wildlife II: New conservation and management challenges in the human-wildlife interactions* (pp. 451–484).

Romanelli, C., Cooper, H. D., & de Souza Dias, B. F. (2014). The integration of biodiversity into One Health. *Revue Scientifique et Technique, 33*(2), 487–496. https://doi.org/10.20506/rst.33.2.2291

Rose, P. E., & Riley, L. M. (2022). Expanding the role of the future zoo: Wellbeing should become the fifth aim for modern zoos. *Frontiers in Psychology*, 13, 1018722. https://doi.org/10.3389/fpsyg.2022.1018722

Routman, E. O., Khalil, K., Wesley Schultz, P., & Keith, R. M. (2022). Beyond inspiration: Translating zoo and aquarium experiences into conservation behavior. *Zoo Biology, 41*(5), 398–408. https://doi.org/10.1002/zoo.21716

Sahrmann, J. M., Niedbalski, A., Bradshaw, L., Johnson, R., & Deem, S. L. (2016). Changes in human health parameters associated with a touch tank experience at a zoological institution. *Zoo Biology, 35*(1), 4–13. https://doi.org/10.1002/zoo.21257

Sakagami, T., & Ohta, M. (2010). The effect of visiting zoos on human health and quality of life. *Animal Science Journal, 81*(1), 129–134. https://doi.org/10.1111/j.1740-0929.2009.00714.x

Sandifer, P. A., Sutton-Grier, A. E., & Ward, B. P. (2015). Exploring connections among nature, biodiversity, ecosystem services, and human health and well-being: Opportunities to enhance health and biodiversity conservation. *Ecosystem Services, 12*, 1–15. https://doi.org/10.1016/j.ecoser.2014.12.007

Schmeller, D. S. (2022). Conservation biology meets medical science. *Nature Conservation, 46*, 39–40. https://doi.org/10.3897/natureconservation.46.79204

Schmeller, D. S., Courchamp, F., & Killeen, G. (2020). Biodiversity loss, emerging pathogens and human health risks. *Biodiversity and Conservation, 29*(11–12), 3095–3102. https://doi.org/10.1007/s10531-020-02021-6

Sheppard, D. J., Moehrenschlager, A., McPherson, J. M., & Mason, J. J. (2010). Ten years of adaptive community-governed conservation: Evaluating biodiversity protection and poverty alleviation in a West African hippopotamus reserve. *Environmental Conservation, 37*(3), 270–282. https://doi.org/10.1017/S037689291000041X

Sulzner, K., Fiorello, C., Ridgley, F., Garelle, D., & Deem, S. L. (2021). Conservation medicine and One Health in zoos: Scope, obstacles, and unrecognized potential. *Zoo Biology, 40*(1), 44–51. https://doi.org/10.1002/zoo.21572

United Nations (2021). *Make the SDGS a reality.* Department of Economic and Social Affairs, United Nations.

Vaasjo, E., Black, S. R., Pastor, A., & Whiteside, D. P. (2021). Assessing the humoral response to and safety of a commercially available equine west nile virus vaccine in a zoo-based conservation breeding population of endangered greater sage-grouse (Centrocercus urophasianus). *Journal of Zoo and Wildlife Medicine, 52*(2), 732–736. https://doi.org/10.1638/2020-0076

Vansteenkiste, J. (2014). Considering the ecohealth approach: Shaping Haitian women's participation in urban agricultural projects. *Development in Practice, 24*(1), 18–29. https://doi.org/10.1080/09614524.2014.867307

Vitali, S., Reiss, A., & Eden, P. (2011). Conservation medicine in and through zoos. *International Zoo Yearbook, 45*, 1–6. https://doi.org/10.1111/j.1748–1090.2010.00127

Walzer, C. (2017). Beyond One Health–Zoological medicine in the Anthropocene. *Frontiers in Veterinary Medicine, 4*, 1–3. https://doi. https://doi.org/10.3389/fvets.2017.00102

Wolf, M. (2015). Is there really such a thing as "one health"? Thinking about a more than human world from the perspective of cultural anthropology. *Social Science and Medicine, 129*, 5–11. https://doi.org/10.1016/j.socscimed.2014.06.018

WWF. (2022). *Building a nature-positive society*. R. E. A. Almond, M. Grooten, D. Juffe Bignoli & T. Petersen (Eds.). https://wwflpr.awsassets.panda.org/downloads/lpr_2022_full_report.pdf, 2022. WWF.

Wyatt, J., & Jennings, J. (2023). A one health initiative in Borneo saving orangutans, transforming community health and promoting sustainable livelihoods. In R. E. Miller, P. P. Calle & N. Lamberski (Eds.), *Fowler's zoo and wild animal medicine current therapy*, 10 (pp. 85–87). Elsevier/Saunders.

Zhang, T., & Rawson, D. M. (2007). The frozen ark: Cryo-banking of genetic and cellular material of endangered species. *Cryobiology, 55*, 355

Zimmermann, A., & Wilkinson, R. (2007). The conservation mission in the wild: Zoos as conservation NGOs. In A. Zimmermann, M. Hatchwell, L. A. Dickie & C. West (Eds.), *Zoos in the 21st Century: Catalysts for Conservation* (pp. 303–321). Cambridge University Press.

Zordan, M. (2021). Progressive zoos and aquariums must be part of the World's response to COVID-19. https://www.iucn.org/crossroads-blog/202101/progressive-zoos-and-aquariums-must-be-part-worlds-response-covid-19

Chapter **3**

Sustaining primate conservation through ecotourism in Uganda

Gladys Kalema-Zikusoka, Allison Carden Hanes, Kimberley Hockings and Nancy Jeanne Stevens

Abstract

Mountain gorilla tourism began at Bwindi Impenetrable National Park (BINP) and Mgahinga National Park (MGNP) in 1993 and within a decade was generating enough revenue to sustain the conservation operations of BINP and MGNP, and also to support other non-great ape protected areas in Uganda Wildlife Authority (UWA)'s estate that were previously unable to cover operational costs. To minimize the risk of disease transmission and behavioral disturbance during tourism encounters, great ape viewing guidelines were developed, specifying limits on the number of tourists per gorilla viewing group, the length of time for a gorilla viewing experience, and the maintenance of a minimum distance between humans and gorillas during tourism encounters (Macfie and Williamson, 2010). In particular, the minimum distance guideline has been difficult to enforce, with both humans and gorillas approaching within 3 m of one another (Sandbrook and Semple, 2006; Weber et al., 2020). Disease transmission between humans and gorillas is a serious concern, as gorillas are susceptible to human pathogens, and they are highly social which leads to pathogens spreading rapidly between individuals, furthermore low population numbers render great apes vulnerable to extinction. The COVID-19 pandemic revealed the significant limitations of relying on tourism to sustain community livelihoods and conservation efforts in Uganda. This chapter provides an overview of initiatives developed by the non-profit organization Conservation Through Public Health (CTPH) to mitigate disease transmission and enhance One Health-informed conservation strategies for Uganda's endangered mountain gorilla population.

3.1 Introduction

The first reported human disease transmitted to Bwindi Impenetrable National Park (BINP) gorillas occurred in 1996, when an outbreak of scabies, caused by the mite *Sarcoptes scabiei*, was traced back to local community interactions with gorillas that forage outside the park. It was found that gorillas were coming into contact with contaminated clothing on scarecrows

Susan C. Cork and Douglas P. Whiteside (eds) **Case Studies in EcoHealth**
DOI: 10.52517/9781789183368.003, © 5m Books Ltd, 2024

CTPH's One Health approach acknowledges that interactions between people, wildlife, livestock and the environment can increase the risk of pathogen transmission between humans and animals, among other negative outcomes for conservation and public health. CTPH works to improve (1) wildlife health, (2) community health, (3) livestock health (4) conservation knowledge and attitudes through Village Health and Conservation Team programs and (5) enhances sustainable livelihoods by offering farmers adjacent to BINP above-market prices for coffee sold mainly to gorilla trekking tourists as Gorilla Conservation Coffee (GCC) (Kalema-Zikusoka et al., 2021a) https://ctph.org/.

Figure 3.1a–d shows CTPH staff and mountain gorillas.

Fig 3.1a Dr. Kalema-Zikusoka conducting a gorilla health check on Rushegura family with CTPH team in BINP, Uganda. © One Health Productions LLC 2021.

Fig 3.1b Uganda Wildlife Authority ranger guide sanitizing tourist and porter hands prior to 1 hour gorilla viewing in BINP, Uganda. © One Health Productions LLC 2021.

Fig 3.1c Adult female gorilla Kibande of Rushegura family in BINP, Uganda. © One Health Productions LLC 2021.

Fig 3.1d Juvenile gorilla of Habinyanja gorilla family in BINP, Uganda. © One Health Productions LLC 2021.

in community gardens (Kalema-Zikusoka et al., 2002). This prompted the establishment of the NGO and nonprofit organization Conservation Through Public Health (CTPH) in 2003, to help prevent human and wildlife disease transmission through education and public programming.

The COVID-19 pandemic revealed the significant limitations of relying on tourism to sustain community livelihoods and conservation efforts in Uganda. A sudden reduction in tourism due to the global lockdowns also contributed to an increase in wild meat hunting by local communities (Kalema-Zikusoka et al., 2021b) In response to these challenges, and with full consultation of local communities, CTPH provided agricultural crop seedlings and identified markets outside of Uganda to buy and sell local farmers' coffee beans at above-market prices. Importantly, the pandemic was a strong impetus for CTPH to work with conservation partners and Uganda Wildlife Authority (UWA) to advocate for sustainable local livelihoods together with updated great ape viewing rules, including mandatory masking during gorilla tourism encounters (Weber et al., 2020) and increasing the minimum distance for human–gorilla viewing from 7 to 10 m (Uganda Wildlife Authority, 2020).

3.2   Mountain gorilla conservation

Mountain gorillas (*Gorilla beringei beringei*) comprise two isolated populations within the Virunga Mountains of eastern Africa, primarily residing in protected areas in the Democratic Republic of the Congo (DRC), Rwanda and Uganda. Both populations are threatened by habitat loss and human encroachment, so much so that it was thought that mountain gorillas imminently faced extinction. But by the latter half of the 20th century, extreme conservation efforts were implemented (Robbins et al., 2011) together with the development of tourism initiatives to provide sustainable community livelihoods to offer alternatives to

habitat destruction. Initiated in 1993 in Uganda, gorilla viewing tourism has offered a means of revenue generation to enable the protection of critical great ape habitats (Muresherwa et al., 2022), and since 1998, mountain gorilla populations in Uganda, Rwanda and DRC have increased from an estimated 650 individuals to a record 1063 individuals in 2018 (Truscott, 2022) (Figure 3.2). This positive trend, documented through intense ecological monitoring, research and censuses conducted approximately every 5 years has resulted in a decision by the International Union for Conservation of Nature (IUCN) to revise the status of mountain gorillas from Critically Endangered to Endangered (i.e. moving from an extremely high risk of extinction to a very high risk of extinction, IUCN [2018]).

Threats to the long-term survival of mountain gorillas persist, including (1) continued habitat loss and conversion to agriculture – with remaining gorilla populations restricted to habitats surrounded by high densities of people exceeding 300 people per square kilometer, (2) war/conflict and hunting of gorillas in DRC, (3) poaching of other animals (e.g. duiker, bush pigs) resulting in accidental injuries of gorillas from snares and spears (Refisch et al., 2018) and (4) zoonotic diseases spread from humans with whom they come into close contact inside protected areas and during foraging bouts outside park boundaries (i.e. a phenomenon exacerbated through habituation to tourist and researcher presence).

3.3   Development of gorilla tourism in Uganda

Owing to the intense habitat pressure on mountain gorilla populations, the International Gorilla Conservation Programme (IGCP) supported Uganda National Parks (UNP) to develop gorilla viewing tourism as a revenue-generating product to support habitat protection. Gorilla

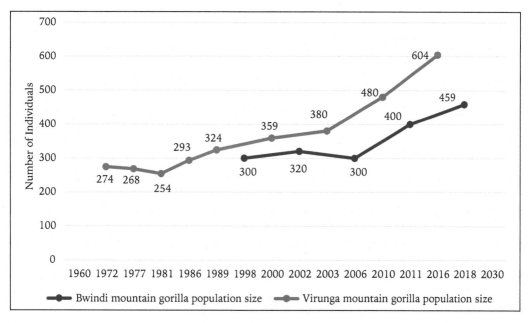

Figure 3.2 Mountain gorilla (*Gorilla beringei beringei*) populations (1972–2018) © Conservation Through Public Health.

tourism began in Uganda in 1993 with just two gorilla groups habituated for viewing encounters. Tourists generated revenue through the purchase of gorilla viewing permits (in 1993 the cost of a gorilla viewing permit was US$250). In 1996, an Act of Parliament stipulated that 20% of park entry fees should be allocated to local communities surrounding protected areas, to be used for development projects to reduce poverty and support social services including the building of schools, health centers and roads, and support to alternative livelihoods (Uganda Wildlife Act, 1996). By 2009, the number of people visiting each gorilla group per day had increased from six to eight people per group, and in 2014, the gorilla viewing permit price increased to US$600. By 2020 the number of habituated gorilla groups had grown to 22 out of an estimated 50 gorilla social groups, and by 2021 the permit fee had increased to US$700. Currently over one-half of the Bwindi gorillas are habituated for visitation by tourists, with one group habituated for research purposes

only. Taken together over the past 30 years, the average number of people visiting BINP mountain gorillas has grown from 12 to 176 per day (Figure 3.3).

3.4 Habituation of great apes

Habituation of great apes has enabled the generation of tourism revenue, and this revenue helps to sustain conservation efforts by providing support for law enforcement and community engagement, as well as improved community health infrastructure and wildlife veterinary care (Blomley et al., 2010). These factors are often cited as contributors to the recent growth in Uganda's mountain gorilla populations. In contrast, gorilla populations experiencing limited or no organized tourism have experienced population declines. However, habituation for tourism can reduce gorillas' fear of people, placing them at greater risk from poaching and emboldening them to travel outside protected areas, increasing the potential for negative

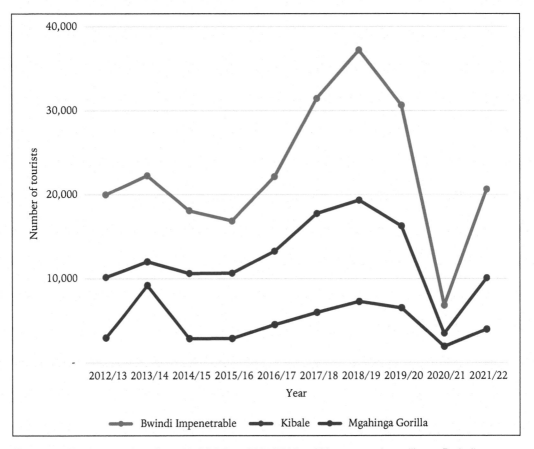

Figure 3.3 Tourism numbers from 2012/2013 to 2021/2022 trekking mountain gorillas at Bwindi Impenetrable and Mgahinga National Parks and chimpanzees at Kibale National Park © Uganda Wildlife Authority.

interactions with people including crop foraging and aggressive encounters, and higher likelihood of disease transmission.

3.5 Understanding and mitigating infectious disease risks

From the outset, it was recognized that great ape tourism could increase the risk of disease transmission from tourists to gorillas, particularly for respiratory diseases as human and nonhuman apes are susceptible to many of the same infectious pathogens and parasites (Woodford et al., 2002). To mitigate concerns

about disease transmission, and to minimize behavioral disturbance to habituated groups, limitations were placed on the number of tourists permitted to visit each gorilla group. Visit duration was limited to 1 hour per gorilla group per day and tourists who were visibly sick, particularly with flu-like symptoms, were not allowed to visit the gorillas. Initially, a minimal viewing distance of 5 m was proposed in order to reduce the likelihood of disease transmission, a distance that was increased to 7 m a few years later. Moreover, people were asked to turn away to cough or sneeze during gorilla viewing, and to dig a hole at least 30 centimeters deep to bury human fecal material while in the forest. A minimum age for gorilla viewing was set at 15 years to minimize risk of human childhood diseases like measles being transmitted to gorillas as well as for child safety. In 1996, a veterinarian was hired to attend to the health of gorillas and other great apes to mitigate the risk of habituated gorillas acquiring human diseases.

Despite guidelines being developed and implemented to protect great ape and human health, numerous occurrences of disease transmission were reported, along with failures in rule adherence during great ape viewing encounters (Weber et al., 2020). This is of particular concern as infectious diseases (including respiratory pathogens) are thought to be responsible for 20% of sudden deaths among mountain gorillas, although the cause of death cannot be ascertained in all cases. Human metapneumovirus (hMPV) caused the death of a mother and her infant in Volcanoes National Park in Rwanda in 2009 (Palacios et al., 2011). Other viruses of concern include respiratory syncytial virus (RSV) (Mazet et al., 2020) and human rhinovirus C (HRV-C), the latter causing a 2013 outbreak among a group of 56 wild chimpanzees in Kibale National Park (KNP), Uganda, resulting in the deaths of five individuals (Scully et al., 2018).

It was clear that the disease transmission risks between humans and gorillas were high, and 7 m spacing violations were being reported at a rate that threatened sustainable great ape tourism. In response to disease risks to habituated gorillas from tourists, research was conducted in 2011 to assess whether tourists would be willing to adopt stricter measures to protect gorilla health. At that time, 51% of tourists reported willingness to wear masks during gorilla viewing encounters (Hanes et al., 2018). A study conducted in 2014 (Weber et al., 2020) documented a higher percentage of tourists (>73%) willing to wear a mask during gorilla viewing encounters, possibly reflecting the fact that

mask wearing had already been implemented in other tourism contexts. The latter study also explored tourist knowledge and attitudes about disease transmission risks, comparing tourist survey results with observational data collected during park briefings and gorilla viewing encounters (Weber et al., 2020). This revealed an interesting pattern: in the overwhelming majority of pre-trek briefings, park rangers informed tourists about the 7 m distance rule (98% of pre-trek briefings), underscoring that the main reason for maintaining distance was to prevent human to gorilla disease transmission (86% of pre-trek briefings). Nonetheless, the 7 m distance rule was violated in *all but one trek in the study* (when gorillas were higher than 7 m in the trees). To complicate matters, gorillas initiated 40% of the exceedingly close (<3 m) encounters with humans (Weber et al., 2020). In 2014, mandatory mask wearing was included in the general management plan of Bwindi Mgahinga Conservation Area, to be instituted within 10 years (Uganda Wildlife Authority, 2014). By 2016, Virunga National Park in DRC had begun mandatory mask wearing during great ape encounters, and a mask taskforce was set up by the International Gorilla Conservation Project to explore implementation of mask mandates during great ape tourism encounters in Rwanda and Uganda. Furthermore, studies showed that viewing distances of closer than 3 m increased stress and human directed behavior in mountain gorillas as well as reduced feeding (Costa et al., 2023).

A primate ecotourism roundtable was held in Uganda in 2019 at the second African Primatological Society (APS) Congress. Primatologists, government and nonprofit leaders, researchers, donors and Ugandan tour companies discussed the opportunities and challenges facing great ape tourism. Stakeholders acknowledged that responsible tourism is needed to sustain conservation efforts and recognized that steps were needed to mitigate disease risk.

Stakeholders also recommended diversifying ecotourism opportunities to include other primates (e.g., red colobus monkeys, *Piliocolobus* sp., baboons, *Papio* sp.) and to develop other wildlife and community ecotourism experiences in order to reduce pressure on gorilla and chimpanzee populations (African Primatological Society, 2019).

3.6 COVID-19 pandemic

Shortly after African Primatological Society Congress discussions in September 2019, the COVID-19 pandemic struck. In January 2020, the World Health Organization (WHO) declared the COVID-19 pandemic a public health emergency of international concern (World Health Organization, 2020). The vulnerabilities of endangered apes to human respiratory diseases was a sobering realization for conservation stakeholders that action needed to be taken sooner rather than later and emphasized the importance of One Health approaches in informing conservation and economic security (Gillespie et al., 2020). As the outbreak spread across the globe, concern for endangered great apes deepened, as human coronavirus OC43 was already known to have caused morbidity in wild chimpanzees in the Ivory Coast in 2016 (Patrono et al., 2018). Studies conducted in 2020 demonstrated that great apes and other Old World primates exhibit similar angiotensin converting enzyme (ACE2) protein receptors as humans, rendering them highly susceptible to the effects of COVID-19 (Damas et al., 2020., Melin et al., 2020). Angiotensin-converting enzyme (ACE) inhibitors help relax the veins and arteries to lower blood pressure. In early 2021, captive gorillas at the San Diego Wild Animal Park and Prague Zoo contracted COVID-19 from asymptomatic keepers, exhibiting symptoms similar to those in humans (Gibbons, 2021; Willoughby, 2021). Multiple members in the gorilla enclosure developed mild to moderate symptoms, but an aged silverback developed more severe symptoms and recovered only after treatment with monoclonal antibodies. After vaccinations were widely available, captive gorillas in Zoo Atlanta contracted the COVID-19 Delta variant from vaccinated keepers (Associated Press, 2021). By this time, experimental vaccination of susceptible zoo animals had begun for great apes and other susceptible species including felids and mustelids, likely protecting the captive gorillas from more severe COVID-19 infections (IUCN, 2021; see also Non-Human Primate COVID-19 Information Hub, https://umnadvet.instructure.com/courses/324).

In response to the pandemic, sudden lockdowns resulted in a 6 month reduction in local tourism. Global lockdowns persisted between 1 to 2 years, devastating the tourism industry and destabilizing local livelihoods (see above link). Desperation contributed to increased poaching by hungry communities, with a silverback gorilla killed by a hungry bushmeat poacher. The situation on the ground prompted CTPH to provide fast growing food crop seedlings and training in sustainable agriculture methods to vulnerable community members to address hunger and diversify income potential during and after the resumption of tourism.

Importantly, published research on human–gorilla distancing during great ape visits enabled CTPH to work with conservation partners and UWA and advocate for visitor masking protocols during gorilla encounters (Gilardi and Uwingeli, 2020; Otsuka and Yamakoshi, 2020; Weber *at al.* 2020). A local tailoring enterprise, Ride for a Woman, was engaged to create the first masks for rangers and trackers at Bwindi to wear to reduce risks of disease transmission during gorilla monitoring visits.

In the previously vibrant tourism destination, receipt of external income was limited to just handful of community members, including workers supported by Ride for a Woman, and farmers supported by Gorilla Conservation Coffee (Kalema-Zikusoka et al., 2021b). As the local lockdown resulted in reduced income for social enterprises, CTPH sought markets outside Uganda, such as those in the UK and USA, to support artisans and farmers. The pandemic revealed clear limitations of relying on tourism alone to sustain conservation and community livelihoods, as has been reported in many other conservation contexts globally (Leendertz and Kalema-Zikusoka, 2021). It also afforded new means to strengthen systems to protect gorillas and communities in the long term (Leendertz & Kalema-Zikusoka, 2021; Kalema-Zikusoka et al., 2021b).

3.7 COVID-19 vaccination requirements

To overcome the challenges arising from the COVID-19 pandemic, and the related economic depression due to global lockdowns, vaccinations were made available and administered within 1 year after the pandemic began to prevent severe human COVID-19 infections and to limit the spread of the virus among people and between people and animals. In response to health concerns, great ape viewing regulations were upgraded in Uganda and Rwanda in March 2022, in line with best practice guidelines provided by the IUCN Primate Specialist Group Section on Great Apes and the Gorilla Pathology Specialist Group. This included mandatory wearing of protective face masks when within 10 m of nonhuman great apes, hand washing and boot disinfection before entering the forest (see Table 3.1) as well as temperature check points using a noncontact infrared thermometer to detect subclinical human illness. The UWA further increased the viewing distance from 7 to 10 m (Uganda Wildlife Authority, 2020). These changes were also championed by the tourism community as the pandemic increased awareness of the seriousness of spreading respiratory illnesses. Tourists also began calling for stronger protective measures (including evidence of vaccination) to be instituted for gorilla and chimpanzee viewing to protect these animals from COVID-19 transmission from humans.

To support these measures and enhance knowledge translation, a national gorilla trekking film was developed and produced by One Health Productions LLC (Thurstan et al.,

Table 3.1 Temporal changes in great ape tourism 'biosecurity' guidelines as a result of pandemic COVID-19.

1993	2020
6 people per group per day for one hour	8 people per group per day for one hour
5 m	10 m
No masks	Protective face masks
No temperature taking	Temperature taking
Hand washing or sanitizing not enforced	Hand washing or sanitizing enforced
Boot disinfection not enforced	Boot disinfection enforced
No tracking when showing clinical signs	No tracking when showing clinical signs
15 years old minimum visitor age	15 years old minimum visitor age
Digging a hole 30 cm deep for human waste	Digging a hole 30 cm deep for human waste
Turn away to cough or sneeze	Turn away to cough or sneeze

2021). The film outlined all updated gorilla trekking protocols, clearly highlighting regulations such as the masking requirement during gorilla viewing encounters, the expanded 10 m distance rule, the requirement to provide proof of COVID-19 vaccination prior to engaging in gorilla viewing, and other essential safety information including how to behave during gorilla viewing encounters. The film also emphasized that mask wearing is mandatory even for people who are fully vaccinated against COVID-19. Importantly, the video highlighted the needs of local staff, emphasizing the need to respect and support sustainable livelihoods for local communities through tourists buying locally made crafts and masks, engaging in community walks and entertainment options, hiring a porter to carry backpacks and to offer support when hiking and trekking, as well as staying at community lodging and eating in community restaurants. Ongoing communications at park headquarters offer information on effective ways to promote conservation by supporting locally owned and operated establishments, contributing to long-term local educational and health initiatives, staying informed about community conservation and sustainable livelihoods projects around the park.

3.8 Broadening the message through policy and education

The African civil society organization, Biodiversity Alliance (ACBA), aims to strengthen the African voice in the Convention of Biological Diversity. In accordance with that goal, CTPH developed an ACBA policy brief with the International Gorilla Conservation Program highlighting 'Challenges and threats amidst the covid-19 pandemic for sustaining conservation through responsible great ape tourism.' This policy brief, designed for African governments, donors, and tour operators, advocated for responsible great ape tourism across African countries with gorilla and chimpanzee populations impacted by primate viewing tourism. The policy brief emphasized adherence to great ape viewing protocols even after the pandemic ends and outlined the role that governments, tour operators and donors play in ensuring health and safety measures. It also highlighted the need to support local communities at great ape sites by engaging tour operators in the development of tourism products that benefit local communities. The policy brief was launched at the first IUCN Africa Protected Areas Congress held in Rwanda in 2022.

The effectiveness of any disease prevention strategy will depend on tourists understanding the risks to the wildlife, and compliance with regulations, as well as facilitating the enforcement of guidelines by tourist guides and relevant regulatory authorities. The development of regulations and the effective communication of those regulations requires a strong evidence base that must be adapted depending on the site context. Nuno et al. (2022) quantified perception of risk, sense of responsibility, social norms, expectations, and willingness to adopt SARS-CoV-2 mitigation measures by tourists and guides to prevent transmission to African great apes and local communities. This research enabled us to create evidence-based, multi-media educational material for ape tourism sites across Africa.

Recommendations are shared on www.protectgreatapesfromdisease.com, a website engaging stakeholders to improve practices and improve sustainability across the sector (Nuno et al., 2022). CTPH trained 152 guides and rangers across five sectors of BINP. To demonstrate improved knowledge of infectious disease and capacity to enforce regulations, and for guides to complete training, they must be able to confidently deliver a pre-trek briefing including all disease mitigation measures as detailed during the training session. For the guides to receive a certificate of completion they must deliver a short

---

Knowledge Translation

Recognizing the need to effectively engage tourists in responsible tourism, an education campaign was developed entitled Protect Great Apes from Disease in Tourism, including COVID-19 and other diseases. Based on a survey of over 1000 travelers who had visited or were planning to visit great ape habitats (Mbayahi and Kalema-Zikusoka, 2022) the campaign engaged travelers in the implementation of responsible great ape tourism practices. Most survey participants indicated a willingness to adhere to upgraded great ape viewing protocols, including mandatory mask wearing during the great ape viewing encounters. Some expressed reluctance to obtain vaccinations, likely in part reflecting the fact that the survey was conducted before vaccines had become widely available, but worth revisiting in an ongoing study. Protect Great Apes from Disease in Tourism also held two ranger training workshops, connecting rangers at BINP with ecoguards at a chimpanzee tourism site in Cantannez National Park in Guinea Bissau. The workshop aimed to evaluate and expand practitioner understanding of disease transmission risks associated with great ape tourism, and to gather input on challenges and opportunities for sustainable tourism across great ape habitats.

---

speech using the education materials provided and demonstrate competence in dealing with a difficult scenario through role play. For example, a tourist refusing to wear their mask properly. All 152 guides and rangers at Bwindi completed their training and demonstrated improved knowledge and were awarded certificates.

3.9   Conclusion and future prospects

A primary goal of responsible tourism is to help support local and national conservation efforts. That said, negative effects on mountain gorillas and other great apes must still be further minimized. Zoonotic disease transmission from people remains one of the greatest threats to habituated great apes. Susceptibility of these animals to novel human infections is unpredictable, and coinfections with COVID-19 and common flu viruses or other pathogens are of ongoing concern. Gorillas are monitored through clinical observations and regular testing of fecal samples collected noninvasively from night nests and trails. Thankfully, as yet there

have been no reported cases of COVID-19 infections in wild mountain gorillas, monitored through observation of clinical signs and non-invasive fecal sample collection and testing (Kalema-Zikusoka et al., 2021b) and the frequency of respiratory disease outbreaks in

---

Following stricter trekking guidelines and improved education for all gorilla viewing tourists in Uganda, stakeholders have widely acknowledged improvements in disease prevention practices. Tourists have indicated willingness to comply with improved guidelines. A concerted effort is now needed to mitigate disease transmission risks during chimpanzee viewing encounters in Uganda and more widely, and provide the same level of visitor education, enforcing adherence to great ape viewing best practice guidelines, and importantly limiting visitor numbers to reduce impacts on chimpanzees, especially when they are located in small groups.

Virunga mountain gorillas has declined in conjunction with lower tourist numbers and mandatory mask wearing (Chesney and Hockings, 2021). This demonstrates the immediate benefits of strong measures for protecting great ape health and encourages rigorous enforcement of protective measures into the future.

Importantly, the COVID-19 pandemic has underscored limitations of relying on tourism to sustain local communities and great ape conservation efforts. Governmental organizations, NGOs, local communities and tour operators alike have acknowledged that alternative livelihood opportunities are needed in conjunction with enhanced responsible tourism practices (see Table 3.2) in order to develop viable solutions in the long term (Leendertz and Kalema-Zikusoka, 2021).

Beyond implementation of best practices to mitigate disease risks during great ape tourism encounters, focused efforts are needed to diversify tourism experiences and reduce pressure on great ape communities (African Primatological Society, 2019). Engaging with broader international markets for agricultural crops and artisanal products can help to secure more diverse and stable sources of income for local communities, including opportunities for income generation at times when tourism is not possible (Weber et al., 2020). The development of sustainable livelihoods for local communities that reduce pressure on the forest and its inhabitants can improve prospects for great ape populations into the future.

Acknowledgments

CTPH acknowledges the following sources and collaborators: UWA, IGCP, Mountain Gorilla Veterinary Project/Gorilla Doctors, Max Planck Institute, Bwindi Community Hospital, Kanungu and Kisoro District Local Governments. CTPH would like to acknowledge the following donors who supported the upgrading of great ape viewing regulations and supported community engagements and livelihoods during the pandemic including IUCN,

Table 3.2 Interventions to improve responsible tourism to the great apes.

1993	Gorilla tourism rules instituted by UNP/UWA with a minimum viewing distance of 5 meters and 6 tourists per gorilla group per day
1999	Distance increased to 7 meters (*on recommendation of IGCP/Homsy report on Ape tourism and human diseases* (Homsy 1999))
2009	Number of tourists increased from 6 to 8 to avoid habituating more gorilla groups as the demand for great ape tourism increased
2010	IUCN guidelines for great ape viewing developed
2014	UWA (BMCA) general management plan included mask wearing in the next ten years
2016	Mask task force established by IGCP
2019	APS Primate ecotourism roundtable recommended strict adherence to ICN great ape viewing guidelines to enable sustainable great ape tourism and diversifying tourism to the primates and wildlife and engaging local communities more meaningfully
2020	Mandatory mask wearing and 10 meters viewing distance instituted during the COVID-19 pandemic
2021	Protect great apes form disease campaign developed for tourists by One Health experts
2022	Responsible tourism to Great Apes policy brief launched at the first IUCN protected area congress IUCN APAC emphasis on mask wearing even after the pandemic ends

Arcus Foundation, British High Commission, Tusk Trust and Wildlife Conservation Network.

The Protect Great Apes from Disease in Tourism project was conducted by the University of Exeter, CTPH, the Robert Koch Institute, Bristol Zoological Society and other partners funded by the Darwin Initiative. We are grateful to all survey participants who kindly provided us with their time and insights. The project Reducing Transmission of SARS-CoV-2 to African Great Apes in Tourism (CV19RR18) was funded by the Darwin Initiative COVID-19 Rapid Response Round linked with the ongoing project 'Promoting Public Health in a Biodiverse Agroforest Landscape in Guinea Bissau' (26–018) to Kimberly J. Hockings.

Additionally, Gladys Kalema-Zikusoka greatly thanks: Alyce Mbayahi and IGCP, and the Africa CSO Biodiversity Alliance and African Primatological Society for their support in advocating for responsible tourism to great apes through an ACBA policy brief and African Primatological Society congress roundtable discussion. Nancy Jeanne Stevens thanks Annalisa Weber for collaboration on rule violations during gorilla viewing tourism, Wren Edwards for collaboration on chimpanzee viewing tourism, and Ohio University's Heritage College of Osteopathic Medicine and Voinovich School for Leadership and Public Service for research support and collaboration.

Photographs and videos courtesy One Health Productions (OHP). OHP designs and develops film and marketing content for projects that are dedicated to improving the health of the planet. OHP specializes in filmmaking that supports human, animal and environmental health. The health of humankind is directly tied to that of animals and our shared environment: One Health. With special focus on wildlife, conservation and human rights issues, OHP utilizes the power of media to thoughtfully showcase what unites us. Science and ethics guide us in highlighting the beauty of our shared natural world to promote stewardship and inspire actionable change. OHP has made several One Health primate conservation films and is available for hire.

References

African Primatological Society. (2019). *Report on the 2019 conference Proceedings*.

Associated Press. (2021). 18 of 20 gorillas at Altanta's Zoo have contracted Covid. https://apnews.com/article/lifestyle-health-environment-and-nature-georgia-coronavirus-pandemic-6a5a30215f8a3f6e49e5d74679791098

Blomley, T., Namara, A., McNeilage, A., Franks, P., Rainer, H., Donaldson, A., Malpas, R., Olupot, W., Baker, J., Sandbrook, C., Bitariho, R., & Infield, M. (2010). Development and gorillas? Assessing fifteen years of integrated conservation and development in south-western Uganda. *Natural Resources Issues No. 23*. IIED.

Chesney, C., & Hockings, K. J. (2021). Protect great apes from disease: Freely available education materials for Research and tourism. *African Primates, 15*, 47–50.

Costa, R., Takeshita, R. S. C., Tomonaga, M., Huffman, M. A., Kalema-Zikusoka, G., Bercovitch, F., & Hayashi, M. (2023). The impact of tourist visits on mountain gorilla behavior in Uganda. *Journal of Ecotourism*, 1–19. https://doi.org/10.1080/14724049.2023.2176507

Damas, J., Hughes, G. M., Keough, K. C., Painter, C. A., Persky, N. S., Corbo, M., Hiller, M., Koepfli, K.-P., Pfenning, A. R., Zhao, H., Genereux, D. P., Swofford, R., Pollard, K. S., Ryder, O. A., Nweeia, M. T., Lindblad-Toh, K., Teeling, E. C., Karlsson, E. K., & Lewin, H. A. (2020). Broad Host Range of SARS-CoV-2 Predicted by Comparative and Structural Analysis of ACE2 in Vertebrates. *BioRxiv preprint doi.* https://www.biorxiv.org/content/10.1101/2020.04.16.045302v1

Gibbons, A. (2021). Captive gorillas test positive for coronavirus. *Science*. https://doi.org/10.1126/science.abg5458

Gilardi, K., & Uwingeli, P. (2022). Keep mountain gorillas free from pandemic virus. *Nature, 602*(7896), 211. https://doi.org/10.1038/d41586-022-00331-z

Gillespie, T. R., & Leendertz, F. H. (2020). COVID-19: Protect great apes during human pandemics. https://www.nature.com/articles/d41586–020–00859-y. *Nature, 579*(7800), 497. https://doi.org/10.1038/d41586-020-00859-y

Hanes, A. C., Kalema-Zikusoka, G., Svensson, M. S., & Hill, C. M. (2018). Assessment of health risks posed by tourists visiting mountain gorillas in Bwindi impenetrable national park, Uganda. *Primate Conservation, 2018*(32): 10 pp.

Homsy, J. (1999). Ape tourism and human disease – how close should we get? A critical review of the rules and regulations governing park management and tourism for the wild mountain gorilla, *Gorilla gorilla beringei*.

IUCN (International Union for Conservation of Nature and Natural Resources). (2018). Redlist update signals hope for the gorillas. https://www.the-scientist.com/newsopinion/iucn-red-list-update-signals-hope-for-gorillas-6510512

IUCN (International Union for Conservation of Nature and Natural Resources. (2021). Primate Specialist Group. SARS-CoV-2 and COVID-19. http://www.primate-sg.org/covid-19

Kalema-Zikusoka, G., Kock, R. A., & Macfie, E. J. (2002). Scabies in free-ranging mountain gorillas (*Gorilla beringei beringei*) in Bwindi Impenetrable National Park, Uganda. *Veterinary Record, 150*(1), 12–15, https://doi.org/10.1136/vr.150.1.12

Kalema-Zikusoka, G., Rubanga, S., Ngabirano, A., & Zikusoka, L. (2021a). Mitigating impacts of the COVID-19 pandemic on gorilla conservation: Lessons from Bwindi Impenetrable Forest, Uganda. *Frontiers in Public Health*, 9, 655175. https://doi.org/10.3389/fpubh.2021.655175

Kalema-Zikusoka, G., Ngabirano, A., & Rubanga, S. (2021b). Gorilla conservation and one health. *Springer nature Switzerland. Agriculturists* C. Underkoffler & H. R. Adams (Eds.), Wildlife Biodiversity Conservation, *2021*(S) (pp. 271–281). https://doi.org/10.1007/978–3-030–64682–0_13

Leendertz, F. H., & Kalema-Zikusoka, G. (2021). Vaccinate in biodiversity hotspots to protect people and wildlife from each other. *Nature, 591*(7850), 369. https://doi.org/10.1038/d41586-021-00690-z.

Macfie, E. J., & Williamson, E. A. (2010). Best practice guidelines for great ape tourism. International Union for Conservation of Nature and Natural Resources SSC Primate Specialist Group. https://portals.iucn.org/library/node/9636

Mazet, J. A. K., Genovese, B. N., Harris, L. A., Cranfield, M., Noheri, J. B., Kinani, J. F., Zimmerman, D., Bahizi, M., Mudakikwa, A., Goldstein, T., & Gilardi, K. V. K. (2020). Human respiratory syncytial virus detected in mountain gorilla respiratory outbreaks. *EcoHealth, 17*(4), 449–460. https://doi.org/10.1007/s10393-020-01506-8

Mbayahi, A., & Kalema-Zikusoka, G. (2022). *COVID-19 and Africa's great apes. Challenges and threats amidst the COVID-19 pandemic for sustaining conservation through responsible great ape tourism.* Policy Press Brief of the Africa CSO Biodiversity Alliance.

Melin, A. D., Janiak, M. C., Marrone, F., Arora, P. S., & Higham, J. P. (2020). Comparative ACE2 variation and primate COVID-19 risk. *BiorxiV. The Preprint server for Biology.* https://www.biorxiv.org/content/10.1101/2020.04.09.034967v2.

Muresherwa, G., Makuzva, W., Dube, C. N., & Amony, I. (2022). The management of mountain gorilla tourism in Uganda: Are the socio-economic benefits realized? Journal of Transdisciplinary Research S Afr., *18*(1), a1136. https://doi.org/10.4102/td.v18i1.1136

Non-Human Primate Covid-19 Information Hub. https://umnadvet.instructure.com/courses/324

Nuno, A., Chesney, C., Wellbelove, M., Bersacola, B., Kalema-Zikusoka, G., Leendertz, F., Webber, A. D., & Hockings, K. J. (2022). Protecting great apes from disease: Compliance with measures to reduce anthroponotic disease transmission. *Peopleand.* Nature Publishing [Research article]. https://doi.org/10.1002/pan3.10396

One Health Productions. http://onehealthproductions.com/gorillatrekkingfilm

Otsuka, R., & Yamakoshi, G. (2020). Analysing the popularity of YouTube videos that vio late mountain gorilla tourism regulations. *PLOS ONE*, 15(5), e0232085. https://doi.org/10.1371/journal.pone.0232085

Palacios, G., Lowenstine, L. J., Cranfield, M. R., Gilardi, K. V. K., Spelman, L., Lukasik-Braum, M., Kinani, J. F., Mudakikwa, A., Nyirakaragire, E., Bussetti, A. V., Savji, N., Hutchison, S., Egholm, M., & Lipkin, W. I. (2011). Human Metapneumovirus infection in wild mountain gorillas, Rwanda. *Emerging Infectious Diseases, 17*(4), 711–713. https://doi.org/10.3201/eid1704.100883

Patrono, L. V., Samuni, L., Corman, V. M., Nourifar, L., Röthemeier, C., Wittig, R. M., Drosten, C., Calvignac-Spencer, S., & Leendertz, F. H. (2018). Human coronavirus OC43 outbreak in wild chimpanzees, Côte d'Ivoire. *Emerging Microbes and Infections, 7*, 118 Emerging Microbes & Infections. https://doi.org/10.1038/s41426–018–0121–2

Refisch, J., Wich, S. A., & Williamson, E. A. (Eds.) (2018). Great apes status report. United Nations Environment Programme Great Apes Survival Partnership, and International Union for Conservation of Nature. https://cites.org/sites/default/files/eng/com/sc/70/E-SC70-52.pdf

Robbins, M. M., Gray, M., Fawcett, K. A., Nutter, F. B., Uwingeli, P., Mburanumwe, I., Kagoda, E., Basabose, A., Stoinski, T. S., Cranfield, M. R., Byamukama, J., Spelman, L. H., & Robbins, A. M. (2011). Extreme conservation leads to recovery of the Virunga mountain gorillas. *PloS One, 6*(6), e19788. https://doi.org/10.1371/journal.pone.0019788

Sandbrook, C., & Semple, S. (2006). The rules and the reality of mountain gorilla *Gorilla beringei beringei* tracking: How close do tourists get? Oryx vol. *Oryx, 40*(4, October), 428–433. https://doi.org/10.1017/S0030605306001323

Scully, E. J., Basnet, S., Wrangham, R. W., Muller, M. N., Otali, E., Hyeroba, D., Grindle, K. A., Pappas, T. E., Thompson, M. E., Machanda, Z., Watters, K. E., Palmenberg, A. C., Gern, J. E., & Goldberg, T. L. (2018) Lethal Respiratory Disease Associated with Human Rhinovirus C in Wild Chimpanzees, Uganda, 2013. *Emerging Infectious Diseases, 24*(2), 267–274. https://doi.org/10.3201/eid2402.170778

Tenywa, G. (2019). Gorillas biggest tourism foreign exchange earner. https://www.newvision.co.ug/news/1506406/gorillas-biggest-tourism-foreign-exchange-earner.

Thurstan, R. H., Hockings, K. J., Hedlund, J. S. U., Bersacola, E., Collins, C., Early, R., Ermiasi, Y., Fleischer-Dogley, F., Gilkes, G., Harrison, M. E., Imron, M. A., Kaiser-Bunbury, C. N., Refly Katoppo, D., Marriott, C., Muzungaile, M. M., Nuno, A., Regalla de Barros, A., van Veen, F., . . . Bunbury, N. (2021). Envisioning a resilient future for biodiversity conservation in the wake of the COVID-19 pandemic. *People and Nature.* https://besjournals.onlinelibrary.wiley.com/doi/full/10.1002/pan3.10262, 3(5), 990–1013. https://doi.org/10.1002/pan3.10262

Truscott, R. (2022). Call for COVID rules that reduced infections in gorilla parks to remain. https://news.mongabay.com/2022/02/call-for-covid-rules-that-reduced-infections-in-gorilla-parks-to-remain

Uganda Wildlife Act. (1996). Chapter 200. In *Legislation as at 16 June 2006. FRBR URI: /akn/ug/act/statute/1996/14/eng@2006–06–16.*

Uganda Wildlife Authority. (2014). *Bwindi and Mgahinga conservation area general management plan.*

Uganda Wildlife Authority (2020). *Standard operating procedure for tourism services and research activities in UWA estates and the reopening of the protected areas to the General Public During COVID-19 pandemic.* https://www.ugandawildlife.org/phocadownload/press_releases/SOPs-for-opening-tourism-and-research-post-covid-protected-areas.pdf

Van Hamme, G., Svensson, M. S., Morcatty, T. Q., Nekaris, K. A. I., & Nijman, V. (2021). Keep your distance: Using Instagram posts to evaluate the risk of zoonotic disease transmission in gorilla ecotourism. *Journal of Zoo and Wildlife Medicine, 44*, 1027–1035. https://doi.org/10.1002/pan3.10187

Weber, A., Kalema-Zikusoka, G., & Stevens, N. J. (2020). Lack of rule-adherence during mountain gorilla tourism encounters in Bwindi impenetrable national park, Uganda, places gorillas at risk from human disease. Frontiers in public health. Original Research published: 13 February 2020 https://doi.org/10.3389/fpubh.2020.0

Willoughby, I. (2021). Two more gorillas at Prague Zoo test positive for Covid. *Huffington Post.* https://english.radio.cz/two-more-gorillas-prague-zoo-test-positive-covid-8712362

Woodford, M. H., Butynski, T. M., & Karesh, W. B. (2002). Habituating the great apes: The disease risks. *Oryx vol, 36*(2, April).

World Health Organization. (2020). Emergencies. Coronavirus disease (COVID-19) pandemic. https://www.who.int/emergencies/diseases/novel-coronavirus-2019

Chapter 4

Wood bison management in northern Canada: The importance of considering people in managing wildlife disease

Kyle Plotsky, Louise Caplan and David C. Hall

Abstract

Wood Buffalo National Park (WBNP) straddles the border of Alberta and the Northwest Territories in northern Canada and is the largest national park in Canada. The park contains a large proportion of the remaining Canadian wood bison population (*Bison bison athabascae*), a keystone herbivore species and the largest terrestrial mammal in North America. The recovery of this species in Canada is hampered by infection of the WBNP bison with two zoonotic diseases, bovine tuberculosis (*Mycobacterium bovis*) and brucellosis (*Brucella abortus*). These diseases have the potential to spread to Canadian livestock and humans with major economic and health implications. Little has been done to reduce the prevalence of these diseases in the WBNP bison though recent management has focused on mitigating the spread of the diseases to bison outside WBNP. This chapter describes how these diseases have been managed over the last century. We focus on how the management and decision-making process has neglected important stakeholders and rightsholders in the past. We use the case of wood bison disease management in WBNP to show that consideration of people, their histories, and attitudes can improve wildlife disease management. The chapter concludes with a brief discussion of our current work to understand management preferences and how that information can inform future policy.

4.1 Wood bison in North America over the last 150 years

Wood bison (*Bison bison athabascae*) are the largest terrestrial mammal in North America standing up to 6 feet (1.83 m) tall and weighing 2400 lb (1088 kg) (Environment and Climate Change Canada, 2018). Wood bison are a cousin species to the smaller plains bison (*Bison bison bison*) and are in the same zoological family as cattle (Bovidae). Whereas plains bison roamed across most of North America, wood bison were limited to Alaska, the Yukon, the Northwest Territories, and northern

Susan C. Cork and Douglas P. Whiteside (eds) **Case Studies in EcoHealth**
DOI: 10.52517/9781789183368.004, © 5m Books Ltd, 2024

British Columbia, Alberta, and Saskatchewan (Stephenson et al., 2001). Although reliable wood bison population numbers prior to European arrival are not known, Indigenous management kept wood bison populations high and their numbers were considered plentiful (McCormack, 1992).

Wood bison were a desirable species to hunt as their fur was in high demand and their hides were used for industrial belts in European and North American factories (McCormack, 1992). These activities reduced the wood bison population to 150–300 individuals by the late 19th century (Sandlos, 2007). Most of these remaining wood bison were in northern Alberta and southern Northwest Territories. Recognizing that wood bison were near extinction, the Canadian government prohibited hunting of the species in 1894. Wood Buffalo National Park (WBNP) was subsequently created in 1922 to help recover the wood bison population and was placed under the purview of the Department of the Interior. The park's boundaries expanded southward in 1926 to encompass 44,807 km^2 and was transferred to the National Parks Branch in 1964 (McCormack, 1992). Wood bison populations were found to have increased to more than 10,000 individuals by 1971 (Carbyn et al., 1998). These populations have fluctuated over the last few decades with the total Canadian wood bison population now estimated at only 9000 individuals. More than half of the Canadian wood bison population is found in the greater Wood Buffalo National Park area (GWBNPA) but self-sustaining herds are found across Canada (Figure 4.1) and Alaska (Doney et al., 2018; Environment and Climate Change Canada, 2018).

Although cattle are now the dominant grazing species in the Great Plains, bison served this role for nearly 10,000 years until their near extirpation in the 19th century (Allred et al., 2011). Biodiversity is the variety of life in an ecosystem or in a more specific habitat. The protection of biodiversity supports ecosystems, provides resources for humans, and is fundamental to environmental health at the macro and microscopic ecological scales. Grazing by large mammals, such as bison, across grasslands helps maintain biodiversity through their widespread influence on other trophic levels, including plant, bird, insect, and fish life. These widespread influences are one of the reasons bison and other herbivores are described as keystone species. Although many North American landscapes are now grazed by cattle, the influence of domestic grazing on the environment is not equal to that of wild grazers (Allred et al., 2011; Nickell et al., 2018). One way in which bison grazing has a large effect on grassland biodiversity is through their foraging, which focuses on graminoid plants (i.e. grasses, sedges) and avoids forbs (i.e. herbaceous flowering plants) (Knapp et al., 1999). Preferential grazing of graminoids allows forbs to flourish with the additional effect of creating new habitats for other species and increasing the overall biodiversity of the grassland. Grazing also affects bird populations with some species showing increased populations with bison grazing and others reduced populations (Powell, 2006); bison grazing may help maintain shortgrass prairies and the bird species that thrive in those shortgrass systems compared to medium and tall grass systems (Askins et al., 2007). Bison also are considered ecosystem engineers due to the effects of their non-grazing behaviours, such as the wallows created by their rolling and pawing at the ground. These wallow patches result in increased arthropod abundance and diversity (Nickell et al., 2018). The influence of bison on their environment is not limited to biodiversity.

Of particular importance in WBNP, due to the importance of the Peace-Athabasca Delta and other tributaries in the area, is the potential impact that wood bison can have on aquatic and riparian environments. The recovery of wood

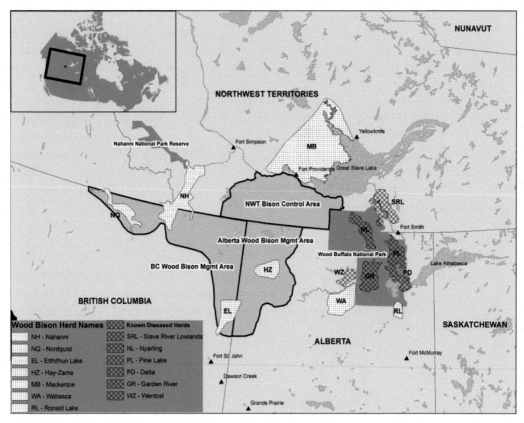

Figure 4.1 Diseased and disease-free bison herds and management areas in northern Canada. From Shury TK, Nishi JS, Elkin BT, Wobeser GA. (2015) Tuberculosis and brucellosis in wood bison (*Bison bison athabascae*) in Northern Canada: A renewed need to develop options for future management. *Journal of Wildlife Diseases* 51:543–554. Used with permission of the authors and of the Journal of Wildlife Diseases, Wildlife Disease Association.

bison may have habitat implications for the wetlands the bison inhabit (DeMars et al., 2016). Grazing in aquatic environments has not been well studied (Wood et al., 2017), but work has related grazing to reduced impacts of climate change on wetlands (McKenna et al., 2021), increased time until ephemeral pools dry (Pyke and Marty, 2005), higher water levels and more dynamic vegetation (Voldseth et al., 2007), and can benefit certain plant species (Falk et al., 2015; Kitti et al., 2009). Overall, a restored bison population could help mitigate the effects of climate change on the Peace-Athabasca Delta (McKenna et al., 2021) and enhance ecosystem

hydrology and vegetation (Falk et al., 2015; Kitti et al., 2009; Voldseth et al., 2007).

4.2 Zoonotic diseases in wood bison

Wood bison populations have had to contend with a variety of issues over the last century stemming from economic resource development, climate change, and the impacts of disease. This last issue is the largest limit on the recovery of wood bison populations in Canada as management has had to balance increasing populations with addressing bovine tuberculosis

(*Mycobacterium bovis* or bTB) and brucellosis (*Brucella abortus*) (Environment and Climate Change Canada, 2018). The WBNP wood bison population became infected around 1925 after 6600 plains bison were transferred into the park. These plains bison were rounded up in a bison park near Wainwright, Alberta after the herd became too large for the fenced park (Figure 4.2). The Wainwright bison tested positive for bTB and brucellosis in 1917 after the animals were likely infected by cattle prior to being shipped to Alberta from Montana. The presence of these diseases led to the transfer to WBNP being initially opposed by various parties, such as veterinarians and biologists. Eventually the transfer was agreed to with the expectation that each plains bison being transferred would be tested for bTB and brucellosis. This testing requirement was eventually dropped due to the

cost (McCormack, 1992; Sandlos, 2003). As expected, the plains bison interbred and hybridized with the wood bison in WBNP resulting in the park bison becoming infected as well.

Brucellosis and bTB are both bacterial diseases that can result in chronic infections. Brucellosis in particular results in fever and abortion, whereas bTB results in slow growth of masses or 'tubercles', particularly in the lungs of the infected animal (Box 4.1). In the GWBNPA wood bison population, the prevalence of bTB has been estimated at 21–49% while brucellosis has been estimated at 25–31% (Environment and Climate Change Canada, 2018; Joly & Messier, 2004a; Tessaro et al., 1990). The prevalence of both diseases occurring in an individual wood bison has been estimated at 42% (Tessaro et al., 1990). Differences in disease prevalence have been found between

Figure 4.2 Rounding up of plains bison at the Wainwright Buffalo National Park for transfer to Wood Buffalo National Park in the 1920s. From the Provincial Archives of Alberta (no known copyright restrictions).

Box 4.1 Tuberculosis and brucellosis

Tuberculosis is an infectious bacterial disease, found in multiple species including cattle, pigs, dogs, and humans. It is caused by a wide range of bacteria of the family Mycobacteriaceae. Clinical signs of pulmonary tuberculosis include coughing, fever, progressive weight loss, and eventually death. Bovine tuberculosis is caused by the airborne aerosol transmission of *Mycobacterium bovis* and is still a problem in many countries. The bovine sub-species of the disease is still found in cattle in other areas of the world but has been essentially eliminated in Canadian cattle. Nevertheless, small pockets of bovine tuberculosis still exist in wild bison (e.g. WBNP) and, rarely, other wildlife. In the rare instances of spillback and identification of the disease in cattle, the herds are depopulated to protect other healthy cattle, humans, and the economic interests of the Canadian beef industry.

Brucellosis is an infectious bacterial disease caused by *Brucella abortus* and other related Brucella species. It also is primarily found in bovines, including bison, but can infect humans and other species. In cattle and bison, the disease typically causes fever, abortion, weight loss, and death. Brucellosis is transmitted primarily through contact with infected placenta material or through ingestion of infected milk or milk products. It can also be transmitted to animals while still in the uterus, resulting in animals being born infected. Canadian cattle are officially brucellosis free, but the disease is endemic in some wood bison populations and has been found in other wildlife species.

---

age and sex classes but not between herds with different bison densities (Joly and Messier, 2004a). Reduction in the park wolf population in the first half of the 20th century may have helped these diseases spread among the WBNP wood bison population (McCormack, 1992; Parks Canada, 2010; Sandlos, 2003).

The WBNP wood bison population has changed drastically over the last five decades. Park bison populations were estimated to be over 10,000 around 1970, below 2500 in 1998, and 3300 in 2014 (Carbyn et al., 1998; Environment and Climate Change Canada, 2018). Infection with bTB and brucellosis may have played a role in the large population decrease between 1970 and 1998 (Joly and Messier, 2004b). Infection with bTB and brucellosis can negatively affect winter survival, pregnancy, and birth rates (Joly and Messier, 2005). Infection also has been linked with an increased likelihood of depredation by wolves with impacts on population growth (Environment and Climate Change Canada, 2018; Joly and Messier, 2004b).

Nevertheless, bison populations have increased without known changes to disease prevalence, showing there are other important influences (Bradley and Wilmshurst, 2005). For example, work on plains bison in Yellowstone National Park in the United States (US) has shown that nutritional quality can affect infection and disease prevalence and bison populations (Treanor, 2012). Juveniles also are least likely to be affected by bTB or brucellosis and reduced juvenile survival was an important factor in the wood bison population changes (Bradley & Wilmshurst, 2005).

Brucellosis and bTB are zoonotic diseases that can infect other wildlife, particularly gregarious or congregating species, domestic animals, and humans. Deer (*Odocoileus virginianus*) and elk (*Cervus canadensis*) around Riding Mountain National Park (Manitoba, Canada) as well as deer in Minnesota in the US have been found to be infected with bTB (Nishi et al., 2006; VerCauteren et al., 2018); the disease was effectively controlled and eliminated in both of these cases. Various wildlife species in Michigan, including deer, elk, bears (*Ursus americanus*), coyotes (*Canis latrans*), opossums (*Didelphis virginiana*), raccoons (*Procyon lotor*), and foxes (*Vulpes vulpes*), have been found infected with bTB (Michigan Department of Natural Resources, 2022). Diseases in wildlife have often been exacerbated by artificial feeding areas for wildlife and livestock that encourage animals to congregate. Artificial feeding grounds also likely played a role in plains bison and elk around Yellowstone National Park being persistently infected with brucellosis (Schumaker et al., 2012; Scurlock & Edwards, 2010). Some isolated muskox (*Ovibos moschatus*) and caribou (*Rangifer tarandus*) in northern Canada and Alaska have been found to be infected with brucellosis (Afema et al., 2017; Allen et al., 2020). There have been recorded instances of spillback into domestic cattle in Manitoba and Montana with major economic implications for the agricultural industry. Although unlikely, some Canadians have been infected with the diseases after consumption of wildlife meat (Chan et al., 1989; Forbes and Tessaro, 1993; Kanji and Saxinger, 2018).

4.3 Management of wood bison in Wood Buffalo National Park

Eliminating zoonotic diseases from a wildlife reservoir can be quite difficult, especially in gregarious species like wood bison (Miller and Sweeney, 2013; Tessaro et al., 1993). Typical wildlife disease management strategies are based on utilitarian agricultural approaches that emphasize the removal of animals that are infected or could be infected (Sandlos, 2003). A few years after the introduction of the Wainwright bison to WBNP, park management

began harvesting wood bison for use by residential schools in the area due to food shortages (McCormack, 1992; Sandlos, 2003). Around the time of the food shortages in the residential schools, there were food shortages in the communities in and around the park as well. Although Indigenous people have harvested other animals in the park as part of their traditional activities, wardens increased the number of bison harvested by park staff to supply people with food rather than allow the residents to hunt the bison themselves.

The hunting of bison by park wardens set the precedent for bison to be used in a potential commercial enterprise while attempting to reduce disease prevalence (McCormack, 1992). Around 1950, bison management began to shift towards commercial production of bison meat rather than as a local food source (McCormack, 1992). The government published information on a test and slaughter strategy in 1954 and completed meat production infrastructure (e.g. corrals, abattoirs) around the same time. Bison harvesting peaked in 1954 with around 900 bison killed and more than 300,000 pounds of bison meat produced that year (Figure 4.3). The second half of the 1950s often saw 100,000–250,000 lb (45–113 tonnes) of bison meat produced each year. This amount continued decreasing and most years in the 1960s, when the slaughter program was active, had less than 100,000 lb (45 tonnes) of bison meat produced. This commercial slaughter strategy was not effective at eliminating either bTB or brucellosis in the wood bison population. Management of wood bison through commercial harvest discontinued when WBNP was transferred to the National Parks Branch in 1964 and the final hunt for local food occurred in 1972 (McCormack, 1992).

In addition to bTB and brucellosis, bison in the GWBNPA can suffer from outbreaks of anthrax (*Bacillus anthracis*) that can be brought on by environmental conditions. These outbreaks often are short-lived with high

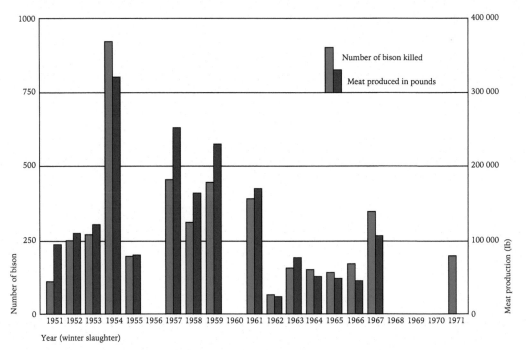

Figure 4.3 Bison slaughtered and meat produced from 1951–1971 when commercial harvesting was used to manage WBNP wood bison populations and health. 100,000 lb = 45 tonnes. Credit: Arctic Institute of North America.

mortality rates. An anthrax outbreak in 1962 led to a management plan that influenced disease management practices through 1977 (Nishi et al., 2002). Short term management included proper carcass disposal, corralling and vaccinating bison, removing other bison (e.g. old, unfit, remote individuals), and creating control zones to limit bison movement (Nishi et al., 2002; Sandlos, 2003). Long-term management focused on simultaneously eliminating bTB, brucellosis, and anthrax through large round-ups and vaccination (Nishi et al., 2002). These round ups began in 1965 but were discontinued in 1977 due to the public being concerned about bison mortalities, costs, and ineffectiveness. At least 1400 wood bison mortalities were attributable to the management actions in this period with no known lasting impact on the prevalence of bTB and brucellosis. Importantly, some of the bison deaths were due to bison handling

practices based on limited knowledge (T. Shury, personal communication). Lessons learned from these past experiences and through the commercial bison industry have made current handling practices more refined and humane with much lower mortality rates.

Bison management in WBNP became a strategy of non-interference after 1977 bringing WBNP policy into conformity with broader national park policy (Parks Canada, 2010). Brucellosis and bTB remained endemic in the park wood bison during this time. Two decades of bison committees and boards began in 1986 with a Disease Task Force (Table 4.1). The Task Force concluded in 1988 and referred the wood bison disease issue to the Federal Environmental Assessment and Review Office. The subsequent 1990 recommendation was to remove and replace all bison in WBNP (Carbyn et al., 1998; Chisholm et al., 1997). Some communities

Table 4.1 The committees and boards focused on the WBNP wood bison disease issue from 1986 to 2005.

Name	Years	Purpose/outcome/issues
Wood Bison/Disease Task Force[1]	1986–1988	• Reviewed existing information, developed and evaluated management recommendations • No representatives from local communities or environmental groups included • Only supported recommendation was depopulation and replacement • Recommendation became the basis of the Environmental Assessment Review Panel in the Federal Environmental Assessment and Review Office
Northern Diseased Bison Environmental Assessment Panel[1]	1989–1990	• Required action when a proposed action (e.g. depopulation) may have environmental impact on federal land • Five members (all but chair from biology) with none from the communities • Two sets of public hearings • Agriculture Canada included as a proponent in second hearing to focus discussion • A fully developed depopulation and replacement plan was displayed throughout the hearing process • Non-agricultural submissions to the panel included concerns that were not addressed, including proposals from concerned Indigenous groups that opposed depopulation • Recommended depopulation and replacement of WBNP bison: disease too great a risk and non-hybridized wood bison should be in the park
Northern Buffalo Management/Recovery Board[1,2]	1991–1993	• Result of opposition to 1990 depopulation proposal • Members from nine WBNP communities and seven government departments • One Indigenous and one government co-chair • Concluded there are important knowledge gaps and recommended gaps be addressed before an action plan
Bison Research Containment Program[2,3]	1995–2001	• Established to make decisions based on science, traditional knowledge, community participation • Discontinued after extension denied
Research Advisory Committee[3]	1996–2001	• Goal to increase community participation • Discontinued with the Bison Research Containment Program after extension denied
Interim Measures Working Group[3,4]	2003–2004	• Collaboration between Canadian Food Inspection Agency, Parks Canada, Governments of Alberta and the Northwest Territories • Goal was to create bison disease control areas, increase public knowledge, and monitoring of domestic bison/cattle • Goals were not implemented until funding secured and public consulted
Disease Task Group of the Canadian Wildlife Director's Committee[3,5]	2005	• Goal to create long-term bison disease management strategy • Bison Diseases Technical Workshop concluded depopulation and repopulation would take 15–20 years and cost CAN$62–78 million

Sources: [1]Information from (McCormack, 1992); [2]Information from (Chisholm et al., 1997); [3]Information from (Environment and Climate Change Canada, 2018); [4]Information from (Nishi et al., 2006); [5]Information from (Shury et al., 2006).

around WBNP felt left out of the decision-making process and the subsequent negative public reaction to the replacement plan led to the plan not being implemented (Chisholm et al., 1997). In order to include the diverse perspectives of rights holders and stakeholders, the Northern Buffalo Management Board was formed in 1991 with most members of the board being from communities around the park (Carbyn et al., 1998; Chisholm et al., 1997).

The decade after the 1990 depopulation proposal saw various committees begin and conclude without an implemented bison management plan (Table 4.1). A proposed 1992/93 WBNP management plan was not approved and the Northern Buffalo Management Board concluded there are knowledge gaps around bison, brucellosis, and bTB (Chisholm et al., 1997; Parks Canada, 2010). These gaps limited the board's ability to develop and implement management programs. A Bison Research Containment Program was established in 1995 to make decisions based on science, traditional knowledge, and community participation (Chisholm et al., 1997; Environment and Climate Change Canada, 2018). Subsequently, a multi-stakeholder Research Advisory Committee began in 1996 to increase community participation. Although the management board attempted to initiate a community participatory approach to bison management, the Research Advisory Committee and Containment program concluded in 2001 after a 4 year extension was denied (Environment and Climate Change Canada, 2018). A subsequent Interim Measures Working Group was formed in 2003 as a collaboration between the Canadian Food Inspection Agency, Parks Canada, and the governments of Alberta and the Northwest Territories. The goal of the Interim Measures Working Group was to create bison disease control areas while increasing public education and the monitoring of cattle (Nishi et al., 2006). It was decided that these goals would not be implemented until funding was secured

and the public consulted. The Interim Measures Working Group proposal was set aside in favour of developing a long-term strategy (Environment and Climate Change Canada, 2018). A long-term strategy was to be created by the Disease Task Group of the Canadian Wildlife Director's Committee in 2005. Participants at a workshop in the same year concluded that although depopulation and repopulation of the park bison may be the best option, it would take two decades and cost more than CAN$62 million (Environment and Climate Change Canada, 2018; Shury et al., 2006). The 2010 WBNP Management Plan focused on reducing transmission of brucellosis and tuberculosis to cattle and wood bison outside the park through a containment strategy that would be implemented by 2012 (Parks Canada, 2010). To our knowledge, these recommendations were not implemented.

Recent bison management plans allow for the WBNP bison population to be reduced to as few as 1000 individuals (Environment and Climate Change Canada, 2018). Public acceptance of bison management and the perceived value of wood bison also would be increased through increased hunting opportunities. Although the park wood bison population would be reduced, the overall Canadian wood bison population would be more than 1000 individuals in at least five free-ranging disease-free herds outside WBNP. The importance of addressing the bison disease issue has increased in importance recently. The World Heritage Status of WBNP was classified as threatened after a Mikisew Cree First Nation petition and the status of wood bison was changed to threatened in northern Alberta (Government of Alberta, 2021; Parks Canada, 2019).

4.4 Ongoing bison management

Overall, management of wood bison within WBNP has been relatively hands off since the park came under the purview of Parks Canada.

The infections have spread to bison in the Slave River Lowlands area northeast of WBNP and to an Alberta herd along the western border of WBNP (Figure 4.1). The governments of the Northwest Territories (NWT) and Alberta have taken direct management action on wood bison to reduce the likelihood that uninfected individuals will interact with infected individuals. These control actions contribute to the diseases being the largest limiting factor on the recovery of Canadian wood bison.

The NWT began operating a control zone north of WBNP in 1987 to stop the northward expansion of the diseases (Nishi, 2002). The control zone now encompasses the area from the north boundary of WBNP to the Mackenzie area and west past the British Columbia border (Conference of Management Authorities, 2019). Higher risk areas have regular government surveillance and bison spotted within the control area are removed. Lower risk areas rely on public reporting of bison and any NWT resident can kill bison in the control area. The NWT control zone costs an estimated CAN$100,000/year in 2010 and is not expected to be rescinded until the WBNP bison are disease free (Conference of Management Authorities, 2019; Environment and Natural Resources, 2010).

The Alberta control zone focused on the area west of WBNP with some aerial surveys to remove bison and bison hunting without a license being permissible (Government of Alberta, 2011, 2017b). The health of wood bison herds around the park are actively monitored by the government (Nishi, 2017). The population of the Hay-Zama herd, a large disease-free wood bison herd in the northwest of Alberta, is maintained at 400–600 bison through hunting to limit the eastward expansion of the herd towards WBNP (Government of Alberta, 2017b; Nishi, 2017). The status of wood bison in Alberta changed in November 2021 to threatened; bison in some areas west of WBNP can still be harvested without a license

while bison in other areas adjacent to the park are now protected due to the dwindling size of some herds (Government of Alberta, 2021).

Efforts to conserve wood bison in Alberta may be further complicated by evidence that some herds show stronger genetic differentiation than other herds from the WBNP population (Ball et al., 2016). No Canadian wood bison are considered genetically pure as all wood bison were recovered from the WBNP populations after the 1925 introduction of plains bison. Although most wood bison herds are genetically similar to the WBNP population, the Harper Creek bison herd is much more genetically diverse than other herds evaluated by Ball et al. (2016). However, the reasons for this herd's high genetic diversity are unknown. The lack of genetic diversity in the wood bison population is a focus of a recent CAN$5 million project, the Bison Integrated Genomics project (Genome Canada, 2022). This project has four pillars, including the preservation of bison genetic material and increasing genetic diversity across the Canadian bison population as well as the development of a reliable test and a combination vaccine for bTB and brucellosis.

4.5 People are integral to management going forward

Although wildlife disease management has typically taken an agricultural approach to wildlife management by removing infected or potentially infected animals. The acceptance of removing animals varies depending on context but often is not deemed acceptable by members of the public (Miguel et al., 2020). Removal of the WBNP bison population has been proposed multiple times without being implemented. Public backlash to the 1990 proposal led to multiple committees and management boards over the following two decades without a concrete management plan being implemented. Similarly, negative public reaction to the cost and bison deaths were part of the reason the corralling and vaccination program in the 1960s was discontinued (Nishi et al., 2002). This history shows the importance of considering people when developing and implementing wildlife management policies (Dickman, 2010; Redpath et al., 2013; Vaske et al., 2008) and WBNP will likely require a different approach than depopulation (Shury et al., 2015).

In the case of WBNP and the wood bison health issue, people from different backgrounds are expected to be interested in the way bison health is addressed. These groups include rights holders (e.g. First Nations, Métis) and other stakeholder groups, such as the livestock industry, recreationists (e.g. hunters), and broader civil society. These groups are expected to have diverse perspectives on wood bison management and those perspectives may conflict with one another (Bath et al., 2022; Schroeder et al., 2021; Vaske & Miller, 2018). Here we will outline some of these expected perspectives to highlight the variability and potential for conflict.

4.6 Indigenous people and bison management

Indigenous people were managing and harvesting wood bison populations for thousands of years prior to European arrival (McCormack, 1992). Wood bison are an important cultural, spiritual, and subsistence animal for many Indigenous groups in the GWBNPA. The relationship between the government and Indigenous groups regarding bison began to deteriorate soon after the protection of wood bison. The first person convicted of poaching north of the 60th parallel was an Indigenous man caught hunting wood bison around WBNP (Sandlos, 2007). Francois Biscaye was found guilty of having hunted wood bison in the spirit of mischief after the police did not believe his

claim of starvation. Protecting the bison from hunting by Indigenous and other residents of the WBNP area was a focus of management after the park was created (McCormack, 1992). Park management allowed themselves to hunt wood bison to supply Indigenous people and residential schools around the park with food. Indigenous people were not allowed to hunt bison for their own food while these park management hunts occurred.

Beyond being hired to assist with the anthrax control programs in the 1960s (e.g. corralling bison), the role of Indigenous groups in the management of the WBNP has been limited. This exclusion of Indigenous perspectives from park management reflects the North American Model (NAM) of Conservation that began in the late 19th century and is discussed below in more detail (Hessami et al., 2021; Jacob et al., 2020; Sandlos, 2014). The 1984 and 2010 WBNP management plans included statements of intent to work with Indigenous groups (Parks Canada, 2010). The 1984 plan included the equitable settlement of land claims and establishing new reserves in the park. Some of these land claims were settled; for example, the Mikisew Cree First Nation signed a Treaty Land Entitlement in 1986 that created reserves in and around Fort Chipewyan, Alberta

Recent legal decisions reinforce that governments are obligated to include Indigenous people in wildlife management and research shows that involvement to be integral to positive outcomes (Holmes et al., 2016; Langdon et al., 2010; Massey et al., 2021; Sandlos, 2014). Many Indigenous communities were vocal critics of the 1990 proposal to depopulate the park bison and feel they do not have control over their traditional territories nor a role in the decision-making process (McCormack, 1992). Indigenous leaders also were disappointed that the Bison Research Containment Program left management decision making in the hands of Parks Canada (Chisholm et al., 1997).

Indigenous groups were approached in 2006 about changes to WBNP's Game Regulations to help rebuild relationships between the groups and park management. A 2009 report on the state of WBNP and the 2010 management plan emphasized the importance of engaging with local Indigenous groups on bison health (Parks Canada, 2010). A 2022 Parks Canada managed clean-up of a WBNP anthrax outbreak (e.g. burning of carcasses) included representation of Indigenous spirituality (Ulrich, 2022). Although Indigenous groups now have representation on WBNP co-management committees, groups still feel their voices do not carry weight in the decision-making process (Willow Spring Strategic Solutions, 2021).

Wood bison, the relationship of wood bison with wolves, and the Indigenous history in WBNP were reasons WBNP was granted World Heritage Site status in 1983. The Mikisew Cree First Nation began a petition in 2014 to have the universal value of WBNP considered under threat (Government of Canada, 2018). This petition was successful and resulted in a wide ranging 2019 Action Plan and an ongoing World Heritage reactive monitoring mission (Parks Canada, 2019). The Action Plan includes 17 recommendations around the ecological integrity of the GWBNPA and four recommendations focused on improving the partnership between Indigenous groups and park management.

4.7 Livestock production and zoonotic diseases

Brucellosis and bTB can be transmitted to domestic livestock with major economic implications. Agriculture Canada was given a prominent role in the 1990 depopulation decision due to the perceived economic and health risk of the bison health issue (McCormack, 1992). Livestock ranches and game ranches exist within ~70 km of WBNP. The Government of Alberta created a provincial grazing area in 1981 that is approximately 120 km from the southwest border of WBNP (Government of Alberta, 2017a). This is the most northerly livestock grazing reserve in Alberta and can be grazed May through October by 5300 animal unit months (i.e 5300 cattle weighing 1000 lb [454 kg] each). Transmission of one of these diseases from wildlife to cattle has occurred in Manitoba, Montana, and Michigan (Carstensen & Doncarlos, 2011; Kauffman et al., 2016; Nishi et al., 2006; VerCauteren et al., 2018). To date there has not been a known case of transmission to livestock around WBNP. This may be due to active efforts by the Alberta government to minimize disease spread or to cattle being less dense or farther from infected animals compared to other areas (Nishi et al., 2006).

Cattle and commercial bison producers are concerned about diseases entering their livestock given the risk of herd losses and damage to their livelihoods (Smith et al., 2012; White et al., 2011). The Canadian livestock industry, particularly the cattle industry, relies on the ability to export their product outside Canada. There have been sporadic outbreaks of bTB in Canadian cattle which have been rapidly contained. In general, if additional uncontrolled outbreaks of bTB occur in Canadian cattle on independent premises within the course of 18–24 months, producers might not be able to export their product internationally (CFIA, 2019) resulting in losses of at least CAN$4.5 billion per annum (Canfax, 2021; Toor & Hamit-Haggar, 2021). The possible tremendous economic impact escalating from a single outbreak of bTB or brucellosis in Canadian livestock are expected to make livestock producers particularly interested in the issue.

4.8 Hunting for conservation

Indigenous people have been hunting wildlife in North America for millennia. Yet it only took a few centuries after European arrival in North America for populations of multiple wildlife species to be decimated by commercialization and open harvest (Wright, 2022). The banning of wildlife markets and the encouragement of sport hunting and scientific management is purported to be the reason many wildlife species recovered. This focus on sport hunting rather than for subsistence or economic gain is fundamental to the NAM of wildlife conservation. Under this system, wildlife is not owned by any individual but are held in the public trust for use by anyone, with scientific process being the way to identify appropriate actions (Organ et al., 2012). The NAM is generally considered successful but there are ethical criticisms of the model; for example, recent criticisms include the lack of Indigenous involvement in management and whether putting hunters at the centre of wildlife management is beneficial for everyone (Eichler & Baumeister, 2018; Nelson et al., 2011).

Hunting in North America is considered a primary mechanism by which wildlife management is conducted. Sportspeople provide observations, samples for research, much of the funding for conservation, and political lobbying (Organ et al., 2012; Schaefer, 2019). This means that the views of hunters are often given priority and included in management decisions (Eichler and Baumeister, 2018; Nelson et al., 2011). A recent press release on the status of wood bison in Alberta included hunting opportunities as a goal: 'The Government of Alberta wants sustainable wild wood bison populations that provide ecological benefits, support food security for Indigenous communities, and increased hunting, guiding and outfitting opportunities for Albertans' (Government of Alberta, 2021, p. 2). Although WBNP does not allow hunting by non-Indigenous people, the past, ongoing, and future role of hunters in wildlife conservation makes the views of this stakeholder group important to the implementation and public acceptance of any management policy. Other jurisdictions include public hunting as part of wildlife disease management strategies (e.g. Nishi et al., 2006; Ramsey et al., 2016). Hunting fees also are a primary funding source for conservation across Alberta, including areas around WBNP (e.g. Alberta Conservation Association, 2022). In other areas of North America, funding for wildlife conservation has been affected by reduced hunting after identification of diseases (Heberlein, 2004).

Indigenous perspectives of wildlife emphasize different aspects of cultural, food, and economic security than have been included in the NAM, which has emphasized sport hunting (Hessami et al., 2021; Wright, 2022). These different perspectives and goals could lead to opposing management preferences. Conflict also may arise between non-Indigenous and Indigenous hunters as only Indigenous people can hunt within WBNP; Indigenous people also may hunt bison within WBNP in the future to ensure diseased herds do not come into contact with disease-free herds. Equal access to wildlife hunting opportunities is considered a fundamental principle of the NAM so non-Indigenous hunters may oppose an Indigenous focused hunting policy. The results of this potential conflict, if it were to occur, are unknown.

4.9 Civil society

Parks Canada is an agency within the federal government making it the purview of an elected official (i.e the Minister of Environment and Climate Change Canada). Thus, decisions and expenses are under political scrutiny with the views of taxpayers and voters being important components of the situation.

Although many members of broader civil society may not be directly affected by the disease issue, their support or opposition to wildlife management policy affects the implementation of management plans. For instance, the termination of the corralling program in the 1960s and not implementing the 1990 depopulation plan were both affected by the reactions of the broader Canadian public. Understanding the attitudes of this large group can decrease the likelihood that management decisions face widespread public criticism. Because WBNP is a national park, the people we include in civil society are intentionally broad; for instance, we advertise questionnaires to people from across western Canada and Ontario.

4.10 Determining the best way forward

There are a variety of potential ways to manage wood bison health in the GWBNPA. Some have been tried (e.g. corralling) or proposed (e.g. depopulation) for the park. Optimal strategies for managing diseases in wildlife are often those strategies that are acceptable and supported as these characteristics contribute to increased effectiveness (Vaske et al., 2008). Identifying the optimal strategy from a social science perspective includes consideration of inter-related issues regarding people's beliefs and preferences: (1) are people knowledgeable and concerned about the disease issue?; (2) how important is addressing the issue to people?; (3) which management strategies are supported?; and (4) are strategies similarly supported across stakeholders and rightsholders? We are currently collecting information on these and related questions to understand people's beliefs and preferences. To show the importance of this type of information to addressing a wildlife disease issue, we report on the results of a preliminary analysis

of questionnaire results from civil society (CS) and self-identifying hunters (SH).

To assess concern about zoonotic diseases from bison, our questionnaire asked respondents to indicate their agreement with four statements (i.e. bTB/brucellosis in bison poses a threat to my health/economic livelihood) using a five-point Likert scale. Neither group indicated that they were overly concerned about the effect of the diseases on their economic livelihoods. But SH were significantly more likely than the CS respondents to see a disease as a threat to their health (e.g. bTB: $n = 423$ (151 CS, 272 SH), $U = 2.169, p = 0.030$). The most common CS response was Strongly disagree that bTB is a threat to their health; SH responses were more evenly spread between Somewhat agree, Neutral, Somewhat disagree, and Strongly disagree.

Respondents also were asked about their agreement with statements that the diseases should be eliminated from wood bison as a measure of the importance of the disease issue. When asked about the diseases in general, the CS and SH respondents were not different from each other. But the CS and SH groups were significantly different than one another when asked about the diseases individually (e.g. bTB: $n = 408$ (150 CS, 272 SH), $U = -2.933$, $p = 0.003$) with CS more frequently Strongly agreeing than SH.

Wildlife management strategies differ in their characteristics, including effectiveness, necessary timeframe, and cost. We expect the level of importance of each strategy characteristic to be different for stakeholders and rightsholders. Both the CS and SH samples stated that a vaccination strategy was the most preferred out of a selection of strategies. Another highly ranked option was to capture, test, remove infected animals, and vaccinate uninfected individuals. Taking no action to address the bison disease issue was the least preferred strategy in both the CS and SH respondent groups.

We also asked respondents to declare their level of support for individual strategies. Although vaccination was the most preferred strategy for both CS and SH, there were differences in each groups' average level of support for the strategy ($n = 445$ (158 CS, 287 SH), $U = -5.720$, $p < 0.001$). CS more commonly indicated that they extremely support a vaccination strategy; SH respondents were more evenly distributed across the moderate, high, and extremely support categories.

The differences in the attitudes within and between the groups can be visualized via Potential for Conflict Index$_2$ (PCI) illustrations (Figure 4.4) (Manfredo et al., 2003; Vaske et al., 2010). A PCI value and associated graphic bubbles simultaneously show the central tendency, dispersion, and skewness of responses (Manfredo et al., 2003). A PCI value quantifies the amount of disagreement within a group of responses; a PCI value of 0 means there is little conflict within a group of responses (i.e. all responses are similar) while a PCI value of 1 shows maximum conflict and contrast between

respondents (i.e. a bimodal distribution). Bubble size is determined by the PCI value and bubble position along the y-axis shows the average response as well as the skewness of those responses. In the case of vaccination, the CS group shows more consistently positive responses than the SH group (Figure 4.4).

Respondent support for two other strategies display instances where the CS and SH groups are similar and relatively further apart (Figure 4.4). Support for a Hunting Unhealthy strategy (i.e allowing hunting of the unhealthy wood bison within WBNP) differed between the two respondent groups ($n = 438$ (153 CS, 285 SH), $U = 4.158$, $p < 0.001$). The highest frequency category in the CS group was moderate support followed by low and not at all support categories. In the SH respondents, moderate support was the highest frequency category followed by high and then low support. No category in either group contained more than 50% of responses but the mean response of the CS group was lower than the SH group. The PCI (Figure 4.4) for the Hunt Unhealthy strategy

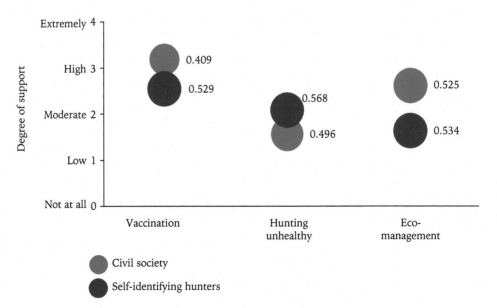

Figure 4.4 Potential for Conflict Index$_2$ for respondent indicated support for three potential wood bison management strategies with separate values for CS and SH respondent groups.

highlights that the groups had similar amounts of variability, but that SH were overall more supportive of the potential strategy.

The CS and SH groups are farther away from one another in their support for an eco-management program (e.g. predator, habitat management) ($n = 413$ (146 CS, 267 SH), $U = -7.511$, $p < 0.001$). These results show a higher likelihood of conflict between the two groups. CS respondents were more supportive than SH and this can be seen in the associated PCI visualizations (Figure 4.4). The groups are farther away from one another on this option than either of the other two strategies. This shows that managers and policy makers can expect different responses from these groups if an eco-management strategy were implemented; but the groups would likely respond more similarly to the other two strategies, particularly a Hunting Unhealthy strategy.

As mentioned, WBNP also allows Indigenous people to hunt as part of their traditional lifestyle.

Bison hunting within the park is currently not allowed, but could be as part of a management strategy. Given that the NAM emphasizes equal access to hunting opportunities, we inquired about respondents' agreement with Indigenous people hunting in WBNP for the purposes of management. The CS and SH responses to this question were significantly different ($n = 459$ (161 CS, 298 SH), $U = -4.567$, $p < 0.001$). PCI visualizations of these responses highlight that SH respondents agreed less and were much more variable as a group than CS respondents (Figure 4.5).

4.11 Conclusion

The bTB and brucellosis disease issue is currently the largest limit on the recovery of the wood bison population in northern Canada (Environment and Climate Change Canada, 2018). The 20th century saw a variety of proposed and undertaken actions to address the

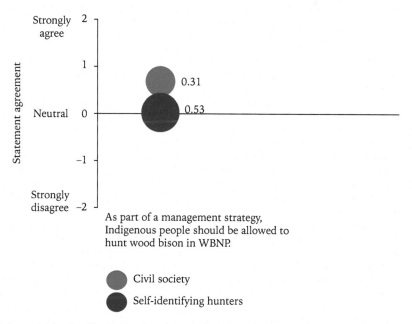

Figure 4.5 Potential for Conflict Index$_2$ for stated agreement with statement about Indigenous hunting within WBNP as part of a management strategy with separate values for CS and SH respondent groups.

issue. Some actions, such as harvesting for meat in the mid-20th century, were not effective at reducing the prevalence of bTB and brucellosis in the wood bison population. Public reaction also led strategies to be abandoned, such as the 1960s bison round ups, or not implemented, such as the 1990 depopulation proposal. The wood bison disease issue has a complicated past that often has neglected the importance of people to wildlife management.

We have provided a brief discussion showing how a single wildlife disease issue can result in diverse and potentially conflicting perspectives from stakeholders and rightsholders. Groups may align on some aspects of an issue while having other contrasting attitudes that could lead to conflict. Identifying those attitudes and where they may come from is an important component of designing management policy. Human dimensions information can inform on the ground strategies as well as the education, communication, and public involvement components of a policy. For instance, our preliminary results show there may be a higher potential for conflict with some strategies than others (e.g. eco-management versus hunting unhealthy). This may necessitate different education and communication for the different groups. In the case of Indigenous hunting within the park, communication to hunters could focus on the rights of Indigenous people to guide and participate in management on their traditional territories. The results also could be used by the Alberta government to guide management in the area around the park; Alberta could investigate allowing non-Indigenous and Indigenous hunters to take bison outside the park to keep herds apart. This combined approach could help address some potential conflict by being more in line with the equal access principle of the NAM, but may require including the attitudes of the local Indigenous groups and non-Indigenous hunters in management.

Wood bison are an important species ecologically, historically, culturally, and for future food security. Wildlife disease management is a complicated picture that requires a multidisciplinary approach that incorporates knowledge and judgment founded in the biological, social, and humanities disciplines.

References

Afema, J. A., Beckmen, K. B., Arthur, S. M., Huntington, K. B., & Mazet, J. A. K. (2017). Disease complexity in a declining Alaskan muskox (Ovibos moschatus) population. *Journal of Wildlife Diseases, 53*(2), 311–329. https://doi.org/10.7589/2016-02-035

Alberta Conservation Association. (2022). Revenue sources 2021/2. https://www.ab-conservation.com/about/revenue-sources/

Allen, S. E., Vogt, N. A., Stevens, B., Ruder, M. G., Jardine, C. M., & Nemeth, N. M. (2020). A retrospective summary of cervid morbidity and mortality in Ontario and Nunavut regions of Canada (1991–2017). *Journal of Wildlife Diseases, 56*(4), 884–895. https://doi.org/10.7589/JWD-D-19-00018

Allred, B. W., Fuhlendorf, S. D., & Hamilton, R. G. (2011). The role of herbivores in Great Plains conservation: Comparative ecology of bison and cattle. *Ecosphere, 2*(3). https://doi.org/10.1890/ES10-00152.1

Askins, R. A., Chávez-Ramírez, F., Dale, B. C., Haas, C. A., Herkert, J. R., Askins, R. A., Chávez-Ramírez, F., Dale, B. C., Haas, C. A., Herkert, J. R., Knopf, F. L., & Vickery, P. D. (2007). Conservation of grassland birds in North America: Understanding ecological processes in different regions Recommended Citation. *Ornithological Monographs, 64*, 1–46. https://doi.org/10.1525/0078-6594(2007)64[1:COGBIN]2.0.CO;2

Ball, M. C., Fulton, T. L., & Wilson, G. A. (2016). Genetic analyses of wild bison in Alberta, Canada: Implications for recovery and disease management. *Journal of Mammalogy, 97*(6), 1525–1534. https://doi.org/10.1093/jmammal/gyw110

Bath, A. J., Engel, M. T., van der Marel, R. C., Kuhn, T. S., & Jung, T. S. (2022). Comparative views of the public, hunters, and wildlife managers on the management of reintroduced bison (Bison bison). *Global Ecology and Conservation, 34*, e02015. https://doi.org/10.1016/j.gecco.2022.e02015

Bradley, M., & Wilmshurst, J. (2005). The fall and rise of bison populations in Wood Buffalo National Park. *Canadian Journal of Zoology, 83*(9), 1195–1205, 1971–2003. https://doi.org/10.1139/z05–106

Canfax. (2021). *The beef industry's contribution to the Canadian economy.* https://www.beefresearch.ca/files/pdf/Multiplier-Summary-May-2021.pdf

Carbyn, L. N., Lunn, N. J., & Timoney, K. (1998). Trends in the distribution and abundance of bison in Wood Buffalo National Park. *Wildlife Society Bulletin, 26*(3), 463–470.

Carstensen, M., & Doncarlos, M. W. (2011). Preventing the establishment of a wildlife disease reservoir: A case study of bovine tuberculosis in wild deer in Minnesota, USA. *Veterinary Medicine International, 2011, 413240.* https://doi.org/10.4061/2011/413240

CFIA. (2019). Bovine tuberculosis (bovine TB) trace-in activities. https://inspection.canada.ca/animal-health/terrestrial-animals/diseases/reportable/bovine-tuberculosis/investigation-british-columbia/trace-in-activities/eng/155553777 2049/1555537831661

Chan, J., Baxter, C., & Wenman, W. M. (1989). Brucellosis in an Inuit child, probably related to caribou meat consumption. *Scandinavian Journal of Infectious Diseases, 21*(3), 337–338. https://doi.org/10.3109/00365548909035706

Chisholm, J., Comin, L., & Unka, T. (1997). Consensus-based research to assist with bison management in wood Buffalo national park. In L. R. Irby & J. E. Knight (Eds.), International Symposium on Bison Ecology and Management in North America (pp. 199–204).

Conference of Management Authorities. (2019). Recovery strategy for wood bison (Bison bison athabascae) in the Northwest Territories. https://www.nwtspeciesatrisk.ca/sites/enr-species-at-risk/files/wood_bison_recovery_strategy_final_formatted_jul1619.pdf

DeMars, C. A., Nielsen, S. E., & Edwards, M. A. (2016). Range use, habitat selection, and the influence of natural and human disturbance on wood bison (Bison bison athabascae) in the Ronald Lake Area of Northeastern Alberta. Report to the Ronald Lake Bison Herd Technical Team, March 31, 2017. University of Alberta, Edmonton. https://ace-lab.ca/assets_b/2017_RLBH_Update_Report_March.pdf

Dickman, A. J. (2010). Complexities of conflict: The importance of considering social factors for effectively resolving human-wildlife conflict. *Animal Conservation, 13*(5), 458–466. https://doi.org/10.1111/j.1469-1795.2010.00368.x

Doney, E. D., Bath, A. J., & Vaske, J. J. (2018). Understanding conflict and consensus regarding wood bison management in Alaska, USA. *Wildlife Research, 45*(3), 229–236. https://doi.org/10.1071/WR17056

Eichler, L., & Baumeister, D. (2018). Hunting for Justice: An indigenous critique of the North American Model of Wildlife Conservation. *Environment and Society: Advances in Research, 9*(1), 75–90. https://doi.org/10.3167/ares.2018.090106

Environment and Climate Change Canada. (2018). Recovery strategy for the wood bison (Bison bison athabascae) in Canada. https://www.canada.ca/en/environment-climate-change/services/species-risk-public-registry/recovery-strategies/wood-bison-2018.html

Environment and Natural Resources. (2010). Wood bison management strategy for the Northwest Territories. https://www.wildlifecollisions.ca/docs/wood_bison_management_strategynwt2010.pdf

Falk, J. M., Schmidt, N. M., Christensen, T. R., & Ström, L. (2015). Large herbivore grazing affects the vegetation structure and greenhouse gas balance in a high arctic mire. *Environmental Research Letters, 10*(4). https://doi.org/10.1088/1748-9326/10/4/045001

Forbes, L. B., & Tessaro, S. V. (1993). Transmission of brucellosis from reindeer to cattle. *Journal of the American Veterinary Medical Association, 203*(2), 289–294.

GenomeCanada. (2022). Genome Canada announces $5.1 million genomics project to protect threatened Canadian bison population. https://genomecanada.ca/genome-canada-announces-5-1-million-genomics-project-to-protect-threatened-canadian-bison-population/

Government of Alberta. (2011). Disease detected in wood bison outside Wood Buffalo National Park. Information Bulletin. https://www.alberta.ca/release.cfm?xID=310047213C5D3-F4A9-7084-95883ED04DE8830B

Government of Alberta. (2017a). Fort Vermilion provincial grazing reserve. https://open.alberta.ca/dataset/5eb489b8-e6b3-4e92-87e4-8372e1a2ba62/resource/512f4d85-469d-4802-8688-6158ac8f08b4/download/fortvermilion-pgr-2017-12.pdf

Government of Alberta. (2017b). *Managing disease risk in Northern Alberta Wood Bison – Outside of Wood Buffalo National Park: 2015–2016 progress report.*

Government of Alberta. (2021). Wood bison status changes in Alberta (Issue November, pp. 1–3). https://open.alberta.ca/dataset/7e7a3c4e-f2a6–4b21-b329–8d8cebc5ea34/resource/1203bf27-fdec-4b97-a5bc-50fe2d77a6d6/download/aep-wood-bison-status-changes-2021–11.pdf

Government of Canada. (2018) Strategic environmental assessment of wood Buffalo national park. Executive summary. https://parks.canada.ca/pn-np/nt/woodbuffalo/info/action/strategie-env-assessment

Heberlein, T. A. (2004). 'fire in the sistine chapel': How Wisconsin responded to chronic wasting disease. *Human Dimensions of Wildlife, 9*(3), 165–179. https://doi.org/10.1080/10871200490479954

Hessami, M. A., Bowles, E., Popp, J. N., & Ford, A. T. (2021). Indigenizing the North American model of wildlife conservation. *FACETS, 6,* 1285–1306. https://doi.org/10.1139/facets-2020-0088

Holmes, A. P., Grimwood, B. S. R., & King, L. J. (2016). Creating an Indigenized visitor code of conduct: The development of Denesoline self-determination for sustainable tourism. *Journal of Sustainable Tourism, 24*(8–9), 1177–1193. https://doi.org/10.1080/09669582.2016.1158828

Jacob, M. M., Gonzales, K. L., Chappell Belcher, D., Ruef, J. L., & RunningHawk Johnson, S. (2021). Indigenous cultural values counter the damages of white settler colonialism. Environmental Sociology, 7(2), 134–146. https://doi.org/10.1080/23251042.2020.1841370

Joly, D. O., & Messier, F. (2004a). Factors affecting apparent prevalence of tuberculosis and brucellosis in wood bison. *Journal of Animal Ecology, 73*(4), 623–631. https://doi.org/10.1111/j.0021-8790.2004.00836.x

Joly, D. O., & Messier, F. (2004b). Testing hypotheses of bison population decline (1970–1999) in Wood Buffalo National Park: Synergism between exotic disease and predation. *Canadian Journal of Zoology, 82*(7), 1165–1176. https://doi.org/10.1139/z04-072

Joly, D. O., & Messier, F. (2005). The effect of bovine tuberculosis and brucellosis on reproduction and survival of wood bison in Wood Buffalo National Park. *Journal of Animal Ecology, 74*(3), 543–551. https://doi.org/10.1111/j.1365-2656.2005.00953.x

Kanji, J. N., & Saxinger, L. (2018). Brucella infection at cardiac pacemaker site in a patient who had consumed raw caribou meat in Northern Canada. *CMAJ, 190*(37), E1108–E1110. https://doi.org/10.1503/cmaj.170234

Kauffman, M., Peck, D., Scurlock, B., Logan, J., Robinson, T., Cook, W., Boroff, K., & Schumaker, B. (2016). Risk assessment and management of brucellosis in the southern greater Yellowstone area (I): A citizen-science based risk model for bovine brucellosis transmission from elk to cattle. *Preventive Veterinary Medicine, 132,* 88–97. https://doi.org/10.1016/j.prevetmed.2016.08.004

Kitti, H., Forbes, B. C., & Oksanen, J. (2009). Long- and short-term effects of reindeer grazing on tundra wetland vegetation. *Polar Biology, 32*(2), 253–261. https://doi.org/10.1007/s00300-008-0526-9

Knapp, A. K., Blair, J. M., Briggs, J. M., Collins, S. L., Hartnett, D. C., Johnson, L. C., & Towne, E. G. (1999). The keystone role of bison in North American tallgrass prairie. *BioScience, 49*(1), 39–50. https://doi.org/10.2307/1313492

Langdon, S., Prosper, R., & Gagnon, N. (2010). Two paths one direction: Parks Canada and aboriginal peoples working together. *George Wright Forum, 27*(2), 222–233.

Manfredo, M. J., Vaske, J. J., & Teel, T. L. (2003). The potential for conflict index: A graphic approach to practical significance of human dimensions research. *Human Dimensions of Wildlife, 8*(3), 219–228. https://doi.org/10.1080/10871200304310

Massey, C. D., Vayro, J. V., & Mason, C. W. (2021). Conservation values and actor networks that shape the Adams River salmon run in Tsútswecw provincial Park, British Columbia. *Society and Natural Resources, 34*(9), 1174–1193. https://doi.org/10.1080/08941920.2021.1946225

McCormack, P. A. (1992). The political economy of bison management in Wood Buffalo National Park. *Arctic, 45*(4), 367–380. https://doi.org/10.14430/arctic1416

McKenna, O. P., Renton, D. A., Mushet, D. M., & DeKeyser, E. S. (2021). Upland burning and grazing as strategies to offset climate-change effects on wetlands. *Wetlands Ecology and Management, 29*(2), 193–208. https://doi.org/10.1007/s11273-020-09778-1

Michigan Department of Natural Resources. (2022). Bovine tuberculosis. https://www.michigan.gov/dnr/managing-resources/wildlife/wildlife-disease/bovine-tuberculosis

Miguel, E., Grosbois, V., Caron, A., Pople, D., Roche, B., & Donnelly, C. A. (2020). A systemic approach to assess the potential and risks of wildlife culling for infectious disease control. Communications Biology, 3(1), 353. https://doi.org/10.1038/s42003-020-1032-z

Miller, R. S., & Sweeney, S. J. (2013). Mycobacterium bovis (bovine tuberculosis) infection in North American wildlife: Current status and opportunities for mitigation of risks of further infection in wildlife populations. *Epidemiology and Infection, 141*(7), 1357–1370. https://doi.org/10.1017/S0950268813000976

Nelson, M. P., Vucetich, J. A., Paquet, P. C., & Bump, J. K. (2011). *An inadequate construct?* (pp. 58–60). Wildlife Communications Professional.

Nickell, Z., Varriano, S., Plemmons, E., & Moran, M. D. (2018). Ecosystem engineering by bison (Bison bison) wallowing increases arthropod community heterogeneity in space and time. *Ecosphere, 9*(9). https://doi.org/10.1002/ecs2.2436

Nishi, J. S. (2002). Surveillance activities under the Northwest Territories bison control area program (1987–2000). https://www.gov.nt.ca/ecc/sites/ecc/files/bison_control_area_1987-2000pdf.pdf

Nishi, J. S. (2017). Status of the American bison (Bison bison) in Alberta. https://open.alberta.ca/publications/9781460140901.

Nishi, J. S., Dragon, D. C., Elkin, B. T., Mitchell, J., Ellsworth, T. R., & Hugh-Jones, M. E. (2002). Emergency response planning for anthrax outbreaks in bison herds of Northern Canada: A balance between policy and science. *Annals of the New York Academy of Sciences, 969*, 245–250. https://doi.org/10.1111/j.1749-6632.2002.tb04386.x

Nishi, J. S., Shury, T., & Elkin, B. T. (2006). Wildlife reservoirs for bovine tuberculosis (Mycobacterium bovis) in Canada: Strategies for management and research. *Veterinary Microbiology, 112*(2–4), 325–338. https://doi.org/10.1016/j.vetmic.2005.11.013

Organ, J. F., Geist, V., Manohey, S. P., Williams, S., Krausman, P. R., Batcheller, G. R., Decker, T., Carmichael, R., Nanjappa, P., Regan, R., Medellin, R., Cantu, R., McCabe, R. E., Craven, S., Vecellio, G. M., & Decker, D. J. (2012). *The North American model of wildlife conservation.*

Parks Canada. (2010). Wood Buffalo national park of Canada management plan. https://parks.canada.ca/pn-np/nt/woodbuffalo/info/plan/plan1

Parks Canada. (2019). Wood Buffalo national park world heritage site action plan. https://publications.gc.ca/site/eng/9.866672/publication.html

Powell, A. F. L. A. (2006). Effects of prescribed burns and bison (Bos bison) grazing on breeding bird abundances in tallgrass prairie. *Auk, 123*(1), 183–197. https://academic.oup.com/auk/article/123/1/183/5562659. https://doi.org/10.1093/auk/123.1.183

Pyke, C. R., & Marty, J. (2005). Cattle grazing mediates climate change impacts on ephemeral wetlands. *Conservation Biology, 19*(5), 1619–1625. https://doi.org/10.1111/j.1523-1739.2005.00233.x

Ramsey, D. S. L., O'Brien, D. J., Smith, R. W., Cosgrove, M. K., Schmitt, S. M., & Rudolph, B. A. (2016). Management of on-farm risk to livestock from bovine tuberculosis in Michigan, USA, white-tailed deer: Predictions from a spatially explicit stochastic model. *Preventive Veterinary Medicine, 134*, 26–38. https://doi.org/10.1016/j.prevetmed.2016.09.022

Redpath, S. M., Young, J., Evely, A., Adams, W. M., Sutherland, W. J., Whitehouse, A., Amar, A., Lambert, R. A., Linnell, J. D. C., Watt, A., & Gutiérrez, R. J. (2013). Understanding and managing conservation conflicts. *Trends in Ecology and Evolution, 28*(2), 100–109. https://doi.org/10.1016/j.tree.2012.08.021

Sandlos, J. (2002). Where the scientists roam: Ecology, management and bison in northern Canada. *Journal of Canadian Studies, 37*(2), 93–129. https://doi.org/10.3138/jcs.37.2.93

Sandlos, J. (2007). Bison. In *Hunters at the margin* (pp. 21–108). UBC Press.

Sandlos, J. (2014). National parks in the Canadian North: Comanagement or colonialism revisited? In S. Stevens (Ed.), *Indigenous peoples, national parks, and protected areas: A new paradigm linking conservation, culture, and rights* (pp. 133–149). University of Arizona Press.

Schaefer, J. A. (2019). Science and the North American Model: Edifice of knowledge, exemplar for conservation. In S. P. Mahoney and V. Geist (Eds.), *The North American model of wildlife conservation* (pp. 95–205). Johns Hopkins University Press.

Schroeder, S. A., Landon, A. C., Fulton, D. C., & McInenly, L. E. (2021). Social identity, values, and trust in government: How stakeholder group, ideology, and wildlife value orientations relate to trust in a state agency for wildlife management. *Biological Conservation, 261*(July), 109285. https://doi.org/10.1016/j.biocon.2021.109285

Schumaker, B. A., Peck, D. E., & Kauffman, M. E. (2012). Brucellosis in the Greater Yellowstone area: Disease management at the wildlife-livestock interface. *Human-Wildlife Interactions, 6*(1), 48–63. https://doi.org/10.26077/95s0-ah13

Scurlock, B. M., & Edwards, W. H. (2010). Status of brucellosis in free-ranging elk and Bison in Wyoming. *Journal of Wildlife Diseases, 46*(2), 442–449. https://doi.org/10.7589/0090-3558-46.2.442

Shury, T. K., Nishi, J. S., Elkin, B. T., & Wobeser, G. A. (2015). Tuberculosis and brucellosis in wood bison (Bison bison athabascae) in Northern Canada: A renewed need to develop options for future management. *Journal of Wildlife Diseases, 51*(3), 543–554. https://doi.org/10.7589/2014-06-167

Shury, T. K., Woodley, S., & Reynolds, H. W. (2006). Proceedings of the bison diseases technical workshop. Parks Canada Agency.

Smith, G. C., McDonald, R. A., & Wilkinson, D. (2012). Comparing Badger (Meles meles) management strategies for reducing tuberculosis incidence in cattle. *PLOS ONE, 7*(6), e39250. https://doi.org/10.1371/journal.pone.0039250

Stephenson, R. O., Gerlach, S. C., Guthrie, R. D., Harington, C. R., Mills, R. O., & Hare, G. (2001). Wood bison in Late Holocene Alaska and adjacent Canada: Paleontological, archaeological and historical records. In S. C. Gerlach (Ed.), *People and wildlife in North America: Essays in Honor of R. Dale Guthrie* (pp. 124–158). British Archaeological Reports.

Tessaro, S. V., Forbes, L. B., & Turcotte, C. (1990). A survey of brucellosis and tuberculosis in bison in and around Wood Buffalo National Park, Canada. *Canadian Veterinary Journal, 31*(3), 174–180.

Tessaro, S. V., Gates, C. C., & Forbes, L. B. (1993). The brucellosis and tuberculosis status of wood bison in the Mackenzie Bison Sanctuary, Northwest Territories, Canada. *Canadian Journal of Veterinary Research, 57*(4), 231–235.

Toor, K., & Hamit-Haggar, M. (2021). Reports on special business projects: Analysis of the beef supply chain. Statistics Canada. https://www150.statcan.gc.ca/n1/pub/18-001-x/18-001-x2021002-eng.htm

Treanor, J. J. (2012). The biology and management of brucellosis in Yellowstone bison. University of Kentucky. http://uknowledge.uky.edu/biology_etds/7

Ulrich, C. (2022). Wood Buffalo National Park anthrax outbreak appears to slow, no new carcasses in past week. CBC News. https://www.cbc.ca/news/canada/north/wood-buffalo-national-park-anthrax-outbreak-appears-to-slow-no-new-carcasses-in-past-week-1.6540723

Vaske, J. J., Beaman, J., Barreto, H., & Shelby, L. B. (2010). An extension and further validation of the potential for conflict index. *Leisure Sciences, 32*(3), 240–254. https://doi.org/10.1080/01490401003712648

Vaske, J. J., & Miller, C. A. (2018). Hunters and non-hunters chronic wasting disease risk perceptions over time. *Society and Natural Resources, 31*(12), 1379–1388. https://doi.org/10.1080/08941920.2018.1463424

Vaske, J. J., Shelby, L. B., & Needham, M. D. (2008). Preparing for the next disease: The human-wildlife connection. In M. J. Manfredo, J. J. Vaske, P. J. Brown, D. J. Decker & E. A. Duke (Eds.), *Wildlife and society: The science of human dimensions* (pp. 244–261). Island Press.

VerCauteren, K. C., Lavelle, M. J., & Campa, H. (2018). Persistent spillback of bovine tuberculosis from white-tailed deer to cattle in Michigan, USA: Status, Strategies, and Needs. *Frontiers in Veterinary Science, 5*(V), 301. https://doi.org/10.3389/fvets.2018.00301

Voldseth, R. A., Johnson, W. C., Gilmanov, T., Guntenspergen, G. R., & Millett, B. V. (2007). Model estimation of land-use effects on water levels of northern Prairie wetlands. *Ecological Applications, 17*(2), 527–540. https://doi.org/10.1890/05-1195

White, P. J., Wallen, R. L., Geremia, C., Treanor, J. J., & Blanton, D. W. (2011). Management of Yellowstone bison and brucellosis transmission risk – Implications for conservation and restoration. *Biological Conservation, 144*(5), 1322–1334. https://doi.org/10.1016/j.biocon.2011.01.003

Willow Spring Strategic Solutions. (2021). A history of wood Buffalo national parks relations with the Denésuliné. https://cabinradio.ca/wp-content/uploads/2021/07/WBNP-Willow-Springs.pdf

Wood, K. A., O'Hare, M. T., McDonald, C., Searle, K. R., Daunt, F., & Stillman, R. A. (2017). Herbivore regulation of plant abundance in aquatic ecosystems. *Biological Reviews of the Cambridge Philosophical Society, 92*(2), 1128–1141. https://doi.org/10.1111/brv.12272

Wright, R. E. (2022). *The history and evolution of the North American wildlife conservation model.* Palgrave Macmillan. https://doi.org/10.1007/978-3-031-06163-9

Chapter 5

Reptiles as sentinels of ecosystem health: Case example in eastern box turtles

Matthew C. Allender and Laura A. Adamovicz

Abstract

Eastern box turtles (*Terrapene carolina carolina*) are the primarily terrestrial chelonian in the eastern and midwestern USA. They inhabit both terrestrial and semi-aquatic environments, are long-lived, and have relatively small home ranges making them ideal sentinels for ecosystem health. Over a series of previously published studies on health assessments, hematology, acute phase responses, pathogen detection, and statistical modeling in the eastern box turtle, we describe characteristics of how this sentinel species was established. This modeling approach provides a framework for identifying and simplifying wellness assessments in reptiles. Furthermore, the framework can and should be applied to other taxa, including amphibians, fish, mammals, and birds. Ultimately, assessing health in ecological sentinel assemblages may provide the most robust and comprehensive understanding of ecosystem health.

5.1 Importance

The ability to detect changes in ecosystem health requires transdisciplinary cooperation and holistic approaches, and wildlife sentinels have been proposed as early monitors of ecosystem wellness (Carson et al., 2014; Childs et al., 2007; Guilette et al., 1995; Hamer et al., 2012; Mazet et al., 2000; Sleeman, 2008). Monitoring the health of sentinel species allows early detection of ecosystem change and directly benefits species health and recovery efforts. Hematologic, plasma biochemical, and pathogen prevalence data have been utilized as a means of assessing health in free-ranging reptile populations (Anderson et al., 1997; Archer et al., 2017; Brown and Sleeman, 2002; Chaffin et al., 2008; Hidalgo-Vila et al., 2011; Kimble and Williams, 2012; Rose and Allender, 2011; Schrader et al., 2010; Sleeman, 2008; Winter et al., 2019; Wright and Skeba, 1992). However, the association between these biomarkers and overall health status has not been critically evaluated in most species. In addition, the close ties between temperature, reproductive status, and physiology in ectotherms

Susan C. Cork and Douglas P. Whiteside (eds) **Case Studies in EcoHealth**
DOI: 10.52517/9781789183368.005, © 5m Books Ltd, 2024

(Lillywhite, 1987; Peterson et al., 1993) can complicate interpretation of health metrics compared to endotherms.

Recently, several reptile species have been proposed as sentinels for contamination with heavy metals (Blanvillain et al., 2007; Santos et al., 2021), atrazine (Douros et al., 2015; Tavalieri et al., 2020), organochlorines (Keller et al., 2004), endocrine disruptors (Jandegian et al., 2015; Rie et al., 2001), environmental radiation (Ulsh et al., 2000), and *Salmonella* spp. (Hernandez et al., 2021). However, when the role of a sentinel is extended to other wellness variables, opportunities arise to identify complex interactions that exist in the natural world between contaminants, pathogens, habitat quality, and immune function, among others. The ability to determine the wellness of wild reptiles in this context could prove invaluable for species and ecosystem conservation. Longitudinal monitoring of clinical pathology, contaminant exposure, pathogen prevalence, and mortality investigation data across broad spatiotemporal scales has been proposed as an effective means to both understand the complex drivers of individual health status and evaluate the wellness of turtle populations (Rose and Allender, 2011). Considering these data in association with specific stressors provides an even more nuanced understanding of ecosystem wellness. In mammals, wellness has been characterized using hematologic or plasma biochemical parameters in the presence of habitat fragmentation (Johnstone et al., 2012), contamination (Mazet et al., 2000), and climate change (Kirk et al., 2010), thus providing a scaffold for studying reptiles. To accomplish this long-term objective in reptiles, the natural history, physiology, and wellness parameters of individual species may be required prior to validating the species use during these ecosystem threats. Rigorous baseline studies that establish criteria for assessment of overall reptile health will be addressed in this chapter.

5.2 Introduction

Eastern box turtles (*Terrapene carolina carolina*) are charismatic inhabitants of forests, shrublands, and wetlands in the eastern half of the United States (Dodd, 2001; Kiester and Willey, 2015). This species was historically abundant but is now experiencing range-wide declines and is considered vulnerable by the IUCN Red List (van Dijk, 2011). Threats to eastern box turtles include habitat destruction and fragmentation, direct mortality via automobiles, agricultural equipment, and lawn mowers, overcollection for turtle races and the pet trade, and increasing nest predation by subsidized mesopredators (van Dijk, 2011). Disease outbreaks have also resulted in significant mortality, highlighting the importance of health for box turtle conservation. This species is a proposed sentinel for ecosystem health due to its omnivorous nature, use of terrestrial and aquatic habitats, long lifespan, relatively small home range size, and high site fidelity (Lloyd et al., 2016). The Wildlife Epidemiology Laboratory at the University of Illinois has performed health surveillance studies on wild eastern box turtles for over a decade in order to characterize biological and analytical variables which support box turtle and ecosystem wellness.

5.3 Methods

Owing to the inherent lack of studies investigating holistic health of eastern box turtles prior to 2010, we initially performed retrospective (Schrader et al., 2010; Rivas et al., 2014; Adamovicz et al., 2018a) and prospective studies characterizing physical examination (dePersio et al., 2019), hematology (Rose and Allender, 2011; Lloyd et al., 2016), plasma biochemistries (Lloyd et al., 2016; Adamovicz et al., 2019), blood gas analysis (Adamovicz et al., 2018b), protein electrophoresis (Flower et al., 2014), acute phase protein detection (Flower et al., 2014; Adamovicz et al., 2020b), innate immune response (Adamovicz et al., 2020a; Adamovicz et al., 2020c), circulating hormones and vitamins (Watson et al., 2017; Boers et al., 2019), heavy metal exposure (Allender et al., 2015), pathogen prevalence (Allender et al., 2011; Allender et al., 2013; Kane et al., 2017; Archer et al., 2017; Rasmussen et al., 2017; Doden et al., 2021), and non-infectious disease (Rayl et al., 2020) in free-living populations. Various statistical methods were used for each study including descriptive characterizations, univariate comparisons between factors (sex, site, age class, location), multivariable modeling, and spatial analysis. Due to differences in established methods between mammals and reptiles, several additional studies were performed to validate or standardize methods to assess the above characteristics (Butkus et al., 2017; Adamovicz et al., 2019; Winter et al., 2019; Klein et al., 2020).

For all studies, individual box turtles were captured using human and canine search teams as previously described (Boers et al., 2019) between one to three times per year from 2010 through 2021. At capture, data were collected on turtle habitat and microhabitat and GPS coordinates were recorded. Turtles then received a complete physical examination including assessment of mentation and responsiveness and visual evaluation of the eyes, tympanic membranes, nares, oral cavity, shell, integument, musculoskeletal system, and urogenital system (Figure 5.1). Attitude was assessed as bright (moving or active) or quiet (minimal movement). Mentation was assessed as alert (aware of surroundings, head held in normal position) or depressed (held low or unable to retract in the shell). Responsiveness was assessed as responsive (movement when stimulated) or unresponsive (no movement when examined). Morphometric data including mass and standardized shell measurements were also recorded.

Blood samples (<0.8% body weight) and combined swabs of the oral cavity and cloaca were collected. These samples were tested for up to 132 variables including hematology, plasma biochemistries, protein electrophoresis, acute phase proteins, blood gas, heavy metal screening, and presence of up to 20 pathogens. Each turtle was given a permanent identifier, then released at its original site of capture. Mortality events were investigated opportunistically as they occurred. Live turtles and carcasses were located using visual encounter surveys. Live animals were examined and sampled as previously described. Carcasses were collected for necropsy and histopathology, if the remains were relatively fresh, or for bone marrow sampling and subsequent ranavirus testing if no soft tissue remained (Butkus et al., 2017).

As an insensitive measure of population health, population sizes were estimated for each of five study sites within Illinois using mark-recapture data. Population viability analysis (PVA) was performed to assess the probability of extinction for each population over the next century and several scenarios were tested to examine the effect of removing one to five adult female turtles on each population, representing increased losses due to predation, disease, or human collection. This provides an estimate of population resiliency.

5.4 Characterizing parameters of a sentinel

As outlined below, many studies have been performed recently that describe normal variation in eastern box turtle health parameters. This information gives biologists an opportunity to determine whether variation outside these baseline values is a response to changes in environmental or habitat parameters.

A physical examination can be the most valuable diagnostic tool for a clinician, and physical

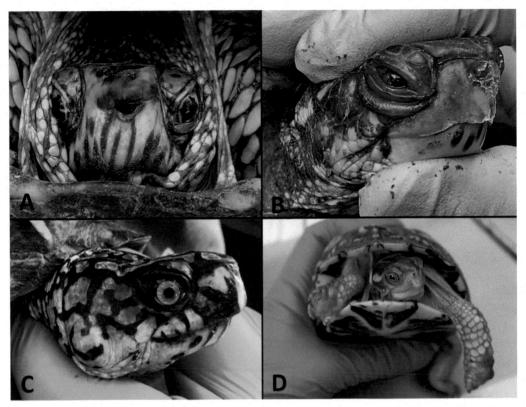

Figure 5.1 Common physical examination abnormalities in eastern box turtles (*Terrapene carolina carolina*). (A & B) Clinical signs of upper respiratory disease including nasal asymmetry or erosion (A), blepharoedema (B), and nasal discharge (B). (C) Aural abscess. (D) Trauma – this individual previously lost both front feet.

exam abnormalities are associated with clinical pathology changes in box turtles. For example, turtles with a quiet attitude (vs. bright) had a decrease in venous blood pH and partial pressure of oxygen and increases in partial pressure of carbon dioxide (Adamovicz et al., 2018b). Additionally, erythrocyte sedimentation rate was higher in animals with several physical examination abnormalities, such as injuries to the shell or presence of upper respiratory disease (Adamovicz et al., 2020b), which is consistent with active non-specific inflammation.

Sex can have dramatic effects on health parameters in all species. Reproductive hormones influence physiology, behavior, and disease conditions, including in eastern box

turtles. Specific sex-based differences in eastern box turtle clinical pathology include increases in aspartate aminotransferase, calcium, phosphorus, albumin, total protein, and basophils in females (Lloyd et al., 2016; Rose and Allender, 2011). Plasma protein fractions (alpha 1, alpha 2, beta, and gamma globulins) and acute phase proteins (haptoglobin) are also subsequently increased in females (Flower et al., 2014). Many of these changes would be predicted based on knowledge of the reproductive cycle and the need for adult females to metabolize minerals and proteins to support egg development. The increases in total calcium in females are likely associated with increased circulating concentrations of total calcium and not due to differences in habitat use as vitamin D concentrations are similar between males and females (Watson et al., 2017).

A robust population of healthy individuals of all age classes is needed for population sustainability. However, experiences and developmental processes differ between adults and juveniles and subsequently many health parameters also differ. Juveniles have been observed with higher creatine kinase, albumin : globulin ratio, phosphorus, and adenovirus detection (Rose and Allender, 2011), and lower haptoglobin (Adamovicz et al., 2020b; Flower et al., 2014) compared to adults. Despite several other factors being tested, no other differences in age class have been observed.

Possibly the largest source of variation in eastern box turtle health parameters is the temporal variation that occurs within and between years. It is hypothesized that fluctuating health metrics are driven largely by temperature but changes due to day length and precipitation cannot be ruled out. Turtles sampled in the summer have the highest albumin, beta globulins, plasma zinc concentrations, *Terrapene* herpesvirus 1 prevalence, partial pressure of carbon dioxide, and lactate concentrations (Adamovicz et al., 2018b; Allender et al., 2015; Flower et al., 2014; Kane et al., 2017). In the fall, albumin is the lowest, alpha 1 and gamma globulins the highest, plasma chromium concentrations the highest, and prevalence of *Terrapene* herpesvirus 1 and *Mycoplasmopsis* sp. are elevated (Allender et al., 2015; Archer et al., 2017; Flower et al., 2014). In the spring, pre-albumin, plasma copper concentrations, and *Terrapene* adenovirus 1 prevalence are the highest (Allender et al., 2015; Archer et al., 2017; Flower et al., 2014). Temporal variation also occurs between years as white blood cell count, packed cell volume, and *Terrapene* herpesvirus 1 prevalence varied from year to year (Archer et al., 2017; Lloyd et al., 2016).

Eastern box turtles range from Georgia north through Michigan and from the Mississippi River to the Atlantic Ocean, encompassing a broad range of latitude and longitude that may affect health. Studies have investigated differences in health factors for eastern box turtles between Illinois and Tennessee, and between sites within these states. In free-ranging individuals in Tennessee, haptoglobin, total protein, albumin, alpha 2 globulins, beta globulins, gamma globulins, plasma chromium and lead concentrations, and *Terrapene* herpesvirus 1 prevalence are higher than Illinois turtles (Allender et al., 2015; Flower et al., 2014; Kane et al., 2017; Lloyd et al., 2016). In contrast, calcium, phosphorus, and plasma selenium and zinc were higher in Illinois turtles (Allender et al., 2015; Lloyd et al., 2016). No differences were observed in vitamin D concentration between turtles in each state. Owing to the large species range and terrestrial nature of EBTs, many individuals intersect with urban and suburban habitats. In these locations, trauma is the most common cause of poor health (Schrader et al., 2010), with infectious diseases (ranavirus) more common in rehabilitation centers than in free-ranging populations nearby (Allender et al., 2013). Recurrent ranavirus infections every 1–3 years have been documented at the same site (Adamovicz et al., 2018a).

Changes in box turtle health parameters based on habitat characteristics are best described from sites within the same state. Five study sites in Illinois range in estimated population size from 161 to 1037 turtles. Management and habitat differ from contiguous forest managed intensively (fire, invasive species removal) to less aggressive, invasive plant dominant habitat. Not surprisingly, the probability of extinction at the least intensively managed site is 97% with annual loss of a single adult female beyond baseline mortality rates (Adamovicz, 2018). In contrast, the population at the most heavily managed site can tolerate loss of up to three adult females per year with only a 23% probability of extinction within the next century. In the context of wellness, there is much less room for error at certain sites due to overall deterioration of health, introduction or establishment of high mortality infectious diseases (e.g. ranavirus), and delays in growth or reproductive success due to poor nutrition. Unfortunately, over the gradient of poor habitats, mortality events have been documented at several locations (Adamovicz et al., 2018a). Two sites had mortality events attributable to ranavirus, a pathogen causing significant disease and death in many species of chelonians, but notably eastern box turtles (Adamovicz et al., 2018a, Allender et al., 2011; DeVoe et al., 2004; Johnson et al., 2008; Kimble et al., 2017; Sim et al., 2016). The poorest quality site also appeared to have a high mortality rate due to non-infectious diseases, with 22/36 individuals in a radiotelemetered group succumbing to winter mortality and a variety of pathologic processes over the course of 3 years (Rayl et al., 2020). The occurrence of a poorer plane of health in poor habitats is not surprising, but the inability to support a viable population of box turtles signals that the habitat is poor for other species that rely on the natural resources at or near these sites. This is further emphasized by turtles at the poorest site demonstrating

increased stress and inflammation (higher heterophil and eosinophil counts, lower albumin : globulin ratios and lymphocyte counts), higher levels of tissue damage (higher aspartate aminotransferase), and more prevalent liver dysfunction (higher bile acids) compared to turtles from the best site. The turtles at the best habitat site also have superior plasma antibacterial capacity against *Escherichia coli* and higher plasma complement activity (as assessed using an ovine erythrocyte hemolysis assay) compared to turtles from other sites ($p < 0.05$), demonstrating that not only is a robust immune response observed at better habitats, but also a better overall condition.

5.5 Modeling wellness in box turtles

Owing to the complexity identified above, we developed a statistical approach to characterize the most important drivers of wellness in box turtles. First, relationships between all health factors were depicted using a directed acyclic diagram, which was used to identify potential confounding variables and structure statistical models (Figure 5.2). Next, reference intervals

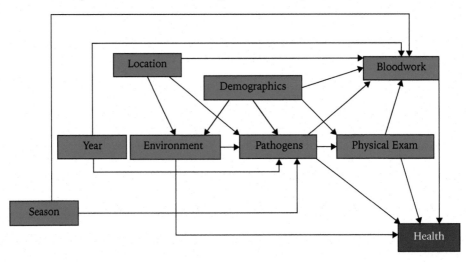

Figure 5.2 Directed acyclic diagram depicting the relationships between health status and its predictors in eastern box turtles (*Terrapene carolina carolina*).

were developed for each clinical pathology parameter and partitioned based on year, season, site, sex, and age class, if appropriate based on a priori data. Each turtle's clinical pathology values were compared to these reference intervals and the number of values falling outside the reference intervals were counted. Finally, turtles were assigned to a health category ('apparently healthy' vs. 'unhealthy'). Apparently healthy turtles met the following three criteria (Figure 5.3).

• No clinically significant physical examination abnormalities. A clinically significant abnormality is one expected to affect the animal's ability to navigate, prehend food, mate, and protect itself from predation.
• No more than three clinical pathology parameters outside of reference intervals. By definition, reference intervals comprise 95% of the values of a 'healthy' population, so 5% of 'healthy' animals can be expected to have a value outside of the reference interval (Friedrichs et al., 2012). Increasing the

number of abnormal values required to classify an animal as 'abnormal' decreases the risk of misclassifying an otherwise 'healthy' animal.
• No clinically apparent qPCR detection of infectious disease. The presence of a pathogen does not necessarily equate to disease. We only considered pathogen-positive animals 'unhealthy' if they had concurrent clinical signs of illness or if they had greater than three abnormal clinical pathology values, indicating a negative physiologic effect of the pathogen.

Models predicting individual health were then constructed using all available health metrics to identify the most clinically useful diagnostic tests. Information-theoretic methods were utilized to identify the most parsimonious model. This model was internally validated via bootstrapping and externally validated using a new dataset.

Three statistical approaches (structural equation modeling, Bayesian network modeling, and

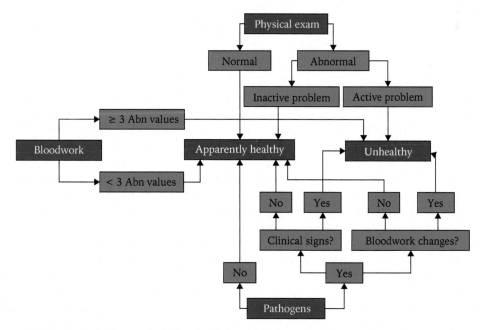

Figure 5.3 Health classification decision tree for eastern box turtles (*Terrapene carolina carolina*).

generalized linear models) produced the same top model for predicting health in eastern box turtles, which contained the additive effects of total leukocyte count (WBC), heterophil : lymphocyte ratio (H : L), condition of the shell (normal vs. active lesions), and presence of upper respiratory disease (nasal discharge, blepharoedema, and/or ocular discharge) (Adamovicz, 2018). Specifically, turtles with WBC and H : L values furthest away from population median values, those with active shell lesions, and those with clinical signs of upper respiratory disease were significantly more likely to be classified as 'unhealthy'. Detection of Terrapene herpesvirus 1, Terrapene adenovirus 1, and *Mycoplasmopsis* sp. was not associated with overall health status; too few ranavirus positive turtles were included for this pathogen to be significantly associated with health (Adamovicz, 2018). This information enables us to refine our health assessment protocols to prioritize the most useful diagnostics for health assessment in this species: hematology and physical examination. Modeling provides a useful framework for objectively assessing health status in challenging species such as reptiles.

5.4   Conclusion

An abundance of wellness data exists on the eastern box turtle and compelling reasons presented in this chapter identify its use as a biosentinel across its range. Fortunately, the complex relationships between health and season, year, sex, and age class have been characterized for this species. This facilitated the development of a modeling approach to elucidate the most important drivers of health and the most clinically useful diagnostic tests for health assessment. This modeling approach provides a framework for identifying and simplifying wellness assessments in reptiles in the very near future. Furthermore, the framework

can and should be applied to other taxa, including amphibians, fish, mammals, and birds. And when applied generally, trends in wellness across multiple species in a habitat can signal deteriorating natural resources and opportunities for intervention. Ultimately, assessing health in ecological sentinel assemblages may provide the most robust and comprehensive understanding of ecosystem health.

References

Adamovicz, L. (2018). Modeling the health of free-living Illinois herptiles: An integrated approach incorporating environmental, physiologic, spatiotemporal, and pathogen factors [PhD Dissertation]. University of Illinois.

Adamovicz, L., Allender, M. C., Archer, G., Rzadkowska, M., Boers, K., Phillips, C., Driskell, E., Kinsel, M. J., & Chu, C. (2018a). Investigation of multiple mortality events in eastern box turtles (*Terrapene carolina carolina*). *PLOS ONE*, *13*(4), e0195617. https://doi.org/10.1371/journal.pone.0195617

Adamovicz, L., Leister, K., Byrd, J., Phillips, C. A., & Allender, M. C. (2018b). Venous blood gas in free-living eastern box turtles (*Terrapene carolina carolina*) and effects of physiologic, demographic, and environmental factors. *Conservation Physiology*, *6*(1), coy041. https://doi.org/10.1093/conphys/coy041

Adamovicz, L., Griffioen, J., Cerreta, A., Lewbart, G. A., & Allender, M. C. (2019). Tissue enzyme activities in free-living eastern box turtles (*Terrapene carolina carolina*). *Journal of Zoo and Wildlife Medicine*, *50*(1), 45–54. https://doi.org/10.1638/2018-0079

Adamovicz, L., Baker, S. J., Merchant, M., Darville, L., & Allender, M. C. (2020). Plasma complement activation mechanisms differ in ornate (Terrapene ornata ornata) and eastern box turtles (Terrapene carolina carolina). Journal of Experimental Zoology. Part A, Ecological and Integrative Physiology, 333(10), 720–731. https://doi.org/10.1002/jez.2423.

Adamovicz, L., Baker, S. J., Kessler, E., Kelly, M., Johnson, S., Winter, J., Phillips, C. A., & Allender, M. C. (2020b). Erythrocyte sedimentation rate and hemoglobin-binding protein in free-living box turtles

(*Terrapene* spp.). *PLOS ONE*, *15*(6), e0234805. https://doi.org/10.1371/journal.pone.0234805

Adamovicz, L., Baker, S. J., Merchant, M., & Allender, M. C. (2020c). Plasma antibacterial activities in ornate (*Terrapene ornata ornata*) and Eastern box turtles (*Terrapene carolina carolina*). *Journal of Experimental Zoology Part A*, *333*(5), 295–305. https://doi.org/10.1002/jez.2352

Allender, M. C., Abd-Eldaim, M., Schumacher, J., McRuer, D., Christian, L. S., & Kennedy, M. (2011). Prevalence of *Ranavirus* Causing morbidity and mortality in eastern box turtles (*Terrapene carolina carolina*) in three Southeastern US states. *Journal of Wildlife Diseases*, *47*(3), 759–764. https://doi.org/10.7589/0090-3558-47.3.759

Allender, M. C., Mitchell, M. A., McRuer, D., Christian, S., & Byrd, J. (2013). Prevalence, clinical signs, and natural history characteristics of frog virus 3-like infections in eastern box turtles (*Terrapene carolina carolina*). *Herpetological Conservation and Biology*, *8*, 308–320.

Allender, M. C., Dreslik, M. J., Patel, B., Luber, E. L., Byrd, J., Phillips, C. A., & Scott, J. W. (2015). Select metal and metalloid surveillance of free-ranging eastern box turtles from Illinois and Tennessee (*Terrapene carolina carolina*). *Ecotoxicology*, *24*(6), 1269–1278. https://doi.org/10.1007/s10646-015-1498-5

Anderson, N. L., Wack, R. F., & Hatcher, R. (1997). Hematology and clinical chemistry reference ranges for clinically normal, captive New Guinea snapping turtle (*Elseya novaeguineae*) and the effects of temperature, sex, and sample type. *Journal of Zoo and Wildlife Medicine*, *28*(4), 394–403.

Archer, G. A., Phillips, C. A., Adamovicz, L., Band, M., Byrd, J., & Allender, M. C. (2017). Detection of co-pathogens in free-ranging eastern box turtles (*Terrapene carolina carolina*) in Illinois and Tennessee. *Journal of Zoo and Wildlife Medicine*, *48*(4), 1127–1134. https://doi.org/10.1638/2017-0148R.1

Blanvillain, G., Schwenter, J. A., Day, R. D., Point, D., Christopher, S. J., Roumillat, W. A., & Owens, D. W. (2007). Diamondback terrapins, *Malaclemys terrapin*, as a sentinel species for monitoring mercury pollution of estuarine systems in South Carolina and Georgia, USA. *Environmental Toxicology and Chemistry*, *26*(7), 1441–1450. https://doi.org/10.1897/06-532r.1

Boers, K. L., Allender, M. C., Novak, L. J., Palmer, J., Adamovicz, L., & Deem, S. L. (2020). Assessment of hematologic and corticosterone response in free-living eastern box turtles (*Terrapene carolina carolina*) at capture and after handling. *Zoo Biology*, *39*(1), 13–22. https://doi.org/10.1002/zoo.21518

Brown, J. D., & Sleeman, J. M. (2002). Morbidity and mortality of reptiles admitted to the wildlife center of Virginia, 1991 to 2000. *Journal of Wildlife Diseases*, *38*(4), 699–705. https://doi.org/10.7589/0090-3558-38.4.699

Butkus, C. E., Allender, M. C., Phillips, C. A., & Adamovicz, L. A. (2017). Diagnosis of Ranavirus using bone marrow harvested from mortality events in eastern box turtles (*Terrapene carolina carolina*). *Journal of Zoo and Wildlife Medicine*, *48*(4), 1210–1214. https://doi.org/10.1638/2017-0098.1

Carson, C., Lavender, C. J., Handasyde, K. A., O'Brien, C. R., Hewitt, N., Johnson, P. D. R., & Fyfe, J. A. M. (2014). Potential wildlife sentinels for monitoring the endemic spread of human Burui ulcer in South-East Australia. *PLOS Neglected Tropical Diseases*, *8*(1), e2668. https://doi.org/10.1371/journal.pntd.0002668

Chaffin, K., Norton, T. M., Gilardi, K., Poppenga, R., Jensen, J. B., Moler, P., Cray, C., Dierenfeld, E. S., Chen, T., Oliva, M., Origgi, F. C., Gibbs, S., Mazzaro, L., & Mazet, J. (2008). Health assessment of free-ranging alligator snapping turtles (*Macrochelys temminckii*) in Georgia and Florida. *Journal of Wildlife Diseases*, *44*(3), 670–686. https://doi.org/10.7589/0090-3558-44.3.670

Childs, J. E., Krebs, J. W., Real, L. A., & Gordon, E. R. (2007). Animal-based national surveillance for zoonotic disease: Quality, limitations, and implications of a model system for monitoring rabies. *Preventive Veterinary Medicine*, *78*(3–4), 246–261. https://doi.org/10.1016/j.prevetmed.2006.10.014

dePersio, S., Allender, M. C., Dreslik, M. J., Adamovicz, L., Phillips, C. A., Willeford, B., Kane, L. P., Joslyn, S., & O'Brien, R. T. (2019). Body condition of eastern box turtles (*Terrapene carolina carolina*) evaluated by computed tomography. *Journal of Zoo and Wildlife Medicine*, *50*(2), 295–302. https://doi.org/10.1638/2018-0201

DeVoe, R., Geissler, K., Elmore, S., Rotstein, D., Lewbart, G., & Guy, J. (2004). Ranavirus-associated morbidity and mortality in a group of captive eastern box turtles (*Terrapene carolina carolina*). *Journal of Zoo and Wildlife Medicine*, *35*(4), 534–543. https://doi.org/10.1638/03-037

Dodd, K. (2001). *North American box turtles: A natural history*. University of Oklahoma Press.

Doden, G., Gartlan, B., Klein, K., Maddox, C. W., Adamovicz, L. A., & Allender, M. C. (2021). Prevalence and antimicrobial resistance patterns of *Salmonella* spp. in two free-ranging populations of eastern box turtles (*Terrapene carolina carolina*). *Journal of Zoo and Wildlife Medicine, 52*(3), 863–871. https://doi.org/10.1638/2020-0061

Douros, D. L., Gaines, K. F., & Novak, J. M. (2015). Atrazine and glyphosate dynamics in a lotic ecosystem: The common snapping turtle as a sentinel species. *Environmental Monitoring and Assessment, 187*(3), 114. https://doi.org/10.1007/s10661-015-4336-6

Flower, J. E., Byrd, J., Cray, C., & Allender, M. C. (2014). Plasma electrophoretic profiles and acute phase protein reference intervals in the eastern box turtle (*Terrapene carolina carolina*) and influences of age, sex, season, and location. *Journal of Zoo and Wildlife Medicine, 45*(4), 836–842. https://doi.org/10.1638/2014-0035.1

Friedrichs, K. R., Harr, K. E., Freeman, K. P., Szladovits, B., Walton, R. M., Barnhart, K. F., Blanco-Chavez, J., & American Society for Veterinary Clinical Pathology. (2012). ASVCP reference interval guidelines: Determination of de novo reference intervals in veterinary species and other related topics. *Veterinary Clinical Pathology, 41*(4), 441–453. https://doi.org/10.1111/vcp.12006

Guillette, L. J., Crain, D. A., Rooney, A. A., & Pickford, D. B. (1995). Organization versus activation: The role of endocrine-disrupting contaminants (EDCs) during embryonic development in wildlife. *Environmental Health Perspectives, 103,* Suppl. 7, 157–164. https://doi.org/10.1289/ehp.95103s7157

Hamer, S. A., Lehrer, E., & Magle, S. B. (2012). Wild birds as sentinels for multiple zoonotic pathogens along an urban to rural gradient in greater Chicago, Illinois. *Zoonoses and Public Health, 59*(5), 355–364. https://doi.org/10.1111/j.1863-2378.2012.01462.x

Hernandez, S. M., Maurer, J. J., Yabsley, M. J., Peters, V. E., Presotto, A., Murray, M. H., Curry, S., Sanchez, S., Gerner-Smidt, P., Hise, K., Huang, J., Johnson, K., Kwan, T., & Lipp, E. K. (2021). Free-Living Aquatic Turtles as Sentinels of *Salmonella* spp. for water Bodies. *Frontiers in Veterinary Science.* Jul 22;8:674973, 8, 674973. https://doi.org/10.3389/fvets.2021.674973

Hidalgo-Vila, J., Martínez-Silvestre, A., Ribas, A., Casanova, J. C., Pérez-Santigosa, N., & Díaz-Paniagua, C. (2011). Pancreatitis associated with the helminth Serpinema microcephalus (Nematoda: *Camallanidae*) in exotic Red-Eared slider turtles (*Trachemys scripta elegans*). *Journal of Wildlife Diseases, 47*(1), 201–205. https://doi.org/10.7589/0090-3558-47.1.201

Jandegian, C. M., Deem, S. L., Bhandari, R. K., Holliday, C. M., Nicks, D., Rosenfeld, C. S., Selcer, K. W., Tillitt, D. E., Vom Saal, F. S., Vélez-Rivera, V., Yang, Y., & Holliday, D. K. (2015). Developmental exposure to bisphenol A (BPA) alters sexual differentiation in painted turtles (*Chrysemys picta*). *General and Comparative Endocrinology, 216,* 77–85. https://doi.org/10.1016/j.ygcen.2015.04.003

Johnson, A. J., Pessier, A. P., Wellehan, J. F., Childress, A., Norton, T. M., Stedman, N. L., Bloom, D. C., Belzer, W., Titus, V. R., Wagner, R., Brooks, J. W., Spratt, J., & Jacobson, E. R. (2008). Ranavirus infection of free-ranging and captive box turtles and tortoizes in the United States. *Journal of Wildlife Diseases, 44*(4), 851–863. https://doi.org/10.7589/0090-3558-44.4.851

Johnstone, C. P., Lill, A., & Reina, R. D. (2012). Does habitat fragmentation cause stress in the agile Antechinus? A haematological approach. *Journal of Comparative Physiology. Part B, 182*(1), 139–155. https://doi.org/10.1007/s00360-011-0598-7

Junge, R. E., & Louis, E. E. (2005). Biomedical evaluation of two sympatric lemur species (*Propithecus verreauxi deckeni* and *Eulemur fulvus rufus*) in Tsiombokibo Classified Forest, Madagascar. *Journal of Zoo and Wildlife Medicine, 36*(4), 581–589. https://doi.org/10.1638/05-025.1

Kane, L. P., Allender, M. C., Archer, G., Dzhaman, E., Pauley, J., Moore, A. R., Ruiz, M. O., Smith, R. L., Byrd, J., & Phillips, C. A. (2017). Prevalence of Terrapene herpesvirus 1 in free-ranging eastern box turtles (*Terrapene carolina carolina*) in Tennessee and Illinois, USA. *Journal of Wildlife Diseases, 53*(2), 285–295. https://doi.org/10.7589/2016-06-138

Keller, J. M., Kucklick, J. R., Stamper, M. A., Harms, C. A., & McClellan-Green, P. D. (2004). Associations between organochlorine contaminant concentrations and clinical health parameters in loggerhead sea turtles from North Carolina, USA. *Environmental Health Perspectives, 112*(10), 1074–1079. https://doi.org/10.1289/ehp.6923

Kiester, A. R., & Willey, L. L. (2015). *Terrapene carolina* (Linnaeus 1758) – Eastern box turtle, common box turtle. In A. G. J. Rhodin et al. (Eds.). Conservation Biology of Freshwater Turtles and Tortoises:

A Compilation Project of the IUCN/SSC Tortoise and Freshwater Turtle Specialist Group. *Chelonian Research Monographs, 5*(8), 085.1–25. https://doi.org/10.3854/crm.5.085.carolina.v1.2015

Kimble, S. J. A., & Williams, R. N. (2012). Temporal variance in hematologic and plasma biochemical reference intervals for free-ranging eastern box turtles (*Terrapene carolina carolina*). *Journal of Wildlife Diseases, 48*(3), 799–802. https://doi.org/10.7589/0090-3558-48.3.799

Kimble, S. J. A., Johnson, A. J., Williams, R. N., & Hoverman, J. T. (2017). A severe Ranavirus outbreak in captive, wild-caught box turtles. *EcoHealth, 14*(4), 810–815. https://doi.org/10.1007/s10393-017-1263-8

Kirk, C. M., Amstrup, S., Swor, R., Holcomb, D., & O'Hara, T. M. (2010). Hematology of southern Beaufort Sea polar bears (2005–2007): Biomarker for an Arctic system health sentinel. *EcoHealth, 7*(3), 307–320. https://doi.org/10.1007/s10393-010-0322-1

Klein, K., Gartlan, B., Doden, G., Fredrickson, K., Adamovicz, L., & Allender, M. C. (2021). Comparing the effects of lithium heparin and dipotassium ethylenediaminetetraacetic acid on hematologic values in eastern box turtles (*Terrapene carolina carolina*). *Journal of Zoo and Wildlife Medicine, 51*(4), 999–1006. https://doi.org/10.1638/2020-0109

Lillywhite, H. N. (1987). Temperature, energetics, and physiological ecology. In R. A. Seigel, J. T. Collins & S. Novak (Eds.), *Snakes: Ecology and evolutionary biology* (pp. 422–477). Macmillan Publishing Company.

Lloyd, T. C., Allender, M. C., Archer, G., Phillips, C. A., Byrd, J., & Moore, A. R. (2016). Modeling Hematologic and Biochemical Parameters with Spatiotemporal Analysis for the Free-Ranging Eastern Box Turtle (Terrapene carolina carolina) in Illinois and Tennessee, a Potential Biosentinel. EcoHealth, 13(3), 467–479. https://doi.org/10.1007/s10393-016-1142-8

Mazet, J. K., Gardner, I. A., Jessup, D. A., Lowenstine, L. J., & Boyce, W. M. (2000). Evaluation of changes in hematologic and clinical biochemical values after exposure to petroleum products in mink (*Mustela vison*) as a model for assessment of sea otters (*Enhydra lutris*). *American Journal of Veterinary Research, 61*(10), 1197–1203. https://doi.org/10.2460/ajvr.2000.61.1197

Peterson, C. R., Gibson, A. R., & Dorcas, M. E. (1993). Snake thermal ecology: The causes and consequences of body-temperature variation. In R. A. Seigel & J. T. Collins (Eds.), *Snakes: Ecology and behavior* (pp. 241–314). McGraw-Hill, Inc.

Rasmussen, C., Allender, M. C., Phillips, C. A., Byrd, J., Lloyd, T., & Maddox, C. (2017). Multidrug resistance patterns of enteric bacteria in two populations of free-ranging eastern box turtles (*Terrapene carolina carolina*). *Journal of Zoo and Wildlife Medicine, 48*(3), 708–715. https://doi.org/10.1638/2016-0194.1

Rayl, J. M., Adamovicz, L., Stern, A. W., Vieson, M. D., Phillips, C. A., Kelly, M., Beermann, M., & Allender, M. C. (2020). Mortality investigation of radio telemetered Eastern box turtles (*Terrapene carolina carolina*) in central Illinois from 2016–2018. *Journal of Wildlife Diseases, 56*(2), 306–315. https://doi.org/10.7589/2019-01-016

Rie, M. T., Lendas, K. A., & Callard, I. P. (2001). Cadmium: Tissue distribution and binding protein induction in the painted turtle, *Chrysemys picta*. *Comparative Biochemistry and Physiology. Toxicology and Pharmacology, 130*(1), 41–51. https://doi.org/10.1016/s1532-0456(01)00219-8

Rivas, A. R., Allender, M. C., Mitchell, M., & Whittington, J. K. (2014). Morbidity and mortality in reptiles presented to a wildlife care facility in central Illinois. *Human-Wildlife Interactions, 8,* 78–87.

Santos, R. L. D., Correia, J. M. D. S., & Santos, E. (2021). Freshwater aquatic reptiles (*Testudines* and *Crocodylia*) as biomonitor models in assessing environmental contamination by inorganic elements and the main analytical techniques used: A review. *Environmental Monitoring and Assessment, 193,* 498. https://doi.org/10.1007/s10661-021-09212-w

Schrader, G. M., Allender, M. C., & Odoi, A. (2010). Diagnosis, treatment, and outcome of eastern box turtles (*Terrapene carolina carolina*) presented to a wildlife clinic in Tennessee, USA, 1995–2007. *Journal of Wildlife Diseases, 46*(4), 1079–1085. https://doi.org/10.7589/0090-3558-46.4.1079

Sim, R. R., Allender, M. C., Crawford, L. K., Wack, A. N., Murphy, K. J., Mankowski, J. L., & Bronson, E. (2016). Ranaviral epizootic in captive eastern box turtles (*Terrapene carolina carolina*) with concurrent herpesvirus and Mycoplasma infection: Management and Monitoring. *Journal of Zoo and Wildlife Medicine, 47*(1), 256–270. https://doi.org/10.1638/2015-0048.1

Sleeman, J. (2008). Use of wildlife rehabilitation centers as monitors of ecosystem health. In M. E.

Fowler, & R. E. Miller (Eds.) *Zoo and wild animal medicine, current therapy* (pp. 97–104). Saunders Elsevier.

Tavalieri, Y. E., Galoppo, G. H., Canesini, G., Luque, E. H., & Muñoz-de-Toro, M. M. (2020). Effects of agricultural pesticides on the reproductive system of aquatic wildlife species, with crocodilians as sentinel species. *Molecular and Cellular Endocrinology, 518*(518), 110918. https://doi.org/10.1016/j.mce.2020.110918

Ulsh, B. A., Mühlmann-Díaz, M. C., Whicker, F. W., Hinton, T. G., Congdon, J. D., & Bedford, J. S. (2000). Chromosome translocations in turtles: A biomarker in a sentinel animal for ecological dosimetry. *Radiation Research, 153*(6), 752–759. https://doi.org/10.1667/0033-7587(2000)153[0752:ctitab]2.0.co;2

van Dijk, P. P. (2011). *Terrapene carolina* (errata version published in 2016). IUCN Red List of Threatened Species, 2011: e.T21641A97428179. https://doi.org/10.2305/IUCN.UK.2011–1.RLTS.T21641A9303747.en

Watson, M. K., Byrd, J., Phillips, C. A., & Allender, M. C. (2017). Characterizing the 25-hydroxyvitamin D status of two populations of free-ranging eastern box turtles (*Terrapene carolina carolina*). *Journal of Zoo and Wildlife Medicine, 48*(3), 742–747. https://doi.org/10.1638/2016-0236.1

Way Rose, B. M. W., & Allender, M. C. (2011). Health assessment of wild eastern box turtles (*Terrapene carolina carolina*) in east Tennessee. *Journal of Herpetological Medicine and Surgery, 21*(4), 107–112. https://doi.org/10.5818/1529-9651-21.4.107

Winter, J. M., Stacy, N. I., Adamovicz, L. A., & Allender, M. C. (2019). Investigating the analytical variability and agreement of manual leukocyte quantification methods in Eastern box turtles (*Terrapene carolina carolina*). *Frontiers in Veterinary Science, 6*, 398. https://doi.org/10.3389/fvets.2019.00398

Wright, K. M., & Skeba, S. (1992). Hematology and plasma chemistries of captive prehensile – tailed skinks (*Corucia zebrata*). *Journal of Zoo and Wildlife Medicine, 23*, 429–432.

Chapter 6

River turtle conservation, Australia: Conservation management planning for a critically endangered species

Richard Jakob-Hoff, Shane Ruming, Caroline Lees, Gerry McGilvray, Melissa Giese, Michael McFadden, Adam Skidmore, Karrie Rose, Ricky Spencer and Bruce Chessman

Abstract

This case study of effective conservation management planning describes a collaborative, multi-stakeholder and multi-sectoral planning process, the steps taken and their results for an endemic Australian freshwater turtle critically endangered by a mass-mortality event. Following an initial emergency response, including the capture and transfer of a group of healthy adult turtles to establish a captive insurance population, conservation planning principles, developed over 40 years by the IUCN-SSC's Conservation Planning Specialist Group (CPSG), were applied in a workshop to conduct a species status review, disease risk analysis and conservation action plan. Guided by this plan over the following 5 years, the coordinated, combined efforts of a large and diverse team of individuals and organizations – including local landholders and citizens – have made substantial progress towards the recovery of this species. Strong leadership, ongoing engagement with all stakeholders and an inclusive decision-making process have been critical to achieving the successes to date. The case study provides a model that could readily be applied to other species conservation projects.

6.1 Introduction

With over 1300 native threatened species, Australia has one of the highest extinction rates in the world – the result of a combination of factors including habitat loss, changed fire regimes, introduced invasive species and disease (Australian Government, 2021). However, this pattern is repeated globally and is now recognized as the sixth mass extinction event in Earth's history, primarily driven by human activity (Kolbert, 2014; WWF, 2020). Increased human intervention through conservation action is necessary to prevent further extinction

Susan C. Cork and Douglas P. Whiteside (eds) **Case Studies in EcoHealth**
DOI: 10.52517/9781789183368.006, © 5m Books Ltd, 2024

and, to understand and effectively mitigate the drivers of these threats, transdisciplinary and trans-sectoral planning is essential. Such an approach enables key stakeholders (government, organizations, communities and individuals) to develop a shared understanding of the systems under threat while integrating their combined knowledge and perspectives into the development of an effective conservation plan (Westley and Miller, 2003).

While human encroachment on wilderness and over-harvesting of wildlife are globally the biggest drivers of biodiversity loss (WWF, 2020), disease is increasingly recognized as an additional significant threat for many species (Daszak et al., 2000; MacPhee and Greenwood, 2013). Although disease is a natural process, it can have severe population impacts on threatened species where remaining populations are small, fragmented or restricted in range. In this case study we describe the transdisciplinary, multi-stakeholder approach taken in response to a mass-mortality event impacting an Australian endemic freshwater turtle with a restricted geographic range in northern New South Wales (NSW). Disease was identified as the immediate driver of the mortality event but its cause and origin were obscure and a system-wide approach to planning was necessary to canvas the wider ecological and social landscape in which it occurred and to identify the mitigating actions most likely to be effective.

6.2 The Bellinger River snapping turtle

The Bellinger River snapping turtle (BRST) (*Myuchelys georgesi*) (Figure 6.1) is a medium-sized Australian freshwater turtle endemic to roughly 80 km of the Bellinger River and its tributaries in northern NSW (Figure 6.2). The species is highly aquatic, leaving the water only to bask or lay eggs (King and Heatwole, 1994), and

Figure 6.1 Bellinger River snapping turtle (*Myuchelys georgesi*). Photo: R. Spencer.

Figure 6.2 View of turtle habitat in the Bellinger River. Photo: S. Ruming

generally prefers waterholes greater than 5 m deep (Blamires and Spencer, 2013). Juveniles, however, can be found in shallow water near banks and in riffle zones (Shane Ruming and Gerry McGilvray pers. obs.). Its restricted range makes *M. georgesi* highly vulnerable to threats.

6.3 The event

On 18 February 2015, several *M. georgesi* were observed dead or dying in the Bellinger River. An immediate investigation by the NSW Environment Protection Authority (EPA) found no evidence of toxicity or water contamination. A disease outbreak was declared and a multi-agency, emergency response was initiated by the former NSW Office of Environment and Heritage (OEH), now superseded by the NSW Department of Planning and Environment (DPE). A total of 432 dead or dying turtles were found during the subsequent 2 months, mostly adults (Moloney et al., 2015). However, a higher number of deaths is likely because initial investigations were hampered by flooding (Moloney et al., 2015).

Prior to 2015, *M. georgesi* was described as common and locally abundant within the Bellinger River (Blamires et al., 2005; Cann et al., 2015; Georges et al., 2007), with an estimated population of 4468 individuals derived from data obtained up to 2004 (Blamires et al., 2005), and an unpublished estimate of 1600–3200 in early 2015 noted by the NSW Scientific Committee. This mass-mortality event is estimated to have caused a >90% decrease in the total population from historical levels (Chessman et al., 2020). In 2016, *M. georgesi* was re-classified as Critically Endangered on both the NSW and Australian threatened species lists and the IUCN/SSC Tortoise and Freshwater Turtle Specialist Group Draft Red List (Rhodin et al, 2018).

6.4 Initial response

Immediately following the mass-mortality event planning focused on biosecurity, containment, communications and surveillance. Euthanasia protocols had to be developed for severely affected animals (see in situ mortality investigation plan below). Key knowledge gaps were targeted for investigation: principally to identify the cause(s) of mortality and the epidemiology and ecology of the disease. Of crucial importance was a plan to capture as many healthy adult *M. georgesi* as possible and transfer them to quarantine facilities to secure an insurance population. Subsequent captive breeding could then support recovery of the remaining in situ population. Early collaborators included

the NSW National Parks and Wildlife Service (NPWS), the Department of Primary Industries (DPI), Western Sydney University (WSU) and the Taronga Conservation Society Australia (TCSA), including its Australian Registry of Wildlife Health (ARWH). The local community was also consulted, as it maintains an intense interest in, and concern for, both the turtles and the health of the Bellinger River. Working in close collaboration with DPIE, partner organizations and individuals drew on their specialist expertise and resources to plan the investigation of the mass mortality *in situ* and the intensive management of *M. georgesi* ex situ.

6.5 In situ mortality investigation plan

The initial response involved tending to the welfare of affected animals while gathering as much information as possible to inform a systematic investigation plan. The significance of the event was clear from initial reports documenting an unusual mass mortality in a geographically restricted, and potentially vulnerable species. Immediate provision of local veterinary care to affected animals failed to provide positive health outcomes. Every effort was then made to transport 10 severely ill turtles and five dead turtles to Taronga's ARWH for immediate veterinary assessment. An additional 11 profoundly sick animals were transported for examination 2 weeks into the outbreak. Clinical findings included swollen or ulcerated eyelids, corneal oedema, pawing at the eyes, clear nasal discharge, and occasionally hind limb paresis or small tan skin lesions (Figure 6.3). Owing to the painful nature of the lesions, debility of the animals, and a grave prognosis for recovery, euthanasia was determined to be the most humane option.

Armed with the knowledge that we were investigating a significant disease event, gross post-mortem examinations were planned and

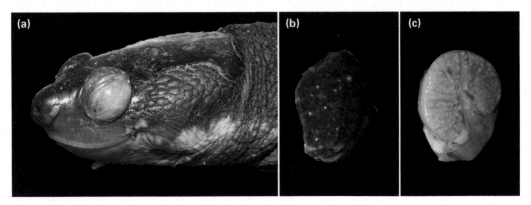

Figure 6.3 Representative lesions of the emergent disease syndrome in Bellinger River snapping turtles. (a) An affected turtle with a severely swollen and pale lower eyelid associated with necrosis and inflammation in the underlying lacrimal gland. (b) Multiple, small, tan foci of necrosis throughout the spleen. (c) Diffuse pallor of the kidney due to severe necrosis and inflammation. Photos: Jane Hall, Australian Registry of Wildlife Pathology.

conducted within a meticulous data and sample collection framework. Morphometric data, photographs and video were collected from each animal. Representative samples of each organ were placed into 10% buffered formalin for histological processing and microscopic examination (Figure 6.3). Multiple sets of fresh tissue samples (liver, kidney, spleen, spinal cord, lung, ingesta, ectoparasites, whole blood) were collected from each animal to allow several lines of investigation to be conducted simultaneously.

These gross post-mortem examinations revealed a consistent pattern of tissue damage and inflammation in the spleen, kidney and tissues around the eyes. The preliminary pattern was inconsistent with intoxication and most consistent with the presence of an infectious or parasitic agent. This knowledge, 2 days after the outbreak onset, underpinned the early establishment of biosecurity actions.

A thorough and broad list of potential diagnoses was established to guide the disease investigation (Box 6.1). This differential diagnosis list was then prioritized according to likelihood, and level of perceived risk posed to the turtle

> **Box 6.1   Potential causes of the mass mortality of Bellinger River snapping turtles**
>
> - Viral infection – ranavirus, iridovirus, adenovirus, herpesvirus, paramyxovirus, picornovirus.
> - Bacterial infection – Mycoplasma species, Chlamydophila species.
> - Bacterial infection secondary to nutritional stress, caustic agents, algal toxins, other environmental factors, trichomonas, viral infection.
> - Exposure to caustic agents, herbicides, pesticides, heavy metals, algal or other toxins.
> - Fungal infection.
> - Trypanosomes, trichomoniasis, amoebiasis, plasmodium, myxosporidiosis, intranuclear coccidiosis.

population and other animals. The list included agents known to cause acute mass mortalities in aquatic species or reptiles, agents known to cause severe eye lesions, and intoxicants.

Microscopic examination of the fixed tissues from affected animals supported the earlier findings of severe inflammation targeting the lacrimal gland, spleen, kidney and blood vessels (Figure 6.3). No organisms were evident within the lesions when examined microscopically and no bacteria, fungi and viral organisms were identified in turtle tissue samples through culture or specific polymerase chain reaction (PCR) tests. Although the clinical and pathological presentation seemed identical to a previous report of ranavirus-associated illness in free-ranging and captive box tortoises in the United States, this virus was not detected by specific PCR testing (Johnson et al., 2008). Analysis of river water samples and turtle tissues failed to detect evidence of intoxicants. After approximately 6 weeks, tests for all known pathogens were negative and the focus of the investigation turned to the potential presence of a novel pathogen. By this time the disease event had spread significantly along the known habitat of the species and the race was on to establish a diagnosis before the population was potentially driven to extinction.

If bacteria, fungi or parasites were causing the outbreak we expected that organisms should have been visible microscopically within at least some of the 26 turtles examined. Thus, we began to more carefully consider the likely presence of an unusual or novel viral agent. With the collaboration of NSW DPI's Elizabeth Macarthur Agricultural Institute, transmission electron microscopy was conducted targeting severe lesions, but no organisms were identified. A second round of viral culture was established on a more diverse array of cell cultures, and when cell damage was detected in one of these lines, electron microscopy identified viral particles with a size and shape consistent with a Nidovirus. Genomic examination of the virus established that it was a novel organism (Zhang et al, 2018) now named the Bellinger River virus (BRV). With the genomic data in hand,

NSW DPI developed a sensitive PCR test specific to the virus. This test enabled extensive screening of animals along the river to attempt to establish the origins, geographic and host range of the virus and better understand the ecology of the organism. The specific PCR test also allowed repeated testing of animals as they were taken into quarantine to enter conservation breeding programs.

6.6   Captive management plan

Because of the scale and speed of the mortality event, there was an urgent need to establish a captive insurance population to prevent the possible extinction of the species. Captive breeding of turtles and 'head-starting' of the resulting hatchlings, i.e. rearing them to a more predator-resistant size before release into the wild, have been viewed as a symptomatic treatment of a larger conservation problem, or 'halfway technology' (Frazer, 1992). Captive breeding from a limited number of individuals may reduce genomic variation and individual fitness (Willoughby et al., 2017), and captive rearing may alter subsequent behaviour (Bowen et al., 1994). However, because there were few adult *M. georgesi* remaining in the wild, the conservation program initially relied heavily on captive breeding. Population modelling for another Australian freshwater turtle (*Chelodina longicollis*) has suggested that only management scenarios including head-starting eliminate all risks of extinction and allow population growth (Spencer et al., 2017). This modelling predicts that even small increases in adult mortality (2%) will have a great effect on population growth and extinction risk, and that supplementation with hatchlings derived from a separate source population can compensate for adult mortality and nest predation to avoid local extinction. A captive breeding and head-starting program for *M. georgesi* was expected to provide such supplementation, at least until juvenile turtles remaining in the wild matured and began to breed.

Initially, in the absence of suitable zoo facilities, WSU drew on its extensive experience with *M. georgesi* to establish a temporary quarantine facility in biosecure conditions for turtles collected from the wild. They remained in this facility for 12 months. Taronga Zoo was asked to co-ordinate the captive breeding program in view of its strong record of ex situ management of other threatened reptiles and amphibians. A dedicated facility was planned and, with funding from the NSW Government's signature threatened species conservation program, Saving Our Species (SoS), established at Taronga Zoo. In April 2016, nine male and seven female *M. georgesi* were transferred from WSU. The facility design (Figure 6.4) combined the biological needs of the species with strict biosecurity conditions to prevent the transfer of any potential pathogens between the captive and wild populations. The enclosures consisted of nine large 5000 l aquaculture tubs, filled with 4000 l water and maintained on individual life-support systems, comprising water filtration, UV sterilization and heating – the last providing for the needs of

juvenile and any sick turtles. Within the middle of each tub was a raised, sand-filled nesting tub and two basking platforms that also provided retreats, while a mesh roof permitted natural UVB exposure for the turtles. Husbandry routines incorporated strict biosecurity protocols including use of personal protective equipment (PPE), dedicated tools for each pond and, as the cause of the mass mortality was initially unclear, animals were fed frozen Fish Fuel Co turtle cubes, Hikari Tropical carnivore pellets and algal wafers[1] and lab-cultured invertebrates. The turtles were housed as male-female pairs or trios of two males and one female. To minimize the potential for inbreeding, pairings were based on relatedness values determined through genetic analysis by the University of Canberra. All turtles were closely monitored, including with ultrasound scans, to determine whether females were gravid. Eggs were collected after laying and placed in dedicated incubators until hatching, whereupon juveniles were transferred to a separate biosecure room in glass aquaria (Figure 6.8). A studbook was established to guide long-term genetic management of the ex situ population and detailed records were maintained of water temperatures, breeding behaviour, reproductive output, incubation results and juvenile growth.

Figure 6.4 Biosecure captive management facility situated in an off-display site at Taronga Zoo, Sydney. Photo: M. McFadden.

6.7 Multi-stakeholder planning workshop

The listing of *M. georgesi* as Critically Endangered prompted the preparation of a site-managed species conservation plan through the SoS program. To assist this process, the Conservation Planning Specialist Group (CPSG), part of the IUCN's[2] Species Survival Commission (IUCN-SSC), was engaged to facilitate a status review, disease risk analysis (DRA) and species action plan. A 2 day conservation planning workshop was held in November 2016, bringing together relevant expertise, alongside decision makers and representatives of impacted stakeholders.

Founded in 1980, with a mission to 'save threatened species by increasing the effectiveness of conservation efforts worldwide' the CPSG combines science, grounded in small population biology, with group facilitation tools, to enable people from diverse backgrounds to freely share their perspectives and knowledge for the benefit of threatened species. A recent review of its work demonstrated that these CPSG-facilitated planning events provide a significant positive turning point in the conservation trajectory of threatened species (Lees et al., 2021).

6.8 Preparing to plan

Following the CPSG'S planning principles and steps (CPSG, 2020; Tables 6.1 and 6.2), the workshop organizing group worked with the facilitators over the 6 months leading up to the workshop, clarifying the purpose and objectives of the workshop and reviewing key stakeholders whose knowledge and perspectives would be needed. These were identified as DPE (OEH), Bellingen Shire Council, NPWS, a Bellinger River snapping turtle local stakeholders group, DPI, TCSA, ARWH, Wildlife Health Australia, WSU and the University of Canberra. Representatives brought a combination of

backgrounds in Australian freshwater turtles and their ecology, in situ and ex situ turtle management, research, freshwater ecology, disease investigation and management, and small population recovery. A set of briefing notes was circulated to all participants as a ready reference prior to, and during, the workshop. These notes included a synopsis of the turtle's natural history, the mortality event and actions taken to protect and recover the species to date. A comprehensive review of all available published and unpublished information on the infectious and non-infectious diseases of Australian freshwater turtles was compiled. The two aims for the workshop were:

1. to agree, using the expertise available, on the current state of knowledge regarding the BRST, the BRV, and other disease and non-disease issues relevant to the conservation of this turtle
2. to use this information to draft a plan of action for the sustained recovery of this turtle.

Guided by CPSG facilitators, the organizing group developed and circulated a detailed workshop program to achieve these aims. The workshop venue at Taronga Zoo enabled participants to fully focus on the task at hand and gain a first-hand appreciation of the complexities of husbandry, health management and biosecurity of the captive turtles.

6.9 Workshop Planning Process

6.9.1 Defining and measuring success

As shown in Figure 6.5, the planning workshop followed the standard CPSG conservation planning steps with the addition of a comprehensive DRA (Jakob-Hoff et al., 2014). A critical early

Table 6.1 CPSG Planning Principles (adapted from CPSG, 2020).

1.	Plan to act	The intent of planning is to promote and guide effective action to save species. This principle underpins everything we do.
2.	Promote inclusive participation	People with relevant knowledge, those who direct conservation action, and those who are affected by that action are all key to defining conservation challenges and deciding how those challenges will be addressed. Inclusivity refers not only to who is included in the planning process, but also to how their voices are valued and incorporated.
3.	Use sound science	Working from the best available information—whether that be established facts, well supported assumptions or informed judgments—is crucial to good conservation planning. Using science-based approaches to integrate, analyse and evaluate this information supports effective decision-making.
4.	Ensure good design and neutral facilitation	Good species planning is designed to move diverse groups of people through a structured conversation in a way that supports them to coalesce around a common vision for the species and to transform this into an achievable, effective plan. Facilitators skilled in planning are essential in guiding these processes. Critically, neutral facilitation eliminates potential or perceived bias in the planning process, helping participants to contribute their ideas and perspectives freely and equally.
5.	Reach decisions through consensus	Effective species conservation planning results in decisions that all participants can support or accept. Recognizing shared goals, seeing the perspective of others, and proceeding by consensus helps galvanize participants behind a single plan of action that is more likely to be implemented.
6.	Generate and share products quickly	Producing and sharing the products of a conservation planning process quickly, freely and widely are important factors in its success. Delays carry a cost in terms of lost momentum, duplicated or conflicting effort or missed opportunities for action.
7.	Adapt to changing circumstances	Effective plans are those that evolve in response to new information and to changing circumstances—biological, political, socio-economic, and cultural—that influence conservation efforts. Plans are considered living documents that are reviewed, updated and improved over time.

step in aligning workshop participants towards a common objective was to jointly develop a hypothetical future scenario, or vision, in which the BRST recovery project had been entirely successful and was being reflected upon by others. A smaller group synthesized contributions from all workshop participants and, following discussion, the aspirational vision shown in Box 6.2 was agreed on. This vision provided the basis on which to consider operational definitions of success and how these would be measured over time (Table 6.3).

> **Box 6.2   Aspirational vision**
>
> It is 2030. The BRST project is a model conservation program for supporting critically endangered native fauna, facilitated by multi-agency collaboration and community engagement. This program has ultimately led to river health restoration and a sustainable turtle population that is disease free.

Table 6.2 CPSG Planning Steps (adapted from CPSG, 2020).

1.	Prepare to plan	Agree on the scope, rationale and required product of planning. Design and prepare a planning process that will meet these requirements.
2.	Define success	Define the core elements of a future state for the species that represents the desired outcome both for conservation and for other relevant stakeholder needs or values.
3.	Understand the system	Assemble the best available information on the biology, history, management, status and threats to the species, the obstacles to addressing those threats, and the opportunities or options for successful intervention.
4.	Decide where to intervene	Determine where in the system to intervene and recommend and prioritize the changes needed to achieve the desired future state.
5.	Agree on how to intervene	Identify alternative approaches to achieving the recommended changes, compare their relative costs, benefits and feasibility, and choose which one(s) to pursue.
6.	Specify what is to be done	Agree on what will be done, when and by whom, to implement the chosen approach, and which measures will be used to indicate progress or completion of specific tasks.
7.	Prepare to implement	Agree on how key individuals and organizations will communicate, coordinate, make decisions, and track and report on progress as they move forward together to implement the plan.
8.	Share, learn and improve	Produce the plan swiftly, share it widely and strategically to maximize conservation impact, and capture lessons learned in order to develop more effective conservation planning processes.

Table 6.3 Operational definitions and measures of success.

Definition	Measures
Bellinger River virus does not pose a threat to species in the wild.	The virus is not detectable via testing or immunity or protection provided to the species by vaccine or otherwise.
Emydura macquarii does not pose a threat to species in the wild.	Control methods have ensured elimination of the hybridization threat [more precise measures to be determined once acceptable control methods have been studied and evaluated].
The species is abundant in the Bellinger River.	Achieve an adult population of at least 1500 turtles by 2030.
Restoration of the species and its ecosystem are sufficient for ongoing resistance to known threats.	Restored population size is stable over time and recovers swiftly from occasional declines.
The community supports the recovery program and is actively engaged in the long-term health of the Bellinger River system.	Landholders are involved in rehabilitation of at least 15 km of riparian zone by Year 5, and there is significant community participation (more than 70 people) in citizen science projects on river health.
Multi-agency collaboration is in place and working positively for the program.	Key institutions have continued active involvement.

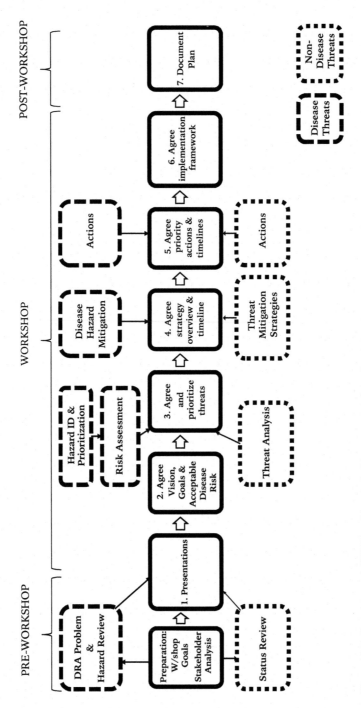

Figure 6.5 Sequence of CPSG multi-stakeholder conservation planning process.

6.9.2 Understanding the system

Bellinger River snapping turtles and the living and non-living components of their environment are part of a complex ecosystem with linkages that are not fully known or may not be readily apparent. As it can be difficult to evaluate interactions in such complex systems (Jones, 1998) workshop participants pooled their collective knowledge using a process of threats analysis. Individual threats to achieving the agreed vision, and in particular the viability or recovery of the species, were identified, described and clustered. The threats considered included not only disease and non-disease impacts on turtle mortality and reproduction, but also the risks associated with delayed or inadequate conservation action. The group then divided into two: one to consider disease and the other non-disease threats. Each group explored in detail each threat in turn, considering its likelihood of occurrence, potential impact, and root causes, with effort taken at each stage to distinguish fact from assumption and to agree key information gaps. Findings were discussed in the combined group for further input.

6.9.3 Deciding where and when to intervene

Critical control points (CCPs) were identified as points in the threat pathways in which actions could be taken to mitigate the threat. Figure 6.6 provides an example of a threat pathway graphic developed for the BRV. The group was encouraged to brainstorm as many interventions as possible for each CCP. The listed interventions were then prioritized based on perceived feasibility and effectiveness, noting that priorities were expected to change over time.

6.9.4 Specifying what is to be done

Actions were developed for all high priority strategies detailing what needed to be done, who would take the lead, potential collaborators, the timeline, success measures and the high-level goal expected to be moved forward as a result (see Box 6.3 for an example).

Box 6.3   Example goal, strategy and success measures

Goal: Establish a serological test for BRV.
Detail: Establish and deploy the test as part of the epidemiological investigation of BRV disease. Establishing a serological testing method will help identify the virus recognized as a primary pathogen. We would expect that during the outbreak, affected animals died so quickly that they did not have time to produce antibodies. Therefore, if antibodies are identified in those animals, it is likely that the virus was present prior to the disease event. We are also currently uncertain whether the juvenile animals currently alive in the river are resistant to the virus or have not been exposed to it. The serological test can provide answers to this.
Lead agency: Australian Registry of Wildlife Health.
Potential collaborators: Dept. Primary Industries, Office of Environment & Heritage.
Timeline/frequency: Commence Year 1 (June 2016–2017).
Success measure(s): Serological test with high sensitivity and specificity is developed and enables the detection of animals that have been exposed to BRV.

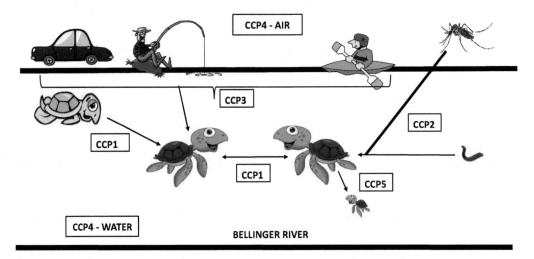

Figure 6.6 Potential transmission pathways for Bellinger River Virus (BRV) and Critical Control Points (CCP).

6.9.5   Prepare to implement

With multiple collaborators needed to implement the conservation management plan, an organizational framework was developed to specify the various players and their roles. A programme coordinator, based in the DPE and supported through SoS, would oversee the project and provide a central point of coordination and communication, to ensure all parties were kept abreast of progress.

6.9.6   Share, learn and improve

To ensure that relevant new information was made available to stakeholders throughout the program, the conservation management plan included the development of an explicit communication strategy. Also, as there were several experimental elements to the program, adapting to lessons learned would be essential. Regular reviews were built into the project's timeline, with major reviews beginning after the first release of captive-bred animals (Figure 6.7).

6.7   Plan Implementation

Implementation of the conservation management plan has been achieved through cooperation and engagement across disciplines and sectors. The program is now led by DPE as the successor to OEH, providing coordination and communication. Most on-ground actions are funded by the NSW Government, through the SoS program, supplemented with cash or in-kind contributions from partner organizations and individuals. The DPE is supported by the Bellinger River Snapping Turtle Expert Reference Group with members contributing expertise in ecology, management, captive breeding, husbandry, genetics, virology, wildlife medicine and wildlife pathology. The implementation program is planned and costed each year around seven central components:

1. in situ population and individual monitoring
2. captive breeding, husbandry and release
3. virus research
4. ecological research (into predation, hybridization, competition and climatic impacts)
5. ecosystem restoration and rehabilitation

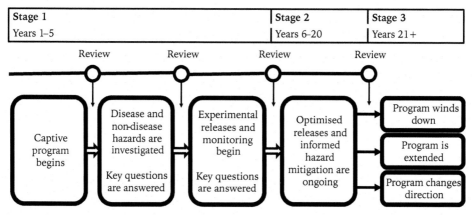

Figure 6.7 Three stages of the recovery program for Bellinger River snapping turtles. In the initial stage, the captive breeding program is developed and key information gaps are filled relating to the disease and to the situation in the river. The second stage involves pursuing recovery while continuing to gather information and test and hone strategies. The third stage involves evaluating program progress and either winding down the program or changing direction.

6. water quality monitoring
7. community and stakeholder communication.

6.8 In situ population and individual monitoring

Annual population surveys have been undertaken in November–December of each year since 2015. These surveys enable assessment of trends in population size and structure for both *M. georgesi* and *E. macquarii* (Chessman et al., 2020) and suggest a possible additional mortality event for *M. georgesi* between December 2016 and November 2017. While these surveys track long-term trends (as they are only annual and subject to statistical error margins), they do not provide rapid feedback on possible mortality events. Therefore, in 2020 it was decided to track a number of 'sentinel' adult *M. georgesi* to monitor survival with the aid of VHF radio-tracking devices (Figure 6.9). Initially, all captured *M. georgesi* were swabbed for virus testing but now only tracked animals are tested unless an animal has disease symptoms, or another

indicator suggests that BRV may be present. The spread of tracked animals through the river is considered a suitable method for virus surveillance.

6.9 Captive breeding, husbandry and release

The ex situ program has been exceptionally successful to date. All founder turtles are still alive within the program 6 years after arrival at Taronga Zoo, and all but one female have produced offspring and are represented within the F1 population. With one exception, all mature females have produced fertile eggs every year. Egg incubation and juvenile rearing have resulted in the expansion of the Taronga population to over 100 and permitted the release of 52 head-started juveniles in the first 3 years of the experimental translocation program (Figure 6.8). Many of these juveniles have been monitored after release with the aid of VHF radio-tracking devices to determine survival, movement, health, and growth. Additionally, the ex situ population has been further reinforced

Figure 6.8 (a) Hatching Bellinger River snapping turtle. (b) Juvenile turtle in a rearing tank. (c) Release of a young, captive-bred turtle into the Bellinger River. Photos: M. McFadden.

Figure 6.9 'Sentinel' *M. georgesi* fitted with VHF radio-tracking device. Photo: G. McGilvray.

by the establishment of a second population in August 2017 at Symbio Wildlife Park, NSW, where 19 healthy young wild-caught turtles are maintained. These immature turtles have yet to breed but are in excellent health.

6.10   Virus research

According to Zhang et al. (2018), BRV was probably the principal or sole cause of the mass mortality that has reduced the total population of *M. georgesi* to about 3% of its historical level (Chessman et al., 2020). A suggestion that *M. georgesi* may have been predisposed to the 2015 disease outbreak by poor body condition and consequently reduced immune function (Spencer et al., 2018) was not supported by analysis of variation in a body condition index (Chessman et al., 2020). The low genomic diversity of *M. georgesi*, with only a single mitochondrial haplotype reported by Georges et al. (2011), may have been a contributing factor to the severity of the 2015 epidemic.

Surveillance in the river for the presence of virus has been undertaken, but three laboratory trials have been unable, to date, to conclusively determine susceptibility and transmission pathways.

Research, led by the DPI in collaboration with the ARWH is ongoing to establish viral transmission routes, the pathogenicity of the virus, and to recreate clinical disease as a first step towards vaccine development. A serological test for BRV antibodies is currently being developed. This would allow us to better understand whether survivors of the outbreak were exposed to the virus, but test development has been challenging.

6.11 Ecological research

Among the non-disease threats identified in the planning workshop were predation, hybridization with the Murray River turtle, *Emydura macquarii* (introduced into the Bellinger River system), competition and climate change.

6.11.1 Predation

Blamires et al. (2005) reported that red foxes (*Vulpes vulpes*) and goannas (*Varanus varius*) destroyed 9 of 13 *M. georgesi* nests. There appear to be no published observations of predation on

juvenile or adult *M. georgesi*. However, Blamires and Spencer (2013) estimated lower survival probabilities of juvenile *M. georgesi* in waterholes with catfish (*Tandanus bellingerensis*, reported as *Tandanus tandanus*) than in waterholes without catfish.

6.11.2 Hybridization

Hybridization between *E. macquarii* and *M. georgesi* has been reported in both the Bellinger and Kalang rivers, and the detection of occasional backcrosses of F1 hybrids to both parental species indicates that at least some F1 hybrids are fertile (Georges et al., 2018). Hybrids totalled only 2% of 474 turtles analysed from the Bellinger River, but 26% of 19 turtles analysed from the Kalang River (Georges et al., 2018). Recent monitoring from November 2015 to November 2018 detected only five morphologically identified F1 hybrids in the Bellinger River (<1% of all turtles) and none in the Kalang River (Chessman et al., 2020). Thus, there is no evidence that the incidence of F1 hybrids is increasing.

6.11.3 Competition

There is potential for competition for food between *M. georgesi* and *E. macquarii* because densities of the two species are positively correlated at the waterhole scale, and they have similar omnivorous diets (Allanson and Georges, 1999; Spencer et al., 2014). However, total turtle densities in the Bellinger River now average only about 15% of densities that the river supported historically (Chessman et al., 2020) and are about 1–2 orders of magnitude below densities at which interspecific competitive impacts have been detected for omnivorous freshwater turtle species (Chessman, 2021). *M. georgesi* may also compete for food with organisms other than turtles, such as fish, water birds, water rats and platypus, but no relevant data appear to be available.

6.11.4 Climate change

Myuchelys georgesi does not appear to be especially sensitive to climatic and hydrological fluctuations. The 2015 mortality event was preceded by a period of low discharge in the Bellinger River (Spencer et al., 2018), but lower discharges have occurred before and since that time without reported mortality. Chessman et al. (2020) found that temporal changes in the body condition of *M. georgesi* were not significantly related to air temperature or river discharge over various antecedent periods.

Spencer et al. (2014) reported that five *M. georgesi* were found to have empty stomachs after extreme flooding removed aquatic vegetation from the Bellinger River in March 2001. In addition, no gravid *M. georgesi* were captured in October–November 2001 (Spencer et al., 2014). Runoff projections for the mid-north coast of NSW vary considerably among climate models but overall suggest an increase in summer runoff but a decrease in winter and total annual runoff (Vaze and Teng, 2011). Nevertheless, flood frequency in coastal southeastern Australia has increased since 1860 (Power and Callaghan, 2016), and floods may pose some risk to *M. georgesi* via scouring of food resources. Climatic warming might benefit *M. georgesi* by extending its growing season, but its thermal ecology has not been assessed.

6.12 Ecosystem restoration and rehabilitation

Blamires et al. (2005) reported that cattle trampled some artificial nests on the Bellinger River

containing hens' eggs although they did not report any impact of cattle on actual turtle nests. Blamires et al. (2005) reported that *M. georgesi* nested in 'heavily vegetated areas', but located only 13 nests and did not specify the type of vegetation (e.g. canopy cover or ground cover). Cohen (2003) documented extensive changes in the channel morphology and riparian vegetation of the Bellinger River since the 1840s, but how these changes may have affected nesting of *M. georgesi* is unknown.

A riparian habitat restoration and rehabilitation program is now ongoing in collaboration with Landcare in the upper sections of the Bellinger River. This restoration has been designed to work upstream to downstream to improve habitat in the riparian zone. Recently, further funding to support the expansion of the program has been provided by partners at local land services. Weed species might impact on the macroinvertebrate assemblage in the river (Llewellyn, 2005), and loss of riparian vegetation can increase erosion and reduce the depth of pools (Prosser et al., 2001). These impacts are likely to be detrimental to *M. georgesi*, given its reliance on deep pools (Blamires and Spencer, 2013) and invertebrate prey (Allanson and Georges, 1999). This program connects with the long-term vision of river health restoration, and it is hoped that the entire length of turtle habitat can be restored in future.

6.13 Water quality monitoring

Water turbidity may affect *M. georgesi* via its impact on availability of various food resources (Spencer et al., 2014). In addition, suspended sediment can influence dive times of other Australian freshwater turtles with cloacal respiration (Schaffer et al., 2016). However, there appears to be no published evidence of adverse effects of turbidity or sedimentation on

M. georgesi. DPE Science has conducted biannual water quality monitoring at 15 sites in the Bellinger River system over 3 years, detecting only moderate and localized land-use impacts. This program has been run concurrently with a citizen science program coordinated by the non-governmental organization *OZGreen*, which is continuing to test water quality with volunteers and has embarked on a macroinvertebrate monitoring program with support from DPE.

6.14 Community and stakeholder communication

The BRST program has generated a lot of interest from the local community and beyond. Intense public meetings were held during and shortly after the mortality event concerning the possible introduction of disease to the river. Community pressure catalysed the highly successful BRST conservation program including the timely rescue of adults for captive breeding. Support from local landholders enabled access to the river for surveys and contributed local knowledge of turtle habitat. Informal communication with landholders has been maintained by DPE project officers during survey and monitoring visits. Strong relationships with those landholders have underpinned and strengthened the program. Critical to this success has been an inclusive and collaborative decision-making approach. That is, all voices were considered equally valid, and contributions were often actively sought prior to meetings to ensure that all ideas could be tested.

This open approach to engagement has been extended with the addition of non-academic representation to the Expert Reference Group. The practical knowledge of amateur herpetologists has been invaluable and provided critical information.

6.15 Reflections

1. There is no doubt that multi-stakeholder and multi-sectoral collaboration has been, and will continue to be, key to the successful recovery of this species. This collaboration has been the result of good communication, cooperative practice and the development of strong relationships with the local community and landholders. There is a wealth of knowledge among the Expert Reference Group members, but local knowledge and advice sought from other parties has been equally important.

2. Partners in the program have been instrumental in its success. Linkages with university research and in-kind husbandry services provided by Taronga Zoo and Symbio Wildlife Park have been invaluable and have meant that costs for the delivery of the program have been smaller than for comparable programs.

3. The success of the diagnostic investigation was also built on the foundations of strong inter-agency collaboration and a pre-existing wildlife health diagnostic program. The investigation was led by Taronga's ARWH, in collaboration with the NSW DPI Elizabeth Macarthur Agricultural Institute, NSW EPA, University of Sydney and Murdoch University research laboratories and the Australian Centre for Disease Preparedness.

4. Thorough research and planning with input from the Expert Reference Group yielded a trial translocation plan that could be a model for other programs around the world. The plan developed at the 2016 workshop, facilitated by the CPSG, provided a shared vision of success and a clear roadmap whereby all stakeholders understood their roles in the context of the wider recovery program.

5. Having a team manage this program contributed significantly to the results. The local project officers coordinating the program

were able to maintain the workload required and provide complementary skills. Being able to work closely, bounce ideas and delve deeply into the program from often opposing views helped to test ideas further and yielded better results. The team extended to local contractors who assisted in monitoring, adding to program successes.

Abbreviations used in the text

ARWH	Australian Registry of Wildlife Health
BRST	Bellinger River Snapping Turtle
BRV	Bellinger River Virus
CPSG	Conservation Planning Specialist Group of the IUCN-SSC
DPE	NSW Department of Planning and Environment (formerly OEH)
DPI	NSW Department of Primary Industries
DRA	Disease Risk Analysis
IUCN-SSC	International Union for the Conservation of Nature – Species Survival Commission
NPWS	NSW National Parks and Wildlife
NSW	New South Wales, Australia
OEH	NSW Office of Environment and Heritage (superseded by DPIE)
SoS	NSW Save our Species program
TCSA	Taronga Conservation Society Australia
WHA	Wildlife Health Australia
WSU	Western Sydney University

Endnotes

1   Fish Fuel Co Turtle Food, Fish Fuel Co Turtle Veg supplied by Fish Fuel Co, Wingfield, South Australia; Hikari Tropical Massivore Delite, Hikari Tropical Sinking Carnivore Pellets, Hikari Tropical

Algal Wafers Supplied by Pet Alliance Pty Ltd, Caringbah, NSW.

2   International Union for the Conservation of Nature, headquartered in Gland, Switzerland comprises 1400 member organizations representing States and government agencies, Non-Government Organizations, Indigenous Peoples' organizations, scientific and academic institutions and business associations. (https://www.iucn.org)

References

Allanson, M., & Georges, A. (1999). Diet of *Elseya purvisi* and *Elseya georgesi* (Testudines: Chelidae), a sibling species pair of freshwater turtles from eastern Australia. *Chelonian Conservation and Biology, 3*, 473–477.

Australian Government. (2021). Australian threatened species strategy 2021–2031. https://www.awe.gov.au/environment/biodiversity/threatened/publications/threatened-species-strategy-2021–2031

Blamires, S. J., & Spencer, R.-J. (2013). Influence of habitat and predation on population dynamics of the freshwater turtle *Myuchelys georgesi. Herpetologica, 69*(1), 46–57. https://doi.org/10.1655/HERPETOLOGICA-D-12-00014

Blamires, S. J., Spencer, R.-J., King, P., & Thompson, M. B. (2005). Population parameters and life-table analysis of two coexisting freshwater turtles: Are the Bellinger River turtle populations threatened? *Wildlife Research, 32*(4), 339–347. https://doi.org/10.1071/WR04083

Bowen, B. W., Conant, T. A., & Hopkins-Murphy, S. R. (1994). Where are they now? The Kemp's ridley headstart project. *Conservation Biology, 8*(3), 853–856. https://doi.org/10.1046/j.1523-1739.1994.08030853.x

Cann, J., Spencer, R.-J., Welsh, M., & Georges, A. (2015). *Myuchelys georgesi* (Cann 1997) – Bellinger River turtle. In A. G. J. Rhodin et al. (Eds.), *Chelonian Research Monographs Conservation biology of freshwater turtles and tortoises*. A compilation project of the IUCN/SSC Tortoise and Freshwater Turtle Specialist Group, *5*, 091.1–9.

Chessman, B. C. (2021). A creeping threat? Introduced Macquarie turtles and the future of endangered helmeted turtles in southern Australia. *Aquatic Conservation: Marine and Freshwater Ecosystems, 31*(12), 3429–3436. https://doi.org/10.1002/aqc.3723

Chessman, B. C., McGilvray, G., Ruming, S., Jones, H. A., Petrov, K., Fielder, D. P., Spencer, R.-J., & Georges, A. (2020). On a razor's edge: Status and prospects of the critically endangered Bellinger River snapping turtle, *Myuchelys georgesi*. *Aquatic Conservation: Marine and Freshwater Ecosystems, 30*(3), 586–600. https://doi.org/10.1002/aqc.3258

Cohen, T. J. (2003). Late Holocene floodplain processes and post-European channel dynamics in a partly confined valley of New South Wales Australia [PhD Thesis]. University of Wollongong.

CPSG. (2020). Species conservation planning principles and steps ver. 1.0. International Union for Conservation of Nature and Natural Resources/ SSC Conservation Planning Specialist Group.

Daszak, P., Cunningham, A. A., & Hyatt, A. D. (2000). Emerging infectious diseases of wildlife— Threats to biodiversity and human health. *Science, 287*(5452), 443–449. https://doi.org/10.1126/science.287.5452.443

Frazer, N. B. (1992). Sea turtle conservation and halfway technology. *Conservation Biology, 6*(2), 179–184. https://doi.org/10.1046/j.1523-1739.1992.620179.x

Georges, A., Spencer, R.-J., Welsh, M., Shaffer, H. B., Walsh, R., & Zhang, X. (2011). Application of the precautionary principle to taxa of uncertain status: The case of the Bellinger River turtle. *Endangered Species Research, 14*(2), 127–134. https://doi.org/10.3354/esr00350

Georges, A., Spencer, R.-J., Kilian, A., Welsh, M., & Zhang, X. (2018). Assault from all sides: Hybridization and introgression threaten the already critically endangered *Myuchelys georgesi* (Chelonia: Chelidae). *Endangered Species Research, 37*, 239–247. https://doi.org/10.3354/esr00928

Georges, A., Walsh, R., & Spencer, R.-J., Welsh M. Shaffer HB. (2007). The Bellinger *Emydura*. Challenges for management. *Final report prepared by the Institute for Applied Ecology, University of Canberra, for the NSW National Parks and Wildlife Service, Sydney*. Institute for Applied Ecology, University of Canberra.

Jakob-Hoff, R. M., MacDiarmid, S. C., Lees, C., Miller, P. S., Travis, D., & Kock, R. (2014). Manual of procedures for wildlife disease risk analysis. World Organization for Animal Health. Published in association with the International Union for Conservation of Nature Species Survival Commission.

Johnson, A. J., Pessier, A. P., Wellehan, J. F., Childress, A., Norton, T. M., Stedman, N. L., Bloom, D. C., Belzer, W., Titus, V. R., Wagner, R., Brooks, J. W., Spratt, J., & Jacobson, E. R. (2008, October). Ranavirus infection of free-ranging and captive box turtles and tortoises in the United States. *Journal of Wildlife Diseases, 44*(4), 851–863. https://doi.org/10.7589/0090-3558-44.4.851

Jones, M. C. (1998). *The thinker's toolkit, 14 powerful techniques for problem solving* (2nd ed). Times Books.

King, P., & Heatwole, H. (1994). Partitioning of aquatic oxygen uptake among different respiratory surfaces in a freely diving pleurodiran turtle, *Elseya latisternum*. *Copeia, 1994*(3), 802–806. https://doi.org/10.2307/1447197

Kolbert, E. (2014). *The sixth extinction: An unnatural history*. Picador

Lees, C. M., Rutschmann, A., Santure, A. W., & Beggs, J. R. (2021). Science-based, stakeholder-inclusive and participatory conservation planning helps reverse the decline of threatened species. *Biological Conservation, 260*, 109194, ISSN 0006-3207. https://doi.org/10.1016/j.biocon.2021.109194

Llewellyn, D. C. (2005). Effect of toxic riparian weeds on the survival of aquatic invertebrates. *Australian Zoologist, 33*(2), 194–209. https://doi.org/10.7882/AZ.2005.016

MacPhee, R. D., & Greenwood, A. D. (2013). Infectious disease, endangerment, and extinction. *International Journal of Evolutionary Biology, 2013*, 571939. https://doi.org/10.1155/2013/571939

Moloney, B., Britton, S., & Matthews, S. (2015). *Bellinger River snapping turtle mortality event 2015*. Epidemiology report. New South Wales Department of Primary Industries.

Power, S. B., & Callaghan, J. (2016). The frequency of major flooding in coastal southeastern Australia has significantly increased since the late 19th century. *Journal of Southern Hemisphere Earth Systems Science, 66*, 2–11. https://doi.org/10.1071/ES16002

Prosser, I. P., Rutherfurd, I. D., Olley, J. M., Young, W. J., Wallbrink, P. J., & Moran, C. J. (2001). Large-scale patterns of erosion and sediment transport in river networks, with examples from Australia. *Marine and Freshwater Research, 52*(1), 81–99. https://doi.org/10.1071/MF00033

Rhodin, A. G. J., Stanford, C. B., van Dijk, P. P. V., Eisemberg, C., Luiselli, L., Mittermeier, R. A., Hudson, R., Horne, B. D., Goode, E. V., Kuchling, G., Walde, A., Baard, E. H. W., Berry, K. H.,

Bertolero, A., Blanck, T. E. G., Bour, R., Buhlmann, K. A., Cayot, L. J., Collett, S., . . . Vogt, R. C. (2018). Global conservation status of turtles and tortoises (Order Testudines). *Chelonian Conservation and Biology, 17*(2), 135–161. https://doi.org/10.2744/CCB-1348.1

Schaffer, J. R., Hamann, M., Rowe, R., & Burrows, D. W. (2016). Muddy waters: The influence of high suspended sediment concentration on the diving behaviour of a bimodally respiring freshwater turtle from north-eastern Australia. *Marine and Freshwater Research, 67*(4), 505–512. https://doi.org/10.1071/MF14117

Spencer, R.-J., Georges, A., Lim, D., Welsh, M., & Reid, A. M. (2014). The risk of inter-specific competition in Australian short-necked turtles. *Ecological Research, 29*(4), 767–777. https://doi.org/10.1007/s11284-014-1169-7

Spencer, R.-J., Van Dyke, J., Petrov, K., Ferronato, B., McDougall, F., Austin, M., Keitel, C., & Georges, A. (2018). Profiling a possible rapid extinction event in a long-lived species. *Biological Conservation, 221,* 190–197. https://doi.org/10.1016/j.biocon.2018.03.009

Spencer, R. J., Van Dyke, J. U., & Thompson, M. B. (2017). Critically evaluating best management practices for preventing freshwater turtle extinctions. *Conservation Biology, 31*(6), 1340–1349. https://doi.org/10.1111/cobi.12930

Vaze, J., & Teng, J. (2011). Future climate and runoff projections across New South Wales, Australia: Results and practical applications. *Hydrological Processes, 25*(1), 18–35. https://doi.org/10.1002/hyp.7812

Westley, F. R., & Miller, P. S. (Eds.). (2003). *Experiments in consilience: Integrating social and scientific responses to save endangered species.* Island Press.

Willoughby, J. R., Ivy, J. A., Lacy, R. C., Doyle, J. M., & DeWoody, J. A. (2017). Inbreeding and selection shape genomic diversity in captive populations: Implications for the conservation of endangered species. *PLOS ONE, 12*(4), e0175996. https://doi.org/10.1371/journal.pone.0175996

World Wildlife Fund. (2020). Living planet report. *Bending the curve of biodiversity loss* R. E. A. Almond, M. Grooten & T. Petersen (Eds.), *2020.* WWF Verlagsgesellschaft.

Zhang, J., Finlaison, D. S., Frost, M. J., Gestier, S., Gu, X., Hall, J., Jenkins, C., Parrish, K., Read, A. J., Srivastava, M., Rose, K., & Kirkland, P. D. (2018). Identification of a novel nidovirus as a potential cause of large scale mortalities in the endangered Bellinger River snapping turtle (*Myuchelys georgesi*). *PLOS ONE, 13*(10), e0205209. https://doi.org/10.1371/journal.pone.0205209

PART II

CLIMATE CHANGE AND LAND USE

Chapter 7

Ecotoxicology: The burgeoning spectre of pharmaceuticals in the environment

Judit E.G. Smits and Kathryn E. Arnold

Abstract

Medications that benefit human and pet health, and which increase production from food animals, are being used at increasingly greater levels worldwide. These pharmaceutical compounds are specifically designed to effect changes in living animals, whether these medications have been taken on purpose, or if exposure occurs through contaminated food or water. Neither wastewater treatment plants nor environmental processes effectively remove or break down these pharmaceuticals. In this chapter we present four dramatic examples of the consequences to wildlife and environmental health, from unplanned, secondary exposures to drugs produced and sold by the pharmaceutical industry, and then prescribed by physicians and veterinarians.

7.1 Introduction

This chapter presents examples in which wildlife provide valuable and honest insight into the state of the environment. 'Proximate causation' is the direct factor that causes death, such as a violent storm, exposure to poisons, or hit by a vehicle. Considering a One Health approach to health problems in animals and humans, what is more important is the 'ultimate causation' of a disease process, in which underlying factors lead to a problem. The example in this chapter is unintended poisoning of wildlife with human medications or agrochemicals (Shore et al., 2014).

This ultimately occurs because we, as a society, undervalue natural resources and fail to consider humans' impacts on the natural world. Rather, we prioritize immediate satisfaction or economic gain instead.

In this chapter we specifically examine the hidden costs (or ultimate causation) associated with medicating ourselves and our animals, with the inevitable, negative impacts on Ecosystem Health and the natural environment (Figure 7.1). Pharmaceuticals are used to improve the quality of life in people, as well as their pets. But in food producing animals the main incentive for antimicrobial use is as growth

Susan C. Cork and Douglas P. Whiteside (eds) **Case Studies in EcoHealth**
DOI: 10.52517/9781789183368.007, © 5m Books Ltd, 2024

Pharmaceuticals are any kind of drug used as a medicine. Historically humans discovered certain plants that could help relieve medical problems. Now, research by scientists and the pharmaceutical industry results in the discovery and development of useful medicines to treat specific illnesses. Common examples are cough syrup, headache pills, antibiotics to treat infections, and drugs to control cancer.

promotants, to increase meat and egg production for the lowest possible cost. The problem is that 20–90% of antibiotics administered to animals are excreted either as the parent compound or as active metabolites in the urine and

faeces (Andersson and Hughes, 2014; Jjemba, 2006). Similarly in humans, total excretion of pharmaceuticals via urine and faeces ranges from 25% to 90% (Lienert et al., 2007).

By working with wild animals in nature, we can learn a great deal about the health of the environment around them. Whether they are fish, birds or mammals, wildlife depends on the natural resources around them for everything, from a plentiful food supply, to secure places to sleep and safe spaces to raise their young. When we detect declines in wild populations, this is a strong indication of possible environmental problems. However, there are huge inherent challenges in working with wildlife. Because it so difficult to detect sick or dead wild animals; (1) we do not detect the vast majority of them,

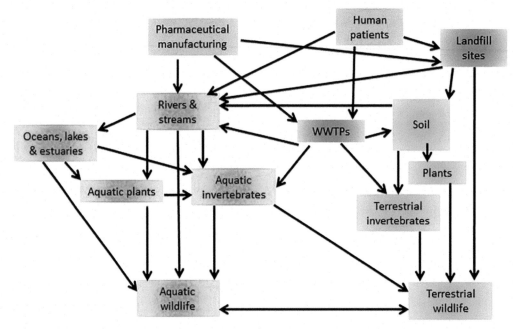

Figure 7.1 Antidepressants disperse through the environment via many and potentially complex pathways some of which are shown here. Sources of pharmaceuticals in the environment (yellow boxes) include pharmaceutical manufacturing and human patients. The drugs then disperse directly into the environment, or via wastewater treatment plants (WWTPs) and landfill (red boxes). Movement of antidepressants in aquatic (blue boxes) and terrestrial (green boxes) environments can result in uptake of these pharmaceuticals into wildlife. Simple food webs are shown to illustrate how antidepressants, particularly those that are lipophilic, bioaccumulate.

and (2) we rarely have high enough numbers, or fresh enough samples to make confident diagnoses. Therefore, the true impact and magnitude of any problem is likely much greater than what can be proven by simply using the hard data in hand (Berny, 2007; Wobeser et al., 2004).

We present examples to illustrate the global extent and magnitude of this threat from pharmaceutical use in livestock and humans. Then we focus on specific cases to illustrate how pharmaceuticals have, and continue to affect humans, animals, and ecosystems. Antimicrobial use in swine and poultry operations in North America, China and ever expanding areas of the world, increases efficiency of production while decreasing costs. Fish, shrimp and shell-fish production (Lulijwa et al., 2020), as well as beef feedlots (Agga et al., 2019), are other major, continuing sources of large volumes of pharmaceuticals entering the environment.

Human use of pharmaceuticals worldwide is increasing at dizzying speed. A largely unknown and highly worrisome fact is that neither sewage treatment plants, nor drinking water plants successfully remove these pharmaceutical compounds! A United Nations database compiled from 71 countries confirms that pharmaceuticals are contaminating not only surface water, but medications and illicit drugs are ending up in groundwater, tap water, and even soil (Center for Behavioural Health Statistics, 2016). Urban wastewater is the primary source. A shocking example of what this means in terms of water contamination, is Patancheru, a town near the city of Hyderabad, southern India, which is considered an 'ecological sacrifice'. It has a cluster of 90, many unregulated, pharmaceutical plants in a small area. When researchers analysed vials of supposedly 'treated' wastewater from one of the plants there, they found the equivalent of 90,000 doses of ciprofloxacin, a valuable antibiotic, were released into the local stream each day (Gunnarsson et al., 2009). 'Drugs in our rivers: Bugs on speed and Prozac in the food chain', is a catchy title by journalist David Trilling (2017), in which he describes another sobering example from the Antarctic, widely thought to be the most pristine, and untouched part of the globe (González-Alonso et al., 2017). Now with as many as 16 thousand tourists, and one to four thousand researchers per year visiting Antarctica, caffeine, ibuprofen, paracetamol, and cocaine are found in the Antarctic Ocean at levels similar to those in Europeans waters.

7.2 Case I Anti-inflammatory drugs and old world vultures

Most people are completely unaware of the great ecological value of scavenger species the world over, basically 'nature's clean-up' crews. No matter what aggressive infectious or parasitic disease has killed an animal, scavengers recycle and reincorporate those carcasses back into nature, supporting a healthy, functioning ecosystem cycle. Additionally, this 'ecosystem service' deals with the unpleasant smells and sights should carcasses linger for extended periods. In this manner, infectious disease is well controlled, while living fauna and flora are naturally nourished. For a long time, both malicious and unintended poisonings have plagued old world vultures which are high trophic level, scavenging birds. Because of this clean-up role, all contaminants that persist or bioaccumulate through the food web will reach highest concentrations in these animals.

A lesser known, historic, and ongoing cultural role of these large vultures in Asia, has been the reincorporation of human remains, into nature. The religion known as Parsi, or Zoroastrian, considers earth and water to be sacred, with their scriptures also commanding that safe disposal of their dead does not result in pollution of air. Therefore, they do not bury or burn the bodies of their dead. Rather they build 'Towers of Silence' upon which their dead

© Erica Sims

Figure 7.2 Gyps vultures (*Gyps* spp.), while performing their 'ecosystem service' of scavenging and cleaning up carcasses present on the landscape, are being poisoned by drug residues in animals that have been treated by humans. In this case an anti-inflammatory medication caused the virtual extinction of several species of old world vultures in a 10 year period.

are placed. In this way vultures, in consuming them and returning them to nature, help release the spirits of their ancestors. With the de facto, rapid extinction of these vultures, a crisis ensued for the Parsi. It took several decades to work out an acceptable and functional solution in the last rites of departed souls. Now in cities in India with big Parsi communities, solar panels have been installed on the 'Towers of Silence' to be able to use energy from the sun to help dispose of their deceased (https://telanganatoday.com/last-rites-parsis-turn-to-tech-as-vulture-numbers-decline).

Dr Vibhu Prakash (Prakash et al., 2003), ornithologist with the Bombay Natural History Society, produced the first scientific publication to alert the world to the biggest population crash of birds of prey in history. In slightly more than a ten-year period, more than 10 million birds were estimated to have died, leaving three species of vultures, the white-rumped (*Gyps bengalensis*), Indian (*G. indicus*) and slender-billed (*G. tenuirostris*) vultures on the brink of extinction. He led a comprehensive survey of various species of the impressive, old world, Gyps vultures in India, comparing survey results from

1990–1993 to those from 2000. The population estimates were carried out along 6000 km of roads in and near 'Protected areas' plus an additional 5000 km of roads for the second survey. The alarming findings indicated a population collapse of 92%.

In the years following this revelation, ornithologists, conservationists, veterinarians, toxicologists and infectious disease specialists were involved in lively debate at conferences, creative thinking, interdisciplinary collaborations, and intense research to determine what, in fact, was responsible for this near extinction of vultures, that is ongoing today, and has expanded to include the entire globe (Smits and Naidoo, 2018).

Ironically, since veterinarians are the profession devoted to the health and welfare of animals, a veterinary anti-inflammatory drug, diclofenac, turned out to be responsible (Oaks et al., 2004) (Figure 7.2).

Observations in the wild, and detailed experimental studies showed that exposed vultures die within 48 hours of feeding on cattle with residual diclofenac (an inexpensive and newly introduced veterinary anti-inflammatory drug) in their tissues when they die within

24 to 48 hours of being treated. These birds are depressed with signs of poisoning within 18 hours of exposure, soon dying from total renal failure. The characteristic pathological finding is visceral gout in which organs in the abdominal cavity are covered by a white chalky material, as the kidneys stop functioning. Microscopically, massive renal tubular damage is evident, which indicates the kidneys have been directly poisoned and destroyed by the drug that they are trying to eliminate from the body. The kidneys have the role of concentrating, and eliminating toxicants from the body, which makes them generally, highly susceptible to damage from toxic compounds.

As well as diclofenac, similar acute toxicity occurs from other anti-inflammatory drugs, namely ketoprofen, flunixin, and aceclofenac in other species of Gyps vultures, including the African white-backed (*G. africanus*), Cape griffon (*G. coprotheres*), the Eurasian griffon (*G. fulvus*) and the Himalayan vultures (*G. himalayensis*) (Das et al., 2011; Galligan et al., 2016; Naidoo et al., 2010; Swan et al., 2006).

Two vulture-safe anti-inflammatory drugs for cattle have been identified. Meloxicam was established as such, about 2006, whereas Tolfenamic acid very recently has been shown to be safe for Gyps vultures (Chandramohan et al., 2022 unpublished results), with 36 of 38 exposed birds surviving with little to no evidence of toxicity during or post-exposure.

A direct factor in this massive poisoning of vultures was the extremely low cost and therefore wide-scale use of diclofenac among veterinary clinics in Pakistan and India. This disastrous impact was exacerbated by the gregarious nature of vultures in which large numbers of birds feed together on each contaminated cattle carcass. Additionally, old world vultures seem unusually sensitive to the drug compared with mammals and other birds (Rattner et al., 2008). Based on population modelling, pharmaceutical contamination of as few as 1% of the total

carcasses available for vultures could cause this massive scale of toxicity (Green et al., 2007).

7.3 Case II Prozac and song birds

Consumption of antidepressants has been increasing globally over the last few decades (González Peña et al., 2021). Despite controversy over the increased pharmaceutical suppression of unhappiness, selective serotonin reuptake inhibitors (SSRIs) have an important role in treating depression, obsessive-compulsive disorder, bulimia and other conditions. As well as their intended effects on mood and behaviour, SSRIs such as Prozac can cause a range of side effects, including changes in appetite and weight, gastrointestinal disruptions, anxiety, sexual dysfunction and lethargy (NICE, 2021). Such side effects in humans are unpleasant rather than life threatening, but in wildlife could alter survival-related traits such as feed intake, reproductive success, and anti-predator, defensive behaviour (reviewed in Saaristo et al., 2018). But why should we even worry about this – people do not leave their Prozac tablets lying around in nature, right?

It is not stray tablets, but human sewage that exposes wild animals to Prozac and other SSRI antidepressants (Figure 7.2). When a person takes a Prozac tablet, for example, roughly one-quarter of the dose is excreted unchanged (Lienert et al., 2007) and flushed away to waste water treatment plants (WWTPs) (Bean et al., 2017). WWTPs and fields fertilized by sewage sludge are a great place to spot birds and other wildlife feeding on the worms, maggots and other invertebrates living in the antidepressant-laced human waste. Fluoxetine, the active ingredient in Prozac, is also very stable in the environment and is not removed by most sewage treatment processes. Consequently, it has been detected downstream of WWTPs in rivers and streams across the world, at concentrations in the ng-µg/l range (Brooks et al., 2003; Castillo-Zacarías et al., 2021; Schultz et al., 2010). Although found at relatively low concentrations in the environment, fluoxetine builds up (aka 'bioaccumulates') in many species eaten by higher predators, for example bivalves (Bringolf et al., 2010), shrimp (*Gammarus pulex*), (Meredith-Williams et al., 2012), earthworms (*Eisenia fetida*) (Bean et al., 2017), as well as fish (Castillo-Zacarías et al., 2021). Thus, there are several pathways by which birds and aquatic predator species can be exposed to fluoxetine and other antidepressants (Arnold et al., 2014) (Figure 7.3).

Antidepressants are designed to change the behaviour and physiology of humans suffering mental health conditions, so how do such potent modifiers of behaviour and physiology affect otherwise healthy, non-depressed, wildlife species? This has been explored by scientists working on Eurasian starlings *Sturnus vulgaris*, a species often spotted feeding at WWTPs. In one experiment, groups of starlings were kept in large outdoor aviaries and hand fed daily a

Figure 7.3 Wastewater treatment plants do not effectively degrade many pharmaceuticals, fluoxetine included, leaving the active drug in the sludge and water released into the environment.

waxworm either containing a low, environmentally relevant concentration of fluoxetine, or control substance (Bean et al., 2014). The fluoxetine concentration was equivalent to one tenth of a therapeutic dose (scaling down from humans to starlings). They found, as in humans, the starlings' appetite changed. Control birds, as is optimal for small birds in the depths of winter, ate a hearty breakfast to recover from a long winter night without food and a big supper to stoke their energy reserves before bedtime. Instead, fluoxetine-treated birds fed less overall, eschewed breakfast and basically lightly snacked throughout the day. So, fluoxetine dosed birds risked starvation, and potentially predation, because they were not balancing their energy reserves properly.

In a similar experiment, evidence of sexual dysfunction was discovered. For females of most species, it is important to pair up with a good quality mate, because eggs are a much bigger investment in reproduction than tiny sperm. So, in starlings, male quality is judged on the length and complexity of their songs, as well as the vigour of their courtship 'moves' (Verheyen et al., 1991). For a male starling, it is important to attract a strong healthy female, not only for her good genes, but also because he will be sharing parental care with her. However, the cues that males use to assess mates are little understood. Scientists found that during courtship, male starlings sang less and were more aggressive towards fluoxetine-treated females than to the normal control females (Whitlock et al., 2018). Somehow the males had 'decided' that fluoxetine dosed females were unattractive, possibly less fertile or potentially would not make good mothers. Unexpectedly, this experiment did not find any evidence that male attractiveness was affected in females exposed to low doses of fluoxetine.

Populations of many wild animal species are declining, often for reasons that we do not fully understand. Low doses of antidepressants and cocktails of different pharmaceuticals in the environment are not causing mass mortality events of wild animals, but the impact is more insidious, subtle and harder to measure. Small changes to physiology or behaviour might seem inconsequential. However, an animal that fails to feed normally during a cold winter, or one that cannot attract a good quality mate will not pass their genes onto the next generation (Saaristo et al., 2018).

Such 'indirect' effects on survival can diminish the resilience of populations in the face of other existing environmental challenges, such as habitat loss and climate change. Under-investment in sanitation, even in developed countries, results in cocktails of pharmaceuticals and other contaminants being released at high levels and contributing to the attrition of our natural environment.

7.4 Case III Dung organisms and insecticidal treatment of livestock

Manure, or dung, from wild animals and livestock is a potential fertilizer containing valuable nutrients such as phosphorous (P), calcium (Ca), and magnesium (Mg). The activity of dung organisms which breakdown dung, is recognized as an 'ecosystem service' by which manure enriches and fertilizes the soil and improves its structure. Dung beetles (Coleoptera: Scarabaeidae) and other organisms increase the rates of dung decomposition and nutrient cycling, such that in healthy, temperate agro-ecosystems, it can be buried and incorporated into the soil within 4–5 days (Beynon et al., 2012). Without the beetles, dung deposits may remain intact for 3–12 months or longer depending on the climatic conditions. Grazing livestock will avoid the patches with dung deposits or the less palatable grass plants that will mature there. The results of undecomposed

manure and subsequent degradation of grazing land is referred to as pasture 'fouling'. However, this beneficial activity of dung organisms is severely threatened by the negative effects of anthelmintics used to control livestock parasites and changes in pasture management.

Avermectins, ivermectin being a well-recognized formula, are antiparasitic drugs used worldwide to control parasites of livestock. Antiparasitic treatment of livestock, whether oral, injected, or pour-on, all end up in the faeces, where they can affect non-target organisms

Although we commonly hear the term 'pesticide', it is a generic reference to any agrochemical or anti-pest compound. These include insecticides for controlling a wide range of insects; rodenticides against rats, mice and other rodents; herbicides used to kill weeds or invasive plants; fungicides which control mould and mildew; and avicides used to kill birds considered to be pests. Toxicologically, these are important subcategories. Worldwide, rodenticides and insecticides are the main pesticides responsible for widespread mortality of non-target avian and mammalian species.

(Figure 7.4). Up to 90% of these types of insecticides are eliminated into the faeces basically unchanged, with their pharmacological activity intact (Alvinerie et al., 1999). This harms the insects that naturally breed in dung, the majority of which are beneficial. Residual antiparasitic drug toxicity has a range of impacts; it may be directly lethal to the beetles and other invertebrates, it may disrupt their life cycle, change the behaviour of the insects, or interfere with other aspects of reproduction in the beneficial dung organisms (Table 7.1). The negative impacts on the dung insects and other non-insect invertebrates in the dung, slows the decomposition of the dung pats from avermectin-treated cattle,

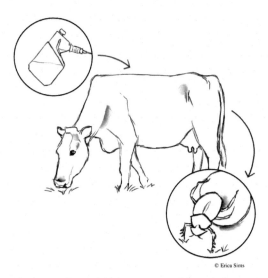

© Erica Sims

Figure 7.4 Insecticidal treatment of livestock, whether injected, poured on, applied through a dip trough, or given orally, decreases the load of harmful parasites, but ends up with a substantial percentage of the insecticide being excreted in the faeces and urine, thereby poisoning dung organisms.

regardless of the route of administration of the drug.

According to Lumaret and Errouissi (2002, and references therein), in experiments in which ivermectin was added directly to dung at levels present in manure from treated cattle, after almost 1 year, the insecticide-treated dung was not appreciably degraded. In contrast, in the same environment, the natural, untreated dung pats deposited at the same time were degraded after 80 days in the field.

A bonus 'ecological service' of coprophagous beetles is that their breaking down of livestock faeces makes the dung unfavourable for the development of gastrointestinal parasites of those livestock, reducing parasitic spread (Sands and Wall, 2017, and references therein). This indirect reduction of gastrointestinal parasites by the beetles is estimated to save UK farmers £188 million per year in production losses. Similarly, in the United States, the value of dung organisms to the cattle industry is estimated at US$380 million annually (Losey and Vaughan, 2006). So, not only do the dung beetles and

Table 7.1 Selected findings from studies on the impacts to non-target organisms, from antiparasitic treatment of livestock. Adapted excerpts from Lumaret et al. (2012, Table 9).

Study author	Feature studied in dung ecosystem	Impacts on dung	Route of administration (livestock treated)
Floate et al. (2002)	Dung community of Coleoptera, Diptera and Hymenoptera	Reduced number of insects in dung voided 2 and 4 weeks post-dosing	Pour-on (cattle)
Floate et al. (2008)	Coprophilous species Coleoptera, Diptera and Hymenoptera	Reduced insect metamorphosis up to 16 weeks post-treatment	Pour-on (cattle)
Iglesias et al. (2006)	Coprophilous arthropods	Reduced number and diversity of arthropods	Injectable (cattle)
Suárez et al. (2009)	Coprophilous arthropods and nematodes	Reduced number of beetle and fly larvae, springtails, mites, nematodes	Injectable and Pour-on (cattle)
Sommer and Bibby (2002)	Decomposition of dung organic matter in soil	Reduced degradation at 8, 12 & 16 weeks	Subcutaneous Injection (cattle)

related organisms decrease parasitism in live-stock, other economic benefits are that they enrich the soil organic matter, prevent fouling of grazing land and they reduce the need for arti-ficial fertilizers (Stevenson and Dindal, 1985), plus the associated real energy costs associated with their application.

7.5 Case IV Role of wildlife in antimicrobial resistance (AMR) in the environment?

Since they started being mass-produced in the 1940s, antimicrobial drugs have saved millions of human lives and improved animal health and wel-fare globally (Davies and Davies, 2010). However, the initial excitement over the power of these wonder drugs was short-lived as we realized that bacteria can quickly develop resistance to them. Once in the environment, antibiotics stimulate the evolution of antibiotic-resistant bacteria, which lead to antibiotic-resistance genes persisting in the environment (Agga et al., 2019). In the One Health framework, antimicrobial-resistant genes are recognized now as major environmental con-taminants. They can remain in the environment longer than 2 years, even after the large-scale animal production has been stopped.

We must wake up to what is going on with humans regarding pharmaceutical use. Based on data available from studies in the United States, it is a heavily medicated country. From 2000 to 2013, the number of Americans using prescrip-tions drugs daily, increased from 44% to 70% of the population according to a Mayo Clinic study. Antibiotics, antidepressants and painkilling nar-cotics (opioids) top this list. Our consumption of these substances is much higher than ever before. In the rest of the world pharmaceutical use is rapidly following the same upward trend.

Over the next 2 decades, antimicrobial use in food production animals (beef, swine, poultry), which has been common for the past 40 years, is expected to increase by almost 70% (Agga et al., 2019) although some countries are beginning to ban this practice (European Parliament, 2018). Antimicrobials in swine and poultry are used because they enhance growth rates and decrease production costs. Poultry harbour coccidial parasites in their gastrointestinal tracts, which decrease their growth rate and egg production (Kant et al., 2013). These parasites can be con-trolled through improved husbandry practices, but most commonly is achieved through rou-tine use of anticoccidial medications in feed. Similarly with swine, antimicrobials are used to increase growth rates, decrease disease prob-lems, and increase meat production per unit feed required per animal (Krishnasamy et al., 2015). This regular use of antibiotics has led to AMR, and the subsequent worrisome threat to human and environmental health. Almost all meat from swine and poultry in North America is produced on a massive scale in concentrated areas. As meat consumption grows along with increasing incomes in countries such as China, antimicrobials are used at ever-expanding levels. High concentrations of antimicrobials from animal waste show up in surface water surrounding all industrial animal feeding opera-tions regardless of the country (e.g., Diarra and Malouin, 2014). Already in 2012, 38.5 million kg were used in China's production of swine and poultry alone (Krishnasamy et al., 2015), not considering other species that are also intensely raised for meat and food production. Beef feed-lots (Agga et al., 2019), plus aquaculture in the marine and freshwater environments (Lulijwa et al., 2020) are other prolific, ever-expanding sources of active pharmaceuticals entering the environment.

The statistics are chilling – the World Health Organization (WHO) estimates that drug-resistant diseases kill more than 700,000 people each year. The evolution and global spread of AMR is one of the major and most complex problems facing medical science and food security (WHO, 2014). Until now, most AMR research has been in hospitals, triggered when a drug treatment fails. However, meaningful solutions to address AMR must be broadly focused, integrating human, livestock, wildlife and environmental aspects, in other words, a One Health framework. Importantly, AMR can evolve anywhere bacteria exist and little is known about the flow and fate of AMR in the natural environment (Arnold et al., 2016).

Like all life forms, bacteria do whatever is necessary to survive and reproduce. They can change their physical or genetic characteristics to adapt to common stresses in their environments. Over-use and improper use of antibiotics is one such increasingly encountered stressor. 'Resistance mechanisms' of bacteria comes from their DNA, in which genes (sections of DNA) create different proteins, combinations of which make them insensitive to specific types of antibiotics. The more commonly encountered the antimicrobial compound, the more likely the bacteria are to develop effective AMR to that specific antibiotic.

So, what can be done to defeat AMR? Reducing antimicrobial use is obviously needed, but the environmental spread of AMR via sewage and manure is as important as original use. A good example is the rapid global spread of blaNDM-1, an AMR gene that makes many bacteria almost untreatable. In 2008 it was first identified in patients in Indian hospitals, but it now has spread to 100+ countries as well as the Arctic. This worldwide spread is unlikely to be from person-to-person transmission. Instead blaNDM-1 and other resistance genes are probably spread by international travel and exposure of wildlife and humans to untreated manure and sewage. The importance of exposure via

the food chain, e.g., eating meat and milk from antibiotic-treated livestock, is poorly understood but could be significant (Jung et al., 2021). Holidaymakers often return home with increased numbers of AMR genes in their guts (D'Souza et al., 2021). Sewage effluent and run-off into streams flowing to the oceans, make coastal waters and beaches a potential critical point of contact where humans, marine animals and migratory seabirds are exposed to AMR (Leonard et al., 2015). Because AMR 'souvenirs' are most common after visits to developing countries, any solutions to AMR need to focus on sustainable development, sanitation and food security, as well as access to health care.

Another big gap in our One Health understanding of AMR is the role of wildlife, particularly migratory animals, which we know are carrying AMR bacteria. Ecologically diverse species such as monkeys, gulls and fish have antimicrobial-resistant bacteria (Ahlstrom et al., 2019). Even wild bottlenose dolphins *Tursiops truncatus* (Stewart et al., 2014) and owls (Miller et al., 2020) living in or near human habitation are more likely to harbour clinically important AMR, than more remote wildlife populations. For species living close to humans, like rodents and some birds, AMR seems to be ubiquitous (Hassell et al., 2019; Swift et al., 2019).

Working out how wildlife acquires AMR and potentially pass it back to humans and livestock is going to be challenging! Novel genomic approaches may help establish directionality of the movement of resistance genes between humans, livestock, wildlife, and the environment. GPS tags for tracking animal movements could be integrated with data on resistance genes to examine in real time, where humans and animals overlap and whether novel AMR genes move between hosts and environmental compartments (Ahlstrom et al., 2019; Arnold et al., 2016). The growing human population and increasing fragmentation of natural habitats forces wildlife into greater contact with humans and their livestock, increasing the potential for transmission of AMR infection between and within wildlife, livestock and human populations.

7.6 Conclusion

The cases presented in this chapter highlight the complex nature, destructive impacts and growing threats of medications entering the global environment. As informed citizens, we all must take responsibility for questioning the value versus cost of over-medicating, looking into alternative solutions to health problems (such as dietary and life-stye changes). Physicians and veterinarians must sincerely seek alternatives to the rampant and unquestioning prescription of antibiotics, even for viral infections although we know it is ineffective (!), and other common drugs (prescribing changes in lifestyle, and improving animal management practices). And very importantly, the pharmaceutical industry must have responsibility for the entire life cycle of the medications they are distributing. This will necessitate redesigning wastewater treatment facilities to effectively degrade pharmaceuticals. Licensing of new drugs must entail not only the safety and effectiveness of that medication in the intended patients, but also must include the effective breakdown and detoxification of those same drugs after they are excreted.

References

Agga, G. E., Cook, K. L., Netthisinghe, A. M. P., Gilfillen, R. A., Woosley, P. B., & Sistani, K. R. (2019). Persistence of antibiotic resistance genes in beef cattle backgrounding environment over two years after cessation of operation. *PLOS ONE*, *14*(2), e0212510. https://doi.org/10.1371/journal.pone.0212510

Ahlstrom, C. A., Bonnedahl, J., Woksepp, H., Hernandez, J., Reed, J. A., Tibbitts, L., Olsen, B., Douglas, D. C., & Ramey, A. M. (2019). Satellite tracking of gulls and genomic characterization of faecal bacteria reveals environmentally mediated acquisition and dispersal of antimicrobial-resistant *Escherichia coli* on the Kenai Peninsula, Alaska. *Molecular Ecology*, *28*(10), 2531–2545. https://doi.org/10.1111/mec.15101

Alvinerie, M., Sutra, J. F., Galtier, P., Lifschitz, A., Virkel, G., Sallovitz, J., & Lanusse, C. (1999). Persistence of ivermectin in plasma and faeces following administration of a sustained-release bolus to cattle. *Research in Veterinary Science*, *66*(1), 57–61. https://doi.org/10.1053/rvsc.1998.0240

Andersson, D. I., & Hughes, D. (2014). Microbiological effects of sublethal levels of antibiotics. *Nature Reviews. Microbiology*, *12*(7), 465–478. https://doi.org/10.1038/nrmicro3270

Arnold, K. E., Brown, A. R., Ankley, G. T., & Sumpter, J. P. (2014). Medicating the environment: Assessing risks of pharmaceuticals to wildlife and ecosystems. *Philosophical Transactions of the Royal Society of London. Series B, Biological Sciences*, *369*(1656), 20130569. https://doi.org/10.1098/rstb.2013.0569

Arnold, K. E., Williams, N. J., & Bennett, M. (2016). 'Disperse abroad in the land': The role of wildlife in the dissemination of antimicrobial resistance. *Biology Letters*, *12*(8), 20160137. https://doi.org/10.1098/rsbl.2016.0137

Bean, T. G., Arnold, K. E., Lane, J. M., Bergström, E., Thomas-Oates, J., Rattner, B. A., & Boxall, A. B. A. (2017). Predictive framework for estimating exposure of birds to pharmaceuticals. *Environmental Toxicology and Chemistry*, *36*(9), 2335–2344. https://doi.org/10.1002/etc.3771

Bean, T. G., Boxall, A. B. A., Lane, J., Herborn, K. A., Pietravalle, S., & Arnold, K. E. (2014). Behavioural and physiological responses of birds to environmentally relevant concentrations of an antidepressant. *Philosophical Transactions of the Royal Society of London. Series B, Biological Sciences*, *369*(1656), 20130575. https://doi.org/10.1098/rstb.2013.0575

Berny, P. (2007). Pesticides and the intoxication of wild animals. *Journal of Veterinary Pharmacology and Therapeutics*, *30*(2), 93–100. https://doi.org/10.1111/j.1365-2885.2007.00836.x

Beynon, S. A., Mann, D. J., Slade, E. M., & Lewis, O. T. (2012). Species-rich dung beetle communities buffer ecosystem services in perturbed agro-ecosystems. *Journal of Applied Ecology*, *49*(6), 1365–1372. https://doi.org/10.1111/j.1365-2664.2012.02210.x

Bringolf, R. B., Heltsley, R. M., Newton, T. J., Eads, C. B., Fraley, S. J., Shea, D., & Cope, W. G. (2010). Environmental occurrence and reproductive effects of the pharmaceutical fluoxetine in native freshwater mussels. *Environmental Toxicology and Chemistry, 29*(6), 1311–1318. https://doi.org/10.1002/etc.157

Brooks, B. W., Foran, C. M., Richards, S. M., Weston, J., Turner, P. K., Stanley, J. K., Solomon, K. R., Slattery, M., & La Point, T. W. (2003). Aquatic ecotoxicology of fluoxetine. *Toxicology Letters, 142*(3), 169–183. https://doi.org/10.1016/s0378-4274(03)00066-3

Chandramohan, S., Mallord, J. W., Mathesh, K., Sharma, A. K., Mahendran, K., Kesavan, M., Gupta, R., Chutia, K., Pawde, A., Prakash, N. V., Ravichandran, P., Saikia, D., Shringarpure, R., Timung, A., Galligan, T. H., Green, R. E., & Prakash, V. M. (2022). Experimental safety testing shows that the NSAID tolfenamic acid is not toxic to Gyps vultures in India at concentrations likely to be encountered in cattle carcasses. Science of the Total Environment, 809. https://doi.org/10.1016/j.scitotenv.2021.152088.

Das, D., Cuthbert, R. J., Jakati, R. D., & Prakash, V. (2011). Diclofenac is toxic to the Himalayan vulture Gyps himalayensis. *Bird Conservation International, 21*(1), 72–75. https://doi.org/10.1017/S0959270910000171

Davies, J., & Davies, D. (2010). Origins and evolution of antibiotic resistance. *Microbiology and Molecular Biology Reviews, 74*(3), 417–433. https://doi.org/10.1128/MMBR.00016-10

D'Souza, A. W., Boolchandani, M., Patel, S., Galazzo, G., van Hattem, J. M., Arcilla, M. S., Melles, D. C., de Jong, M. D., Schultsz, C., COMBAT Consortium, Dantas, G., & Penders, J. (2021). Destination shapes antibiotic resistance gene acquisitions, abundance increases, and diversity changes in Dutch travelers. Genome Medicine, 13(1), 79. https://doi.org/10.1186/s13073-021-00893-z

Castillo-Zacarías, C., Barocio, M. E., Hidalgo-Vázquez, E., Sosa-Hernández, J. E., Parra-Arroyo, L., López-Pacheco, I. Y., Barceló, D., Iqbal, H. N. M., & Parra-Saldívar, R. (2021). Antidepressant drugs as emerging contaminants: Occurrence in urban and non-urban waters and analytical methods for their detection. *Science of the Total Environment, 757,* 143722. https://doi.org/10.1016/j.scitotenv.2020.143722

Center for Behavioral Health Statistics and Quality. (2016). Key substance use and mental health indicators in the United States: Results from the 2015 National Survey on Drug Use and Health (HHS Publication No. SMA 16–4984, NSDUH Series H-51). http://www.samhsa.gov/data/

Diarra, M. S., & Malouin, F. (2014). Antibiotics in Canadian poultry productions and anticipated alternatives. *Frontiers in Microbiology, 5,* 282. https://doi.org/10.3389/fmicb.2014.00282

European Parliament (2018). MEPs back plans to halt spread of drug resistance from animals to humans [Press release]. https://www.europarl.europa.eu/news/en/press-room/20181018IPR16526/meps-back-plans-to-halt-spread-of-drug-resistance-from-animals-to-humans

Floate, K. D., Colwell, D. D., & Fox, A. S. (2002). Reductions of non-pest insects in dung of cattle treated with endectocides: A comparison of four products. *Bulletin of Entomological Research, 92*(6), 471–481. https://doi.org/10.1079/BER2002201

Floate, K. D., Bouchard, P., Holroyd, G., Poulin, R., & Wellicome, T. I. (2008). Does doramectin use on cattle indirectly affect the endangered burrowing owl. *Rangeland Ecology and Management, 61*(5), 543–553. https://doi.org/10.2111/08-099.1

Galligan, T. H., Taggart, M. A., Cuthbert, R. J., Svobodova, D., Chipangura, J., Alderson, D., Prakash, V. M., & Naidoo, V. (2016). Metabolism of aceclofenac in cattle to vulture-killing diclofenac. *Conservation Biology, 30*(5), 1122–1127. https://doi.org/10.1111/cobi.12711

González-Alonso, S., Merino, L. M., Esteban, S., López de Alda, M. L., Barceló, D., Durán, J. J., López-Martínez, J., Aceña, J., Pérez, S., Mastroianni, N., Silva, A., Catalá, M., & Valcárcel, Y. (2017). Occurrence of pharmaceutical, recreational and psychotropic drug residues in surface water on the northern Antarctic Peninsula region. *Environmental Pollution, 229,* 241–254. https://doi.org/10.1016/j.envpol.2017.05.060

González Peña, O. I., López Zavala, M. Á., & Cabral Ruelas, H. (2021). Pharmaceuticals market, consumption trends and disease incidence are not driving the pharmaceutical research on water and wastewater. *International Journal of Environmental Research and Public Health, 18*(5). https://doi.org/10.3390/ijerph18052532

Green, R. E., Taggart, M. A., Senacha, K. R., Raghavan, B., Pain, D. J., Jhala, Y., & Cuthbert, R. (2007). Rate of decline of the oriental white-backed vulture population in India estimated from a survey of diclofenac residues in carcasses of ungulates. *PLOS*

ONE, *2*(8), e686. https://doi.org/10.1371/journal.pone.0000686

Gunnarsson, L., Kristiansson, E., Rutgersson, C., Sturve, J., Fick, J., Förlin, L., & Larsson, D. G. (2009). Pharmaceutical industry effluent diluted 1:500 affects global gene expression, cytochrome P450 1A activity, and plasma phosphate in fish. *Environmental Toxicology and Chemistry*, *28*(12), 2639–2647. https://doi.org/10.1897/09-120.1.

Hassell, J. M., Ward, M. J., Muloi, D., Bettridge, J. M., Robinson, T. P., Kariuki, S., Ogendo, A., Kiiru, J., Imboma, T., Kang'ethe, E. K., Öghren, E. M., Williams, N. J., Begon, M., Woolhouse, M. E. J., & Fèvre, E. M. (2019). Clinically relevant antimicrobial resistance at the wildlife-livestock-human interface in Nairobi: An epidemiological study. *Lancet. Planetary Health*, *3*(6), e259–e269. https://doi.org/10.1016/S2542-5196(19)30083-X

Iglesias, L. E., Saumell, C. A., Fernández, A. S., Fusé, L. A., Lifschitz, A. L., Rodríguez, E. M., Steffan, P. E., & Fiel, C. A. (2006). Environmental impact of ivermectin excreted by cattle treated in autumn on dung fauna and degradation of faeces on pasture. *Parasitology Research*, *100*(1), 93–102. https://doi.org/10.1007/s00436-006-0240-x

Jjemba, P. K. (2006). Excretion and ecotoxicity of pharmaceutical and personal care products in the environment. *Ecotoxicology and Environmental Safety*, *63*(1), 113–130. https://doi.org/10.1016/j.ecoenv.2004.11.011

Jung, D., Morrison, B. J., & Rubin, J. E. (2021). A review of antimicrobial resistance in imported foods. *Canadian Journal of Microbiology*, *0*, 1–15. https://doi.org/10.1139/cjm-2021-0234

Kant, V., Singh, P., Verma, P. K., Bais, I., Parmar, M. S., Gopal, A., & Gupta, V. (2013). Anticoccidial Drugs Used in the Poultry: An Overview. *Science International*, *1*(7), 261–265. https://doi.org/10.17311/sciintl.2013.261.265

Krishnasamy, V., Otte, J., & Silbergeld, E. (2015). Antimicrobial use in Chinese swine and broiler poultry production. *Antimicrobial Resistance and Infection Control*, *4*, 17. https://doi.org/10.1186/s13756-015-0050-y

Leonard, A. F. C., Zhang, L., Balfour, A. J., Garside, R., & Gaze, W. H. (2015). Human recreational exposure to antibiotic resistant bacteria in coastal bathing waters. *Environment International*, *82*, 92–100. https://doi.org/10.1016/j.envint.2015.02.013

Lienert, J., Bürki, T., & Escher, B. I. (2007). Reducing micropollutants with source control: Substance flow analysis of 212 pharmaceuticals in faeces and urine. *Water Science and Technology*, *56*(5), 87–96. https://doi.org/10.2166/wst.2007.560

Lienert, J., Güdel, K., & Escher, B. I. (2007). Screening method for ecotoxicological hazard assessment of 42 pharmaceuticals considering human metabolism and excretory routes. *Environmental Science and Technology*, *41*(12), 4471–4478. https://doi.org/10.1021/es0627693

Losey, J. E., & Vaughan, M. (2006). The economic value of ecological services provided by insects, *BioScience* 56(4), *3568*, 311–323. https://doi.org/10.1641/0006

Lulijwa, R. E., Rupia, E. J., & Alfaro, A. C. (2020). Antibiotic use in aquaculture, policies and regulation, health and environmental risks: A review of the top 15 major producers. *Reviews in Aquaculture*, *12*(2), 640–663. https://doi.org/10.1111/raq.12344

Lumaret, J. P., & Errouissi, F. (2002). Use of anthelmintics in herbivores and evaluation of risks for the non-target fauna of pastures. *Veterinary Research*, *33*(5), 547–562. https://doi.org/10.1051/vetres:2002038

Lumaret, J. P., Errouissi, F., Floate, K., Römbke, J., & Wardhaugh, K. (2012). A review on the toxicity and non-target effects of macrocyclic lactones in terrestrial and aquatic environments. *Current Pharmaceutical Biotechnology*, *13*(6), 1004–1060. https://doi.org/10.2174/138920112800399257

Mayo Clinic. https://www.mayoclinic.org/diseases-conditions/depression/in-depth/antidepressants/art-20046273

Meredith-Williams, M., Carter, L. J., Fussell, R., Raffaelli, D., Ashauer, R., & Boxall, A. B. A. (2012). Uptake and depuration of pharmaceuticals in aquatic invertebrates. *Environmental Pollution*, *165*, 250–258. https://doi.org/10.1016/j.envpol.2011.11.029

Miller, E. A., Ponder, J. B., Willette, M., Johnson, T. J., & VanderWaal, K. L. (2020). Merging metagenomics and spatial epidemiology to understand the distribution of antimicrobial resistance genes from Enterobacteriaceae in wild owls. *Applied and Environmental Microbiology*, *86*(20). https://doi.org/10.1128/AEM.00571-20

Naidoo, V., Wolter, K., Cromarty, D., Diekmann, M., Duncan, N., Meharg, A. A., Taggart, M. A., Venter, L., & Cuthbert, R. (2010). Toxicity of non-steroidal anti-inflammatory drugs to Gyps vultures: A new threat from ketoprofen. *Biology Letters, 6*(3), 339–341. https://doi.org/10.1098/rsbl.2009.0818

Oaks, J. L., Gilbert, M., Virani, M. Z., Watson, R. T., Meteyer, C. U., Rideout, B. A., Shivaprasad, H. L., Ahmed, S., Chaudhry, M. J., Arshad, M., Mahmood, S., Ali, A., & Khan, A. A. (2004). Diclofenac residues as the cause of vulture population decline in Pakistan. *Nature, 427*(6975), 630–633. https://doi.org/10.1038/nature02317

Prakash, V., Pain, D. J., Cunningham, A. A., Donald, P. F., Prakash, N., Verma, A., Gargi, R., Sivakumar, S., & Rahmani, A. R. (2003). Catastrophic collapse of Indian white-backed *Gyps bengalensis* and long-billed *Gyps indicus* vulture populations. *Biological Conservation, 109*(3), 381–390. https://doi.org/10.1016/S0006-3207(02)00164-7

Rattner, B. A., Whitehead, M. A., Gasper, G., Meteyer, C. U., Link, W. A., Taggart, M. A., Meharg, A. A., Pattee, O. H., & Pain, D. J. (2008). Apparent tolerance of turkey vultures (Cathartes aura) to the non-steroidal anti-inflammatory drug diclofenac. *Environmental Toxicology and Chemistry, 27*(11), 2341–2345. *https://doi.org/10.1897/08-123.1*

Saaristo, M., Brodin, T., Balshine, S., Bertram, M. G., Brooks, B. W., Ehlman, S. M., McCallum, E. S., Sih, A., Sundin, J., Wong, B. B. M., & Arnold, K. E. (2018). Direct and indirect effects of chemical contaminants on the behaviour, ecology and evolution of wildlife. *Proceedings. Biological Sciences, 285*(1885), 20181297. https://doi.org/10.1098/rspb.2018.1297

Sands, B., & Wall, R. (2017). Dung beetles reduce livestock gastrointestinal parasite availability on pasture. *Journal of Applied Ecology, 54*(4), 1180–1189. https://doi.org/10.1111/1365-2664.12821

Schultz, M. M., Furlong, E. T., Kolpin, D. W., Logue, A., Painter, M. M., & Schoenfuss, H. L. (2010). Selective uptake of antidepressant pharmaceuticals in fish neural tissue. *Abstracts of Papers of the American Chemical Society, 239*.

Shore, R. F., Taggart, M. A., Smits, J. E., Mateo, R., Richards, N. L., & Fryday, S. (2014). Detection and drivers of exposure and effects of pharmaceuticals in higher vertebrates. *Philosophical Transactions of the Royal Society of London. Series B, Biological Sciences, 369*(1656). https://doi.org/10.1098/rstb.2013.0570

Smits, J. E., & Naidoo, V. (2018). Chapter 10. Toxicology of birds of Prey. In J. H. Sarasola, J. M. Grande & J. J. Negro (Eds.), *Birds of Prey; Biology and conservation in the XXI century. Springer International*

Publishing AG, part of Springer Nature. Cham, Switzerland. https://doi.org/10.1007/978-3-319-73745-4

Sommer, C., & Bibby, B. M. (2002). The influence of veterinary medicines on the decomposition of dung organic matter in soil. *European Journal of Soil Biology, 38*(2), 155–159. https://doi.org/10.1016/S1164-5563(02)01138-X

Stevenson, B. G., & Dindal, D. L. (1985). Growth and development of Aphodius beetles (Scarabaeidae) in laboratory microcosms of cow dung. *Coleopterists' Bulletin, 39*(3), 215–220. *http://www.jstor.org/stable/4008274*

Stewart, J. R., Townsend, F. I., Lane, S. M., Dyar, E., Hohn, A. A., Rowles, T. K., Staggs, L. A., Wells, R. S., Balmer, B. C., & Schwacke, L. H. (2014). Survey of antibiotic-resistant bacteria isolated from bottlenose dolphins *Tursiops truncatus* in the southeastern USA. *Diseases of Aquatic Organisms, 108*(2), 91–102. https://doi.org/10.3354/dao02705

Suárez, V. H., Lifschitz, A. L., Sallovitz, J. M., & Lanusse, C. E. (2009). Effects of faecal residues of moxidectin and doramectin on the activity of arthropods in cattle dung. *Ecotoxicology and Environmental Safety, 72*(5), 1551–1558. https://doi.org/10.1016/j.ecoenv.2007.11.009

Swan, G. E., Cuthbert, R., Quevedo, M., Green, R. E., Pain, D. J., Bartels, P., Cunningham, A. A., Duncan, N., Meharg, A. A., Oaks, J. L., Parry-Jones, J., Shultz, S., Taggart, M. A., Verdoorn, G., & Wolter, K. (2006). Toxicity of diclofenac to Gyps vultures. *Biology Letters, 2*(2), 279–282. https://doi.org/10.1098/rsbl.2005.0425

Swan, G., Naidoo, V., Cuthbert, R., Green, R. E., Pain, D. J., Swarup, D., Prakash, V., Taggart, M., Bekker, L., Das, D., Diekmann, J., Diekmann, M., Killian, E., Meharg, A., Patra, R. C., Saini, M., & Wolter, K. (2006). Removing the threat of diclofenac to critically endangered Asian vultures. *PLOS Biology, 4*(3), e66. https://doi.org/10.1371/journal.pbio.0040066

Swift, B. M. C., Bennett, M., Waller, K., Dodd, C., Murray, A., Gomes, R. L., Humphreys, B., Hobman, J. L., Jones, M. A., Whitlock, S. E., Mitchell, L. J., Lennon, R. J., & Arnold, K. E. (2019). Anthropogenic environmental drivers of antimicrobial resistance in wildlife. *Science of the Total Environment, 649*, 12–20. https://doi.org/10.1016/j.scitotenv.2018.08.180

Trilling, D. (May 22, 2017). *Drugs in our rivers: Bugs on speed and Prozac in the food chain. The journalist's resource.* Harvard Kennedy School, Shorenstein Center on Media, Politics and Public Policy.

Verheyen, R., Eens, M., & Pinxten, R. (1991). Male song as a cue for mate choice in the European starling. *Behaviour, 116*(3–4), 210–238.

Whitlock, S. E., Pereira, M. G., Shore, R. F., Lane, J., & Arnold, K. E. (2018). Environmentally relevant exposure to an antidepressant alters courtship behaviours in a songbird. *Chemosphere, 211*, 17–24. https://doi.org/10.1016/j.chemosphere.2018.07.074

World Health Organization. (2014). Antimicrobial resistance: Global report on surveillance 2014. https://www.who.int/publications/i/item/978 9241564748

Wobeser, G., Bollinger, T., Leighton, F. A., Blakley, B., & Mineau, P. (2004). Secondary poisoning of eagles following intentional poisoning of coyotes with anticholinesterase pesticides in western Canada. *Journal of Wildlife Diseases, 40*(2), 163–172. https://doi.org/10.7589/0090-3558-40.2.163

Chapter 8

Impacts of land use and climate change on natural populations: The butterfly perspective

Zachary G. MacDonald, H. Bradley Shaffer and Felix A. H. Sperling

Abstract

The past century has witnessed an explosion of anthropogenic activity, resulting in land use and climate changes on a global scale. The study of butterflies provides a unique window into the biological impacts of these changes. In this chapter, we explore several case studies that demonstrate the power of butterflies, both as model organisms in theory development and as ecological sentinels in conservation practice. These studies demonstrate how research on butterfly phenology, distribution, and diversity has yielded important insights into the interacting effects of habitat loss, fragmentation, and degradation on natural populations, as well as ecological and evolutionary responses to changing climatic conditions. Further, an important avenue for future research harnesses the power of whole-genome sequencing of butterfly populations to better document and help ameliorate biodiversity loss. Continued collaboration and knowledge transfer between dedicated amateurs and professional researchers, facilitated by humanity's innate appreciation of butterflies, will be essential to our continuing efforts to stem the catastrophic loss of biodiversity that is generally associated with modifications of natural habitats and large-scale shifts in climatic conditions.

8.1 Introduction

Anthropogenic habitat loss, habitat fragmentation, and climate change are among the greatest threats to global biodiversity (IPCC, 2021; Tilman et al., 2017; Warren et al., 2001). The scale of these threats and the rate at which they are accelerating have made it paramount that we understand and ameliorate their effects. A necessary first step is to document their impacts on specific groups of plants and animals that serve as sentinels of biodiversity change. Butterflies are among the most intensively monitored of these groups worldwide, and for good reason. Their rich history of study, variation in habitat and degree of host plant specificity, and suitability for both observational and experimental research render them ideal model organisms in conservation biology (Boggs et al., 2003). From island biogeography theory and

Susan C. Cork and Douglas P. Whiteside (eds) **Case Studies in EcoHealth**
DOI: 10.52517/9781789183368.008, © 5m Books Ltd, 2024

metapopulation dynamics (Ehrlich and Hanski, 2004; Hanski, 1994; Hanski and Gilpin, 1991; Munroe, 1948), to mechanisms of evolution and speciation (Ehrlich and Raven, 1964; Mavárez et al., 2006; Sperling, 2003; Wallace, 1865; Watt, 2003), many theoretical advances have originated from the study of butterflies. This has often translated into real-world conservation science (Bellis et al., 2019; Boggs et al., 2003; Kremen et al., 2003). In this chapter, we explore representative case studies that demonstrate how butterflies can be used to assess anthropogenic impacts on biodiversity worldwide.

8.1.1 Why butterflies?

Everyone likes butterflies. Their charismatic nature has inspired astonishing dedication among naturalists, who have published countless books, papers, field guides, taxonomic reports, life history accounts, and detailed notes on species' distributions. This body of knowledge has facilitated deep integration and collaboration with professional scientists, promoting important knowledge transfer in both directions.

In addition to their charismatic nature, butterflies also exhibit multiple ecological and evolutionary traits that render them key sentinels of land use and climate changes. Most butterfly species have short life cycles with one or more generations per year, allowing the genetic composition and demography of populations to respond relatively quickly to changes in local habitat and environmental conditions (MacDonald et al., 2017; Nowicki et al., 2008; van Swaay & Warren, 1999). Many butterflies can live and reproduce within small fragments of habitat on landscapes modified by human activities, and their population dynamics, genetics, and diversity patterns can be used to infer the effects of habitat loss and fragmentation on ecosystem function at very fine spatial scales

(MacDonald et al., 2020; van Swaay et al., 2006). A few butterfly species are migratory, such as the well-known monarch (*Danaus plexippus*) and painted lady (*Vanessa cardui*), and their study can shed light on population-level consequences of land use and climate changes at the continental scale (Flockhart et al., 2013; Miller et al., 2012; Shreeve, 1992; Stefanescu, et al., 2013). The larvae of most butterfly species are dependent on specific species of host plants as food resources, and butterfly host plant co-occurrence patterns can yield important insights into changes into multi-species ecological and trophic interactions (Filazzola et al., 2020; MacDonald et al., 2018a). Finally, the diversity of butterflies often correlates with that of other terrestrial taxa, making them viable indicators of biodiversity at the ecosystem level (MacDonald et al., 2018a; Nowicki et al., 2008; Thomas, 2005).

8.1.2 Butterfly monitoring programs

Butterfly monitoring has a long history, prominently including the 1976 establishment of the United Kingdom Butterfly Monitoring Scheme and the 1974 initiation of annual single-day butterfly counts in North America by the Xerces Society, now continued by the North American Butterfly Association (Acorn, 2017). While long-term butterfly monitoring schemes are becoming more common in a growing number of countries, the popularity of single-day butterfly counts has diminished in recent decades. For example, the number of single-day counts in Alberta, Canada, dropped from more than 40 in 2000 to around 5 in 2019 (Acorn, 2017). This decline stems, in part, from the realization that diversity data from single-day counts cannot be meaningfully compared across years. Particular species may be observed in some years but not others, and often it cannot be determined whether this is due to variation in emergence times related to weather, long-term shifts in phenology related to climate change, or actual changes to the composition of species assemblages. In contrast, other types of monitoring programs, such as 'Pollard transects' (Pollard, 1977) where counts are completed weekly or bi-weekly, provide more detailed data that can differentiate between these mechanisms (MacDonald et al., 2017; Parmesan, 2003; Westwood and Blair, 2010). The resulting data, as well as other observations from both amateurs and professionals, are now reported and organized in massive digital databases such as eButterfly (Prudic et al., 2017; www.e-butterfly.org), the Global Biodiversity Information Facility (GBIF; www.gbif.org), and iNaturalist (www.inaturalist.org). Open access with no paywall to these databases allows researchers to quantify changes in species' phenologies, distributions, and overall diversity patterns at spatial and temporal scales that are otherwise impossible (Acorn, 2017; MacDonald et al., 2017).

8.2   Effects of land use change on butterfly populations

In the Anthropocene, perhaps the most immediate threat to biodiversity is land use change (Brooks et al., 2002; Hanski, 2011). Global extinction rates are estimated to be 100–1000 times more than they were before intensive human activity, and much of this increase can be attributed to modification or destruction of natural habitats through urbanization, agriculture, and forestry (Pimm et al., 2014; Rosenzweig, 1995). In the majority of ecosystems around the world, effects of these anthropogenic activities are negative for native species, and this holds true for butterflies (Boggs et al., 2003; Tabarelli et al., 2012; Warren et al., 2001). However, some butterfly species depend on specific host plants and diverse nectaring plants that occur in non-forested habitats, meaning anthropogenic disturbances actually increase

butterfly abundance and diversity in particular circumstances (reviewed by Dover and Settele, 2009). For example, Riva et al. (2018a) found that clearing boreal forests along seismic lines for oil exploration in Alberta, Canada, generally increases the overall abundance and diversity of butterflies (but see Riva et al., 2018b) for negative effects on the disturbance-sensitive species, the cranberry blue, *Plebejus optilete*). In another example, Thomas (1991) showed that butterfly species occurring throughout the UK are dependent on high-frequency disturbance regimes that facilitate open forest canopies and early successional plant assemblages. These types of habitats were far more prevalent on landscapes 4500–10,000 BP during warmer climatic conditions, but are now primarily maintained by agricultural practices (Dennis, 1993; Thomas, 1991; Vera, 2000). Continuation of some agricultural practices may therefore be necessary for the persistence of particular butterfly species (Singer and Parmesan, 2018; Singer et al., 1993). However, in the majority of cases, butterflies are negatively affected by land use changes that lead to extensive habitat loss, fragmentation, and degradation.

8.2.1   Habitat loss and fragmentation

Examples of worldwide habitat loss are astonishing. More than 94% of Earth's temperate broadleaf forests have been modified or destroyed by human development (Primack, 2006). Approximately 85% of natural habitat throughout Europe has been modified or destroyed (Primack, 2006), more than 97% of North America's tallgrass prairies have been converted to farmland (White et al., 2000), and more than 50% of wetlands in the USA and 60–70% in Europe have been drained (Ravenga et al., 2000). As of 2021, less than 16% of Earth's terrestrial landmass has been explicitly set aside for conservation and legally protected from human development (UNEP-WCMC IUCN, 2021). Given this modest percentage of protected habitat, conservation biologists are particularly concerned with how its spatial configuration (i.e. degree of fragmentation) affects biodiversity. Human activities often leave behind isolated fragments of natural habitat, from small pockets of undeveloped land to larger nature reserves. While habitat loss has unequivocally negative effects on species diversity, the interconnected fields of theoretical ecology and conservation biology are marked by a vigorous debate on whether habitat fragmentation poses additional threats. The study of butterflies has contributed much to this debate.

8.2.2 Some history

Relationships between habitat fragmentation and biodiversity have interested conservation biologists since Levins' (1969) extrapolation of the theory of island biogeography to habitat fragments on terrestrial landscapes (Haila, 2002). Although the theory is generally attributed to a collaboration between Robert MacArthur and Edward Wilson (MacArthur and Wilson, 1963; Wilson and MacArthur, 1967), its basic tenets were preceded by Eugene Munroe (1948) in his doctoral work on butterflies of the West Indies (Brown and Lomolino, 1989). Munroe (1948) observed that the number of butterfly species observed on oceanic islands may represent an equilibrium between the opposing processes of colonization and extinction, each of which is principally determined by an island's area and isolation. This was a harbinger of MacArthur and Wilson's (1963) theory of island biogeography, which predicts that larger and less isolated islands harbor more species due to lower extinction and higher immigration rates, while smaller and more isolated islands harbor fewer species due to higher extinction and lower immigration rates.

The application of the theory of island biogeography to conservation is intuitive and appealing. If edges of terrestrial habitat fragments delimit populations similar to the shores of oceanic islands, small populations occupying small fragments may be more prone to extinction than larger populations occupying unfragmented habitat (Diamond, 1975). Most butterfly species have minimum area requirements for sustaining viable populations (reviews in Crone and Schultz, 2003; Dover and Settele, 2009), suggesting that at least at the lower size limit the analogy holds. For example, Crone and Schultz (2003) calculated that Fender's blue (*Icaricia icarioides fenderii*), an endangered butterfly endemic to upland prairies in Oregon, requires a fragment area of >6 ha for long-term population persistence. The increased ratio of edge-to-area that results from smaller patch sizes may also degrade the quality of remaining habitat for species that are particularly sensitive to habitat edges (Hadley and Betts 2016; Ries and Sisk, 2004).

The theory of island biogeography also predicts that species diversity is affected by the isolation of habitat fragments. As the distance between habitat fragments increases, they will be less likely to receive immigrants belonging to new species or exchange individuals of resident species that may be important for demographic health and avoiding inbreeding depression. Butterflies have served as model organisms for quantifying these isolation effects, with numerous studies investigating effects of inter-fragment distances on metapopulation dynamics and genetic differentiation (e.g. Hanski, 1994, 1999; Keyghobadi et al., 2005; Kuussaari et al., 1996; MacDonald et al., 2020; Matter et al., 2003). For many species with limited mobility, even modest levels of isolation reduce or eliminate inter-fragment movements (e.g. 300 m for the scarce heath (*Coenonympha hero*); Cassel et al., 2001). For long-term species persistence on these landscapes, the area of habitat fragments must be greater than that required to support a minimum viable population (Crone and Schultz, 2003; MacDonald et al. 2018a).

By these and related mechanisms, habitat fragmentation may result in a greater loss of species than that resulting from habitat loss alone. This has many conservation biologists concerned (reviewed by Fletcher et al., 2018). However, it is becoming increasingly recognized that many inferences of negative fragmentation effects are based on observations and study designs that have not decoupled the correlated effects of habitat fragmentation and habitat loss (Fahrig, 2003, 2013, 2017; Hadley and Betts, 2016). Habitat fragmentation is almost always accompanied by habitat loss, which may in fact be the primary driver of diversity declines on landscapes modified by anthropogenic activities. To effectively control for these confounding effects, many researchers have framed habitat fragmentation in terms of the 'SLOSS' debate, which asks whether the configuration of protected lands as 'Single Large Or Several Small' habitat fragments protects a greater number of species (Diamond, 1975). Because many butterflies respond quickly to changes in habitat configuration and are able to live and reproduce within small habitat fragments, they serve as an excellent indictor group for evaluating SLOSS at fine temporal and spatial scales.

8.2.3 Butterfly case studies

Many studies have used butterflies to infer effects of habitat loss, fragmentation, and quality on species diversity. Here we explore two studies that quantify patterns of butterfly diversity on a naturally fragmented landscape of lake islands (MacDonald et al., 2018a, 2021). True islands provide interesting models for investigating the effects of fragment area and isolation on species diversity (Dover and Settele, 2009;

Haila, 2002). From more than 14,500 islands in Lake of the Woods, Canada, MacDonald et al. (2018a) selected 30 islands, in two sets of 15, based on their area (Figure 8.1). The small island set contained one 0.8 ha island, two 0.4 ha islands, four 0.2 ha islands, and eight 0.1 ha islands. Here, the single 0.8 ha island represented 0.8 ha of unfragmented habitat, the two 4.0 ha islands represent 0.8 ha of moderately fragmented habitat, the four 2.0 ha islands represent 0.8 ha of more fragmented habitat, and the eight 1.0 ha islands represent 0.8 ha of highly fragmented habitat. The large island set followed the same basic design, but was comprised of islands 10 times larger; one 8.0 ha island, two 4.0 ha islands, four 2.0 ha islands, and eight 1.0 ha islands. This study design effectively decouples degree of habitat fragmentation from total habitat area, or, in other words, habitat fragmentation from habitat loss. Species richness and the relative abundance of butterfly species on each of the 30 islands were quantified using repeated full-island surveys, with sampling effort standardized per unit area.

Thirty-four butterfly species were observed across all islands. Surprisingly, in both island sets, the combined diversity of several small islands was approximately equal to the diversity of the single large island of equal total area (Figure 8.2). This result was also found with other SLOSS-based methods that assessed fragmentation effects across all 30 islands simultaneously, such as comparisons of species accumulation curves (MacDonald et al., 2018a). In this analysis, cumulative species richness is plotted against cumulative island area in two ways: (1) adding islands, one at a time, from the largest island to smallest island (large-to-small curve); and (2) adding islands, one at a time, from the smallest island to largest island (small-to-large curve) (Figure 8.3). For any value of cumulative island area along the x-axis of the plot, the cumulative species richness of single/few larger islands and

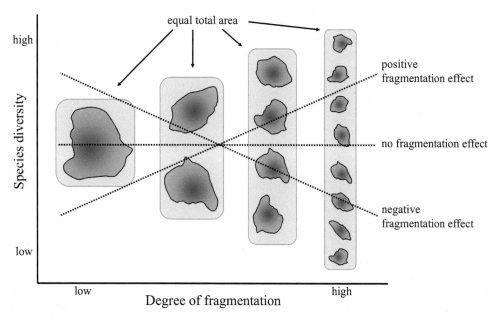

Figure 8.1 Experimental design of MacDonald et al. (2018a, 2018b), showing how sets of islands or habitat fragments, differing in their individual areas and number of replicates, can be used to decouple the effects of habitat fragmentation from habitat loss. Total habitat area is maintained between four groupings of islands, but degree of fragmentation increases from left to right. If species diversity decreases, does not change, or increases across increasing degrees of fragmentation, it can be inferred that fragmentation has a negative effect, no effect, or positive effect, respectively.

several small islands, equal in their total areas, can be compared by examining the difference between the large-to-small and small-to-large curves. Considering all butterflies observed across the 30 study islands, the large-to-small and small-to-large curves are nearly identical, indicating that single/fewer larger and several small islands contain the same number of butterfly species (Figure 8.3a). This equates to a neutral fragmentation effect. Additionally, after controlling for island area, generalized linear models resolved that isolation was not significantly related to species richness. Together, these results suggest that decreasing fragment area and increasing fragment isolation do not affect butterfly diversity after controlling for total habitat area.

These results accord with predictions of Fahrig's (2013) 'habitat amount hypothesis',

which posits that the number of species persisting on fragmented landscapes is solely a function of habitat loss and not its degree of fragmentation. The primary mechanism underlying the habitat amount hypothesis is the sample-area effect, which describes that species richness increases with island/fragment area only because larger islands/fragments randomly sample more individuals from the regional species pool (Connor & McCoy, 1979). In other words, the species-area relationship is just a sampling artefact; if a researcher surveys a larger area, they will find more species simply because of their increasing sampling effort. To date, the majority of SLOSS-based studies addressing a plethora of taxa (e.g., plants, insects, birds, mammals) have found that several small islands or habitat fragments contain an equal or greater number of species compared to single large

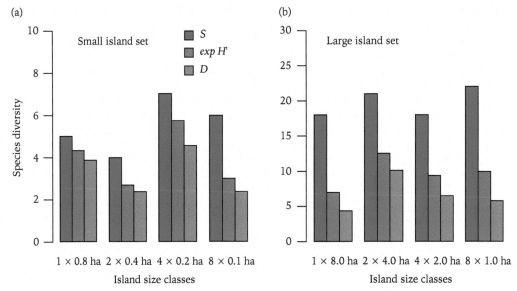

Figure 8.2 Butterfly species richness (*S*), the exponential of the Shannon-Wiener index (exp *H'*), and Simpson's reciprocal index (*D*) for island size classes that represent different degrees of habitat fragmentation while controlling for total habitat area (MacDonald et al., 2018a). Within each of the two island sets (panels a and b), island size classes varied in degree of fragmentation while maintaining an equal total area. There was no obvious change in the diversity of butterflies across island size classes, suggesting that habitat fragmentation does not affect butterfly diversity after controlling for habitat loss.

islands/fragments (reviewed in Fahrig, 2003; 2013; 2017).

Although SLOSS-based studies based on overall patterns of species diversity often report neutral or even positive effects of fragmentation, such inferences may be susceptible to ecological fallacy (*sensu* Robinson, 1950). This addresses biases that may arise when observed effects on aggregated variables (e.g. species richness) differ from causal relationships at finer levels of organization (e.g. occurrences of single species) (MacDonald et al., 2021). Populations respond to external conditions, such as degree of habitat fragmentation, and species richness sometimes emerges in misleading ways. This point is demonstrated by a disparity in the findings of a thorough review of the fragmentation literature. Fahrig (2017) found that 90% (114/127) of studies addressing patterns of species diversity (generally, species richness) inferred a positive

effect of fragmentation. However, studies that addressed responses of individual species were much more varied in their results, with only 68% (158/232) reporting positive fragmentation effects (Fahrig, 2017). Greater congruence in the results of studies addressing species diversity and those addressing individual species would be expected if patterns of species diversity are indeed viable indicators of fragmentation effects on individual species. This suggests that analyses of species diversity may obscure important species-level fragmentation effects.

Looking deeper into butterfly diversity patterns on islands of Lake of the Woods, it became apparent that occurrences of butterfly species with small wingspans were generally restricted to large islands that contained their host plant species (MacDonald et al., 2018a). In other words, small species were excluded from

(a)

(b)

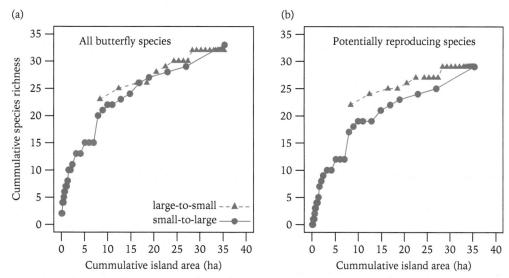

Figure 8.3 Species accumulation curves inform whether single/fewer large islands harbor a greater or lesser number of species than many small islands of equal total area (MacDonald et al., 2018a; 2018b). This was completed for both total butterfly species richness (a) and for the richness of potentially reproducing species only (described below) (b). In each panel, cumulative species richness is plotted against cumulative island area in two ways: 1) by adding islands, one at a time, from the largest to smallest island (large-to-small curve; triangles connected by dashed lines); and 2) adding islands, one at a time, from the smallest to largest island (small-to-large curve; circles connected by solid lines). For any value of cumulative island area along the x-axis, the cumulative species richness of single/fewer large islands and many small islands of equal total area can be compared by examining the difference between the two curves. When all butterfly species were included in the analysis (a), overlap of the two curves indicates that single/fewer large islands and many small islands of equal total area contain the same number of species. However, when transient individuals were removed (b), the large-to-small curve accumulated species more quickly than the small-to-large curve, indicating that single/fewer large islands support a greater number of potentially reproducing species than many small islands of equal total area.

particularly small islands and islands of any size that lacked their host plants. In contrast, butterfly species with large wingspans frequented islands of all sizes, regardless of whether their host plants were present or absent. Wingspan is one of the strongest predictors of butterfly species' mobility and dispersal ability (Burke et al., 2011). Accordingly, MacDonald et al. (2018a) hypothesized that movements of larger and highly mobile 'transient' species from large islands (containing host plants) to smaller islands (lacking host plants) may be obscuring

important effects of habitat fragmentation on species diversity (examples in Figure 8.4). In a re-analysis of the data, MacDonald et al. (2018a) removed from the dataset all individuals occurring on islands that did not contain their host plant species, as these individuals cannot represent reproducing populations. Rerunning SLOSS-based analyses on 'potentially reproducing' species only, negative fragmentation effects became very apparent: comparisons of species accumulation curves showed that several small islands harbored fewer potentially reproducing

a) Large, more mobile species

b) Small, less mobile species

Figure 8.4 Examples of butterflies observed on a naturally fragmented landscape of lake islands; Lake of the Woods, Ontario, Canada. (a) Four large and more mobile butterfly species generally found on islands of all sizes, regardless of the presence of their larval host plants (from top left, clockwise: Canadian tiger swallowtail (*Papilio canadensis*); atlantis fritillary (*Speyeria atlantis*); red admiral (*Vanessa atalanta*); white admiral (*Limenitis arthemis*)). (b) Four small and less mobile species that were generally excluded from both small islands and islands of any size that lacked their host plants (from top left, clockwise: striped hairstreak (*Satyrium liparops*); arctic skipper (*Carterocephalus palaemon*); summer azure (*Celastrina neglecta*); northern crescent (*Phyciodes cocyta*)). Occurrence and abundance patterns of these butterflies across lake islands show that habitat fragmentation disproportionally affects small and less mobile species (MacDonald et al., 2018a, 2021). Photos by Zachary G. MacDonald.

species than single large islands (Figure 8.3b). Additionally, generalized linear models resolved that island isolation had a significant effect on species richness after transient individuals were removed from the dataset. MacDonald et al. (2018a) therefore suggested that inter-fragment movements, but not breeding or establishment of self-sustaining populations, of large, highly mobile butterfly species inflate the total number of species observed on small habitat fragments, obscuring important fragmentation effects.

Differentiating between the richness of transient versus potentially reproducing butterfly species hinted that an autecological (i.e. species-by-species) approach may be most effective for inferring effects of habitat fragmentation on butterflies. Butterfly species can vary considerably in their responses to landscape factors and researchers are particularly interested in morphological, behavioral, and ecological characteristics of butterflies that might explain this variation (Dover and Settele, 2009). Functional traits, including body size, mobility and dispersal ability, perceptual range, degree of ecological specialization, rarity/conservation status, and trophic position are hypothesized to relate to species' sensitivity to habitat fragmentation for a variety of taxa (see references in MacDonald et al., 2019, 2021). However, few studies have tried to relate functional traits to interspecific variation in fragmentation effects, or model how this interspecific variation scales to emergent patterns of species diversity on fragmented landscapes.

To address this knowledge gap, MacDonald et al. (2021) reanalyzed the Lake of the Woods butterfly data with a novel methodological framework, using random placement models (instead of SLOSS-based methods) to control for the sample-area effect while integrating functional trait analyses. These analyses showed that habitat fragmentation may have negligible effects on overall butterfly species richness, in agreement with the previous SLOSS-based analyses. However, applying the new methods to individual butterfly species identified a number of important fragmentation effects that were not previously detected. For many butterfly species, probabilities of occurrence were lower than predicted by the sample-area effect for small islands. This indicates that small islands are of lower habitat quality than larger islands, suggesting an important fragmentation effect. Even more prominent were effects on butterfly species' abundances, which were much lower on both smaller and more isolated islands than predicted by the sample-area. Most importantly, the analyses demonstrated that negative fragmentation effects were significantly greater for smaller and less mobile butterfly species, highlighting the importance of individual species traits in fragmentation research.

In sum, the relative size and mobility of butterfly species can be a strong predictor of their sensitivity to fragmentation (and see Dover and Settele, 2009). Individual habitat fragments may contain resources sufficient for the persistence of some butterfly species, but not others. Large and highly mobile species may be able to move among multiple habitat fragments to meet their resources requirements, rendering them more resilient to habitat fragmentation. To adequately evaluate fragmentation effects on species diversity, it is important to differentiate transient individuals from resident individuals representing established populations. Different butterfly species are affected by habitat fragmentation in different ways, and autecological details need to be considered when we evaluate the effects of habitat fragmentation on butterflies. The habitat amount hypothesis is generally unsupported for butterflies, and we suggest that efforts to minimize habitat fragmentation and abate habitat loss remain as foundations in conservation practice.

8.3 Habitat degradation

Many terrestrial landscapes fragmented by anthropogenic activities are more complex than true island systems (Haila, 2002). For example, the terrestrial matrix that separates isolated fragments is often variable and less hostile than open water, and may in fact be the preferred habitat of some matrix-dwelling species (Riva et al., 2018a). The quality of individual habitat fragments themselves can also vary widely due to habitat degradation. Along with total habitat area, habitat quality is often a principal factor determining the persistence of many butterfly species on landscapes fragmented or otherwise modified by anthropogenic activity. For example, rather than area *per se*, larval habitat quality and isolation were the strongest predictors of fragment occupancy for the Glanville fritillary (*Melitaea cinxia*), Adonis blue (*Polyommatus bellargus*), and Lulworth skipper (*Thymelicus acteon*) in the UK (Thomas et al., 2001). Furthermore, large populations inhabiting high-quality (rather than large) habitat fragments have been inferred to facilitate colonization/recolonization of adjacent, unoccupied fragments, helping to stabilize species' metapopulations (Thomas et al., 2001). Habitat quality was also the principal factor affecting the assemblage of grassland butterfly species inhabiting isolated habitat fragments in an urban-dominated landscape in the UK (Wood & Pullin, 2002), and habitat quality and area, but not isolation, were the principal factors governing the distribution of Alcon blue (*Maculinea alcon*) across 127 habitat fragments in the Netherlands (Wallis DeVries, 2004). Notwithstanding these observations of the importance of habitat quality, both Wenzel et al. (2006) and Öckinger and Smith (2006) observed that species richness and overall abundance of grassland butterflies on fragmented agricultural landscapes were strongly related to their isolation from semi-natural grassland habitats. As with the two studies addressing butterfly assemblages on islands of Lake of the Woods (MacDonald et al., 2018a, 2021), small, less mobile species were most sensitive to fragmentation effects in two these studies.

8.3.1 Interacting effects of ecology and evolution

Land use changes can also affect the evolution of butterfly species, threatening their long-term persistence. A particularly rigorous history of research on Edith's checkerspot (*Euphydryas editha*) has resulted in inferences of eco-evolutionary processes that affect species persistence. For example, Singer and Parmesan (2018) found that an isolated population of *E. editha* in Nevada, USA, shifted its primary host plant association from *Collinsia parviflora*, the historical native host, to *Plantago lanceolata*, a non-native species introduced by cattle ranchers. Such host shifts in *E. editha* were first documented in the 1980s, with female *E. editha* rapidly evolving a preference for ovipositing on *Plantago* in heavily grazed landscapes due to higher larval survival (Singer et al., 1993). Some populations even abandoned *Collinsia* completely and evolved total dependence on *Plantago*.

However, it has become clear that the survival of *Plantago*-feeding populations may be heavily dependent on continued cattle grazing. Due to the sale of private land, grazing was eventually stopped on the landscape inhabited by the isolated Nevada population, allowing tall grasses to grow around the *Plantago* plants. The resulting shading drastically reduced survival of the thermophilic *E. editha* larvae, which would have otherwise survived on *Collinsia* occupying drier microhabitats that were unaffected by grazing. Although *Collinsia* was still present on the landscape, female *E. editha* did not switch back to ovipositing on it. Shortly after grazing ceased, Singer and Parmesan (2018) documented that the entire *E. editha* population, which had completely switched its host association from *Collinsia* to *Plantago*, went extinct. This extinction was the product of an 'eco-evolutionary trap', set by anthropogenic activity. Although the area was eventually recolonized by adjacent *Collinsia*-feeding *E. editha* populations, this is a particularly instructive example of the complex effects of land use changes and the possibility for unintended consequences of reclamation practices. Butterflies adapting to land use changes risk becoming dependent on continuation of the same practices. Indeed, Singer et al. (1993, p. 681) warn that 'This is a serious risk, because human cultural evolution can be even faster than the rapid genetic adaptation that the insects can evidently achieve.'

In sum, effects of land use changes on butterflies are complex and often difficult to predict. Different butterfly species can respond to similar land use changes in very different ways; some species are lucky 'winners' and succeed in changing landscapes, while the unfortunate majority are likely to be 'losers' suffering population declines and possibly extinction (Filgueiras et al., 2021; Tabarelli et al., 2012). Continued study of habitat loss, fragmentation and degradation will be key to understanding and possibly ameliorating these effects. However, the effects of land use changes should not be considered in isolation. Changing climatic conditions, and the rate at which these changes are accelerating, must also be taken into consideration.

8.4 Climate change

Earth's climate is changing, and anthropogenic activities are a large contributor (IPCC, 2021). Generalizable patterns include warming temperatures, increased cloud cover, and increased overall precipitation concentrated in fewer precipitation events (Easterling et al., 2000; Parmesan, 2003; Wang et al., 2016). Faced with these changes, individual populations and entire species will exhibit one of three possible responses (Parmesan, 2003).

1. Track changing climatic conditions by shifting distributions either poleward in latitude or higher in elevation.
2. Persist in the current location via phenotypic/behavioural plasticity or local adaptation to new climatic conditions.
3. Suffer extirpation or extinction.

It is generally unknown what the frequency of these three responses will be, or whether they will vary systematically across space, through time, or among taxa. Notwithstanding, the high suitability of butterflies as sentinel species for detecting, modelling, and predicting effects of climate change has been amply demonstrated in case studies on their phenology, distributions, and diversity.

8.4.1 Butterfly phenology and climate change

Butterflies are highly sensitive to abiotic conditions (Dennis, 1993; Sparks and Yates, 1997). In most mid- to high-latitude regions around the world, where butterfly species complete one or a few generations every summer, the timing of larval development, pupation, and eclosion as adults is advancing due to warming temperatures. A number of mechanisms account for this relationship. For example, faster melting of snowpacks allows butterflies (and their host plants) to begin growth and development earlier, regardless of the life stage in which they overwintered. Additionally, most temperate butterfly species are thermophilic, and the rate at which larvae grow is a function of available resources and heat, with warmer temperatures generally corresponding with faster growth rates (Singer and Parmesan, 2018). 'Growing degree days' is a commonly used measure of thermal accumulation for the development of many insects and plants. For most butterfly species, growing degree days is a better predictor of adult emergence time and overall abundance than date of the year (e.g. Cayton et al., 2015).

Around the world, emergence times of many butterfly species are shifting from historical patterns because growing day degrees are accumulating earlier in the spring as climatic conditions warm. For example, 23 butterfly species inhabiting the Central Valley of California, USA, have advanced their spring emergence times by an average of 24 days since the early 1970s (Forister and Shapiro, 2003). Changing climatic conditions in California, including temperature and precipitation, were significantly related to these phenological shifts. Similarly, data from the United Kingdom Butterfly Monitoring Scheme have shown significant advances in the emergences of multiple UK butterfly species (Sparks and Yates, 1997). Extrapolating the observed trends, regional warming of ~3°C within the next century is expected to advance the emergence of most UK butterfly species by 2 to 3 weeks. Along with earlier emergence times, flight periods of many butterfly species are becoming longer. For example, Westwood and Blair (2010) examined trends in the flight periods of 19 Canadian butterfly species and found that 13 were flying significantly longer into the autumn. With longer flight periods, we may expect some univoltine species to switch to a bivoltine life cycle, with two generations flying every summer (MacDonald et al., 2020).

8.4.2 Butterfly distributions and abundances and climate change

Butterflies were the first taxonomic group for which shifts in species' distributions and abundances were linked to changing climatic conditions. Well before anthropogenic activity was recognized to be affecting our planet's climate, butterfly enthusiasts were documenting poleward range shifts of multiple species, particularly in the UK (e.g. *Limenitus camilla*; Ford, 1945), and it was hypothesized that these range shifts were due to a general warming trend (reviewed by Dennis, 1993; Parmesan, 2003). Mean average temperatures across the UK rose by an average of 0.8°C in the 20th century, and this warming trend is predicted to continue (Murphy et al., 2009). With these types of climatic changes receiving increasing attention, research on how butterfly distributions are responding has proliferated.

A common method for inferring range shifts is to compile species' historical occurrence records and compare them to contemporary distributions. Using this approach, Burton (2003) inferred that 260 European species had experienced some degree of change in the northern limit of their occurrences since 1850. Of these, 190 exhibited a northern range expansion, while only 70 exhibited a northern range contraction. In a more thorough analysis addressing the entire ranges of 35 European butterfly species, northward shifts of both northern and southern range limits were observed for 22 species, with distances varying from 35 to 240 km (Parmesan et al., 1999). In contrast, only two species exhibited southward range shifts. In the northeastern USA, abundances of many butterfly species are increasing throughout the northern extant of their range and decreasing in the southern extent, suggesting that range shifts are underway (Breed et al., 2013).

In the absence of habitat destruction preventing movement, butterflies may continue shifting their ranges poleward as climatic conditions continue to warm (Parmesan, 2003). However, based on the limited data available, it seems that changing climatic conditions, coupled with habitat loss, fragmentation, and degradation, may exceed the resiliency of many butterfly species. In many instances, extensive habitat loss and fragmentation mean that butterfly species may be unable to track changing climatic conditions. This is predicted to result in population declines and possibly extinction. Empirical data support this prediction. For example, Warren et al. (2001) examined changes in the populations of 46 UK butterfly species that, based on previous observations, were expected to respond positively to warming temperatures. However, 34 of the 46 species experienced significant population declines over the previous 30 years. This was attributed to habitat loss and barriers to movement, which outweighed any positive effects of warming temperatures. In the Netherlands, 51 of the 72 butterfly species native to the region have experienced population declines over the 20th century, while 15 have suffered regional extinction (van Swaay and Warren, 1998). Research on Canadian butterfly distributions has also shown that while some species may be able to track changing climatic conditions via poleward shifts in their distributions, these shifts are unlikely to match the pace of changing climatic conditions for the majority of species due to their limited dispersal ability and habitat loss (Lewthwaite et al., 2018). Among these climate threatened species, those with the smallest range sizes seem to be accumulating the greatest 'climate debt', wherein northward range expansions cannot sufficiently offset southward range contractions (Lewthwaite et al., 2018). Changing climatic conditions are also having significant effects on butterfly abundances. For example, analyses of a massive dataset, including 70 survey locations across western North America, suggested that butterfly abundances have exhibited a 1.6% annual reduction over the past four decades

(Forister et al., 2021). This overall trend of fewer butterflies, as well as inter-annual fluctuations in butterfly abundance, was significantly associated with warming temperatures.

8.4.3   An autecological approach

Studying the effects of changing climatic conditions on patterns of species diversity has generated interesting hypotheses that warrant further study. However, effectively evaluating the probability of climate-induced range shifts, local adaptation, and extinction likely requires an autecological approach that considers the unique ecology and life history of individual species. *Euphydryas E. editha*, the victim of the eco-evolutionary trap studied by Singer and Parmesan (2018), has probably received more research attention in this realm than any other butterfly. One study (Parmesan, 1996) has become a landmark for quantitatively demonstrating the effects of climate change on sensitive species and their ecological associations. Examining patterns of *E. editha* population extinctions throughout western North America, from Baja California, Mexico, to British Columbia, Canada, Parmesan (1996) inferred that the species' range had shifted poleward and higher elevation by 92 km and 124 m, respectively, since the beginning of the 20th century. Most interestingly, these figures closely mirrored latitudinal and elevational shifts in temperature isotherms over the same time period, inferred to be 105 km and 105 m, respectively (Parmesan, 2003). In many correlative relationships like this, it is unclear whether butterflies are responding directly to changes in environmental conditions, or whether these relationships are mediated by intermediate variables, such as the occurrences and phenologies of host and nectar plants. However, extensive research on *E. editha* suggests that, at least for this species, the latter scenario is most likely.

For many *E. editha* populations, environmentally driven host plant senescence and its effect on larval mortality is the principal driver of abundance fluctuations (Singer, 1972; reviewed in Parmesan, 2003). In general, warm and dry climatic conditions accelerate the senescence of host plants, leaving larvae to starve if they have not reached their fourth or fifth instar and entered diapause. The effects of this can be disastrous. For example, the 1975–1977 drought in California lead to the extinction of 5 out of 21 *E. editha* populations that were being monitored at the time (Singer and Ehrlich, 1979; Ehrlich et al., 1980). Extreme weather events, which are predicted to increase in frequency with climate change, also have significant effects on abundance fluctuations and population extinctions. For example, the extinction of an entire *E. editha* metapopulation in the Sierra Nevada, California, coincided with three extreme weather events: (1) a very low winter snowpack in 1989, which caused adults to emerge earlier than nectar resources were available, leading to substantial adult mortality; (2) another very low winter snowpack in 1990, causing many early emerging adults to perish in a spring snowstorm; and (3) unseasonably cold temperatures of −5°C on 16 June 1992, which killed an estimated 97% of the metapopulation's host plant, *Collinsia*, leaving larvae to starve (Thomas et al., 1996). Following the extinction of this *Collinsia*-feeding metapopulation, the area was eventually recolonized by adjacent *Pedicularis*-feeding populations, which typically fly 2 to 3 weeks later and were less affected by these extreme weather events (reviewed by Parmesan, 2003).

While warmer and drier climatic conditions threaten the persistence of many *E. editha* populations, cooler and wetter conditions can have opposing, positive effects. In southern California, USA, and Baja California, Mexico, two consecutive years (1976 and 1977) of particularly high precipitation in the dry season led to population explosions of an endangered subspecies of *E. editha*, *E. editha quino* (Murphy and White, 1984). These high abundances were attributed to increased host plant availability and delayed host plant senescence over the two-year period. However, since then, the majority of these *E. editha quino* metapopulations have suffered extinction due to a general trend of warmer and drier conditions. Parmesan (2003) noted that, in her 1996 surveys of *E. editha quino* populations, the phenologies of larvae and local host plants were asynchronous, with host senescence preceding the hatching of larvae. As climatic conditions of this region continue to become warmer and drier, the persistence of remaining *E. editha quino* populations is questionable. Other notable threats to the subspecies include habitat loss related to urban and agricultural development, competitive exclusion of larval host plants (*Plantago*, *Antirrhinum*, and *Collinsia* spp.), and increased fire frequency. The US Fish and Wildlife Quino Checkerspot Working Group and the California Conservation Genomics Project (https://www.ccgproject.org/) are working together to identify specific conservation measures, such as population augmentation and translocation, that might aid in the persistence of *E. editha quino*. However, given the rate at which land use and climate changes are advancing in southern California and Baja California, the efficacy of these measures is uncertain.

8.4.4 Climate-induced habitat loss

An insidious effect of climate change on butterflies is climate-induced habitat loss. This is expected to be a major threat to many range-restricted, high-elevation species (Dirnböck et al., 2011), which together comprise some of the most unique species assemblages on our planet. For example, in California's Sierra Nevada, the greatest diversity of butterflies is found in non-forested habitats, such as subalpine/alpine meadows and fellfields. These unique 'sky island' habitats support three butterflies endemic to the region: the Sierra Nevada Parnassian (*Parnassius behrii*), the Sierra sulfur (*Colias behrii*), and the Ivallda Arctic (*Oeneis chryxus ivallda*). The insular nature of these species' distributions makes them particularly interesting subjects for biogeographic and genetic studies (e.g. Schoville and Roderick, 2009; Schoville et al., 2012). However, their specific habitat and environmental requirements also render them particularly sensitive to climate and habitat changes (Condamine and Sperling, 2018; Sperling et al., 2020). Many populations already occur at or near mountain tops, limiting their potential for elevational range shifts to track changing climatic conditions (Figure 8.5). Within the next century, dispersal and resultant gene flow among isolated populations will be increasingly threatened by habitat loss and fragmentation resulting from forests encroaching into subalpine/alpine habitats (Keyghobadi et al., 2005; Roland and Matter, 2007; Sperling et al., 2020). These changes are likely to threaten the persistence of high-elevation specialist species to a greater extent than sympatric generalist species (Condamine & Sperling, 2018; Dirnböck et al., 2011; Ehrlich and Murphy, 1987; Warren et al., 2001). Changing climatic conditions are also likely to impact the distribution and phenology of butterflies' host plant species, compounding their risk of extinction (Filazzola et al., 2020). In light of these predictions, effective conservation planning for alpine butterflies requires research programs that integrate methods of multiple disciplines to better understand the ecological and evolutionary processes that determine species' persistence and extinction probabilities.

8.5 A path forward: conservation and landscape genomics

A plethora of methods in landscape ecology correlate variation in the occurrences, densities,

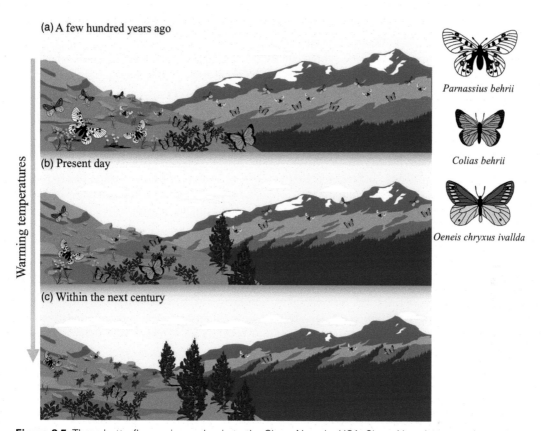

Figure 8.5 Three butterfly species endemic to the Sierra Nevada, USA: Sierra Nevada parnassian (*Parnassius behrii*), Sierra sulfur (*Colias behrii*), and Ivallda Arctic (*Oeneis chryxus ivallda*). Each of these species have highly restricted ranges and are only found in high-elevation environments such as alpine meadows and fellfields. Three panels (a, b, and c) illustrate how climate change and warming temperatures are affecting high-elevation habitats and the butterflies that depend on them. A few hundred years ago (top panel), alpine meadows and fellfields were abundant and populations of each species were well connected. Today (middle panel), warming temperatures are causing treelines to advance up mountain slopes, resulting in climate-induced habitat loss and fragmentation as alpine meadows and fellfields become smaller and further isolated. Within the next century (bottom panel), continued habitat loss, fragmentation, and degradation is expected to lead to drastic population declines and possibly extinction.

and movements of species with environmental and geographic variables to infer factors that affect species persistence. Meanwhile, advances in DNA sequencing technologies have enabled analyses of genome-wide data for almost any species of interest, helping to resolve how gene flow and adaptive genetic variation affect long-term persistence. The integration of these disciplines holds great potential, not only for understanding the effects of habitat loss, habitat fragmentation, and climate change on biodiversity, but also to aid in their amelioration through conservation planning. The emerging field of conservation genomics presents a viable toolkit for informing the conservation of imperiled butterflies and other species. An example of its application is the California Conservation Genomics Project, a state-funded initiative with a single goal: to produce the most comprehensive, multi-species, genomic dataset ever assembled to help manage regional biodiversity. This represents an intersection between genomic, landscape, and environmental methods, which together can be used to quantify the adaptive potential of threatened species and predict how their population structure and genetic health are likely to be affected by land use and climate changes.

Three endemic Sierra Nevada butterflies, along with the pipevine swallowtail (*Battus philenor*), Mormon metalmark (*Apodemia mormo*), and Edith's checkerspot (*Euphydryas editha*), are six of 236 species that are now being surveyed by the California Conservation Genomics Project. Each of these species is receiving a chromosome-level genome assembly, as well as whole-genome DNA sequencing for approximately 150 individuals. These data will be combined with forward-in-time landscape and environmental modelling to evaluate how each species is being/will be affected by land use and climate changes. The larval host plant of *B. philenor*, Dutchman's pipe (*Aristolochia californica*), is also included in this project, allowing scientists to infer how trophic interactions influence genomic patterns. Genome-wide sequence data can also be used to identify adaptive genomic variation corresponding to local environmental conditions and host associations (MacDonald et al., 2020). This information can be particularly informative for conservation practice. For example, if land use or climate changes result in the decline or extinction of populations that are of conservation concern, information on adaptive genomic variation can ensure that future translocation and reintroduction efforts focus on donor populations that are most compatible, both in terms of their ecology and genomic composition (Bellis et al., 2019). These data can also be used to inform captive

breeding programs by maximizing the genetic fitness of introduced individuals.

8.6 Summary

Butterflies are unparalleled sentinels for measuring effects of land use and climate changes on biodiversity worldwide. They exhibit rapid ecological and evolutionary responses to habitat loss, fragmentation, and degradation, as well as changing climatic conditions. Detailed accounts of their phenologies, distributions, and diversity, often led by armies of engaged members of the public, have led to a pool of information available to conservation biologists that is perhaps only exceeded by birds. This knowledge has been used to infer ecological and evolutionary effects of land use and climate changes at a range of spatial scales, from habitat fragments to entire continents, and a range of temporal scales, from inter-annual population fluctuations to extinction dynamics over multiple centuries. Our society's passion for butterflies has created a unique connection between amateur naturalists and professional scientists, accelerating theoretical developments in ecology and evolution and their application to conservation practice. These continuing collaborations will be crucial to understanding and ameliorating the effects of land use and climate changes in the future, especially as we reach potential tipping points and large-scale shifts in climate patterns globally.

Acknowledgements

This work was supported by a Natural Sciences and Engineering Research Council (NSERC) Discovery Grant awarded to FAHS (RGPIN-2018–04920), an NSERC Alexander Graham Bell Canada Graduate Scholarship–Doctoral (CGSD), an NSERC Postdoctoral Fellowship (PDF), and a UCLA La Kretz Center for

California Conservation Science Postdoctoral Fellowship awarded to ZGM and the California Conservation Genomics Project, with funding provided to HBS and the University of California by the State of California, State Budget Act of 2019 [UC Award ID RSI-19-690224]. Figure 8.5 was created by Aidan Sheppard.

References

Acorn, J. H. (2017). Entomological citizen science in Canada. *Canadian Entomologist, 149*(6), 774–785. https://doi.org/10.4039/tce.2017.48

Baz, A., & Garcia-Boyero, A. (1995). The effects of forest fragmentation on butterfly communities in central Spain. *Journal of Biogeography, 22*(1), 129–140. https://doi.org/10.2307/2846077

Bellis, J., Bourke, D., Williams, C., & Dalrymple, S. (2019). Identifying factors associated with the success and failure of terrestrial insect translocations. *Biological Conservation, 236*, 29–36. https://doi.org/10.1016/j.biocon.2019.05.008

Boggs, C. L., Watt, W. B., & Ehrlich, P. R. (2003). *Butterflies: Ecology and evolution taking flight.* University of Chicago Press.

Breed, G. A., Stichter, S., & Crone, E. E. (2013). Climate-driven changes in northeastern US butterfly communities. *Nature Climate Change, 3*(2), 142–145. https://doi.org/10.1038/nclimate1663

Brooks, T. M., Mittermeier, R. A., Mittermeier, C. G., Da Fonseca, G. A. B., Rylands, A. B., Konstant, W. R., Flick, P., Pilgrim, J., Oldfield, S., Magin, G., & Hilton-Taylor, C. (2002). Habitat loss and extinction in the hotspots of biodiversity. *Conservation Biology, 16*(4), 909–923. https://doi.org/10.1046/j.1523-1739.2002.00530.x

Brown, J. H., & Lomolino, M. V. (1989). Independent discovery of the equilibrium theory of island biogeography. *Ecology, 70*(6), 1954–1957. https://doi.org/10.2307/1938125

Burke, R. J., Fitzsimmons, J. M., & Kerr, J. T. (2011). A mobility index for Canadian butterfly species based on naturalists' knowledge. *Biodiversity and Conservation, 20*(10), 2273–2295. https://doi.org/10.1007/s10531-011-0088-y

Burton, J. F. (2003). The apparent influence of climatic change on recent changes of range by European insects (Lepidoptera, Orthoptera). *Société Royale Belge d'Entomologie, 38*, 125–144.

Cassel, A., Windig, J., Nylin, S., & Wiklund, C. (2001). Effects of population size and food stress on fitness-related characters in the scarce heath, a rare butterfly in Western Europe. *Conservation Biology, 15*(6), 1667–1673. https://doi.org/10.1046/j.1523-1739.2001.99557.x

Cayton, H. L., Haddad, N. M., Gross, K., Diamond, S. E., & Ries, L. (2015). Do growing degree days predict phenology across butterfly species? *Ecology, 96*(6), 1473–1479. https://doi.org/10.1890/15-0131.1

Condamine, F. L. (2018). Limited by the roof of the world: Mountain radiations of Apollo swallowtails controlled by diversity-dependence processes. *Biology Letters, 14*(3), 20170622. https://doi.org/10.1098/rsbl.2017.0622

Condamine, F., & Sperling, F. A. H. (2018). Anthropogenic threats to high-altitude Parnassian diversity. *News of the Lepidopterists' Society, 60*, 94–99.

Connor, E. F., & McCoy, E. D. (1979). The statistics and biology of the species-area relationship. *American Naturalist, 113*(6), 791–833. https://doi.org/10.1086/283438

Crone, E. E., & Schultz, C. B. (2003). Movement behaviour and minimum patch size for butterfly population persistence. In C. L. Boggs, W. B. Watt & P. R. Ehrlich (Eds.), *Butterflies: Ecology and evolution taking flight* (pp. 517–540). University of Chicago Press.

Dennis, R. L. H. (1993). *Butterflies and climate change.* Manchester University Press.

Dennis, R. L. H. (2004). Butterfly habitats, broad-scale biotope affiliations, and structural exploitation of vegetation at finer scales: The matrix revisited. *Ecological Entomology, 29*(6), 744–752. https://doi.org/10.1111/j.0307-6946.2004.00646.x

Dennis, R. L. H., & Hardy, P. B. (2007). Support for mending the matrix: Resource seeking by butterflies in apparent non-resource zones. *Journal of Insect Conservation, 11*(2), 157–168. https://doi.org/10.1007/s10841-006-9032-y

Diamond, J. M. (1975). The island dilemma: Lessons of modern biogeographic studies for the design of natural reserves. *Biological Conservation, 7*(2), 129–146. https://doi.org/10.1016/0006-3207(75)90052-X

Dirnböck, T., Essl, F., & Rabitsch, W. (2011). Disproportional risk for habitat loss of high-altitude endemic species under climate change. *Global Change Biology, 17*(2), 990–996. https://doi.org/10.1111/j.1365-2486.2010.02266.x

Dover, J., & Settele, J. (2009). The influences of landscape structure on butterfly distribution and movement: A review. *Journal of Insect Conservation*, *13*(1), 3–27. https://doi.org/10.1007/s10841-008-9135-8

Driscoll, D. A., Banks, S. C., Barton, P. S., Lindenmayer, D. B., & Smith, A. L. (2013). Conceptual domain of the matrix in fragmented landscapes. *Trends in Ecology and Evolution*, *28*(10), 605–613. https://doi.org/10.1016/j.tree.2013.06.010

Easterling, D. R., Evans, J. L., Groisman, P. Y., Karl, T. R., Kunkel, K. E., & Ambenje, P. (2000). Observed variability and trends in extreme climate events: A brief review. *Bulletin of the American Meteorological Society*, *81*(3), 417–425. https://doi.org/10.1175/1520-0477(2000)081<0417:OVATIE>2.3.CO;2

Ehrlich, P. R., & Raven, P. H. (1964). Butterflies and plants: A study in coevolution. *Evolution*, *18*(4), 586–608. https://doi.org/10.1111/j.1558-5646.1964.tb01674.x

Ehrlich, P. R., Murphy, D. D., Singer, M. C., Sherwood, C. B., White, R. R., & Brown, I. L. (1980). Extinction, reduction, stability and increase: The responses of checkerspot butterfly (*Euphydryas*) populations to the California drought. *Oecologia*, *46*(1), 101–105. https://doi.org/10.1007/BF00346973

Ehrlich, P. R., & Murphy, D. D. (1987). Conservation lessons from long-term studies of checkerspot butterflies. *Conservation Biology*, *1*(2), 122–131. https://doi.org/10.1111/j.1523-1739.1987.tb00021.x

Ehrlich, P. R., & Hanski, I. (2004). *On the wings of checkerspots: A model system for population biology*. Oxford University Press.

Fahrig, L. (2003). Effects of habitat fragmentation on biodiversity. *Annual Review of Ecology, Evolution, and Systematics*, *34*(1), 487–515. https://doi.org/10.1146/annurev.ecolsys.34.011802.132419

Fahrig, L. (2013). Rethinking patch size and isolation effects: The habitat amount hypothesis. *Journal of Biogeography*, *40*(9), 1649–1663. https://doi.org/10.1111/jbi.12130

Fahrig, L. (2017). Ecological responses to habitat fragmentation per se. *Annual Review of Ecology, Evolution, and Systematics*, *48*(1), 1–23. https://doi.org/10.1146/annurev-ecolsys-110316-022612

Filazzola, A., Matter, S. F., & Roland, J. (2020). Inclusion of trophic interactions increases the vulnerability of an alpine butterfly species to climate change. *Global Change Biology*, *26*(5), 2867–2877. https://doi.org/10.1111/gcb.15068

Filgueiras, B. K. C., Peres, C. A., Melo, F. P. L., Leal, I. R., & Tabarelli, M. (2021). Winner–loser species replacements in human-modified landscapes. *Trends in Ecology and Evolution*, *36*(6), 545–555. https://doi.org/10.1016/j.tree.2021.02.006

Fletcher, R. J., Didham, R. K., Banks-Leite, C., Barlow, J., Ewers, R. M., Rosindell, J., Holt, R. D., Gonzalez, A., Pardini, R., Damschen, E. I., Melo, F. P. L., Ries, L., Prevedello, J. A., Tscharntke, T., Laurance, W. F., Lovejoy, T., & Haddad, N. M. (2018). Is habitat fragmentation good for biodiversity? *Biological Conservation*, *226*, 9–15. https://doi.org/10.1016/j.biocon.2018.07.022

Flockhart, D. T., Wassenaar, L. I., Martin, T. G., Hobson, K. A., Wunder, M. B., & Norris, D. R. (2013). Tracking multi-generational colonization of the breeding grounds by monarch butterflies in eastern North America. *Proceedings. Biological Sciences*, *280*(1768), 20131087. https://doi.org/10.1098/rspb.2013.1087

Ford, E. B. (1945). *Butterflies* [Reprint], 1962. Collins.

Forister, M. L., & Shapiro, A. M. (2003). Climatic trends and advancing spring flight of butterflies in lowland California. *Global Change Biology*, *9*(7), 1130–1135. https://doi.org/10.1046/j.1365-2486.2003.00643.x

Forister, M. L., Halsch, C. A., Nice, C. C., Fordyce, J. A., Dilts, T. E., Oliver, J. C., Prudic, K. L., Shapiro, A. M., Wilson, J. K., & Glassberg, J. (2021). Fewer butterflies seen by community scientists across the warming and drying landscapes of the American West. *Science*, *371*(6533), 1042–1045. https://doi.org/10.1126/science.abe5585

Hadley, A. S., & Betts, M. G. (2016). Refocusing habitat fragmentation research using lessons from the last decade. *Current Landscape Ecology Reports*, *1*(2), 55–66. https://doi.org/10.1007/s40823-016-0007-8

Haila, Y. (2002). A conceptual genealogy of fragmentation research: From island biogeography to landscape ecology. *Ecological Applications*, *12*(2), 321–334. https://doi.org/10.1890/1051-0761(2002)012[0321:ACGOFR]2.0.CO;2

Hanski, I. (1994). A practical model of metapopulation dynamics. *Journal of Animal Ecology*, *63*(1), 151–162. https://doi.org/10.2307/5591

Hanski, I. (1999). *Metapopulation ecology*. Oxford University Press.

Hanski, I. (2011). Habitat loss, the dynamics of biodiversity, and a perspective on conservation. *Ambio*, *40*(3), 248–255. https://doi.org/10.1007/s13280-011-0147-3

Hanski, I. (2015). Habitat fragmentation and species richness. *Journal of Biogeography*, *42*(5), 989–993. https://doi.org/10.1111/jbi.12478

Hanski, I., & Gilpin, M. (1991). Metapopulation dynamics: Brief history and conceptual domain. *Biological Journal of the Linnean Society*, *42*(1–2), 3–16. https://doi.org/10.1111/j.1095-8312.1991.tb00548.x

IPCC. (2021). *Climate Change 2021: The physical science basis.* https://www.ipcc.ch/report/ar6/wg1/

Keyghobadi, N., Roland, J., & Strobeck, C. (2005). Genetic differentiation and gene flow among populations of the alpine butterfly, *Parnassius smintheus*, vary with landscape connectivity. *Molecular Ecology*, *14*(7), 1897–1909. https://doi.org/10.1111/j.1365-294X.2005.02563.x

Kerr, J. T. (2001). Butterfly species richness patterns in Canada: Energy, heterogeneity, and the potential consequences of climate change. *Conservation Ecology*, *5*(1). https://doi.org/10.5751/ES-00246-050110

Krauss, J., Steffan-Dewenter, I., & Tscharntke, T. (2003). How does landscape context contribute to effects of habitat fragmentation on diversity and population density of butterflies? *Journal of Biogeography*, *30*(6), 889–900. https://doi.org/10.1046/j.1365-2699.2003.00878.x

Kremen, C., Lees, D. C., & Fay, P. (2003). Butterflies and conservation planning in Madagascar: From patterns to practice. In C. L. Boggs, W. B. Watt & P. R. Ehrlich (Eds.), *Butterflies: Ecology and evolution taking flight* (pp. 517–540). University of Chicago Press.

Kuussaari, M., Nieminen, M., & Hanski, I. (1996). An experimental study of migration in the Glanville fritillary butterfly *Melitaea cinxia*. *Journal of Animal Ecology*, *65*(6), 791–801. https://doi.org/10.2307/5677

Levins, R. (1969). Some demographic and genetic consequences of environmental heterogeneity for biological control. *Bulletin of the Entomological Society of America*, *15*(3), 237–240. https://doi.org/10.1093/besa/15.3.237

Lewthwaite, J. M. M., Angert, A. L., Kembel, S. W., Goring, S. J., Davies, T. J., Mooers, A. Ø., Sperling, F. A. H., Vamosi, S. M., Vamosi, J. C., & Kerr, J. T. (2018). Canadian butterfly climate debt is significant and correlated with range size. *Ecography*, *41*(12), 2005–2015. https://doi.org/10.1111/ecog.03534

MacArthur, R. H., & Wilson, E. O. (1963). An equilibrium theory of insular zoogeography. *Evolution*, *17*(4), 373–387. https://doi.org/10.1111/j.1558-5646.1963.tb03295.x

MacDonald, Z. G., Nielsen, S. E., & Acorn, J. H. (2017). Negative relationships between species richness and evenness render common diversity indices inadequate for assessing long-term trends in butterfly diversity. *Biodiversity and Conservation*, *26*(3), 617–629. https://doi.org/10.1007/s10531-016-1261-0

MacDonald, Z. G., Anderson, I. D., Acorn, J. H., & Nielsen, S. E. (2018a). Decoupling habitat fragmentation from habitat loss: Butterfly species mobility obscures fragmentation effects in a naturally fragmented landscape of lake islands. *Oecologia*, *186*(1), 11–27. https://doi.org/10.1007/s00442-017-4005-2

MacDonald, Z. G., Anderson, I. D., Acorn, J. H., & Nielsen, S. E. (2018b). The theory of island biogeography, the sample-area effect, and the habitat diversity hypothesis: Complementarity in a naturally fragmented landscape of lake islands. *Journal of Biogeography*, *45*(12), 2730–2743. https://doi.org/10.1111/jbi.13460

MacDonald, Z. G., Acorn, J. H., Zhang, J., & Nielsen, S. E. (2019). Perceptual Range, Targeting Ability, and Visual Habitat Detection by Greater Fritillary Butterflies *Speyeria cybele* (Lepidoptera: Nymphalidae) and *Speyeria atlantis*. *Journal of Insect Science*, *19*(4), 1–10. https://doi.org/10.1093/jisesa/iez060

MacDonald, Z. G., Dupuis, J. R., Davis, C. S., Acorn, J. H., Nielsen, S. E., & Sperling, F. A. H. (2020). Gene flow and climate-associated genetic variation in a vagile habitat specialist. *Molecular Ecology*, *29*(20), 3889–3906. https://doi.org/10.1111/mec.15604

MacDonald, Z. G., Deane, D. C., He, F., Lamb, C. T., Sperling, F. A. H., Acorn, J. H., & Nielsen, S. E. (2021). Distinguishing effects of area *per se* and isolation from the sample-area effect for true islands and habitat fragments. *Ecography*, *44*(7), 1051–1066. https://doi.org/10.1111/ecog.05563

Matter, S. F., Roland, J., Keyghobadi, N., & Sabourin, K. (2003). The effects of isolation, habitat area and resources on the abundance, density and movement of the butterfly *Parnassius smintheus*. *American Midland Naturalist*,

150(1), 26–36. https://doi.org/10.1674/0003-0031(2003)150[0026:TEOIHA]2.0.CO;2

Mavárez, J., Salazar, C. A., Bermingham, E., Salcedo, C., Jiggins, C. D., & Linares, M. (2006). Speciation by hybridization in *Heliconius* butterflies. *Nature*, 441(7095), 868–871. https://doi.org/10.1038/nature04738

Miller, N. G., Wassenaar, L. I., Hobson, K. A., & Norris, D. R. (2012). Migratory connectivity of the monarch butterfly (*Danaus plexippus*): Patterns of spring re-colonization in eastern North America. *PLOS ONE*, 7(3), e31891. https://doi.org/10.1371/journal.pone.0031891

Munroe, E. G. (1948). *The geographical distribution of butterflies in the West Indies* [PhD Thesis]. Cornell University.

Murphy, D. D., & White, R. R. (1984). Rainfall, resources, and dispersal of southern populations of *Euphydryas editha* (Lepidoptera: Nymphalidae). *Pan-Pacific Entomologist*, 60(4), 350–354.

Murphy, J. M., Sexton, D. M. H., Jenkins, G. J., Booth, B. B., Brown, C. C., Clark, R. T., . . . & Wood, R. A. (2009). *UK climate projections science report: Climate change projections*. Met Office Hadley Centre. http://ukclimateprojections.metoffice.gov.uk/22530

Nowicki, P., Settele, J., Henry, P. Y., & Woyciechowski, M. (2008). Butterfly monitoring methods: The ideal and the real world. *Israel Journal of Ecology and Evolution*, 54(1), 69–88. https://doi.org/10.1560/IJEE.54.1.69

Öckinger, E., & Smith, H. G. (2006). Landscape composition and habitat area affects butterfly species richness in semi-natural grasslands. *Oecologia*, 149(3), 526–534. https://doi.org/10.1007/s00442-006-0464-6

Oliver, T. H., Marshall, H. H., Morecroft, M. D., Brereton, T., Prudhomme, C., & Huntingford, C. (2015). Interacting effects of climate change and habitat fragmentation on drought-sensitive butterflies. *Nature Climate Change*, 5(10), 941–945. https://doi.org/10.1038/nclimate2746

Parmesan, C. (1996). Climate and species' range. *Nature*, 382(6594), 765–766. https://doi.org/10.1038/382765a0

Parmesan, C. (2003). Butterflies as bioindicators for climate change effects. In C. L. Boggs, W. B. Watt & P. R. Ehrlich (Eds.), *Butterflies: Ecology and evolution taking flight* (pp. 541–560). University of Chicago Press.

Parmesan, C., Ryrholm, N., Stefanescu, C., Hill, J. K., Thomas, C. D., Descimon, H., Huntley, B., Kaila, L.,

Kullberg, J., Tammaru, T., Tennent, W. J., Thomas, J. A., & Warren, M. (1999). Poleward shifts in geographical ranges of butterfly species associated with regional warming. *Nature*, 399(6736), 579–583. https://doi.org/10.1038/21181

Pimm, S. L., Jenkins, C. N., Abell, R., Brooks, T. M., Gittleman, J. L., Joppa, L. N., Raven, P. H., Roberts, C. M., & Sexton, J. O. (2014). The biodiversity of species and their rates of extinction, distribution, and protection. *Science*, 344(6187), 1246752. https://doi.org/10.1126/science.1246752

Pollard, E. (1977). A method for assessing changes in the abundance of butterflies. *Biological Conservation*, 12(2), 115–134. https://doi.org/10.1016/0006-3207(77)90065-9

Primack, R. B. (2006). *Essentials of conservation biology*. Sinauer.

Prudic, K. L., McFarland, K. P., Oliver, J. C., Hutchinson, R. A., Long, E. C., Kerr, J. T., & Larrivée, M. (2017). eButterfly: Leveraging massive online citizen science for butterfly conservation. *Insects*, 8(2), 53. https://doi.org/10.3390/insects8020053

Ravenga, C., Brunner, J., Henninger, N., Kassem, K., & Payne, R. (2000). *Pilot analysis of global ecosystems: Wetland ecosystems*. World Resources Institute.

Ries, L., & Sisk, T. D. (2004). A predictive model of edge effects. *Ecology*, 85(11), 2917–2926. https://doi.org/10.1890/03-8021

Riva, F., Acorn, J. H., & Nielsen, S. E. (2018a). Localized disturbances from oil sands developments increase butterfly diversity and abundance in Alberta's boreal forests. *Biological Conservation*, 217, 173–180. https://doi.org/10.1016/j.biocon.2017.10.022

Riva, F., Acorn, J. H., & Nielsen, S. E. (2018b). Distribution of cranberry blue butterflies (*Agriades optilete*) and their responses to forest disturbance from in situ oil sands and wildfires. *Diversity*, 10(4), 112. https://doi.org/10.3390/d10040112

Robinson, W. S. (1950). Ecological correlations and the behavior of individuals. *American Sociological Review*, 15(3), 351–357.

Roland, J., & Matter, S. F. (2007). Encroaching forests decouple alpine butterfly population dynamics. *Proceedings of the National Academy of Sciences of the United States of America*, 104(34), 13702–13704. https://doi.org/10.1073/pnas.0705511104

Rosenzweig, M. L. (1995). *Species diversity in space and time*. Cambridge University Press.

Schmeller, D. S., Henry, P. Y., Julliard, R., Gruber, B., Clobert, J., Dziock, F., Lengyel, S., Nowicki, P.,

Déri, E., Budrys, E., Kull, T., Tali, K., Bauch, B., Settele, J., Van Swaay, C., Kobler, A., Babij, V., Papastergiadou, E., & Henle, K. (2009). Advantages of volunteer-based biodiversity monitoring in Europe. *Conservation Biology, 23*(2), 307–316. https://doi.org/10.1111/j.1523-1739.2008.01125.x

Schoville, S. D., & Roderick, G. K. (2009). Alpine biogeography of Parnassian butterflies during Quaternary climate cycles in North America. *Molecular Ecology, 18*(16), 3471–3485. https://doi.org/10.1111/j.1365-294X.2009.04287.x

Schoville, S. D., Lam, A. W., & Roderick, G. K. (2012). A range-wide genetic bottleneck overwhelms contemporary landscape factors and local abundance in shaping genetic patterns of an alpine butterfly (Lepidoptera: Pieridae: *Colias behrii*). *Molecular Ecology, 21*(17), 4242–4256. https://doi.org/10.1111/j.1365-294X.2012.05696.x

Shapiro, A. M. (1996). Status of butterflies. In *Sierra Nevada Ecosystem Project: Final report to congress, 2* (pp. 743–757). University of California, Centers for Water and Wildland Resources. https://pubs.usgs.gov/dds/dds-43/VOL_II/VII_C27.PDF

Shreeve, T. G. (1992). Monitoring butterfly movements. In R. L. H. Dennis (Ed.), *The ecology of butterflies in Britain* (pp. 120–138). Oxford University Press.

Singer, M. C. (1972). Complex components of habitat suitability within a butterfly colony. *Science, 176*(4030), 75–77. https://doi.org/10.1126/science.176.4030.75

Singer, M. C., & Ehrlich, P. R. (1979). Population dynamics of the checkerspot butterfly *Euphydryas editha*. *Fortschritte der Zoologie, 25*, 53–60.

Singer, M. C., & Parmesan, C. (2018). Lethal trap created by adaptive evolutionary response to an exotic resource. *Nature, 557*(7704), 238–241. https://doi.org/10.1038/s41586-018-0074-6

Singer, M. C., Ng, D., & Thomas, C. D. (1988). Heritability of oviposition preference and its relationship to offspring performance within a single insect population. *Evolution; International Journal of Organic Evolution, 42*(5), 977–985. https://doi.org/10.1111/j.1558-5646.1988.tb02516.x

Singer, M. C., Thomas, C. D., & Parmesan, C. (1993). Rapid human-induced evolution of insect–host associations. *Nature, 366*(6456), 681–683. https://doi.org/10.1038/366681a0

Sparks, T. H., & Yates, T. J. (1997). The effect of spring temperature on the appearance dates of British butterflies 1883–1993. *Ecography, 20*(4), 368–374.

https://doi.org/10.1111/j.1600-0587.1997.tb00381.x

Sperling, F. A. H., Sperling, W., & MacDonald, Z. G. (2020). Canadian Alpine butterflies deserve better monitoring. *State of the mountains report* (pp. 31–34). https://www.stateofthemountains.ca/s/ACC-SotM-2020-MAR22-2021.pdf

Sperling, F. A. H. (2003). Butterfly molecular systematics. In C. L. Boggs, W. B. Watt & P. R. Ehrlich (Eds.), *Butterflies: Ecology and evolution taking flight* (pp. 431–458). University of Chicago Press.

Steffan-Dewenter, I., & Tscharntke, T. (2000). Butterfly community structure in fragmented habitats. *Ecology Letters, 3*(5), 449–456. https://doi.org/10.1111/j.1461-0248.2000.00175.x

Stefanescu, C., Páramo, F., Åkesson, S., Alarcón, M., Ávila, A., Brereton, T., Carnicer, J., Cassar, L. F., Fox, R., Heliölä, J., Hill, J. K., Hirneisen, N., Kjellén, N., Kühn, E., Kuussaari, M., Leskinen, M., Liechti, F., Musche, M., Regan, E. C., ... & Chapman, J. W. (2013). Multi-generational long-distance migration of insects: Studying the painted lady butterfly in the western Palaearctic. *Ecography, 36*(4), 474–486. https://doi.org/10.1111/j.1600-0587.2012.07738.x

Sweaney, N., Lindenmayer, D. B., & Driscoll, D. A. (2014). Is the matrix important to butterflies in fragmented landscapes? *Journal of Insect Conservation, 18*(3), 283–294. https://doi.org/10.1007/s10841-014-9641-9

Tabarelli, M., Peres, C. A., & Melo, F. P. L. (2012). The "few winners and many losers" paradigm revisited: Emerging prospects for tropical forest biodiversity. *Biological Conservation, 155*, 136–140. https://doi.org/10.1016/j.biocon.2012.06.020

Thomas, J. A. (1991). Rare species conservation: Case studies of European butterflies. In I. F. Spellerberg, F. B. Goldsmith & M. G. Morris (Eds.), *The scientific management of temperate communities for conservation* (pp. 149–197). Blackwell Scientific.

Thomas, J. A. (2005). Monitoring change in the abundance and distribution of insects using butterflies and other indicator groups. *Philosophical Transactions of the Royal Society of London. Series B, Biological Sciences, 360*(1454), 339–357. https://doi.org/10.1098/rstb.2004.1585

Thomas, C. D., Singer, M. C., & Boughton, D. A. (1996). Catastrophic extinction of population sources in a butterfly metapopulation. *American Naturalist, 148*(6), 957–975. https://doi.org/10.1086/285966

Thomas, J. A., Bourn, N. A. D., Clarke, R. T., Stewart, K. E., Simcox, D. J., Pearman, G. S., Curtis, R., & Goodger, B. (2001). The quality and isolation of habitat patches both determine where butterflies persist in fragmented landscapes. *Proceedings. Biological Sciences, 268*(1478), 1791–1796. https://doi.org/10.1098/rspb.2001.1693

Tilman, D., Clark, M., Williams, D. R., Kimmel, K., Polasky, S., & Packer, C. (2017). Future threats to biodiversity and pathways to their prevention. *Nature, 546*(7656), 73–81. https://doi.org/10.1038/nature22900

UN Environmental Program. (2021). WCMC IUCN. *Protected planet: The world database on protected areas (WDPA)*. https://www.protectedplanet.net/en

Van Nouhuys, S., & Hanski, I. (2002). Colonization rates and distances of a host butterfly and two specific parasitoids in a fragmented landscape. *Journal of Animal Ecology, 71*(4), 639–650. https://doi.org/10.1046/j.1365-2656.2002.00627.x

van Swaay, C., & Warren, M. (1999). *Red data book of European butterflies (Rhopalocera)*. Council of Europe.

van Swaay, C., Warren, M., & Loïs, G. (2006). Biotope use and trends of European butterflies. *Journal of Insect Conservation, 10*(2), 189–209. https://doi.org/10.1007/s10841-006-6293-4

van Swaay, C., Maes, D., Collins, S., Munguira, M. L., Šašić, M., Settele, J., Verovnik, R., Warren, M., Wiemers, M., Wynhoff, I., & Cuttelod, A. (2011). Applying IUCN criteria to invertebrates: How red is the Red List of European butterflies? *Biological Conservation, 144*(1), 470–478. https://doi.org/10.1016/j.biocon.2010.09.034

Vera, F. W. M. (2000). *Grazing ecology and forest history*. CABI.

Wallace, A. R. (1865). I. On the phenomena of variation and geographical distribution as illustrated by the Papilionidae of the Malayan Region. *Transactions of the Linnean Society of London, 25*(1), 1–71. https://doi.org/10.1111/j.1096-3642.1865.tb00178.x

Wallis DeVries, M. F. (2004). A quantitative conservation approach for the endangered butterfly *Maculinea alcon*. *Conservation Biology, 18*(2), 489–499.

Wang, T., Hamann, A., Spittlehouse, D., & Carroll, C. (2016). Locally downscaled and spatially customizable climate data for historical and future periods for North America. *PLOS ONE, 11*(6),

e0156720. https://doi.org/10.1371/journal.pone.0156720

Warren, M. S., Hill, J. K., Thomas, J. A., Asher, J., Fox, R., Huntley, B., Roy, D. B., Telfer, M. G., Jeffcoate, S., Harding, P., Jeffcoate, G., Willis, S. G., Greatorex-Davies, J. N., Moss, D., & Thomas, C. D. (2001). Rapid responses of British butterflies to opposing forces of climate and habitat change. *Nature, 414*(6859), 65–69. https://doi.org/10.1038/35102054

Watt, W. B. (2003). Mechanistic studies of butterfly adaptation. In C. L. Boggs, W. B. Watt & P. R. Ehrlich (Eds.), *Butterflies: Ecology and evolution taking flight* (pp. 319–352). University of Chicago Press.

Wenzel, M., Schmitt, T., Weitzel, M., & Seitz, A. (2006). The severe decline of butterflies on western German calcareous grasslands during the last 30 years: A conservation problem. *Biological Conservation, 128*(4), 542–552. https://doi.org/10.1016/j.biocon.2005.10.022

Westwood, A. R., & Blair, D. (2010). Effect of regional climate warming on the phenology of butterflies in boreal forests in Manitoba, Canada. *Environmental Entomology, 39*(4), 1122–1133. https://doi.org/10.1603/EN09143

Wettstein, W., & Schmid, B. (1999). Conservation of arthropod diversity in montane wetlands: Effect of altitude, habitat quality and habitat fragmentation on butterflies and grasshoppers. *Journal of Applied Ecology, 36*(3), 363–373. https://doi.org/10.1046/j.1365-2664.1999.00404.x

White, R. P., Murray, S., & Rohweder, M. (2000). *Pilot assessment of global ecosystems: Grassland ecosystems*. World Resources Institute.

Wilcox, B. A., Murphy, D. D., Ehrlich, P. R., & Austin, G. T. (1986). Insular biogeography of the montane butterfly faunas in the Great Basin: Comparison with birds and mammals. *Oecologia, 69*(2), 188–194. https://doi.org/10.1007/BF00377620

Wilson, E. O., & MacArthur, R. H. (1967). *The theory of island biogeography*. Princeton University Press.

Wissel, C., Stephan, T., & Zaschke, S.-H. (1994). Modelling extinction and survival of small populations. In H. Remmert (Ed.), *Minimum animal populations* (pp. 67–103). Springer. https://doi.org/10.1007/978-3-642-78214-5_6

Wood, B. C., & Pullin, A. S. (2002). Persistence of species in a fragmented urban landscape: The importance of dispersal ability and habitat availability for grassland butterflies. *Biodiversity and Conservation*, *11*(8), 1451–1468. https://doi.org/10.1023/A:1016223907962

Zschokke, S., Dolt, C., Rusterholz, H. P., Oggier, P., Braschler, B., Thommen, G. H., Lüdin, E., Erhardt, A., & Baur, B. (2000). Short-term responses of plants and invertebrates to experimental small-scale grassland fragmentation. *Oecologia*, *125*(4), 559–572. https://doi.org/10.1007/s004420000483

Chapter 9

The Asian longhorned tick in North America: An interdisciplinary EcoHealth study

Susan C. Cork, Jamyang Namgyal, Isabelle Couloigner and Sylvia Checkley

Abstract

The Asian longhorned tick (*Haemaphysalis longicornis* Neumann, 1901) is a vector for many pathogenic micro-organisms, of animal and public health importance, in its native range in East Asia. It has been found to be a competent vector for the protozoon, *Theileria orientalis* Ikeda, which has resulted in livestock morbidity and mortality in its introduced range of New Zealand and Australia. In North America, this tick was first formally detected in 2017 in New Jersey on sheep, although it may have actually been present in the United States as early as 2010 but was initially misidentified as a rabbit tick. Currently, this tick has been reported from 19 US states. These are predominantly in the northeast. In this chapter, we examine the habitat suitability for the Asian longhorned tick (ALHT) across North America and consider potential routes of entry into Canada from the northeastern states of the United States. Our best habitat suitability model, using a machine learning application (MaxEnt), predicted that the most suitable North American areas for geographic expansion of the ALHT are from Arkansas–South Carolina to the south of Quebec–Nova Scotia in the east, and from California to the coast of British Columbia in the west. Likely routes of entry into Canada include cross border movement of infested terrestrial wildlife and livestock, dog importations and the movement of migratory birds. Enhanced tick surveillance, including the use of citizen science, is required to ensure early detection of this invasive species and to gain a better understanding of the impact that this tick might have on animal and human health in North America.

Susan C. Cork and Douglas P. Whiteside (eds) **Case Studies in EcoHealth**
DOI: 10.52517/9781789183368.009, © 5m Books Ltd, 2024

9.1   Introduction – the biology of the Asian longhorned tick

The Asian longhorned tick (ALHT) (*Haemaphysalis longicornis* Neumann, 1901) is a three-host hard tick (Acari, Ixodidae), whose native range includes Japan, China, eastern Russia, and the Korean peninsular (Hoogstraal et al., 1968). It has become well established in Australia and New Zealand, as well as many western Pacific islands (e.g., Fiji, New Caledonia, Tonga, Vanuatu, Western Samoa), likely due to accidental introduction with cattle imports (Heath, 2016) (see Figure 9.1). These ticks occur predominantly in temperate regions but are also found in subtropical and tropical

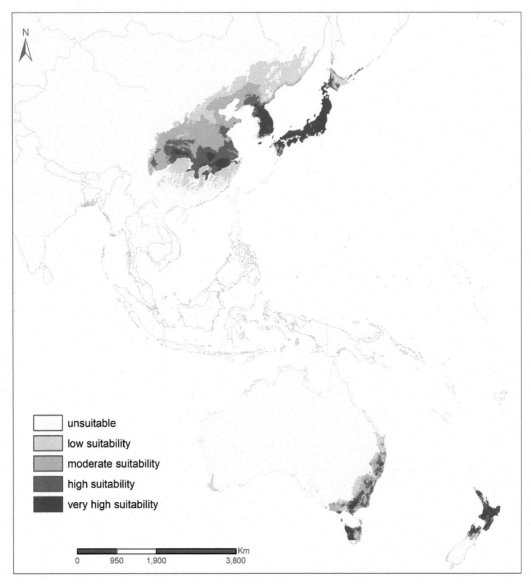

Figure 9.1 Habitat suitability of the ALHT's established populations in its native (Asia) and introduced (Australasia) range.

climates and generally do not tolerate arid conditions.

The female of the ALHT has the unique ability to lay eggs and reproduce without mating with a male (i.e., they are parthenogenetic) (Anon., 2019; Chen et al., 2012). This can result in rapid reproduction and large tick infestations on individual vertebrate hosts. Thousands of ticks may also be found in the environment (e.g., grassland or shrubs) as well as on a wide range of animal hosts (see Figure 9.2, for life cycle). The host range reported for the ALHT is extensive and includes domestic animals and wildlife, including a wide range of mammals and birds. This tick species has also been found to bite humans. Owing to the lack of host specificity, host related factors may not have a large impact when determining suitable habitat in which this tick could thrive if it was to be introduced to new regions.

In East Asia, the ALHT is the main vector for the transmission of severe fever with thrombocytopenia syndrome virus (SFTSV) in humans (Choi et al., 2020). This emerging zoonotic disease, which can be fatal, is caused by a novel bunyavirus. It was first identified in China in 2009 (Yu et al., 2011) and subsequently in South Korea and Japan in 2013 (Kim

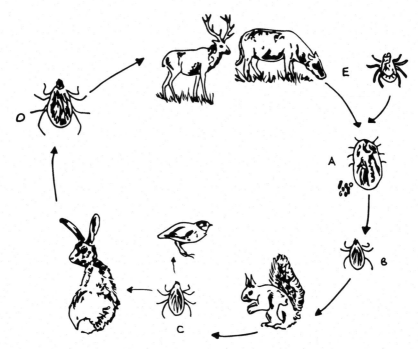

Figure 9.2 Life cycle of *Haemaphysalis longicornis* and the preferred hosts for each developmental life stage. (A) Engorged adult female drops off host and lays eggs in the environment (usually in the autumn). (B) Eggs hatch into six-legged larvae which attach on to a host to feed. (C) Larvae moult into eight-legged nymphs after leaving first host and may overwinter before finding the next host in the spring (often small mammals or birds). (D) Nymphs moult into adults after feeding and leaving the second host. Adults then attach to a third host to feed in the late spring or summer (this includes deer and cattle as well as other species). (E) Male ticks are not necessary for the reproductive cycle to be completed. Female ticks become significantly engorged after feeding and are easily visible unlike the larvae and nymphs which are much smaller.

et al., 2013., Takahashi et al., 2014). In Japan, the ALHT is also a vector of *Rickettsia japonica*, which causes Japanese spotted fever in humans (Mahara, 1997). In Australia and New Zealand, the ALHT is a vector for the protozoan parasite *Theileria orientalis* Ikeda, which causes bovine theileriosis (Hammer et al., 2015; Heath, 2016; Lawrence et al., 2017). Theileriosis can lead to severe anaemia in cattle with subsequent morbidity and mortality resulting in animal welfare concerns and economic losses (Heath, 2016). The ALHT is also a competent vector for other cattle pathogens such as *Babesia ovata*, *B. major*, and *Anaplasma bovis* in New Zealand (Lawrence et al., 2017) and the DNA of *Ehrlichia* and *Borrelia* spp. has been detected in ALHT collected from its native range in East Asia (Chu et al., 2008., Luo et al., 2016). More recently molecular evidence of *Rickettsia* spp. and other potential pathogens has been found in ALHT collected in the USA (Price et al., 2021, 2022; Stanley et al., 2020).

9.1.1 What does this tick look like?

ALHTs (*H. longicornis*) are reddish-brown. Adult females are 2.7–3.4 mm long and 1.4–2.0 mm wide, whereas nymphs are 1.8 × 1.0 mm and larvae are 0.6 × 0.5 mm. When completely blood fed, adult females measure up to 10 mm long.

Two other tick species closely related the ALHT (*H. longicornis*) occur in North America; the rabbit tick (*H. leporispalustris*) and the bird tick (*H. chordeilis*). *H. leporispalustris* parasitizes rabbits (e.g., eastern cottontail rabbits) and hares (e.g., snowshoe hares), but will also parasitize ground-dwelling birds. *H. chordeilis* normally parasitizes birds (e.g., grouse). Owing to the importance of correctly identifying ticks, it is important to engage entomologists in tick submission and surveillance programmes.

The ALHT is a three-host tick which needs to feed at each stage of the life cycle (larva, nymph and adult).

Engorged adult female drops off host and lays eggs in the environment (usually in the autumn).

Eggs hatch into six-legged larvae which attach on to a host to feed.

Larvae moult into eight-legged nymphs after leaving first host and may overwinter before finding the next host in the spring (often small mammals or birds).

Nymphs moult into adults after feeding and leaving the second host. Adults then attach to a third host to feed in the late spring or summer (this includes deer and cattle as well as other species).

Male ticks are not necessary for the reproductive cycle to be completed. Female ticks become significantly engorged after feeding and are easily visible unlike the larvae and nymphs which are much smaller.

This tick can complete its life cycle within 12 months, but typically one generation occurs per year, with most immature ticks entering diapause (dormancy) during winter and other cold periods.

9.2 The Asian longhorned tick in North America

In North America, the ALHT was first formally identified in August 2017, from sheep in New Jersey (Rainey et al., 2018). However, retrospective investigations of archived specimens suggest that this tick was actually present in the United States in 2010 but that the specimens were initially misidentified as the native rabbit tick, *H. leporispalustris* (Packard, 1869) since the latter appears very similar (Oakes et al., 2019). Since 2017, the ALHT has been reported in over 200 counties from 19 eastern states of the United States (USDA, 2022) (see Figure 9.3). The invasive population of ALHT in the United States probably began with three or more self-cloning females from northeastern Asia

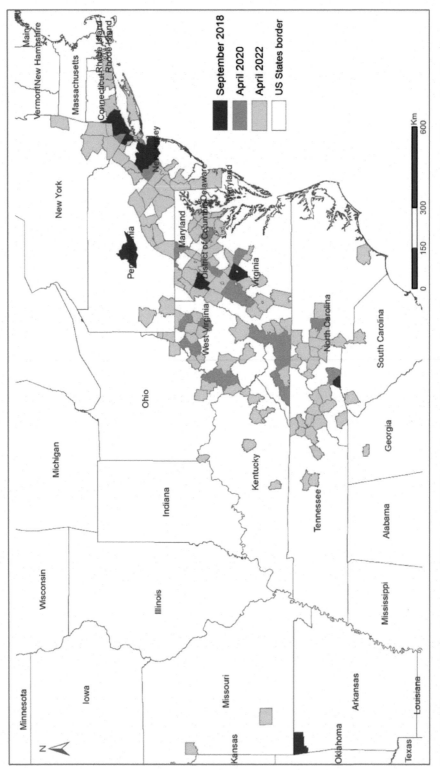

Figure 9.3 Temporal (from September 2018 to April 2022) visualization of the expansion of the confirmed ALHT presence in US counties. See also USDA National *Haemaphysalis longicornis* (Asian longhorned tick) Situation Report [PDF – 8 pages] (updated monthly).

(Egizi et al., 2020). At the time of writing, there has not been any detection of this tick in Canada (Hutcheson et al., 2019).

The first ALHT bite in a human subject in the United States was reported from New York state in 2018 (Wormser et al., 2020). To date, however, there have not been any viable human pathogens detected in ALHT collected in the United States but there remains a concern that this tick has the potential to transmit endemic pathogens such as *Anaplasma*, *Babesia*, and *Rickettsia* species (Hutcheson et al., 2019). However, *Theileria orientalis* Ikeda, a protozoan parasite transmitted by ALHT in East Asia, New Zealand, and Australia, has been detected in cattle in Virginia, USA (Oakes et al., 2019., Dinkel et al., 2021). This has prompted further concerns that this tick species might play a role in the continued transmission of the pathogen causing *Theileria*-associated bovine infectious anaemia (Oakes et al., 2019). This may result in restrictions on cattle movements and subsequent impacts on interstate movements across North America. The potential role of this tick in causing significant morbidity and mortality in livestock in North America cannot be ignored. Aside from the threat of tick-borne diseases, we also know that heavy infestations of this tick species on host animals can cause direct morbidity, and welfare concerns, as a result of excessive tick feeding activity.

9.3 Modelling methods to predict habitat suitability

Habitat suitability models developed by Namgyal et al. (2020) were projected onto North America to indicate regions of predicted habitat suitability for ALHT, and these predictions were examined using statistical tools. For the second step of the analysis, Namgyal et al. (2020) looked at whether using different climatic predictors would improve the predictability of MaxEnt (maximum entropy

MaxEnt is a machine learning method, called maximum entropy modelling, that enables the computation of the potential geographic distribution of species by finding, from its presence-only *georeferenced/GPS* locations, the distribution that is most representative, while considering the limits of the environmental variables of those known locations (Elith et al., 2011). In a recent study, Namgyal et al. (2020) used ALHT presence data derived from Rochlin (2019) augmented with data from Zhang et al. (2019) applying MaxEnt (Phillips et al., 2006) to model the habitat suitability of ALHT in North America. The ALHT presence data were separated into native and introduced range and then MaxEnt models were built for each to compare the habitat suitability predictions for North America. The objectives of this study were to (1) separate the global ALHT occurrence data for different regions in the world and build competing models using environmental predictors identified by Rochlin (2019); (2) build competing models for different regions using environmental variables from WorldClim (Fick and Hijmans, 2017; Hijmans et al., 2005) and ENVIREM (Title and Bemmels, 2018); and (3) finally compare habitat suitability predictions for North America by these competing models and subsequently select the two best models to describe the potential distribution and expansion of ALHT in North America. The maps derived from this study are shown in Figure 9.4.

machine learning algorithm), habitat suitability modelling, for North America.

It was found that temperature and precipitation are the most important climatic factors to influence the distribution of ALHT in both the native (East Asia) and the introduced (Australia/New Zealand) regions where it

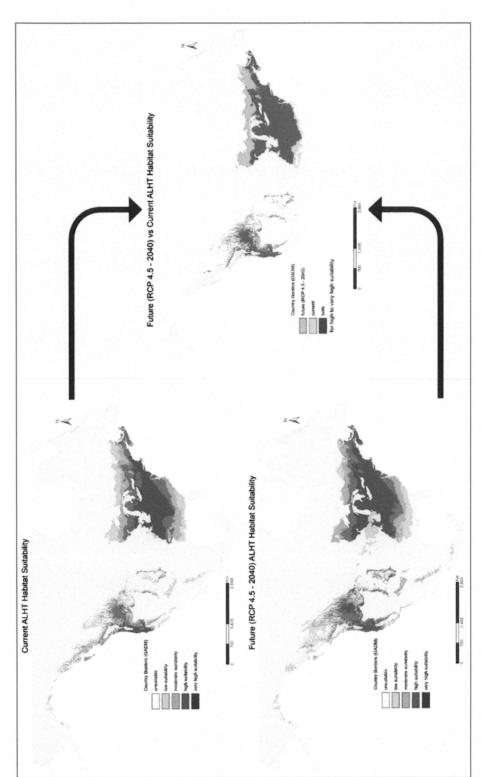

Figure 9.4 Current and future ALHT habitat suitability maps (left) and comparison map (right) to visualize projected expansion of ALHT due to climate change (i.e., median of the future climate projections for 2021–2040 under RCP 4.5, which assumed an estimate of 2°C above pre-industrial period) (using WorldClim).

has become established (Heath, 2016; Zheng et al., 2012). Annual mean temperature greater than 12°C, mean coldest monthly temperature less than 2°C, and annual rainfall above 1000 mm are considered to be optimum for ALHT range expansion in New Zealand (Heath, 2016).

The best habitat suitability models for the ALHT in North America have predicted that the most suitable areas for its geographic expansion (Figure 9.4) are from Arkansas–South Carolina to the south of Quebec–Nova Scotia in the east, and, if introduced, from California to the coast of British Columbia in the west (Namgyal et al., 2020). With current assessment of future climate in North America, it is generally predicted that the Asian long horned tick could establish further north into Canada over the next two decades (Figure 9.4). Access to more accurate publicly available presence and absence data for ALHT would improve the predictive accuracy of these models and, in future, would facilitate more cost-effective targeted surveillance for early detection and subsequent tick control response.

One of the limiting factors to the development of habitat suitability models is the lack of representative location of specific data points. In the case of the ALHT there is a growing data set being generated in the US through active multiyear tick surveillance programmes overseen by government and academic institutions. There are also research teams encouraging public submissions of ticks for identification and testing (Porter et al., 2021), this use of passive surveillance data can add value to formal tick monitoring projects which are quite expensive to run. In Canada, the growing expansion of citizen science platforms such as e-tick, can also add useful data to formal tick research and monitoring projects. See e-tick https://www.etick.ca/

9.4 Risk assessment for the entry of the Asian longhorned tick into Canada

Likely routes of entry into Canada include cross border movement of infested terrestrial wildlife and livestock, dog importations and the movement of migratory birds. Enhanced surveillance and further investigation are required to ensure early detection of this invasive tick and to gain a better understanding of the role that this tick might play in the transmission of diseases to humans and animals in North America. They have been found feeding on a wide range of wild and domesticated animal hosts and will also bite humans. It is evident that the ALHT can adapt rapidly to local fauna, which enhances their ability to survive and spread quickly when introduced to a suitable habitat. Infestation of hosts which have a wide geographic range, such as migratory birds, suggests that these ticks have the opportunity to spread rapidly over long distances. In v, there are records that the ALHT has been found on migratory birds (e.g. thrushes and kingfishers); house sparrows and Australian magpies are reported as hosts for the ALHT in Australia; and geese and doves are commonly considered as hosts in China. In New Zealand, ALHT have been detected on wild and domestic birds (e.g., mynas, domestic and wild ducks, kiwis and house sparrows).

We adapted a qualitative risk analysis methodology, developed by the World Organization for Animal Health (OIE, 2004) for the import of animals and animal products, for our study. Risk assessment (Figure 9.5) is a component of risk analysis and includes hazard identification, entry assessment, exposure assessment, consequence assessment, and risk estimation. In this situation the hazard is the ALHT itself, the susceptible host population includes humans and other vertebrate animal hosts including wildlife; routes of entry include terrestrial and air entry points from the USA into Canada. The

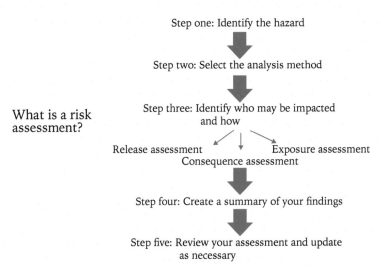

Step one: Identify the hazard

Step two: Select the analysis method

What is a risk assessment?

Step three: Identify who may be impacted and how

Release assessment   Exposure assessment
Consequence assessment

Step four: Create a summary of your findings

Step five: Review your assessment and update as necessary

Figure 9.5 Qualitative risk assessment. Risk assessment is a component of risk analysis and consists of hazard identification, release assessment, exposure assessment and consequence assessment. Findings are summarized as a risk estimation. In this situation, the hazard is the tick itself, the susceptible host population includes vertebrate hosts including humans and wildlife. Qualitative risk assessment is used where specific quantitative data is lacking and relies on a review of published information and expert opinion. Routes of entry include terrestrial and air entry points from the USA into Canada. For a full description of the qualitative risk assessment methodology see Rinchen et al. (2020) for details. https://www.frontiersin.org/articles/10.3389/fvets.2020.00366/full

terrestrial routes are extensive given the ease of moving across the long porous border between these two large countries. Currently the tick has not been found in Canada although there are reports of the tick moving further north in New York state and getting closer to the Great Lakes region. Terrestrial wildlife, e.g., deer, medium-sized mammals and birds, move across the border on a regular basis and migratory birds can travel longer distances from areas where the tick is well established in the northeastern USA into Canada. We know that birds are often infested with ticks such as the blacklegged tick (*Ixodes scapularis*) and have been linked to the spread of *Ixodes* spp. from the USA north and northwest into Canada. Although this is not currently a high-risk pathway into Canada, this could change over the next decade and further examination of the potential role of migratory birds is warranted.

As mentioned earlier, the ALHT is a highly generalist tick species, feeding on whatever host is most abundant in the environment. In the United States, these ticks are currently most commonly submitted from white tailed deer (*Odocoileus virginianus* [USDA, 2022]) as well as the environment (White et al., 2021). Larvae and nymphs often parasitize birds and small mammals whereas adults often parasitize larger animals, such as deer and cattle. Enhanced passive surveillance and collaborative efforts (Trout Fryxell et al., 2021) have facilitated the detection and reporting of ALHT from a wide range of hosts including:

- livestock (e.g., cattle, goat, sheep, pigs, chickens)
- wildlife (e.g., white tailed deer, elk, bears, coyote, gray fox, red fox, eastern cottontail rabbit, racoon, stripped skunk, small rodents and a variety of wild birds including raptors,

Dog imports

In recent estimates of spatial and temporal trends of dog importation into Canada (2013–2019), data on commercial dog imports from the Canadian Border Services Agency (CBSA) was used. This includes animals imported from breeders and from welfare charities. Estimates of personal pet imports via air and land were added to these figures. The total estimate for commercial imports was 13,281 import events involving 1 to 200 dogs in each consignment. These consignments primarily came from the USA, Ukraine, Mexico, and Slovakia. Estimates of personal dog imports are rough but were thought to be between 41,236 and 80,287 with the total number of dogs imported likely to be between 63,085 and 122,827 (2013–2019). Of these, 67% entered Canada by air and 33% over land. It is thought that dog imports into Canada have increased by over 400% since 2013 with 45% coming from the USA. In 2019 alone, there were estimated over 35,000 dogs imported (Public Health Agency of Canada, personal communication 2022).

waterfowl and some migratory passerine species such as the grey cat bird [*Dumetella* sp.] [Pandey et al., 2022])
- humans
- companion animals (e.g., dog, cat, horse).

Citizen science tick submission and testing campaigns can be used to supplement government and other published data sources and have the potential to provide timely estimates of tick, and pathogen, distributions to help characterize and understand tick-borne disease threats to communities (Porter et al., 2021).

9.5 What does all this mean from an EcoHealth perspective?

The ALHT has a wide host range and has the potential to carry and transmit several pathogenic micro-organisms, of animal and public health importance. Likely routes of entry into Canada include cross border movement of infested terrestrial wildlife and livestock, dog importations and the movement of migratory birds. The establishment and potential expansion of ALHT across North America continues to cause concern for animal and human health, particularly in the United States and Canada. From the published literature, we found that this tick species is globally associated with more than 50 different pathogens, of which 30 are potentially pathogenic to humans (Zhao et al., 2020). To date, there have been no reports of any human pathogen transmitted by ALHT in North America (Wormser et al., 2020) but this could change, and laboratory studies have shown that it could transmit a range of pathogens including Heartland virus and *Rickettsia rickettsii*, which causes Rocky Mountain spotted fever (Stanley et al., 2020). There is also molecular evidence of pathogen presence in submitted ticks, e.g., Bourbon virus (Cumbie et al., 2022; Raney et al., 2022). Given the potential for this tick to emerge and spread widely once introduced to a suitable habitat, early detection will be important. Taking a comprehensive interdisciplinary approach, engaging entomologists, epidemiologists and animal and human health practitioners, is important in order to develop and implement surveillance and response plans. The growing expansion of citizen science platforms such as e-tick in Canada, can add useful data to government tick monitoring projects and academic research initiatives.

9.6 Conclusion

The ALHT (*H. longicornis*) is currently distributed in over 217 counties in 19 states of the United States (USDA, 2023). Previously, specimens of the ALHT were misidentified as the native rabbit tick (*H. leporispalustris*), resulting in a delay in an appropriate response to the incursion of this exotic tick. This highlights the need to ensure that entomologists are actively engaged in surveillance programmes to ensure identification of new ticks. It is thought that the ALHT most likely came to North America from East Asia. Based on our habitat suitability models, the geographic distribution of the ALHT will likely continue to expand across North America. Owing to the ability of this tick to transmit pathogens, the potential threat of this tick to animal and human health cannot be ignored. Enhanced tick surveillance to determine the expanding geographical distribution of the ALHT in North America should be continued. There is also a need for animal and human health monitoring systems to work together, along with citizen science initiatives, to determine the potential role this tick might play in the transmission of diseases to animals and humans in North America. Effective monitoring and control methods for this tick in North America should be developed using a broad interdisciplinary approach.

References

Anon. (2019). Public health Ontario. *The Asian longhorned tick: Assessing public health implications for Ontario. On, Canada*. Ontario Agency for Health Protection and Promotion (Public Health Ontario): Toronto.

Cane, R. (2010). *Profile: Haemaphysalis longicornis (Neuman 1901). New Zealand biosecure*. https://www.smsl.co.nz/site/southernmonitoring/files/NZB/Ha%20longicornis%20Profile.pdf

Chen, Z., Yang, X., Bu, F., Yang, X., & Liu, J. (2012). Morphological, biological and molecular characteristics of bisexual and parthenogenetic *Haemaphysalis longicornis*. *Veterinary Parasitology*, *189*(2–4), 344–352. https://doi.org/10.1016/j.vetpar.2012.04.021

Choi, Y., Jiang, Z., Shin, W. J., & Jung, J. U. (2020). Severe fever with thrombocytopenia syndrome virus NSs interacts with TRIM21 to activate the p62-Keap1-Nrf2 pathway. *Journal of Virology*, *94*(6). https://doi.org/10.1128/JVI.01684-19

Chu, C. Y., Jiang, B. G., Liu, W., Zhao, Q. M., Wu, X. M., Zhang, P. H., Zhan, L., Yang, H., & Cao, W. C. (2008). Presence of pathogenic Borrelia burgdorferi sensu lato in ticks and rodents in Zhejiang, south-east China. *Journal of Medical Microbiology*, *57*(8), 980–985. https://doi.org/10.1099/jmm.0.47663-0

Cumbie, A. N., Trimble, R. N., & Eastwood, G. (2022). Pathogen spillover to an Invasive Tick Species: First Detection of Bourbon Virus in *Haemaphysalis longicornis* in the United States. *Pathogens*, *11*(4), 454. https://www.mdpi.com/2076-0817/11/4/454. https://doi.org/10.3390/pathogens11040454

Dinkel, K. D., Herndon, D. R., Noh, S. M., Lahmers, K. K., Todd, S. M., Ueti, M. W., Scoles, G. A., Mason, K. L., & Fry, L. M. (2021). A U.S. isolate of Theileria orientalis, Ikeda genotype, is transmitted to cattle by the invasive Asian longhorned tick, *Haemaphysalis longicornis*. *Parasites and Vectors*, *14*(1), 157. https://doi.org/10.1186/s13071-021-04659-9

Egizi, A. M., Robbins, R. G., Beati, L., Nava, S., Vans, C. R., Occi, J. L., & Fonseca, D. M. (2019). A pictorial key to differentiate the recently detected exotic *Haemaphysalis longicornis* Neumann, 1901 (Acari, Ixodidae) from native congeners in North America. *ZooKeys*, (818), 117–128. https://doi.org/10.3897/zookeys.818.30448

Egizi, A., Bulaga-Seraphin, L., Alt, E., Bajwa, W. I., Bernick, J., Bickerton, M., Campbell, S. R., Connally, N., Doi, K., Falco, R. C., Gaines, D. N., Greay, T. L., Harper, V. L., Heath, A. C. G., Jiang, J., Klein, T. A., Maestas, L., Mather, T. N., Occi, J. L., . . . Fonseca, D. M. (2020). First glimpse into the origin and spread of the Asian longhorned tick, *Haemaphysalis longicornis*, in the United States. *Zoonoses and Public Health*, *67*(6), 637–650. https://doi.org/10.1111/zph.12743

Eisen, L., & Eisen, R. J. (2021). Benefits and drawbacks of citizen science to complement traditional data gathering approaches for medically important hard ticks (acari: Ixodidae) in the United States. *Journal of Medical Entomology*, *58*(1), 1–9. https://doi.org/10.1093/jme/tjaa165

Elith, J., Phillips, S. J., Hastie, T., Dudík, M., Chee, Y. E., & Yates, C. J. (2011). A statistical explanation of MaxEnt for ecologists. *Diversity and Distributions*, *17*(1), 43–57. https://doi.org/10.1111/j.1472-4642.2010.00725.x

Fick, S. E., & Hijmans, R. J. (2017). WorldClim 2: new 1-km spatial resolution climate surfaces for global land areas. *International Journal of Climatology*, *37*(12), 4302–4315. https://doi.org/10.1002/joc.5086

Hammer, J. F., Emery, D., Bogema, D. R., & Jenkins, C. (2015). Detection of *Theileria orientalis* genotypes in *Haemaphysalis longicornis* ticks from southern Australia. *Parasites and Vectors*, *8*, 229. https://doi.org/10.1186/s13071-015-0839-9

Heath, A. C. G. (2016). Biology, ecology and distribution of the tick, *Haemaphysalis longicornis* Neumann (Acari: Ixodidae) in New Zealand. *New Zealand Veterinary Journal*, *64*(1), 10–20. https://doi.org/10.1080/00480169.2015.1035769

Hijmans, R. J., Cameron, S. E., Parra, J. L., Jones, P. G., & Jarvis, A. (2005). Very high resolution interpolated climate surfaces for global land areas. *International Journal of Climatology*, *25*(15), 1965–1978. https://doi.org/10.1002/joc.1276

Hoogstraal, H., Roberts, F. H. S., Kohls, G. M., & Tipton, V. J. (1968). Review of Haemaphysalis (kaiseriana) Longicornis Neumann (resurrected) of Australia, New Zealand, New Caledonia, Fiji, Japan, Korea, and Northeastern China and USSR, and its parthenogenetic and bisexual populations (Ixodoidea, Ixodidae). *Journal of Parasitology*, *54*(6), 1197–1213. https://doi.org/10.2307/3276992

Hutcheson, H. J., Dergousoff, S. J., & Lindsay, L. R. (2019). *Haemaphysalis longicornis*: A tick of considerable veterinary importance, now established in North America. *Canadian Veterinary Journal*, *60*(1), 27–28.

Kim, K. H., Yi, J., Kim, G., Choi, S. J., Jun, K. I., Kim, N. H., Choe, P. G., Kim, N. J., Lee, J. K., & Oh, M. D. (2013). Severe fever with thrombocytopenia syndrome, South Korea, 2012. *Emerging Infectious Diseases*, *19*(11), 1892–1894. https://doi.org/10.3201/eid1911.130792

Lawrence, K. E., Summers, S. R., Heath, A. C. G., McFadden, A. M. J., Pulford, D. J., Tait, A. B., & Pomroy, W. E. (2017). Using a rule-based envelope model to predict the expansion of habitat suitability within New Zealand for the tick *Haemaphysalis longicornis*, with future projections based on two climate change scenarios. *Veterinary Parasitology*, *243*, 226–234. https://doi.org/10.1016/j.vetpar.2017.07.001

Lawrence, K. E., Lawrence, B. L., Hickson, R. E., Hewitt, C. A., Gedye, K. R., Fermin, L. M., & Pomroy, W. E. (2019). Associations between *Theileria orientalis* Ikeda type infection and the growth rates and haematocrit of suckled beef calves in the North Island of New Zealand. *New Zealand Veterinary Journal*, *67*(2), 66–73. https://doi.org/10.1080/00480169.2018.1547227

Luo, L., Sun, J., Yan, J., Wang, C., Zhang, Z., Zhao, L., Han, H., Tong, Z., Liu, M., Wu, Y., Wen, H., Zhang, R., Xue, Z., Sun, X., Li, K., Ma, D., Liu, J., Huang, Y., Ye, L., . . . Yu, X. J. (2016). Detection of a novel Ehrlichia species in *Haemaphysalis longicornis* tick from China. *Vector Borne and Zoonotic Diseases*, *16*(6), 363–367. https://doi.org/10.1089/vbz.2015.1898

Mahara, F. (1997). Japanese spotted fever: Report of 31 cases and review of the literature. *Emerging Infectious Diseases*, *3*(2), 105–111. https://doi.org/10.3201/eid0302.970203

Morshed, M. G., Scott, J. D., Fernando, K., Beati, L., Mazerolle, D. F., Geddes, G., & Durden, L. A. (2005). Migratory songbirds disperse ticks across Canada, and first isolation of the Lyme disease spirochete, *Borrelia burgdorferi*, from the avian tick, Ixodes auritulus. Journal of Parasitology, 91(4), 780–790. https://doi.org/10.1645/GE-3437.1

Namgyal, J., Couloigner, I., Lysyk, T. J., Dergousoff, S. J., & Cork, S. C. (2020). Comparison of Habitat Suitability Models for *Haemaphysalis longicornis* Neumann in North America to Determine Its Potential Geographic Range. *International Journal of Environmental Research and Public Health*, *17*(21), 8285. https://doi.org/10.3390/ijerph17218285

Oakes, V. J., Yabsley, M. J., Schwartz, D., LeRoith, T., Bissett, C., Broaddus, C., Schlater, J. L., Todd, S. M., Boes, K. M., Brookhart, M., & Lahmers, K. K. (2019). *Theileria orientalis* Ikeda genotype in cattle, Virginia, USA. *Emerging Infectious Diseases*, *25*(9), 1653–1659. https://doi.org/10.3201/eid2509.190088

OIE. (2004). Introduction to import risk analysis. In *Handbook on import risk analysis for animals and animal products*, 1. OIE (World Organization for Animal Health).

Pandey, M., Piedmonte, N. P., Vinci, V. C., Falco, R. C., Daniels, T. J., & Clark, J. A. (2022). First detection of the invasive Asian longhorned tick (acari: Ixodidae) on migratory passerines in the Americas.

Journal of Medical Entomology, 59(6, November), 2176–2181. https://doi.org/10.1093/jme/tjac144

Phillips, S. J., Anderson, R. P., & Schapire, R. E. (2006). Maximum entropy modeling of species geographic distributions. *Ecological Modelling, 190*(3–4), 231–259. https://doi.org/10.1016/j.ecolmodel.2005.03.026

Porter, W. T., Wachara, J., Barrand, Z. A., Nieto, N. C., & Salkeld, D. J. (2021). Citizen science provides an efficient method for broad-scale tick-borne pathogen surveillance of *Ixodes pacificus* and Ixodes scapularis across the United States. *mSphere, 6*(5), e0068221. https://doi.org/10.1128/mSphere.00682-21

Price, K. J., Graham, C. B., Witmier, B. J., Chapman, H. A., Coder, B. L., Boyer, C. N., Foster, E., Maes, S. E., Bai, Y., Eisen, R. J., & Kyle, A. D. (2021). *Borrelia burgdorferi* sensu stricto DNA in field-collected *Haemaphysalis longicornis*Ticks, Pennsylvania, United States. *Emerging Infectious Diseases, 27*(2), 608–611. https://doi.org/10.3201/eid2702.201552

Price, K. J., Ayres, B. N., Maes, S. E., Witmier, B. J., Chapman, H. A., Coder, B. L., Boyer, C. N., Eisen, R. J., & Nicholson, W. L. (2022). First detection of human pathogenic variant of *Anaplasma phagocytophilum* in field-collected *Haemaphysalis longicornis*, Pennsylvania, USA. *Zoonoses and Public Health, 69*(2), 143–148. https://doi.org/10.1111/zph.12901

Public Health Ontario. (2019). *The Asian longhorned tick: Assessing public health implications for Ontario*. Ontario Agency for Health Protection and Promotion (Public Health Ontario). https://www.publichealthontario.ca/-/media/documents/F/2019/focus-on-asian-longhorned-tick.pdf?la=en

Rainey, T., Occi, J. L., Robbins, R. G., & Egizi, A. (2018). Discovery of Haemaphysalis longicornis (Ixodida: Ixodidae) parasitizing a sheep in New Jersey, United States. *Journal of Medical Entomology, 55*(3), 757–759. https://doi.org/10.1093/jme/tjy006

Raney, W. R., Perry, J. B., & Hermance, M. E. (2022). Transovarial transmission of heartland virus by invasive Asian longhorned ticks under laboratory conditions. *Emerging Infectious Diseases.* http://www.cdc.gov/eid, *28*(3), *726–729. https://doi.org/10.3201/eid2803.210973*

Rinchen, S., Tenzin, T., Hall, D. C., & Cork, S. C. (2020). A qualitative risk assessment of rabies reintroduction into the rabies low-risk zone of Bhutan. *Frontiers in Veterinary Science, 7,* 366. https://doi.org/10.3389/fvets.2020.00366

Rochlin, I. (2019). Modeling the Asian longhorned tick (Acari: Ixodidae) suitable habitat in North America. *Journal of Medical Entomology, 56*(2), 384–391. https://doi.org/10.1093/jme/tjy210

Stanley, H. M., Ford, S. L., Snellgrove, A. N., Hartzer, K., Smith, E. B., Krapiunaya, I., & Levin, M. L. (2020). The ability of the invasive Asian longhorned tick *Haemaphysalis longicornis* (acari: Ixodidae) to acquire and transmit *Rickettsia rickettsii* (Rickettsiales: Rickettsiaceae), the agent of Rocky Mountain spotted fever, under laboratory conditions. *Journal of Medical Entomology, 57*(5), 1635–1639. https://doi.org/10.1093/jme/tjaa076

Takahashi, T., Maeda, K., Suzuki, T., Ishido, A., Shigeoka, T., Tominaga, T., Kamei, T., Honda, M., Ninomiya, D., Sakai, T., Senba, T., Kaneyuki, S., Sakaguchi, S., Satoh, A., Hosokawa, T., Kawabe, Y., Kurihara, S., Izumikawa, K., Kohno, S., . . . Saijo, M. (2014). The first identification and retrospective study of severe fever with thrombocytopenia syndrome in Japan. *Journal of Infectious Diseases, 209*(6), 816–827. https://doi.org/10.1093/infdis/jit603

Thompson, A. T., Dominguez, K., Cleveland, C. A., Dergousoff, S. J., Doi, K., Falco, R. C., Greay, T., Irwin, P., Lindsay, L. R., Liu, J., Mather, T. N., Oskam, C. L., Rodriguez-Vivas, R. I., Ruder, M. G., Shaw, D., Vigil, S. L., White, S., & Yabsley, M. J. (2020). Molecular characterization of Haemaphysalis species and a molecular genetic key for the identification of Haemaphysalis of North America. *Frontiers in Veterinary Science, 7,* 141. https://doi.org/10.3389/fvets.2020.00141

Title, P. O., & Bemmels, J. B. (2018). ENVIREM: An expanded set of bioclimatic and topographic variables increases flexibility and improves performance of ecological niche modeling. *Ecography, 41*(2), 291–307. https://doi.org/10.1111/ecog.02880

Trout Fryxell, R. T., Vann, D. N., Butler, R. A., Paulsen, D. J., Chandler, J. G., Willis, M. P., Wyrosdick, H. M., Schaefer, J. J., Gerhold, R. W., Grove, D. M., Ivey, J. Z., Thompson, K. W., Applegate, R. D., Sweaney, J., Daniels, S., Beaty, S., Balthaser, D., Freye, J. D., Mertins, J. W., . . . Lahmers, K. (2021). Rapid Discovery and Detection of *Haemaphysalis longicornis* through the Use of Passive Surveillance and Collaboration: Building a State Tick-Surveillance Network. *International Journal of Environmental Research and Public Health, 18*(15), 7980. https://doi.org/10.3390/ijerph18157980

United States Department of Agriculture. (2019). https://www.aphis.usda.gov/animal_health/animal_diseases/tick/downloads/h-longicornis-response-plan_usda.pdf

United States Department of Agriculture. (2022). *ALHT situation report.* https://www.aphis.usda.gov/animal_health/animal_diseases/tick/downloads/longhorned-tick-sitrep.pdf

White, S. A., Bevins, S. N., Ruder, M. G., Shaw, D., Vigil, S. L., Randall, A., Deliberto, T. J., Dominguez, K., Thompson, A. T., Mertins, J. W., Alfred, J. T., & Yabsley, M. J. (2021). Surveys for ticks on wildlife hosts and in the environment at Asian longhorned tick (*Haemaphysalis longicornis*)-positive sites in Virginia and New Jersey, 2018. *Transboundary and Emerging Diseases, 68*(2), 605–614. https://doi.org/10.1111/tbed.13722

Wormser, G. P., McKenna, D., Piedmonte, N., Vinci, V., Egizi, A. M., Backenson, B., & Falco, R. C. (2020). First recognized human bite in the United States by the Asian longhorned tick, *Haemaphysalis longicornis. Clinical Infectious Diseases, 70*(2), 314–316. https://doi.org/10.1093/cid/ciz449

Yu, X. J., Liang, M. F., Zhang, S. Y., Liu, Y., Li, J. D., Sun, Y. L., Zhang, L., Zhang, Q. F., Popov, V. L., Li, C., Qu, J., Li, Q., Zhang, Y. P., Hai, R., Wu, W., Wang, Q., Zhan, F. X., Wang, X. J., Kan, B., . . . Li, D. X. (2011). Fever with thrombocytopenia associated with a novel bunyavirus in China. *New England Journal of Medicine, 364*(16), 1523–1532. https://doi.org/10.1056/NEJMoa1010095

Zhang, G., Zheng, D., Tian, Y. T., & Li, S. (2019). A dataset of distribution and diversity of ticks in China. *Scientific Data, 6*(1), 105. https://doi.org/10.1038/s41597-019-0115-5

Zhang, X., Zhao, C., Cheng, C., Zhang, G., Yu, T., Lawrence, K., Li, H., Sun, J., Yang, Z., Ye, L., Chu, H., Wang, Y., Han, X., Jia, Y., Fan, S., Kanuka, H., Tanaka, T., Jenkins, C., Gedye, K., . . . Zheng, A. (2022). Rapid spread of severe fever with thrombocytopenia syndrome virus by parthenogenetic Asian longhorned ticks. *Emerging Infectious Diseases, 28*(2), 363–372. https://doi.org/10.3201/eid2802.211532

Zheng, H., Yu, Z., Zhou, L., Yang, X., & Liu, J. (2012). Seasonal abundance and activity of the hard tick *Haemaphysalis longicornis* (Acari: Ixodidae) in North China. *Experimental and Applied Acarology, 56*(2), 133–141. https://doi.org/10.1007/s10493-011-9505-x

Zhao, L., Li, J., Cui, X., Jia, N., Wei, J., Xia, L., Wang, H., Zhou, Y., Wang, Q., Liu, X., Yin, C., Pan, Y., Wen, H., Wang, Q., Xue, F., Sun, Y., Jiang, J., Li, S., & Cao, W. (2020). Distribution of *Haemaphysalis longicornis* and associated pathogens: Analysis of pooled data from a China field survey and global published data. *Lancet. Planetary Health, 4*(8), e320–e329. https://doi.org/10.1016/S2542-5196(20)30145-5

Chapter 10

Avian influenza and migratory wild birds: An EcoHealth perspective

Susan C. Cork and Wlodek L. Stanislawek

Abstract

In this chapter we examine the complex ecology and biology of avian influenza viruses and illustrate the importance of interagency cooperation when developing disease monitoring programs. These are required for wild birds as well as for susceptible domestic species. Conservation efforts need to identify environmental factors that might impact the heath of wild birds as well as risk factors which may predispose birds, especially vulnerable migratory species, to be exposed to new strains of avian influenza. The latter requires the engagement of multidisciplinary teams, including ornithologists, virologists, epidemiologists, veterinarians and environmental experts. A core component of avian health monitoring programs is the inclusion of the public through education and the establishment of citizen science programs. In this chapter we consider the potential role of wild birds in the spread of new and emerging strains of avian influenza and also consider the risks to wild birds from domestic poultry. We also examine global conservation efforts to monitor and protect long-distance migrants. In our case study we discuss the avian influenza status of Australia and New Zealand and describe a project that was developed to monitor the health of migratory shore birds arriving from the Arctic through the East Asia-Australasia flyway.

10.1 Introduction

The global spread of a number of new strains of avian influenza virus (AIV) requires organizations working in the human and animal health sectors to carefully examine disease transmission risks between domestic species, humans and wildlife. A key motivator for developing enhanced collaborations between government agencies, and the public, is the fact that approximately 75% of emerging infectious diseases in the human population have originated from domestic and wild animals (Jones et al., 2008; WHO, 2008). Diseases such as avian influenza can also have conservation implications as they can cause significant losses of rare and endangered avian wildlife. Addressing the latter requires careful consideration of species habitat

Susan C. Cork and Douglas P. Whiteside (eds) **Case Studies in EcoHealth**
DOI: 10.52517/9781789183368.010, © 5m Books Ltd, 2024

requirements, protection of conservation areas, monitoring migratory routes and reducing the exposure of wild birds to domestic poultry. At the same time, with climate change and anthropogenic impacts on the environment, many migratory birds are also under significant threat as a result of habitat loss and limited access to historical feeding sites.

A core component of effectively monitoring wildlife health is the inclusion of the public through education campaigns and the establishment of citizen science programs (Schubert et al., 2019; Weisshaupt et al., 2021). The latter includes public participation in ornithology groups, engagement of the public in research projects, and encouraging the submission of bird sighting data to online platforms such as ebird (https://ebird.org/home) and iNaturalist (https://www.inaturalist.org/).

Given the complex ecology of AIVs, developing effective plans for early detection, through disease monitoring, requires an interdisciplinary approach engaging a team of ornithologists, virologists, epidemiologists, veterinarians, environmental experts and others. The EcoHealth approach engages such a team of experts, working with policy makers and the public, to deal with the varied challenges required to identify, prevent and control avian influenza in wild birds and considers factors beyond the zoonotic risk to the human population.

10.2 Avian influenza biology and disease ecology

AIVs are members of the Orthomyxoviridae family which are enveloped viruses that contain a segmented single-stranded RNA genome. There are three genera of influenza virus (A, B and C) but only viruses of the influenza A genus are thought to infect birds. Influenza A viruses are typically classified into different subtypes based on the antigenic properties (i.e. those that stimulate an immune response), and genetic sequences, of their surface haemagglutinin (HA) and neuraminidase (NA) glycoproteins (see Figure 10.1). HA subtypes H1–16 and NA subtypes N1–9 are found in a wide range of avian species, whereas subtypes H17–18 and N10–11 are found in bats (To et al., 2014). These two external glycoproteins (HA & NA) are particularly important for the classification of AIV. The HA glycoprotein is responsible for virus attachment to the host target cell, and the NA glycoprotein is needed for virion maturation and release (Hampson, 2002).

AIVs are naturally found in wild waterfowl, such as ducks and geese, and most of these are low pathogenic strains that typically cause no clinical signs in these birds. However, small changes in the genetic structure of AIVs can result in new strains which may have increased virulence with an enhanced risk of transmission from their natural avian reservoirs to other species. Although humans, and other mammalian species exposed to AIVs can become infected, in most cases this does not cause clinical disease unless the viral strains have become adapted to mammalian species or the viral load is high enough to overcome natural levels of immunity. Poultry can become infected with AIVs through contact with infected wild birds as well as through contaminated environmental sources such as water, litter and feed. Low pathogenicity avian influenza (LPAI) strains of H5Nx and H7Nx AIVs can spread through domestic populations of poultry fairly rapidly and these strains may mutate to become highly pathogenic avian influenza (HPAI) strains resulting in high mortality in poultry and potential spill back to wildlife.

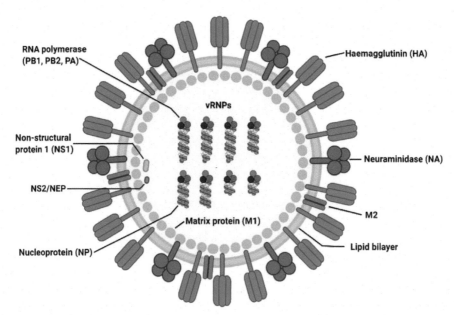

Figure 10.1 The structure of influenza A virus. AIV is a negative-stranded RNA virus belonging to the Orthomyxoviridae family. The AIV genome is divided into eight segments that encode 11 viral proteins in total (HA, NA, M1, M2, NP, NS1, NS2, PA, PB1, PB2, and PB1-F2). The viral envelope of AIV contains the transmembrane proteins HA, NA, and M2. NA = neuraminidase, HA = haemagglutinin, M = matrix protein, NP = nucleoprotein, NS = non-structural protein, LP = lipid bilayer, RNA polymerase (PB1,PB2, PA) vRNPs = viral ribonucleoproteins. Adapted from: Hi Eun Jung and Heung Kyu Lee – https://www.mdpi.com/1999–4915/12/5/504/htm.

Classification of avian influenza viruses

Influenza A viruses are typically classified into different subtypes based on the antigenic prop-erties, and genetic sequences, of their surface haemagglutinin (HA) and neuraminidase (NA) glycoproteins. There are 16 HA subtypes, H1–16 and 9 NA subtypes N1–9, which are found in a wide range of avian species. The HA glycoprotein is responsible for virus attachment to the target cell, and the NA glycoprotein is needed for virion maturation and release. Each influenza strain is known by the HA and NA combination, e.g. H5N8 and H5N1, but further molecular characterization is required to determine virulence. Most of the virulent strains of AIV are H5Nx or H7Nx and sometimes H9Nx subtypes (OIE, 2022; WOAH, 2022). See Figure 10.1.

Wild birds have been found to be infected with most known subtypes of influenza A viruses (Alexander, 2000; Kawaoka et al., 1988; Olsen et al., 2006). Waterbirds, including Anseriformes (includes ducks, geese and swans) and Charadriiformes (includes gulls, terns, knots, plovers), are regarded as the primary natural reservoirs of influenza A viruses (Hurt et al., 2006; Krauss and Webster, 2010; Olsen et al., 2006; Webster et al., 1992). Passerine birds (e.g. perching birds, such as corvids and finches) can serve as local reservoirs for AIVs but the limited number of confirmed outbreaks reported in these species indicates that they are likely to represent spillover infections following exposure to infected poultry or waterbirds (Vandergrift et al., 2010).

10.3 Why is avian influenza a potential public health issue?

AIVs have the potential to lead to pandemic influenza in humans and this was considered serious in 1997 when new HPAI strains of H5N1 virus spread rapidly in poultry, including waterfowl, across Asia. Historically, the AIV subtypes involved in human infection, in addition to H5N1, were H6N1, H7N2, H7N3, H7N7, H7N9, H9N2, H10N7 and H10N8 (To et al., 2014) and more recently H10N3, H5N6 and H5N8 (see also current situation reports on the CDC website at https://www.cdc.gov/flu/avianflu/reported-human-infections.htm). Pandemic emergence remains a concern, especially in Asia, and a new AIV strain, with the potential to infect humans could arise at any time. Pigs can also become infected with AIVs and have the potential to act as a 'mixing vessel' in which AIVs may become more rapidly adapted to become pathogenic to humans (Ma et al., 2008). This is because humans and pigs share the same type of surface receptors in the respiratory tract. Of interest, cases of seasonal influenza in humans (i.e. human adapted strains of H3N2) fell to low levels in the last 2 years, probably due to COVID-19 control programs (i.e. wearing masks, social distancing, hand washing, etc.), however there has been an increase in the number of cases of avian influenza, especially HPAI H5N1 strains, in poultry and wild birds (OIE/WOAH, 2022).

10.4 The evolution of new strains of avian influenza

There are at least two key mechanisms that contribute to the development of highly virulent strains of influenza A: antigenic drift and antigenic shift. Antigenic drift is a gradual process in which there is an accumulation of mutations (or insertions) primarily of HA and NA surface antigens. Mutations in other genes (PB2, PB1, PA and NS) can also play a role in virus pathogenicity (Hampson, 2002; Krauss and Webster, 2010; Neverov et al., 2014). The segmented genome of influenza virus can also exchange genetic segments if different strains of influenza co-infect the same host. This can lead to a new gene constellation and subsequently, a new virus. The latter process is called reassortment or antigenic shift. Most major influenza epidemics of influenza A in humans have been the result of virus mutations caused by reassortment events. This was the case for the 1957 (H2N2) and the 1968 (H3N2) pandemics, as well as for the swine flu in 2009 (H1N1) (Lindstrom et al., 2004, Bastien et al., 2010).

New strains of HPAI H5N1 are currently circulating across several continents and it has been proposed that wild birds have played a role in the spread of these strains within and between countries especially in Europe and North America (OIE/WOAH, 2022). In several countries these new strains have resulted in high mortality in wild birds including snow geese (*Anser caerulescens*) in southern Quebec, Canada (CWHC, 2022), pelicans (*Pelicanus thagus*) in Peru (Anon., 2022) and puffins (*Fratacula arctica*) in the United Kingdom (Natural History Museum, 2022). The high level of mortality in these wild species is unusual.

10.5 Bird migration and conservation initiatives

Bird migration is the seasonal movement of migratory species, many with well-established routes, connecting summer breeding grounds with well-established wintering grounds. Many species migrate north in the spring and return south before the cold winter months. This ensures access to sufficient food supplies available in the milder southern climates. Some species travel very long distances non-stop (e.g. bartailed godwit, *Limosa lapponica)* whereas others have routine stopping off points (e.g. red knot, *Calidris canutus*). Recently, however, these stopping off points have been significantly impacted by human activity, raising concerns about the survival of some long-distance migrants. A key component of monitoring migratory bird movements is the inclusion of public sightings of wild birds through membership of ornithology groups and by encouraging submission of bird sightings to data collection platforms such as iNaturalist (https://www.inaturalist.org/) and ebird (https://ebird.org/home). These citizen science platforms have been valuable in complementing official bird banding (see Figure 10.4b) and conservation programs for shorebirds and for a range of other species (Schubert et al., 2019., Weisshaupt et al., 2021).

10.6 Bird flyways

In this section we will focus on waders and waterfowl, which are thought to be the natural reservoirs for AIVs. The route regularly taken by migratory waders and waterfowl moving between breeding and non-breeding grounds each year is typically known as a flyway (Kirby et al., 2008). The effective monitoring and conservation of migratory birds depends on coordinated efforts engaging countries along the entire length of its flyway. As previously mentioned,

The United Nations Environment Programme (UNEP) estimated that over the past 30 years the number of migratory birds has almost halved (https://www.undp.org/kazakhstan/stories/protection-migratory-birds-and-their-habitats-people-and-planet). The factors contributing to this situation include the destruction of ecosystems and habitat in places used for stops, collisions with high buildings in cities, artificial lighting (which can disorientate migratory birds) and power lines can cause impact trauma and significant mortality. Climate change has also contributed to declines in the number of migratory birds as a result of direct impacts on the biological cycles of birds, and ecosystem changes that impact their breeding, the survival rate of their chicks and destruction of resting places used during migration for adults (https://www.arctic-council.org/news/the-arctic-migratory-birds-initiative/).

the typical migratory routes follow a north–south axis with birds moving to milder climes at lower latitudes for the non-breeding season (Newton, 2007). The significant energetic costs associated with long-distance migration means that birds often use the shortest possible route but other factors such as weather patterns, food resources and geographical features will also influence the route taken. A large number of species may share the same flyways, especially those with similar biological needs. The International Wader Study Group previously described eight multiple-species flyways used by wading birds (Boere and Stroud, 2006; BirdLife International, 2010). These have subsequently been developed into nine generalized global flyways for all fully migratory water bird species (see Figure 10.2). Although these global flyways are an oversimplification, they do provide a useful overview of the complex journeys undertaken by the world's 2274 migratory

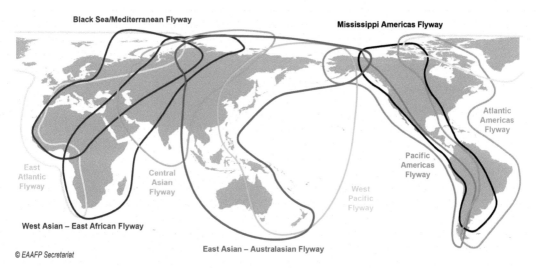

Black Sea/Mediterranean Flyway

Mississippi Americas Flyway

Atlantic
Americas
Flyway

Pacific
Americas
Flyway

East
Atlantic
Flyway

Central
Asian
Flyway

West
Pacific
Flyway

West Asian – East African Flyway

East Asian – Australasian Flyway

© EAAFP Secretariat

Figure 10.2 Global flyways. In the Americas there are three flyways (the **Pacific Americas Flyway**, **Mississippi Americas Flyway** (also known as the central Americas flyway) and **Atlantic Americas Flyway**) connecting the high Arctic to Tierra del Fuego – the southernmost tip of the South American mainland.

The three Palaearctic–African flyways (**East Atlantic Flyway**, **Black Sea/Mediterranean Flyway** and **West Asia–East Africa Flyway**) collectively constitute the world's largest bird migration system with over 2 billion passerines and near-passerines migrating from their breeding grounds in Europe and central and western Asia to winter in tropical Africa each year (Hahn et al., 2009). The **Central Asian Flyway** is the shortest flyway in the world. Lying entirely within the northern hemisphere, it connects a large area of the Palaearctic with the Indian subcontinent. Fewer species use this route due to the formidable barrier presented by the Tibetan plateau and Himalayas (Newton, 2007). The **East Asia–Australasia Flyway** extends from Arctic Russia and North America to New Zealand and is used by over 50 million migratory waterbirds. The **West Pacific Flyway** overlaps with parts of this and the Pacific Americas flyway.

The flyway descriptions are adapted from BirdLife International (2010) http://www.birdlife.org on 20/10/2021. Map credit: East Asian-Australasian Flyway Partnership Secretariat (EAAFP) http://www.eaaflyway.net/.

species (Kirby, 2010). Monitoring the migratory routes of birds and assessing their health is an important component of conservation efforts as well as being key to understanding how AIVs can move around the world.

10.7   East Asia–Australasia Flyway

Large flocks of migratory waders and shorebirds travel north to the Arctic-subarctic to breed in the spring and return south in the autumn

to their wintering grounds. Many species stop off in the area of the Yellow Sea to stock up on food. Different species select different ecosystems in order to access their preferred diet. The Yellow Sea is often the only stopover point for these birds. It is located between the Korean Penninsular and China and is host to plentiful food supplies in the mudflats. The East Asia–Australasia flyway connects 22 countries and has been a key travel route for migratory waders and shorebirds for thousands of years (Figure 10.3).

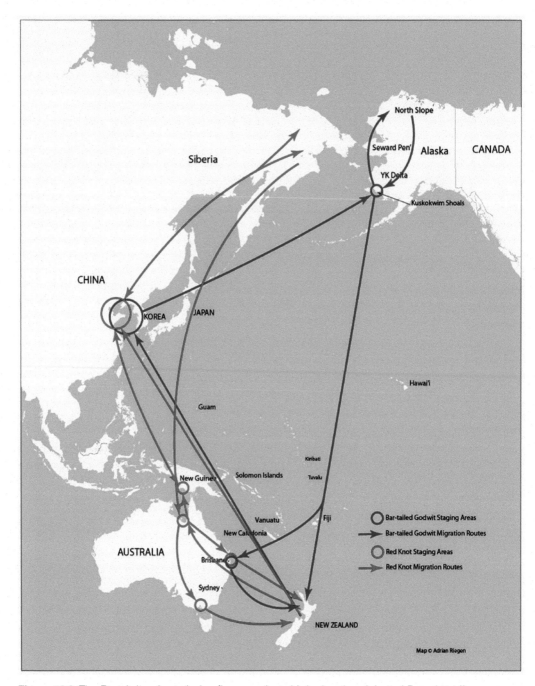

Figure 10.3 The East Asian–Australasian flyway – shore bird migration. Adapted From http://www. teara.govt.nz/en/map/7218/the-east-asian-australasian-flyway. Credit Adrian Riegen- Pūkorokoro Miranda Naturalists' Trust, New Zealand.

Figure 10.4 Bird capture and banding allows ornithologists to monitor the health of migratory birds and to study shore bird populations. A number of projects have followed the migratory paths of birds by applying small transmitting devices which record bird locations via satellite.

10.8 The changing landscape of the East Asia–Australasian flyway

The southern part of the Yellow Sea, including the west coast of Korean peninsula includes a 10 km wide belt of intertidal mudflats covering a total area of 2850 square km. These include highly productive sediments with a rich benthic fauna and are of great importance for migratory waders and shorebirds (Barter, 2002, 2005). It is thought that the area is the most important site for migratory birds on the northward migration in the East Asia–Australasian flyway with more than 35 species occurring in significant numbers. Surveys indicate that at least 2 million birds pass through when migrating north with half that number on the return flight south. The Saemangeum estuary previously had an estimated 300,000 migrating birds passing through until South Korea, as part of major land reclamation project, built a seawall

to separate it from the Yellow Sea (1991–2006), resulting in drying out of the land. Land reclamation also impacted 65% of the intertidal areas in China between the 1950s and 2002 and there are plans to reclaim further estuarine habitat (Murray et al., 2014). This is because the region is home to a large percentage of China's human population and economic development has been a priority. However, more recently there have been efforts to conserve these globally important mudflats (see Cornell Laboratory of Ornithology video, https://www.youtube.com/watch?v=N74zn7bCpq8).

10.9 Bird migration and avian influenza along the East Asian Flyway

East Asia remains an epicenter for the emergence of novel strains of AIV, however, information on the ecology and movements of waterfowl from this region remains limited (Sullivan et al., 2018). Although short distance transmission of virulent strains of AIV by wild birds has been recorded (Grillo et al., 2015; Takekawa et al., 2010) it is still not clear what the role of long-distance migratory birds might be in the spread of highly pathogenic strains.

Active surveillance for AIVs in migratory birds has been undertaken in China since the emergence of virulent H7N9 in early 2013 in China. Two H7N7 strains, from two different wild bird sampling sites were subsequently recovered from fresh faecal samples. Phylogenetic and sequence analyses indicated that several genes of the two H7N7 viruses were closely related to those in AIVs circulating in domestic poultry. Liu et al. (2018) concluded that this strongly suggested that genes from viruses infecting migratory birds had been introduced into poultry. However, the two strains that were found were low pathogenic strains. They concluded that low pathogenic, divergent H7N7 viruses circulate within birds migrating along the East Asian-Australasian flyway and speculated that virus dispersal between migratory birds and domestic poultry may increase the risk of the emergence of novel AIV strains.

To examine the movements of migratory birds on the East Asia–Australasian flyway, Sullivan et al. (2018) marked two species of wild waterfowl, northern pintail *(Anas acuta)* and Eurasian wigeon *(A. penelope)*, with satellite transmitters. They monitored them from their wintering grounds in Hong Kong, China to study the northward spring migration. Northern pintail were found to initiate migration 42 days earlier, travel 2150 km farther, and perform 4.4 more stopovers than Eurasian wigeon. They found that both species used similar stopover locations including areas along the Yangtze River near Shanghai, Bohai Bay and Korea Bay in rapidly developing regions of the Yellow Sea, and the Sea of Okhotsk. They noted that

both species exhibited strong habitat selection for rice paddies during migration stopovers. Due to the presence of domestic poultry in these habitats they concluded that this habitat preference has the potential to influence risks of AIV outbreaks especially as rapid land use and land cover changes are occurring throughout China reducing access to areas previously untouched by anthropogenic activities. Sullivan et al. (2018) noted that both migratory species had greatest association with locations of H5N1 outbreaks during the early stages of migration when they were at lower latitudes. Their work illustrates species-level differences in migration timing and behavior and demonstrates the need to consider temporal and spatial movement ecology when incorporating wild birds into AIV risk modeling and management.

Meng et al. (2019) proposed that the spread of highly pathogenic avian influenza (HPAI) H5N1 virus was associated with wild fowl migration in East Asian–Australasian (EA) and Central Asian (CA) flyways. However, they concluded that the spread of H5N1 virus between the two flyways remained unclear. They examined the movements of wild waterfowl by obtaining satellite tracking data covering seven bar-headed geese and three great black-headed gulls breeding in the Qinghai Lake area (along the EA flyway), and 20 whooper swans wintering in the Sanmenxia Reservoir area (at the CA flyway). They examined 2688 samples from wild birds found at Qinghai Lake after an outbreak of H5N1 in July 2015. Four genomes of H5N1 virus were obtained from bar-headed geese. The results of molecular analysis indicated that these H5N1 viruses belonged to clade 2.3.2.1c and their gene fragments were highly homologous with A/whooper swan/ Henan/SMX1/2015 (H5N1) virus (ranging from 99.76% to 100.00%) isolated from a dead whooper swan from the Sanmenxia Reservoir area along the EA flyway in January 2015 (Meng et al., 2019). They suggest that the coincidental timing of H5N1 outbreaks with spring migration, together with phylogenetic evidence, provided evidence of the east-to-west spread of HPAI H5N1 between the EA and CA migratory flyways of China. However, the source of the infection in the wild birds is still uncertain. It is likely to have been due to contact with infected poultry and the actual role of bar-headed geese in virus distribution remains unclear. This is especially true given the high level of mortality reported in bar-headed geese at the time.

Poyang Lake is situated within the East Asian Flyway, a migratory corridor that also encompasses Guangdong Province, China. This lake is the largest freshwater body in China and is a significant congregation site for wild waterfowl (Takekawa et al., 2010). Surrounding rice fields and grazing of domestic poultry have created a significant overlap with wild water birds increasing interspecies contact which might facilitate avian influenza transmission to and from domestic poultry. Takekawa et al. (2010) explain that reports of pathogenic H5N1 in healthy wild ducks at Poyang Lake raised concerns about the potential of free-ranging birds to disseminate highly pathogenic H5N1 AIVs. During 2007/8 they marked wild ducks at Poyang Lake with satellite transmitters to examine the location and timing of spring migration and to identify any spatio-temporal relationship with H5N1 outbreaks in poultry. Species included the Eurasian wigeon, northern pintail, common teal (*A. crecca*), falcated teal (*A. falcata*), Baikal teal (*A. formosa*), mallard (*A. platyrhynchos*), garganey (*A. querquedula*), and Chinese spotbill (*A. poecilohyncha*). These wild ducks followed the East Asian Flyway along the coast to breeding areas in northern China, eastern Mongolia, and eastern Russia. None migrated west toward Qinghai Lake (site of the largest wild bird H5N1 AIV outbreak), thus failing to demonstrate any migratory connection to the CA Flyway. They used Brownian bridge spatial analysis which indicated that H5N1 AIV outbreaks reported in the flyway were related to

latitude and poultry density but not to the core migration corridor or to wetland habitats. They also found a temporal mismatch between the timing of outbreaks in poultry and wild duck movements. In conclusion, although low pathogenic AIVs have been isolated from migratory birds along the East Asia–Australia Flyway few, if any, studies have demonstrated a clear causal link between these strains and outbreaks of avian influenza in poultry. However, viruses evolve and much remains unknown. Ongoing surveillance is warranted in order to better understand the risks to, and from, migratory species.

Until 1997 HPAI subtypes H5 and H7 were mainly associated with infection and disease outbreaks in poultry. HPAI was very rarely detected in wild birds (Becker, 1966). The situation has changed since the emergence of HPAI H5 viruses of the A/Goose/Guangdong/1/1996 (GsGd) lineage in southeast Asia (Alexander and Brown, 2009; Verhagen et al., 2021). The evolution of the HPAI H5 GsCd viruses continued developing new linages of the haemagglutinin gene. In 2014 a new clade (clade 2.3.4.4) appeared with new genetic and epidemiological characteristics with subtypes H5N1, H5N2, H5N3, H5N4, H5N5, H6N6, and H5N8. These HPAI H5 viruses caused several outbreaks and high mortality in poultry and wild birds in southeast Asia, Siberia, Middle East, Europe, Africa, and North America (Cui et al., 2020).

10.10 Case study: The history of avian influenza in Australia and New Zealand

Although Australia and New Zealand form part of the East Asia–Australasia Flyway, there are very few migratory waterfowl flying directly to Australia and New Zealand from Asia. The location of New Zealand, as an isolated island in the Pacific Ocean, provides a high degree of natural biosecurity from many diseases including HPAI AIVs although commercial and wild birds from Europe were introduced in the 1800s along with their low pathogenic AIVs (Austin and Hinshaw, 1984; Heather and Robertson, 1996). Australia, however, does have closer natural linkages to land masses such as Papua New Guinea with some movement of birds into northern areas of the country (Wille et al., 2022). To reduce the risk of introducing new AIVs, Australia and New Zealand maintain very high levels of biosecurity with regard to the importation of birds and bird products. New AIVs could be introduced in to New Zealand or Australia by a number of routes, these are (1) migratory birds and vagrants (2) importation of contaminated products or equipment (3) people travelling and (4) smuggling illegal live birds or avian products (Watts et al., 2016). In this case study we will examine the situation with regard to avian influenza in Australia and New Zealand in more detail.

10.10.1   Avian influenza in Australia

Despite the relative geographical and ecological isolation of Australia there is evidence of migration of some bird species, mainly waders, arriving in significant numbers. For example, it is estimated that 3 million Charadriiformes (shorebirds, gulls, alcids) make annual trans-hemispheric migrations via southeast Asian countries to spend their non-breeding season in Australia (Grillo et al., 2015). In contrast, however, the majority of waterfowl, i.e. Anseriformes (ducks, geese, swans), in Australia are non-migratory, unlike in the northern hemisphere, although some species are nomadic within the Australian–Papuan region with movements largely determined by rainfall and other resources (Grillo et al., 2015).

Unlike the high prevalence of AIVs in northern hemisphere waterfowl, the prevalence in Australian waterfowl has consistently been less than 2% with no strong seasonal patterns observed (Grillo et al., 2015; Wille et al., 2022). Through a number of studies using samples collected over 15 years from Australian wild birds, 14 different HA subtypes and all 9 different NA subtypes identified in Australia were detected, this comprises 58 HA-NA combinations. The only HA subtypes which were not detected were subtypes H14 and H15, and only a single case each of H13 and H16. The most common subtypes were H1N1, H3N8, H4N6, H5N3, H6N2, H9N2 and H11N9 (Grillo et al., 2015; Willie et al., 2022).

Over a 5-year surveillance period (2011 to 2015), ruddy turnstones (*Arenaria interpres*) that 'overwinter' during the Austral summer in south-eastern Australia showed generally low levels of AIV prevalence (0–2%) with the exception in one year that AIV was detected in 32%, in island population 90 km from the Australian mainland. This epizootic comprised three distinct AIV genotypes (H3N5, H6N8, H10N8) each of which represent a unique reassortment of Australian, recently introduced Eurasian, and recently introduced American lineage gene segments. Together with the diverse geographic origins of the American and Eurasian gene segments, these findings suggest extensive circulation and reassortment of AIVs within Australian wild birds over vast geographic distances (Hoye et al., 2021).

The Australian-lineage gene segments suggested a high similarity to those of H10N7 viruses isolated in 2010 and 2012 from poultry AIV outbreaks. Despite Australia's effective biosecurity and containment measures, the greatest threat to the emergence of an unexpected infectious disease seems to lie with the spread of a novel AIV strain, likely to be introduced by migratory birds (Bisset and Hoyne, 2021).

Historical outbreaks of HPAI H7 in Australia since 1976 (Bisset and Hoyne, 2021)

- 1976 – H7N7 (LPAI, HPAI) In a combined broiler and egg farm in the outer suburbs of Melbourne Victoria
- 1985 – H7N7 (HPAI) Occurred in a chicken farm near Bendigo, Victoria
- 1992 – H7N3 (HPAI) Chicken farm near Bendigo, Victoria
- 1994 – H7N3 (HPAI) Egg farm in Brisbane, Queensland
- 1997 – H7N4 (HPAI) Tamworth, NSW; H7N4 – two chicken broiler-breeder farms and one emu farm near Tamworth, NSW
- 2012 – H7N7 (HPAI) Free range egg laying farm in Maitland, Lower Hunter Valley, NSW
- 2013 – H7N2 (HPAI) Two poultry farms infected with HPAI H7N2 with 100% correlation of the same virus isolated from each property
- 2020 – H7N7 (HPAI), H5N2 (LPAI), H7N6 (LPAI) An outbreak of three different viral strains of AIV cross six different farms in Victoria. Three egg farms (HPAI H7N7), two turkey farms (LPAI H5N2) and one emu farm (LPAI H7N6).

In the 2020 H7N7 AIV outbreak the simultaneous emergence of three different viral strains highlights the likelihood that multiple AIV viral subtypes are naturally circulating in domestic poultry, with viral transmission into domestic poultry likely to have occurred through mixing with wild aquatic bird species (Vijaykrishna et al., 2013). Recurring outbreaks of HPAI H7N7 in Australia have caused significant losses in poultry but are not currently thought to indicate a rising zoonotic risk from this strain.

However, it supports Webster's suggestion that aquatic birds remain an endemic reservoir that includes a genetically diverse range of influenza A viral subtypes that could infect birds, mammals and humans (Webster et al., 1992). In addition, although HPAI H5Nx viruses remain a known threat to the poultry industry and human health (Lee et al., 2017) there is currently no indication that HPAI H5 subtypes are present in Australia, although the risk of introducing new AIV strains from long-distance migratory birds is recognized (Wille et al., 2019). It should be noted, however, that AIVs are constantly evolving and the future potential for HPAI H5N2 to emerge from an endemic LPAI precursor has been demonstrated using ostrich-origin H5N2 progenitor viruses serially passaged in embryonated chicken eggs (Laleye and Abolnik, 2020). Ongoing surveillance for new and re-emerging AIVs in wild birds and poultry is required.

10.10.2 Avian influenza in New Zealand

New Zealand lies at the southeastern extremity of the EA Flyway which was (and still is) of particular relevance for the potential introduction of novel AIVs in the region, especially given the widescale spread of H5N1 across Asia from 2004. The isolation of AIV from migratory shorebirds in Australia (Hurt et al., 2006) highlights the potential risk of introducing novel strains into New Zealand, but this remains unproven (Langstaff et al., 2009).

The first published report on AIV in New Zealand birds was a result of work conducted between 1975 and 1978 by researchers from the Medical Council of New Zealand Virus Research Unit (Austin and Hinshaw, 1984). A number of further surveys of wild birds were conducted by the Ministry of Agriculture and Forestry (MAF) (now Ministry for Primary Industries, MPI), in 1989 (Stanislawek, 1992; Stanislawek et al., 2002, 2020). As mentioned earlier, New Zealand is not on the common migration pathway for key species of waterfowl, although vagrant waterfowl from Australia are occasionally encountered (Williams et al., 2004). Non-migratory waterfowl, predominantly mallard ducks, are sampled in the summer months throughout New Zealand, with a particular focus on coastal areas where they may have had contact with migratory shorebirds, or where large numbers of juvenile ducks congregate.

Owing to New Zealand's geographical isolation relatively few Artic-breeding migratory shorebirds actually reach the country. In total, about 120,000 birds representing 47 species arrive but the vast majority of these comprise only three species: bar-tailed godwits (*Limosa lapponica*), red knots (*Calidris canutus*), and ruddy turnstones (*Arenaria interpres*) (Williams et al., 2004). Godwits are believed to fly directly (non-stop from the Arctic and tundra breeding grounds) to New Zealand, knots have stopovers in East Asia, and turnstones have a number of stops in the Pacific before they reach New Zealand (Stanislawek et al., 2007, 2020). For these birds to introduce AIV to New Zealand, they would have to be infected before or during migration. Given the high levels of energy required for migration it is unlikely that sick birds carrying HPAI AIVs would survive the long journey, but it is possible that low pathogenic AIV strains could be dispersed by healthy birds. For example, on arrival, they could shed new AIV strains while co-habiting with endemic shorebirds, gulls, waterfowl, and other species and thereby introduce new AIVs. From 2004 to 2010, migratory shorebirds were sampled, in particular the bar-tailed godwit and red knot, from late September to November at Miranda, Firth of Thames, their main North Island arrival site. However, surveillance over this period indicated that migratory birds pose a very low risk for the introduction of HPAI AIVs into New Zealand, as no AIV was ever isolated. Therefore,

since 2010 surveillance has focused on resident waterfowl (Stanislawek et al., 2015, 2020).

As mentioned previously, water birds and particularly the family Anatidae (i.e. common ducks, geese and swans), play an important role as natural reservoirs of AIV (Bulach et al., 2010; Gilbert et al., 2006; Hars et al., 2008). The absence of waterfowl migration to New Zealand substantially limits the potential of introduction of new AIVs into New Zealand. All New Zealand AIV isolates to date, have been obtained from wild mallard ducks. These include subtypes: H1, H2, H3, H4, H5, H6, H7, H9, H10, H11 and H12. All nine neuraminidase types associated with these viruses were detected (Stanislawek unpublished data). All of these H5 and H7 isolates were pathotyped and confirmed to be LPAI based on the HA cleavage site assessment, i.e the conformation of this cleavage site is used as an indication of pathogenicity (Langstaff et al., 2009; Stanislawek et al., 2007, 2020).

Globally, there are two well-recognized lineages of AIVs, these are the Eurasian and North American lineages. Eurasian lineage viruses primarily circulate among birds in Eurasia whereas North American lineage viruses circulate among birds in the Americas. However, where there is overlap between migratory flyways, such as in Alaska and Iceland reassortment can occur in viruses shared between wild birds and in future more new strains are likely to be transferred between hemispheres. Bulach et al. (2010) speculated that the pelagic birds and waders that often migrate to North America do not play a role in the movement of influenza viruses to Australia and New Zealand. This conclusion was based on the closer relationship of the endemic Australian and New Zealand H7 isolates to Eurasian AIV isolates than to North American AIV isolates, despite the flyways that link Australia, New Zealand to North America flyways. Further support for this conclusion has come from surveys carried out by the MPI National Animal Health Laboratory during the first 6 years of the avian influenza surveillance programme (2004–2010). These surveys targeted migratory wading birds, in particular the bar-tailed godwit and red (lesser) knot. No AIVs were isolated, indicating that migratory birds currently pose a very low risk for the introduction of AIV to New Zealand (Stanislawek et al., 2020). This conclusion is supported by findings from similar surveillance of migratory birds in Australia (Curran et al., 2014). Additional studies also suggest that AIVs isolated in New Zealand are somewhat geographically isolated compared to other AIVs circulating around the world. This would suggest that AIVs in New Zealand appear to have undergone an extended period of genetic isolation. The genetic differences of the H5 and H7 viruses in New Zealand could be explained by the establishment of resident mallard duck populations from stock brought in from Europe in the 1860s and from North America in the 1930s and 1940s.

10.11   Conclusion

In this chapter we illustrate how the EcoHealth approach to avian influenza includes a focus on disease ecology, bird migration and the conservation of migratory water birds. Wild waterfowl are regarded as the natural reservoir of AIV and predominantly belong to two orders, Anseriformes (mainly ducks, geese, swans) and Charadriiformes (gulls, terns and shorebirds). These birds have the potential to disseminate AIV widely and can introduce new strains to domestic birds (poultry) and mammals including humans. Until 1997 highly pathogenic (HPAI) subtypes of H5 and H7 AIV were mainly associated with infection and disease outbreaks in poultry. HPAI was very rarely detected in wild birds. However, the situation has changed since the emergence of HPAI H5 viruses of the A/Goose/Guangdong/1/1996 (GsGd) lineage in southeast Asia. Tackling this recent fast changing

> ### Conservation concerns
>
> There is a need to increase inter-disciplinary cooperation between the key scientific and professional groups such as disease ecologists, virologists, human and animal health researchers, conservationists, commercial poultry operators, free range poultry keepers, and relevant government authorities, in the management of AIV. Owing to the complex ecology and zoonotic potential of AIV, the study, control and prevention of avian influenza requires an Eco-Health approach. Policies developed for the prevention and control of AIV also have to consider the potential disease risks to wild birds, this is especially important in New Zealand which is home to many unique flightless birds: e.g. the kiwi (*Apteryx* spp.), kakapo (*Strigops habroptilus*) and takahe (*Porphyrio hochstetteri*) and some of the world's most endangered species, including the Campbell island teal (*Anas nesiotis*), black robin (*Petroica traversi*), black stilt (*Himantopus novaezelandiae*), magenta petrel (*Pterodroma magenta*), New Zealand storm petrel (*Fregetta maorianus*) and many more.

outbreak requires the collaborative efforts of virologists, ornithologists, ecologists, pathologists, mathematical modellers, veterinarians and other disciplines to understand the increased risk associated with these emerging viruses. Better management and increased biosecurity in commercial poultry farms along with well-structured passive and active surveillance for AIV in domestic and wild birds could help to break the reoccurring path for the virus to 'spillover' from poultry to wild birds and 'spillback' from wild birds to poultry. At the same time, with climate change and anthropogenic impacts on the environment, many migratory birds are under significant threat as a result of habitat loss and limited access to historical feeding sites. Further efforts to understand the role of migratory birds in the spread of AIVs is needed as well as a coordinated effort to protect species at risk.

References

Alexander, D. J. (2000). A review of avian influenza in different bird species.*Veterinary Microbiology*, *74*(1–2), 3–13. https://doi.org/10.1016/s0378-1135(00)00160-7

Alexander, D. J., & Brown, I. H. (2009). History of highly pathogenic avian influenza. *Revue Scientifique et Technique*, *28*(1), 19–38. https://doi.org/10.20506/rst.28.1.1856

Anon. (2022). https://phys.org/news/2022–11-bird-flu-pelicans-seabirds-peru.html

Austin, F. J., & Hinshaw, V. S. (1984). The isolation of influenza A viruses and paramyxoviruses from feral ducks in New Zealand. *Australian Journal of Experimental Biology and Medical Science*, *62*(3), 355–360. https://doi.org/10.1038/icb.1984.35

Barter, M. A. (2002). Shorebirds of the Yellow Sea – Importance, threats and conservation status. *Wetlands International global series vol. 9. International wader studies*, 12. Canberra.

Barter, M. A. (2005). Yellow Sea – Driven priorities for Australian shorebird researchers. In *Status and conservation of shorebirds in the East Asian – Australasian flyway*. Proceedings of the Australasian Shorebird Conference, 13–15 December 2003 (pp. 158–160). Canberra. International Wader Studies 17. Sydney.

Bastien, N., Antonishyn, N. A., Brandt, K., Wong, C. E., Chokani, K., Vegh, N., Horsman, G. B., Tyler, S., Graham, M. R., Plummer, F. A., Levett, P. N., & Li, Y. (2010). Human infection with a triple-reassortant swine influenza A(H1N1) virus containing the hemagglutinin and neuraminidase genes of seasonal influenza virus.Journal of Infectious Diseases, 201(8), 1178–1182.

Becker, W. B. (1966, September). The isolation and classification of tern virus: Influenza A-tern South Africa—1961. *Journal of Hygiene (Lond)*, *64*(3), 309–320. https://doi.org/10.1017/s0022172400040596

BirdLife International (2010). http://www.birdlife.org viewed on 20/10/2021

Bisset, A. T., & Hoyne, G. F. (2021). An outbreak of highly pathogenic Avian Influenza (H7n7) in Australia and the potential for Novel Influenza A viruses to emerge. *Microorganisms*, *9*(8). https://doi.org/10.3390/microorganisms9081639

Boere, G. C., & Stroud, D. A. (2006). The flyway concept: What it is and what it isn't. In G. C. Boere, C. A. Galbraith & D. A. Stroud (Eds.), *Waterbirds around the world* (pp. 40–47). Stationery Office.

Bulach, D., Halpin, R., Spiro, D., Pomeroy, L., Janies, D., & Boyle, D. B. (2010). Molecular analysis of H7 avian influenza viruses from Australia and New Zealand: Genetic diversity and relationships from 1976 to 2007. *Journal of Virology*, *84*(19), 9957–9966. https://doi.org/10.1128/JVI.00930-10

Cui, Y., Li, Y., Li, M., Zhao, L., Wang, D., Tian, J., Bai, X., Ci, Y., Wu, S., Wang, F., Chen, X., Ma, S., Qu, Z., Yang, C., Liu, L., Shi, J., Guan, Y., Zeng, X., Tian, G., . . . Chen, H.(2020). Evolution and extensive reassortment of H5 influenza viruses isolated from wild birds in China over the past decade. *Emerging Microbes and Infections*, *9*(1), 1793–1803. https://doi.org/10.1080/22221751.2020.1797542

Curran, J. M., Ellis, T. M., & Robertson, I. D. (2014). Surveillance of Charadriiformes in Northern Australia shows species variations in exposure to avian influenza virus and suggests negligible virus prevalence. *Avian Diseases*, *58*(2), 199–204. https://doi.org/10.1637/10634-080913

CWHC (Canadian Wildlife Health Cooperative). (2022). http://blog.healthywildlife.ca/the-highly pathogenic-h5n1-avian-influenza-virus-is-still-present-mass-mortality-of-snow-geese-during-fall-migration-in-southern-quebec/

Gilbert, M., Xiao, X., Domenech, J., Lubroth, J., Martin, V., & Slingenbergh, J. (2006, November). Anatidae migration in the western Palearctic and spread of highly pathogenic avian influenza H5NI virus. *Emerging Infectious Diseases*, *12*(11), 1650–1656. https://doi.org/10.3201/eid1211.060223

Grillo, V. L., Arzey, K. E., Hansbro, P. M., Hurt, A. C., Warner, S., Bergfeld, J., Burgess, G., Cookson, B., Dickason, C., Ferenczi, M., Hollingsworth, T., Hoque, M., Jackson, R., Klaassen, M., Kirkland, P., Kung, N., Lisovski, S., O'Dea, M., O'Riley, K., . . . Post, L. (2015). Avian influenza in Australia: A summary of 5 years of wild bird surveillance. *Australian Veterinary Journal*, *93*(11), 387–393. https://doi.org/10.1111/avj.12379

Hahn, S., Bauer, S., & Liechti, F. (2009). The natural link between Europe and Africa – 2.1 billion birds on migration. *Oikos, 118*(4), 624–626. https://doi.org/10.1111/j.1600-0706.2008.17309.x

Hampson, A. W. (2002). Influenza virus antigens and "antigenic drift". In C. W. Potter (Ed.), *Influenza. Perspectives in medical virology* (7th ed) (pp. 49–85). Elsevier. https://doi.org/10.1016/S0168-7069(02)07004-0

Hars, J. et al. (2008). J Wildl Dis: The epidemiology of the highly pathogenic H5N1 avian influenza in mute swan (*Cygnus olor*) and other Anatidae in the Dombes region (France), 2006. *Journal of Avian Medicine and Surgery, 22*(4, December), p. 365.

Heather, B. D., & Robertson, H. A. (1996). *Field guide to the birds of New Zealand.* Penguin.

Hill, N. J., Bishop, M. A., Trovão, N. S., Ineson, K. M., Schaefer, A. L., Puryear, W. B., Zhou, K., Foss, A. D., Clark, D. E., MacKenzie, K. G., Gass, J. D., Borkenhagen, L. K., Hall, J. S., & Runstadler, J. A.(2022). Ecological divergence of wild birds drives avian influenza spillover and global spread. *PLOS Pathogens, 18*(5), e1010062.https://doi.org/10.1371/journal.ppat.1010062

Hoye, B. J., Donato, C. M., Lisovski, S., Deng, Y.-M., Warner, S., Hurt, A. C., Klaassen, M., & Vijaykrishna, D. (2021). Reassortment and persistence of influenza A viruses from diverse geographic origins within Australian wild birds: Evidence from a small, isolated population of ruddy turnstones. *Journal of Virology, 95*(9), 1–18. https://doi.org/10.1128/JVI.02193-20

Hurt, A. C., Hansbro, P. M., Selleck, P., Olsen, B., Minton, C., Hampson, A. W., & Barr, I. G. (2006). Isolation of avian influenza viruses from two different trans hemispheric migratory shorebird species in Australia. *Archives of Virology,151*(11), 2301–2309. https://doi.org/10.1007/s00705-006-0784-1

Jones, K. E., Patel, N. G., Levy, M. A., Storeygard, A., Balk, D., Gittleman, J. L., & Daszak, P. (2008). Global trends in emerging infectious diseases. *Nature, 451*(7181), 990–993. https://doi.org/10.1038/nature06536

Kawaoka, Y., Chambers, T. M., Sladen, W. L., & Webster, R. G. (1988). Is the gene pool of influenza viruses in shorebirds and gulls different from that in wild ducks? *Virology, 163*(1), 247–250. https://doi.org/10.1016/0042-6822(88)90260-7

Kirby, J. (2010). Review of current knowledge of bird flyways, principal knowledge gaps and conservation priorities. CMS Scientific Council: Flyway Working Group Review. Unpublished.

Kirby, J. S., Stattersfield, A. J., Butchart, S. H. M., Evans, M. I., Grimmett, R. F. A., Jones, V. R., O'Sullivan, J., Tucker, G. M., & Newton, I. (2008). Key conservation issues for migratory land- and waterbird species on the world's major flyways. *Bird Conservation International,18*(S1), S49–S73. https://doi.org/10.1017/S0959270908000439

Krauss, S., & Webster, R. G. (2010, March). Avian influenza virus surveillance and wild birds: Past and present. *Avian Diseases, 54*(1), Suppl., 394–398. https://doi.org/10.1637/8703-031609-Review.1

Laleye, A. T., & Abolnik, C.(2020). Emergence of highly pathogenic H5N2 and H7N1 influenza A viruses from low pathogenic precursors by serial passage in ovo. *PLOS ONE, 15*(10), e0240290. https://doi.org/10.1371/journal.pone.0240290

Langstaff, I. G., McKenzie, J. S., Stanislawek, W. L., Reed, C. E. M., Poland, R., & Cork, S. C. (2009). Surveillance for highly pathogenic avian influenza in migratory shorebirds at the terminus of the East Asian-Australasian Flyway. *New Zealand Veterinary Journal, 57*(3), 160–165. https://doi.org/10.1080/00480169.2009.36896

Lee, D. H., Bertran, K., Kwon, J. H., & Swayne, D. E. (2017). Evolution, global spread, and pathogenicity of highly pathogenic avian influenza H5Nx clade 2.3.4.4. *Journal of Veterinary Science, 18*(S1), 269–280. https://doi.org/10.4142/jvs.2017.18.S1.269

Li, K. S., Guan, Y., Wang, J., Smith, G. J., Xu, K. M., Duan, L., Rahardjo, A. P., Puthavathana, P., Buranathai, C., Nguyen, T. D., Estoepangestie, A. T., Chaisingh, A., Auewarakul, P., Long, H. T., Hanh, N. T., Webby, R. J., Poon, L. L., Chen, H., . . . Peiris, J. S. (2004). Genesis of a highly pathogenic and potentially pandemic H5N1 influenza virus in eastern Asia. *Nature, 430*(6996), 209–213. https://doi.org/10.1038/nature02746

Lindstrom, S. E., Cox, N. J., & Klimov, A. (2004). Genetic analysis of human H2N2 and early H3N2 influenza viruses, 1957–1972: Evidence for genetic divergence and multiple reassortment events. *Virology, 328*(1), 101–119. https://doi.org/10.1016/j.virol.2004.06.009

Liu, H., Xiong, C., Chen, J., Chen, G., Zhang, J., Li, Y., Xiong, Y., Wang, R., Cao, Y., Chen, Q., Liu, D., Wang, H., & Chen, J. (2018, April 11). Two genetically diverse H7N7 avian influenza viruses isolated from migratory birds in central China.

Emerging Microbes and Infections, 7(1), 62. https://doi.org/10.1038/s41426-018-0064-7

Ma, W., Kahn, R. E., & Richt, J. A. (2008). The pig as a mixing vessel for influenza viruses: Human and veterinary implications. *Journal of Molecular and Genetic Medicine, 3*(1), 158–166. https://doi.org/10.4172/1747-0862.1000028

Meng, W., Yang, Q., Vrancken, B., Chen, Z., Liu, D., Chen, L., Zhao, X., François, S., Ma, T., Gao, R., Ru, W., Li, Y., He, H., Zhang, G., Tian, H., & Lu, J. (2019). New evidence for the east–west spread of the highly pathogenic avian influenza H5N1 virus between Central Asian and east Asian-Australasian flyways in China. *Emerging Microbes and Infections, 8*(1), 823–826. https://doi.org/10.1080/22221751.2019.1623719

Murray, N. J., Clemens, R. S., Phinn, S. R., Possingham, H. P., & Fuller, R. A. (2014). Tracking the rapid loss of tidal wetlands in the Yellow Sea. *Frontiers in Ecology and the Environment*. Parkinson's Disease Foundation, 12(5), 267–272. https://doi.org/10.1890/130260

Neverov A. D., Lezhnina K. V., Kondrashov A. S., Bazykin G. A. (2014). Intrasubtype reassortments cause adaptive amino acid replacements in H3N2 influenza genes. PLoS Genet. 2014 Jan;10(1):e1004037. https://doi.org/10.1371/journal.pgen.1004037

National History Museum. (2022). https://www.nhm.ac.uk/discover/news/2022/june/bird-flu-outbreak-devastates-uk-seabird-colonies.html

Newton, I. (2007). *The ecology of bird migration* (1st ed). Elsevier.

OIE. (2021). Updates on the spread of H5N8 avian influenza from an African perspective.https://rr-africa.oie.int/en/news/updates-on-the-spread-of-h5n8-avian-influenza-from-an-african-perspective/

OIE. (2022). https://www.oie.int/en/disease/avian-influenza/

Olsen, B., Munster, V. J., Wallensten, A., Waldenström, J., Osterhaus, A. D. M. E., & Fouchier, R. A. M. (2006). Global patterns of influenza A virus in wild birds. *Science, 312*(5772), 384–388. https://doi.org/10.1126/science.1122438

Semangeum Shorebird Monitoring Program (SSMP). (2006–8). *Birds Korea*. https://awsg.org.au/pdfs/Saemangeum-Report.pdf

Schubert, S. C., Manica, L. T., & Guaraldo, A. D. C. (2019). Revealing the potential of a huge citizen-science platform to study bird migration. *Emu – Austral Ornithology, 119*(4), 364–373. https://doi.org/10.1080/01584197.2019.1609340

Stanislawek, W. L. (1992). Survey of wild ducks for evidence of avian influenza virus, 1988–90. *Surveillance, 19*(1), 21–22.

Stanislawek, W. L., Wilks, C. R., Meers, J., Horner, G. W., Alexander, D. J., Manvell, R. J., Kattenbelt, J. A., & Gould, A. R. (2002). Avian paramyxoviruses and influenza viruses isolated from mallard ducks (*Anas patyrhynchos*) in New Zealand. *Archives of Virology, 147*(7), 1287–1302. https://doi.org/10.1007/s00705-002-0818-2

Stanislawek, W. L., Cork, S. C., Rawdon, T., Thornton, R. N., & Melville, D. S. (2007). Surveillance for Avian Influenza in wild birds in New Zealand. In *Option for the Control. of Influenza Vi Toronto Canada* (pp. 358–360). Conference proceedings.

Stanislawek, W. L., McFadden, A., & Tana, T. (2015). Avian influenza surveillance programme. *Surveillance,40*(3), 19–22.

Stanislawek, W., Rawdon, T., & Tana, T. (2020). Avian influenza surveillance programme. *Surveillance,47*(3), 29–32.

Sullivan, J. D., Takekawa, J. Y., Spragens, K. A., Newman, S. H., Xiao, X., Leader, P. J., Smith, B., & Prosser, D. J. (2018). Waterfowl spring migratory behavior and avian influenza transmission risk in the changing landscape of the East Asian-Australasian flyway. *Frontiers in Ecology and Evolution, 6*. https://doi.org/10.3389/fevo.2018.00206

Takekawa, J. Y., Newman, S. H., Xiao, X., Prosser, D. J., Spragens, K. A., Palm, E. C., Yan, B., Li, T., Lei, F., Zhao, D., Douglas, D. C., Muzaffar, S. B., & Ji, W. (2010). Migration of waterfowl in the East Asian flyway and spatial relationship to HPAI H5N1 outbreaks. *Avian Diseases, 54*(1), Suppl., 466–476. https://doi.org/10.1637/8914-043009-Reg.1

To, K. K., Tsang, A. K. L., Chan, J. F. W., Cheng, V. C. C., Chen, H., & Yuen, K. Y. Y. (2014). Emergence in China of human disease due to avian influenza A (H10N8) – Cause for concern? *Journal of Infection, 68*(3), 205–215. https://doi.org/10.1016/j.jinf.2013.12.014

Vandegrift, K. J., Sokolow, S. H., Daszak, P., & Kilpatrick, A. M. (2010). Ecology of avian influenza viruses in a changing world. *Annals of the New York Academy of Sciences,1195*, 113–128. https://doi.org/10.1111/j.1749-6632.2010.05451.x

Verhagen, J. H., Fouchier, R. A. M., & Lewis, N. (2021). Highly pathogenic avian influenza viruses at the wild–domestic bird interface in Europe: Future directions for research and surveillance. *Viruses, 13*(2), 212.https://doi.org/10.3390/v13020212

Vijaykrishna, D., Deng, Y. M., Su, Y. C. F., Fourment, M., Iannello, P., Arzey, G. G., Hansbro, P. M., Arzey, K. E., Kirkland, P. D., Warner, S., O'Riley, K., Barr, I. G., Smith, G. J., & Hurt, A. C. (2013). The recent establishment of North American H10 lineage influenza viruses in Australian wild waterfowl and the evolution of Australian avian influenza viruses. *Journal of Virology, 87*(18), 10182–10189. https://doi.org/10.1128/JVI.03437-12

Watts, J., Rawdon, T., Stanislawek, W., Tana, T., Cobb, S., & Mulqueen, K. (2016). Avian influenza: Epidemiology and surveillance in New Zealand. *Surveillance, 43*(4), 6–14.

Webster, R. G., Bean, W. J., Gorman, O. T., Chambers, T. M., & Kawaoka, Y. (1992). Evolution and ecology of influenza A viruses. *Microbiological Reviews, 56*(1), 152–179. https://doi.org/10.1128/mr.56.1.152-179.1992

Weisshaupt, N., Lehikoinen, A., Mäkinen, T. M., & Koistinen, J. (2021). Challenges and benefits of using unstructured citizen science data to estimate seasonal timing of bird migration across large scales. *PLOS ONE, 16*(2), e0246572. https://doi.org/10.1371/journal.pone.0246572

WHO (World Health Organization). (2008). *The global burden of disease: 2004 update*. World Health Association Press.

Wille, M., Lisovski, S., Risely, A., Ferenczi, M., Roshier, D., Wong, F. Y. K., Breed, A. C., Klaassen, M., & Hurt, A. C. (2019). Serologic evidence of exposure to highly pathogenic avian influenza H5 viruses in migratory shorebirds, Australia. *Emerging Infectious Diseases, 25*(10), 1903–1910. https://doi.org/10.3201/eid2510.190699

Wille, M., Grillo, V., Ban de Gouvea Pedroso, S., Burgess, G. W., Crawley, A., Dickason, C., Hansbro, P. M., Hoque, M. A., Horwood, P. F., Kirkland, P. D., Kung, N. Y., Lynch, S. E., Martin, S., McArthur, M., O'Riley, K., Read, A. J., Warner, S., Hoye, B. J., Lisovski, S., . . . Wong, F. Y. K. (2022). Australia as a global sink for the genetic diversity of avian influenza A virus. *PLOS Pathogens, 18*(5), e1010150. https://doi.org/10.1371/journal.ppat.1010150

Williams, M., Gummer, H., Powlesland, R., & Taylor, G. (2004). *Migrations and movements of birds to New Zealand and surrounding seas*. Department of Conservation Science and Research Unit. https://www.doc.govt.nz/globalassets/documents/science-and-technical/sap232.pdf

WOAH. (2022). Avian influenza. https://www.woah.org/en/disease/avian-influenza/

WOAH. (2022). Updates on the spread of H5N8 avian influenza from an African perspective. https://rr-africa.woah.org/en/news/updates-on-the-spread-of-h5n8-avian-influenza-from-an-african-perspective/

World Health Organization, FAO, & OIE. (2008). Zoonotic diseases: A guide to establishing collaboration between animal and human health sectors at the country level. http://www.oie.int/doc/ged/D12060.PDF

Chapter 11

Elk, forest and fire management: Balancing wildlife and human needs in Jasper National Park, Alberta, Canada

Colleen Arnison and Marco Musiani

Abstract

Forest fires have always shaped natural landscapes. In Jasper National Park (Alberta, Canada), historical fires have helped maintain a healthy mixture of young and old forests, shrublands and open meadows. However, since the 1930s, fire suppression has been implemented to reduce the loss of what was perceived to be 'forest resources', and these practices created an unnaturally old forest with, ironically, an increased risk of large, catastrophic fires. Recently, a 'FireSmart' program was designed to mimic natural disturbance, such as fire, and consisted of timber removal to protect the community of Jasper from wildfires and improve ecological conditions for wildlife. Our findings indicate that, following FireSmart, forage availability for ungulates increased including grass, forb and shrub biomass, along with biodiversity of plant species. We also fitted with GPS collars the areas' elk to determine the spatial distribution of the species and its resource selection. Here we show that elk selected for areas that would be enhanced by natural disturbances because they preferred herbaceous, shrub and open coniferous habitat types as well as burn sites. Consistent with this interpretation, we also found that the FireSmart areas were used by elk and other herbivores more than 'control' areas. This work therefore demonstrates that disturbances such as fires can benefit plants and herbivores, and likely other species in healthy food webs. More broadly, natural disturbances including fire, drought, wind, parasite and disease outbreaks, herbivory and predation should all be accounted for, as they may contribute to the biodiversity of healthy ecosystems.

Susan C. Cork and Douglas P. Whiteside (eds) **Case Studies in EcoHealth**
DOI: 10.52517/9781789183368.011, © 5m Books Ltd, 2024

11.1 Restoring natural ecological balance through fire management

Forest fires have shaped the landscape of Jasper National Park (hereafter referred to as JNP) for centuries, maintaining a healthy mixture of young and old forests, shrublands and open meadows, and providing habitat for abundant wildlife. Since the 1930s, fire suppression has been implemented to reduce the loss of what were at that time perceived to be 'forest resources' (Chavardès et al., 2018). However, these effective fire suppression practices created an unnaturally old forest with reduced biodiversity and, ironically, they also increased the risk of a large, catastrophic fire. Additionally, fire suppression can have negative consequences on wildlife and vegetation in fire-dependent ecosystems, such as a decrease in biodiversity (Kelly and Brotons, 2017). Artificially old forests produced by decades of fire suppression are not unique to JNP. Such landscapes dominate large regions of North America, raising questions about how best to allow fires in order to improve biodiversity, without risking human communities and facilities inside and outside protected areas.

In the early 2000s, the impacts of fire suppression on wildlife and vegetation became well known. Researchers were worried about in-growth of forests, as well as a decline of fire-dependent ecosystems and species that depend on them (Risbrudt, 1995; Westhaver, 2003). Consequently, land managers started to use new measures to counteract the impacts of fire suppression, using such methods as prescribed burning or manual thinning of forests. In 2004, JNP started a FireSmart–ForestWise Program (hereafter referred to as FireSmart) for the dual purposes of protecting the residential community of the municipality of Jasper from wildfires, and of improving ecological conditions, specifically ungulate habitat

(Westhaver, 2003). This program simultaneously applied forest-thinning treatments and prescribed burns to protect the town of Jasper (FireSmart), as well as to return the forest to a stand structure that was consistent with its historical conditions (ForestWise).

11.2 Habitat selection by elk in forested ecosystems

Rocky Mountain elk (*Cervus elaphus*) is a dominant herbivore species in the Canadian and United States (US) Rocky Mountains (Rockies). They are iconic megafauna and are esteemed for their conservation value (van Dyke et al., 2012). Elk populations provide ecological, social and economic benefits to local communities. For example, wildlife viewing and hunting (conducted outside protected areas) generate millions of dollars annually in Alberta, Canada (Limited, 2008). Ecologically, elk contribute to maintaining meadows in open habitat – i.e., this requires conditions that have declined in some areas due to anthropogenic influences such as fire suppression activities. However, elk can also cause concerns as reservoirs for livestock diseases and from potential damage to young trees. Within protected areas, damage caused by elk includes overbrowsing of timber resources, vehicle collisions, nuisance and safety concerns due to habituation close to human developments (Walter et al., 2010).

Elk are large and fairly common herbivores, which makes them an ideal species to study whole-forest ecosystems. They are known to respond to timber-removal treatments and are sufficiently abundant in JPN to be adequately sampled (Shepherd and Whittington, 2006). In particular, understanding how elk (and other wildlife) use their surrounding habitats is essential to ensure sustainable environmental management. Habitat selection studies usually compare habitat use with habitat availability, showing how animals actively select the environments where they spend most of their time (Manly, 1993). Resource selection models can then be used to predict the geographical distribution of a particular species and to analyse the intensity of resource use (Boyce et al., 2002). As a result, both animal distribution and their utilization of resources can then be managed, to accomplish conservation and management goals.

Figure 11.1 Herbivores, such as this female elk that we monitored with a GPS collar in JNP, may prefer to use the same flat areas that are also used for infrastructure. Photo credit: Colleen Arnison.

11.3 Case study: Monitoring elk to understand forest resources selected by herbivores

For this case study we used location data from 10 elk that were fitted with GPS collars in JNP from 2008 to 2011 by Parks Canada (Figure 11.1). Twelve locations per day were recorded for each collared elk. The location data were collated and divided into two seasons (winter and summer), which are dramatically different for resource availability – i.e., North American winter from 15 October to 15 April and summer from 16 April to 14 October. We used a statistical method (95% kernel density) to create individual elk home ranges for the summer and winter seasons. Multivariable logistic regression was used to create several candidate models. These models were then compared using Aikaike's information criteria (AIC) to determine the best and most representative model for each season. The best models were then validated to test how each was able to predict the resource use of each monitored elk (Arnison, 2014).

We examined elk response to landscape variables in JNP using resource selection functions (RSFs). Multivariable logistic regression is typically used to create resource selection models. These models estimate the probability of the use of a resource by animals: essentially they compare resource use with the availability of those resources on the landscape (Austin,

2002; Manly et al., 2002). Often, RSFs are developed to understand critical resources used by animals and to determine whether the current landscape is providing the needed components. As such, RSFs are increasingly used as a tool in natural resource management, assessment of cumulative environmental effects, land-management planning and population viability analysis (Boyce et al., 2002; Johnson et al., 2004).

To evaluate elk selection or avoidance of burned areas of forest, datasets representing fire were obtained from Parks Canada. Post-fire vegetation biomass is known to fluctuate over time (Schimmel and Granström, 1996), and elk have been found to select burned areas as they consume plant species that increase following fire (Bailey and Whitham, 2002). Because we wanted to determine whether elk selected burn sites based on the number of years since the fire occurred, a continuous 'year since burn' dataset was derived (BrnAge). Interestingly, herb-dominated vegetation can persist on burned forest sites for up to 25 years (Sachro et al., 2005), therefore, we created a categorical dataset of different burn ages to determine whether elk select for recent, past and historic burns differently (BurnCat).

Variables considered in model development

The environmental variables that we analysed (i.e., those potentially selected by elk) included elevation, slope and aspect derived from a digital elevation model (DEM). We also looked at a composite model generated from land-cover data, forest canopy closure and species composition of trees (for more details refer to McDermid (2006) and DeCesare (2012)). The outputs from the model were reclassified into a number of categories and an average normalized difference vegetation index (NDVI) was used as a measure of productivity of green forage (DeCesare, 2012; Hebblewhite et al., 2008; Pettorelli et al., 2005). We obtained geodatabases of roads, railways and official and unofficial trails from Parks Canada and used these to evaluate elk's potential avoidance or selection of these resources – i.e., a distance to linear feature variable (DistLinear). Further, a categorical spatial dataset was developed to determine what type of linear feature (i.e., trail, road or railway – linked to human activity) was closest to the elk location (LinearType). This was used to determine whether elk selection varied by the type of linear feature it was close to, because elk and a variety of other species perceive road types and vehicle use differently (Ager et al., 2003; Clark et al., 2001; Dickson and Beier, 2002; Dodd et al., 2007). Additionally, we used data estimated by Parks Canada representing the number of humans or vehicles per month along each linear feature (SummerUse).

Most animals look for food, which justifies our focus on the resources mentioned above. Animals also avoid their predators. We therefore used a predation risk spatial dataset (prisk) formulated for wolves, which are the elk's major predators (DeCesare, 2012). This dataset was not necessarily specific to elk, and considered an averaged pool of ungulates killed by wolves in the area. However, taking a broad EcoHealth view, we maintain that the dataset provides a realistic prediction of predation risk for elk in JNP, because wolves are opportunistic in their prey selection and are not limited to hunting only elk in the area.

11.4 Key factors influencing elk distribution in summer and winter

Following the development of multiple candidate models to explain elk distribution, the best model for each season was determined using AIC, which estimates prediction errors (Table 11.1). We found that the best summer and winter models included resources avoided or selected by elk: forest cover types (cover), elevation, slope, predation risk (prisk), year since FireSmart treatment (FireSmt), year since burn (BrnAge), burn category (BurnCat) and an interaction of closest linear feature type (LinearType) and distance to nearest linear feature (DistLinear). In addition, the best summer model also contained greenness (NDVI) and distance to water (Dwater), while the top winter model also included a water body category (WatrBndy), aspect (Cosaspect) and human activity level (SummerUse).

Our findings demonstrate that, in summer, elk highly selected areas with broadleaf forests, herbaceous and open wetland as cover types (Cover), as well as higher elevation (Elevation) and with flatter terrain (Slope). Surprisingly, they also selected areas that had higher wolf predation risk (Prisk_s), and higher human activity levels (SummerUse), very likely because these same areas had other essential resources for elk. In the summer, elk selected former burn areas

Table 11.1 Elk Resource Selection Models in JNP: summer and winter preferences with AIC comparison, including ΔAIC (difference between best motel and each model in set) and w$_i$ values (probability that each model is the best approximating model).

Summer models	Rank	AIC	ΔAIC	w$_i$
Cover + Elevation + Slope + Prisk_s + SummerUse + LinearType * DistLinear + BrnAge + NDVI + FireSmt + BurnCat + DistWater	1	268682.2	0	0.323
Cover + Elevation + Slope + Prisk_s + SummerUse + LinearType * DistLinear + BrnAge + NDVI + BurnCat + DistWater	2	268685.6	3.4	0.0590
Cover + Elevation + Slope + Prisk_s + SummerUse + LinearType * DistLinear + BrnAge + NDVI + BurnCat + DistWater + Cosaspect	3	268686.2	4	0.0437
Cover + Elevation + Slope + Prisk_s + SummerUse + LinearType * DistLinear + BrnAge + FireSmart + BurnCat + DistWater	4	268703.1	20.9	<0.0001

Winter models	Rank	AIC	ΔAIC	w$_i$
Cover + Elevation + Slope + Prisk_w + Snow_w + BrnAge + BurnCat + FireSmt + LinearType * DistLinear + WatrBndy + Cosaspect	1	276119.9	0	0.429
Cover + Elevation + Slope + Snow_w + BrnAge + BurnCat + FireSmt + LinearType * Distlinear + WatrBndy + Cosaspect	2	276122.1	2.2	0.143
Cover + Elevation + Slope + Prisk_w + Snow_w + BrnAge + BurnCat + FireSmt + LinearType + DistLinear + WatrBndy + Cosaspect	3	276425	305.1	<0.0001

(BrnAge and BrnCat) and FireSmart treatment areas (FireSmt), most likely due to increased forage availability there, including an increase in the diversity and biomass of grass and herbaceous plants (forbs) (see section 11.7). Also related to food and water availability, we found that in summer elk selected for areas with a higher level of plant productivity or biomass (NDVI) and those closer to water bodies (Dwater).

Our findings also demonstrate that, in winter, elk did not avoid any land cover type (Cover), with herbaceous, shrub, open conifer and barren land being the strongest selected types. Similar to the findings for summer, elk preferred flatter terrain (Slope) in the winter but preferred lower elevation areas (Elevation) with a southern aspect (Cosaspect) and lower snow depths (Snow_w). Areas that were at risk to wolf predation were also selected (Prisk_w); however, this pattern was less pronounced than

in the summer season. Like in summer, elk also selected former burn areas (BrnAge and BrnCat) and FireSmart treatment areas (FireSmt). Additionally, elk preferred areas closer to linear features (DistLinear), especially to roads, but not to trails and railways (LinearType).

Our study generated maps with relative (high to low) probability of the presence of elk throughout JNP based on park-wide rRSF models for (a) summer and (b) winter distributions (Figure 11.2). These models were refined and validated using data locations from the ten elk that were GPS collared from 2008 to 2011. The maps indicated that elk are restricted to valley bottoms in winter and more evenly distributed across JNP in summer, likely because vegetation is also more evenly distributed following the spring 'green-up' in higher-elevation areas. However, it is noticeable that elk avoid areas used intensively by people, including roads and trails, especially in the summer.

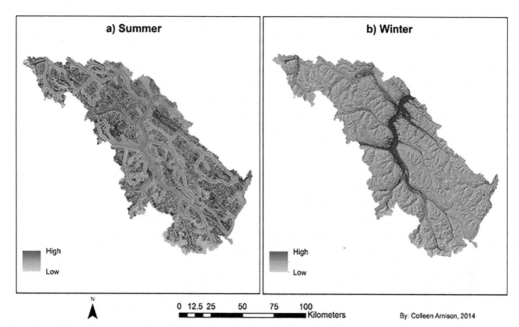

Figure 11.2 Relative (high to low) probability of occurrence of elk throughout JNP based on park-wide RSF models for (a) summer and (b) winter distributions. Distributional models were developed and validated using locations from 10 elk that were GPS collared from 2008 to 2011.

11.5 Additional observations: elk preferred burned or thinned forest

Our findings indicated that, as elk are considered a generalist herbivore and are often flexible in their choice of foods and habitat (Christianson and Creel, 2007; Cook, 2002; Geist, 2002), they may select a variety of land-cover types to acquire forage. Herbaceous land cover (i.e., grass and non-grass-like herbaceous plants or forbs) was highly selected in each season because it contains the highest forage availability compared with all other land-cover types. In winter this forage might be particularly important, because 90% of elk diet can consist of grass-like herbaceous plants (also known as graminoids) (Hebblewhite, 2006). Other land-cover types were also selected by elk in this study. In particular, open conifer forests may be attractive to elk in winter, as we observed,

because snow depth tends to be lower under coniferous forest canopy than in open areas (Huot, 1974) and elk are known to avoid areas of deep snow (Pauley et al., 1993; Sweeney and Sweeney, 1984). Finally, the broadleaf forests that were also selected may include more than 80% of deciduous species, including aspen, which is a major browse species for elk (White et al., 1998).

Dense and moderately dense coniferous forest patches were selected less often by elk in both winter and summer. Both of these land-cover types are typical of landscapes that have been exposed to fire-suppression management practices (Mitchell, 2006; Rhemtulla, 1999).

Regarding natural features, elk preferred lower elevation and low snow cover in the winter and a higher elevation in the summer. It is common in herbivore species to follow retreating snow cover and greater food availability in higher elevations in the summer, and to choose lower elevation for their winter range (Fryxell and Sinclair, 1988; Sawyer et al., 2009). Moving to lower elevations in the winter is a strategy to find wintering areas with shallow snow depth (Boyce, 1991; White et al., 2010). Further, in winter elk were found to prefer south-facing slopes, which are often snow-free due to the effect of wind and solar radiation (D'eon and Delparte, 2005).

Regardless of the season, the collared elk selected areas where they were at risk from wolf predation; however, this was more so in summer than in winter. This is contrary to studies that have indicated that elk avoid areas with high predation risk in general (Hebblewhite and Merrill, 2009), perhaps indicating that, in JNP, elk movements are explained by forage availability and wolves subsequently followed elk.

We also found from our study that elk preferred areas close to linear features in both summer and winter, and that this selection often varied with traffic volume, road density and adjacent habitat type. In winter in JNP the traffic volume is less due to a lower number of tourists visiting the area, and elk may be more attracted to roads for forage availability or important adjacent habitat during this time, as demonstrated in other studies (Dodd et al., 2007; Gagnon et al., 2007). In summer elk may avoid linear features due to the higher traffic volume on roads and trails and may choose suitable habitat with adequate forage in other areas. Additionally, elk avoidance of trails in the winter may be due to wolves using such trails during this season (Hebblewhite et al., 2005; Whittington et al., 2005).

The collared elk in this study preferred burned or thinned forest areas, likely due to the better forage found in these areas. Natural forest fires, and also (human-made) prescribed burns, have been shown to strongly increase herbaceous forage biomass (Hebblewhite et al., 2008; Sachro et al., 2005), primarily as a result of increased light reaching the ground in these cleared areas. Because the elk diet typically consists of herbaceous vegetation, they often select for such burns (Hebblewhite et al., 2008; Irwin and Peek, 1983; Peck and Peek, 1991). In our study we also found that there was a seasonal difference in elk selection of burned areas. In summer, while elk selected for all ages of burns, they preferred areas that had been burned more recently. In winter elk selected for more recent burn (<5 years) sites and even avoided older burn sites (5–59 years). An increase in herbaceous forage biomass occurs within the first few years following a fire, which is when the area is most attractive to elk (Hebblewhite et al., 2008; Irwin and Peek, 1983; Peck and Peek, 1991). Since overall forage resources are limited in winter, elk may concentrate on these more recently burned areas, which contain the most forage, and avoid older burn areas, which have lower forage availability and still elevated snow cover due to the open canopy.

Year round, our elk selected the 'FireSmart program' timber-removal (thinning) areas and preferred those that were older. Openings created by logging and thinning operations are considered to be beneficial to ungulates, such as elk, because they generally increase forage production and suitable edge habitat (Basile and Jensen, 1971; Harper, 1971; Krefting, 1962). However, other studies have indicated that open areas could increase elk visibility, and hence vulnerability to predation (Beall, 1976; Hershey and Leege, 1976; Lyon and Jensen, 1980; Marcum, 1976; Pengelly, 1972). The size of logged areas in the JNP FireSmart program was limited, with an average of 10.91 ha in size for each treatment site, which would limit ungulate vulnerability by allowing elk adequate cover provided by forest patches close by. Overall, elk's preference for former burn areas and FireSmart treatment areas in our study was most likely due to increased forage availability, including an increase in the diversity and biomass of grasses and forbs (see below).

11.6 Assessing the impact of the FireSmart on vegetation and wildlife

The JNP 'FireSmart program' was implemented around the town of Jasper and a nearby cottage subdivision at Lake Edith, approximately 3 km north-east of town. This area is in the Alberta Montane ecoregion, which is mainly composed of closed forests with fragments of grassland and open and trembling aspen forests. It is also considered the most productive and biologically diverse ecoregion within the region because it contains unique and precious wildlife habitats (Holroyd and van Tighem, 1983).

FireSmart programs have been implemented across North America and Australia; however, a detailed account on the impact of vegetation characteristics following these programs has not previously been documented. This case study is the first detailed assessment of vegetation response to a FireSmart Program in Canada, and the results are intended to inform future conservation and management approaches within the Canadian Rockies and to provide a comparative measure for future research in the area. We hope that the knowledge gathered will enhance the efficiency of habitat management in human-altered landscapes, as well as providing better predictions of the effects of future development in previously unaltered environments.

Before the FireSmart program was implemented the structure of the forest was an artifact of recent fire-suppression management practices, rather than a reflection of natural fire cycle processes. Due to fire-suppression management practices, formerly open, savannah-like Douglas fir and lodgepole pine forests in the area changed significantly. Westhaver (2003) found that the dense tree understory present in the area recently was younger than 75 years, and originated from the start of fire-suppression policies within the park.

To achieve FireSmart program's objectives, manual and mechanical vegetation treatments such as selective thinning, pruning and burning took place on approximately 350 ha of forest surrounding the town of Jasper and nearby Lake Edith (Figure 11.3). Although the size of each treatment varied, it averaged 10.91 ha in size. FireSmart treatments occured between 2004 and 2009.

To evaluate the effectiveness of the FireSmart program we established a before-after-control-impact (BACI; Green, 1979) research design. Permanent monitoring sites (grids) were established in treatment and (untreated) control areas (Figure 11.4). Note that surveys of both plants and ungulate use (through pellets) were conducted in 2003 and 2012 within 30×30 m^2 plots, forming a grid for each site. We focused on responses to FireSmart in terms of both plant species present and wildlife use. The parameters measured included stand density, vegetation cover, diversity of plants, biomass production and ungulate relative abundance.

11.7 The use of FireSmart to enhance plant diversity and attract ungulates

Following the FireSmart program, in the treatment areas there was an obvious decrease in both over- and understory tree density (due to the treatment of tree removal). However, densities of sapling and regeneration trees increased (Figure 11.5). Such increases mainly involved lodgepole pine and Douglas fir (65.6 and 57.6% average cover, respectively). Lodgepole pine is shade intolerant, benefitting from openings created by natural disturbances (Daniel

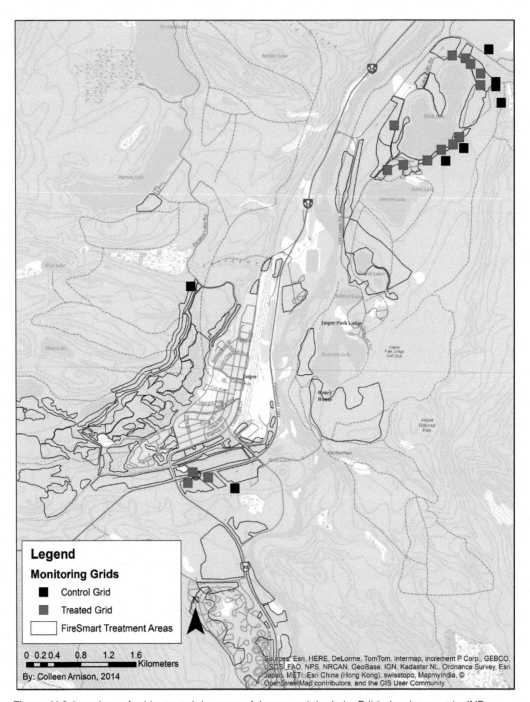

Legend

Monitoring Grids

- ■ Control Grid
- ■ Treated Grid
- ▢ FireSmart Treatment Areas

0 0.2 0.4 0.8 1.2 1.6
━━━━━━━━━ Kilometers
By: Colleen Arnison, 2014

Sources: Esri, HERE, DeLorme, TomTom, Intermap, increment P Corp., GEBCO, USGS, FAO, NPS, NRCAN, GeoBase, IGN, Kadaster NL, Ordnance Survey, Esri Japan, METI, Esri China (Hong Kong), swisstopo, Mapmyindia, © OpenStreetMap contributors, and the GIS User Community

Figure 11.3 Locations of grids around the town of Jasper and the Lake Edith development in JNP, Alberta used to monitor those areas treated by the FireSmart program in 2003 and 2012. Topography and roads are also indicated.

Figure 11.4 To evaluate the effectiveness of the FireSmart program we established a BACI research design. The areas were treated with FireSmart between 2004 and 2009. Permanent monitoring sites (grids) were established in treatment areas (a) and (untreated) control areas (b). Photo credits: Colleen Arnison.

et al., 1979), while Douglas fir is considered to be moderately shade intolerant (Burns and Honkala, 1990). Therefore, the sudden increase in light from the opening of the canopy likely promoted the rapid increase in seedling growth, especially for shade-intolerant species such as lodgepole pine (Kemball et al., 2005; Wright et al., 2000).

Following the FireSmart program, plant biodiversity increased in all stand types that we monitored (Pine (*Pinus* spp.), Douglas fir (*Pseudotsuga menziesii*) and others). Increases were consistent for shrub and regeneration trees, as well as for grasses and forbs (Figure 11.6, left and right panels). In addition, both grass and forb biomass increased following FireSmart when compared with the period before, and such increases could not be attributed to other causes because they were more pronounced in treatment areas than in control sites (Figure 11.7). Herbaceous forage species, especially grasses, are physiologically adapted to grow rapidly under high light and are rapid pioneer species in forest gaps (Keenan and Kimmins, 1993). Other studies have shown that, following tree removal, grass and forbs increase in biomass for approximately 10 years

then slowly decline as the cut block ages and canopy closure increases (Visscher and Merrill, 2009).

Although we focused on specific stand types that are characteristic of the mountainous environment of the study area, the applicability of our findings is broader. Other studies have reported a similar response of vegetation to environmental disturbances (Bergeron et al., 1999; Crawford, 1976; Keenan and Kimmins, 1993). Regardless of stand type, forest thinning, logging, fires and other disturbances are likely to increase forage availability for ungulates, which is consistent with the foraging patterns we observed for the collared elk. We observed that, following FireSmart removal, elk usage of treated sites increased both during the summer and the winter season (Figure 11.8).

Elk in the study area may need to access those openings where forage abundance is higher in the winter season when forage is scarce elsewhere (Robinson et al., 2010). Additionally, the increases in plant biodiversity and biomass that we documented (see text box below and earlier sections) were most significant for grasses and forbs, which are the most important component

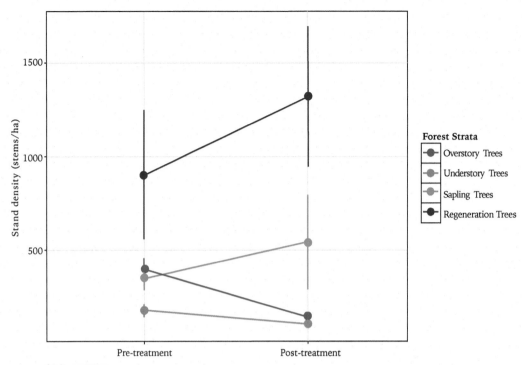

Figure 11.5 Stand density (stems/ha, with mean ± standard deviation (SD) values) in the experimental periods before (pre-treatment) and after timber removal (post-treatment) assessed in vegetation layers (Forest Strata), including overstory, understory, sapling and regeneration trees in JNP, Alberta, 2003–2012.

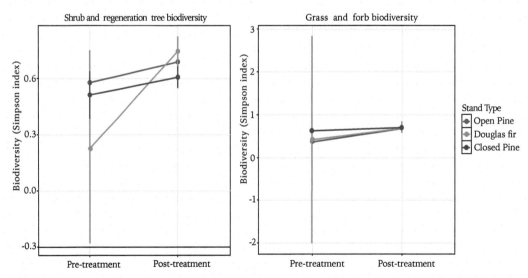

Figure 11.6 Shrub and regeneration tree diversity, and grass and forb diversity (calculated with Simpson biodiversity index, with mean ± SD values) before and after the FireSmart treatment according to stand type (open pine, Douglas fir and closed pine) in JNP, Alberta, 2003–2012.

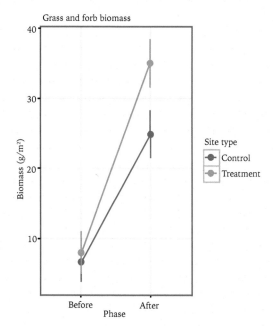

Figure 11.7 Total biomass of grasses and forbs (g /m², with mean ± SD values) in the experimental phases before and after FireSmart timber removal on treatment and control sites in JNP, Alberta, 2003–2012.

of elk forage (van Dyke et al., 2012), especially in winter (Hebblewhite, 2006). Ungulate populations have been known to disperse or even decline when forests mature and increase in canopy closure (Peek et al., 2002). However, ungulate populations can be maintained or increased when forage is supplemented during times of limited resource in the winter period (Boyce, 1988). Therefore, an increase in grass and forb biomass, like that we observed, may be critical (Irwin and Peek, 1983; Visscher and Merrill, 2009), especially for ungulates living in environments that have been adversely impacted by previous fire-suppression methods.

11.8 Conclusion

This case study indicates that careful manipulations used to manage forest fires could

Following FireSmart removal, elk usage of treated sites increased during both summer and winter seasons (Figure 11.8). However, the season affected elk usage (x^2 = 15.262, $P < 0.001$), increasing it by 21.25 ± 5.40 pellet groups/ha in winter compared with summer. Elk, like those of this study, were therefore likely influenced by the distribution and quality of forage resources in the landscape (Langvatn and Hanley, 1993; Thomas and Toweill, 1982; van Dyke et al., 2012; Wilmshurst et al., 1995) – resources that were more abundant in FireSmart-treated sites. Logging, like the thinning provided by FireSmart, is considered to be beneficial to ungulates such as deer, elk and moose, because such disturbances increase forage production and create more edge habitat typically selected by ungulates (Basile and Jensen, 1971; Harper, 1971; Krefting, 1962). Consistent with this interpretation, Lyon and Jensen (1980) also found, like us, that ungulates use logged forest openings in search of better quality or greater quantities of forage. However, this selection is often influenced in ungulates by security requirements during feeding. Therefore, this study's GPS-collared elk, as demonstrated by our resource selection analysis, were also found to prefer little openings with forest cover nearby.

mimic the positive effects of natural disturbance. The goal of the FireSmart program in JNP was to reduce the risk of wildfire losses to infrastructure in a local commuity, and adjacent human developments, as well as to restore the ecological condition of fire-dependent areas and enhance wildlife habitats (Parks Canada & Municipality of Jasper, 2011). Overall, and in keeping with the stated objectives, we found that the FireSmart program

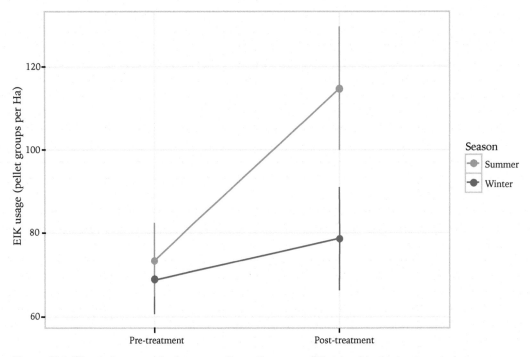

Figure 11.8 Elk relative usage (pellet groups/ha, with mean ± SD values) in the treatment phases before and after FireSmart program's timber removal in summer and winter seasons in JNP, Alberta, 2003–2012.

most likely increased forage availability for elk and other ungulates in general by increasing shrub regeneration and tree diversity, as well as increasing the diversity and biomass of grasses and herbaceous plants, which are all key components of elk diet (Arnison, 2014). Most ecologists and environmental planners now agree on the fundamental role played by fires in shaping and reshaping natural landscapes (Noss et al., 2006). However, several areas, including our study area, are still shaped by decades-long regimes of fire suppression that have created an unnaturally old forest. This study demonstrates that programs such as FireSmart, which are designed to mimic natural disturbance, can contribute to the biodiversity of ecosystems while also likely diminishing the risk for local communities of catastrophic fires.

References

Ager, A. A., Johnson, B. K., Kern, J. W., & Kie, J. G. (2003). Daily and seasonal movements and habitat use by female Rocky Mountain elk and mule deer. *Journal of Mammalogy, 84*(3), 1076–1088. https://doi.org/10.1644/BBa-020

Arnison, C. (2014). *Forest stand management and implications for elk selection in Jasper National Park, Alberta.*

Austin, M. P. (2002). Spatial prediction of species distribution: An interface between ecological theory and statistical modelling. *Ecological Modelling, 157*(2–3), 101–118. https://doi.org/10.1016/S0304-3800(02)00205-3

Bailey, J. K., & Whitham, T. G. (2002). Interactions among fire, aspen, and elk affect insect diversity: Reversal of a community response. *Ecology, 83*(6), 1701–1712. https://doi.org/10.1890/0012-9658(2002)083[1701:IAFAAE]2.0.CO;2

Basile, J., & Jensen, C. E. (1971). *Grazing potential on lodgepole pine clearcuts in Montana.*

Beall, R. C. (1976). Elk habitat selection in relation to thermal radiation. In *Proceedings of the Elk-Logging Roads Symposium* (pp. 97–100). University Idaho.

Bergeron, Y., Harvey, B., Leduc, A., & Gauthier, S. (1999). Forest management guidelines based on natural disturbance dynamics: Stand- and forest-level considerations. *Forestry Chronicle, 75*(1), 49–54. https://doi.org/10.5558/tfc75049-1

Bettinger, P. (2010). An overview of methods for incorporating wildfires into forest planning models. *Mathematical and Computational Forestry and Natural Resource Sciences, 2*(1).

Boyce, M. S. (1988). Elk winter feeding dampens population fluctuations at the national elk refuge. In *Western States and Provinces Elk Workshop Proceedings*.

Boyce, M. S. (1991). Migratory behavior and management of elk (*Cervus elaphus*). *Applied Animal Behaviour Science, 29*(1), 239–250. https://doi.org/10.1016/0168-1591(91)90251-R

Boyce, M. S., Vernier, P. R., Nielsen, S. E., & Schmiegelow, F. K. A. (2002). Evaluating resource selection functions. *Ecological Modelling, 157*(2–3), 281–300. https://doi.org/10.1016/S0304-3800(02)00200-4

Burns, R. M., & Honkala, B. H. (1990). *Silvics of North America, 1. Conifers*. United States Department of Agriculture. https://www.srs.fs.usda.gov/pubs/misc/ag_654_vol1.pdf

Chavardès, R. D., Daniels, L. D., Gedalof, Z. E., & Andison, D. W. (2018). Human influences superseded climate to disrupt the 20th century fire regime in Jasper National Park, Canada. *Dendrochronologia, 48*, 10–19. https://doi.org/10.1016/j.dendro.2018.01.002

Christianson, D. A., & Creel, S. (2007). A review of environmental factors affecting elk winter diets. *Journal of Wildlife Management, 71*(1), 164–176. https://doi.org/10.2193/2005-575

Clark, B. K., Clark, B. S., Johnson, L. A., & Haynie, M. T. (2001). Influence of roads on movements of small mammals. *Southwestern Naturalist, 46*(3), 338–344. https://doi.org/10.2307/3672430

Cook, J. G. (2002). Nutrition and food. *North American elk: Ecology and management*. Smithsonian Institution Press, 259–349.

Crawford, H. S. (1976). *Relationships between forest cutting and understory vegetation: An overview of eastern hardwood stands*. USDA Forest Service Reserach Paper.

Daniel, T. W., Helms, J. A., & Baker, F. S. (1979). *Principles of silviculture* (2nd ed). McGraw-Hill.

DeCesare, N. J. (2012). Separating spatial search and efficiency rates as components of predation risk. *Proceedings. Biological Sciences, 279*(1747), 4626–4633. https://doi.org/10.1098/rspb.2012.1698

D'eon, R. G. D., & Delparte, D. (2005). Effects of radio-collar position and orientation on GPS radio-collar performance, and the implications of PDOP in data screening. *Journal of Applied Ecology, 42*(2), 383–388. https://doi.org/10.1111/j.1365-2664.2005.01010.x

Dickson, B. G., & Beier, P. (2002). Home-range and habitat selection by adult cougars in Southern California. *Journal of Wildlife Management, 66*(4), 1235–1245. https://doi.org/10.2307/3802956

Dodd, N. L., Gagnon, J. W., Boe, S., & Schweinsburg, R. E. (2007). Assessment of elk highway permeability by using global positioning system telemetry. *Journal of Wildlife Management, 71*(4), 1107–1117. https://doi.org/10.2193/2006-106

Fryxell, J. M., & Sinclair, A. R. E. (1988). Causes and consequences of migration by large herbivores. *Trends in Ecology and Evolution, 3*(9), 237–241. https://doi.org/10.1016/0169-5347(88)90166-8

Gagnon, J. W., Theimer, T. C., Dodd, N. L., Boe, S., & Schweinsburg, R. E. (2007). Traffic volume alters elk distribution and highway crossings in Arizona. *Journal of Wildlife Management, 71*(7), 2318. https://doi.org/10.2193/2006-224

Geist, V. (2002). Adaptive behavioral strategies. *North American elk: Ecology and management*. Smithsonian Institution Press.

Green, R. H. (1979). *Sampling design and statistical methods for environmental biologists*. John Wiley & Sons.

Harper, J. A. (1971). *Ecology of Roosevelt elk*. Oregon State Game Commission.

Hebblewhite, M. (2006). Linking predation risk and forage to ungulate population dynamics [Dissertation]. University of Alberta.

Hebblewhite, M., & Merrill, E. H. (2009). Trade-offs between predation risk and forage differ between migrant strategies in a migratory ungulate. *Ecology, 90*(12), 3445–3454. https://doi.org/10.1890/08-2090.1

Hebblewhite, M., Merrill, E. H., & Mcdonald, T. L. (2005). Spatial decomposition of predation risk using resource selection functions: An example in a wolf-elk predator–prey system. *Oikos, 111*(1), 101–111. https://doi.org/10.1111/j.0030-1299.2005.13858.x

Hebblewhite, M., Merrill, E., & McDermid, G. (2008). A multi-scale test of the forage maturation hypothesis in a partially migratory ungulate population. *Ecological Monographs, 78*(2), 141–166. https://doi.org/10.1890/06-1708.1

Hershey, T. J., & Leege, T. A. (1976). Influences of logging on elk on summer range in northcentral Idaho. In *Proceedings, Elk Logging Roads Symposium* (pp. 73–80).

Hillis, J. M., Thompson, M. J., Canfield, J. E., Lyon, L. J., & Lonner, T. N. (1991). Defining elk security: The Hillis paradigm. In *Proceedings of the Elk Volunerability Symposium* (pp. 38–43).

Holroyd, G. L., & van Tighem, K. J. (1983). *Ecological (biophysical) land classification of Banff and Jasper National Parks, 3. The wildlife inventory. Canadian Wildlife Service.*

Huot, J. (1974). Winter habitat of white-tailed deer at thirty-one mile Lake, Quebec. *Canadian Field-Naturalist, 88,* 293–301.

Irwin, L. L., & Peek, J. M. (1983). Elk habitat use relative to forest succession in Idaho. *Journal of Wildlife Management, 47*(3), 664–672. https://doi.org/10.2307/3808602

Johnson, C. J., Seip, D. R., & Boyce, M. S. (2004). A quantitative approach to conservation planning: Using resource selection functions to map the distribution of mountain caribou at multiple spatial scales. *Journal of Applied Ecology, 41*(2), 238–251. https://doi.org/10.1111/j.0021-8901.2004.00899.x

Keenan, R. J., & Kimmins, J. P. (1993). The ecological effects of clear-cutting. *Environmental Reviews, 1*(2), 121–144. https://doi.org/10.1139/a93-010

Kelly, L. T., & Brotons, L. (2017). Using fire to promote biodiversity. *Science, 355*(6331), 1264–1265. https://doi.org/10.1126/science.aam7672

Kemball, K. J., Wang, G. G., & Dang, Q.-L. (2005). Response of understory plant community of boreal mixedwood stands to fire, logging, and spruce budworm outbreak. *Canadian Journal of Botany, 83*(12), 1550–1560. https://doi.org/10.1139/b05-134

Kittle, A. M., Fryxell, J. M., Desy, G. E., & Hamr, J. (2008). The scale-dependent impact of wolf predation risk on resource selection by three sympatric ungulates. *Oecologia, 157*(1), 163–175. https://doi.org/10.1007/s00442-008-1051-9

Krefting, L. W. (1962). Use of silvicultural techniques for improving deer habitat in the Lake States. *Journal of Forestry, 60*(1), 40–42.

Langvatn, R., & Hanley, T. A. (1993). Feeding-patch choice by red deer in relation to foraging efficiency:

An experiment. *Oecologia, 95*(2), 164–170. https://doi.org/10.1007/BF00323486

Limited, E. R. (2008). Hunting in Alberta in 2008: Performance, value and socioeconomic impacts. http://www1.agric.gov.ab.ca/$Department/deptdocs.nsf/all/csi12823/$FILE/Volume-II-Hunting-May-15.pdf

Lyon, L. J., & Jensen, C. E. (1980). Management implications of elk and deer use of clear-cuts in Montana. *Journal of Wildlife Management, 44*(2), 352–362. https://doi.org/10.2307/3807965

Manly, B. F. J. (1993). *Resource selection by animals.* Chapman & Hall.

Manly, B. F. J., Mcdonald, L. L., Thomas, D. L., Mcdonald, T. L., & Erickson, W. P. (2002). In *Resource selection by animals statistical design and analysis for field* (2nd ed). Technology Publishing.

Marcum, C. (1976). Habitat selection and use during summer and fall months by a western Montana elk herd. In *Proceedings of the Elk-Logging-Roads Symposium* (pp. 91–96). Forestry Wildlife and Range Experiment Station, University of Idaho.

McDermid, G. J. (2006). *Remote sensing for large-area, multi-jurisdictional habitat mapping: Vol. PhD.* University of Waterloo.

Mitchell, M. P. (2006). *Montane landscape heterogeneity and vegetation change in jasper National Park, Alberta (1949–1997)* (Vol. MR28374). University of Northern British Columbia.

Montgomery, R. A., Roloff, G. J., & Millspaugh, J. J. (2012). Importance of visibility when evaluating animal response to roads. *Wildlife Biology, 18*(4), 393–405. https://doi.org/10.2981/11-123

Noss, R. F., Franklin, J. F., Baker, W. L., Schoennagel, T., & Moyle, P. B. (2006). Managing fire-prone forests in the western United States. *Frontiers in Ecology and the Environment, 4*(9), 481–487. https://doi.org/10.1890/1540-9295(2006)4[481:MFFITW]2.0.CO;2

Parks Canada, and Municipality of Jasper. (2011). *Jasper community sustainability plan.*

Pauley, G. R., Peek, J. M., & Zager, P. (1993). Predicting white-tailed deer habitat use in Northern Idaho. *Journal of Wildlife Management, 57*(4), 904–913. https://doi.org/10.2307/3809096

Peck, V. R., & Peek, J. M. (1991). Elk, Cervus elaphus, habitat use related to prescribed fire, Tuchodi River, British Columbia. *Canadian Field-Naturalist, 105*(3), 354–362.

Peek, J. M., Dennis, B., & Hershey, T. (2002). Predicting population trends of mule deer. *Journal* of *Wildlife Management*, *66*(3), 729–736. https://doi.org/10.2307/3803138

Pengelly, W. L. (1972). Clearcutting: Detrimental aspects for wildlife resources. *Journal of Soil and Water Conservation*.

Pettorelli, N., Vik, J. O., Mysterud, A., Gaillard, J. M., Tucker, C. J., & Stenseth, N. C. (2005). Using the satellite-derived NDVI to assess ecological responses to environmental change. *Trends in Ecology and Evolution*, *20*(9), 503–510. https://doi.org/10.1016/j.tree.2005.05.011

Preisler, H. K., Ager, A. A., & Wisdom, M. J. (2006). Statistical methods for analysing responses of wildlife to human disturbance. *Journal of Applied Ecology*, *43*(1), 164–172. https://doi.org/10.1111/j.1365-2664.2005.01123.x

Rhemtulla, J. M. (1999). *Eighty years of change: The montane vegetation of Jasper National Park*. University of Alberta.

Risbrudt, C. (1995). Ecosystem management: A framework for management of our National Forests. *Natural Resources and Environmental Issues, 5*(1), 15.

Robinson, B. G., Hebblewhite, M., & Merrill, E. H. (2010). Are migrant and resident elk (Cervus elaphus) exposed to similar forage and predation risk on their sympatric winter range? *Oecologia, 164*(1), 265–275. https://doi.org/10.1007/s00442-010-1620-6

Rogala, J. K., Hebblewhite, M., Whittington, J., White, C. A., Coleshill, J., & Musiani, M. (2011). Human activity differentially redistributes large mammals in the Canadian Rockies national parks. *Ecology and Society, 16*(3). https://doi.org/10.5751/ES-04251-160316

Sachro, L. L., Strong, W. L., & Gates, C. C. (2005). Prescribed burning effects on summer elk forage availability in the subalpine zone, Banff National Park, Canada. *Journal of Environmental Management, 77*(3), 183–193. https://doi.org/10.1016/j.jenvman.2005.04.003

Sawyer, H., Kauffman, M. J., Nielson, R. M., & Horne, J. S. (2009). Identifying and prioritizing ungulate migration routes for landscape-level conservation. *Ecological Applications, 19*(8), 2016–2025. https://doi.org/10.1890/08-2034.1

Schimmel, J., & Granström, A. (1996). Fire severity and vegetation response in the boreal Swedish forest. *Ecology, 77*(5), 1436–1450. https://doi.org/10.2307/2265541

Seidl, R., Fernandes, P. M., Fonseca, T. F., Gillet, F., Jönsson, A. M., Merganičová, K., Netherer, S., Arpaci, A., Bontemps, J., Bugmann, H., González-Olabarria, J. R., Lasch, P., Meredieu, C., Moreira, F., Schelhaas, M., & Mohren, F. (2011). Modelling natural disturbances in forest ecosystems: A review. *Ecological Modelling, 222*(4), 903–924. https://doi.org/10.1016/j.ecolmodel.2010.09.040

Shepherd, B., & Whittington, J. (2006). Response of wolves to corridor restoration and human use management. *Ecology and Society, 11*(2). https://doi.org/10.5751/ES-01813-110201

Sweeney, J. M., & Sweeney, J. R. (1984). Snow depths influencing winter movements of elk. *Journal of Mammalogy*. American Society of Mammalogists, *65*(3), 524–526. https://doi.org/10.2307/1381113

Thomas, J. W., & Toweill, D. E. (1982). Elk of North America: Ecology and management. *Elk of North America: Ecology and Management*.

van Dyke, F., Fox, A., Harju, S. M., Dzialak, M. R., Hayden-Wing, L. D., & Winstead, J. B. (2012). Response of elk to habitat modification near natural gas development. *Environmental Management, 50*(5), 942–955. https://doi.org/10.1007/s00267-012-9927-1

Visscher, D. R., & Merrill, E. H. (2009). Temporal dynamics of forage succession for elk at two scales: Implications of forest management. *Forest Ecology and Management, 257*(1), 96–106. https://doi.org/10.1016/j.foreco.2008.08.018

Walter, W. D., Lavelle, M. J., Fischer, J. W., Johnson, T. L., Hygnstrom, S. E., & VerCauteren, K. C. (2010). Management of damage by elk (Cervus elaphus) in North America: A review. *Wildlife Research, 37*(8), 630–646. https://doi.org/10.1071/WR10021

Westhaver, A. L. (2003). Environmental Screening Report: FireSmart – ForestWise community protection and forest restoration project. Foothills Model Forest, Jasper National Park. Park Registry File J03–004. Jasper National Park.

White, C. A., Olmsted, C. E., & Kay, C. E. (1998). Aspen, elk, and fire in the Rocky Mountain national parks of North America. *Wildlife Society Bulletin*, 449–462.

White, P. J., Proffitt, K. M., Mech, L. D., Evans, S. B., Cunningham, J. A., & Hamlin, K. L. (2010). Migration of northern Yellowstone elk: Implications of spatial structuring. *Journal of Mammalogy, 91*(4), 827–837. https://doi.org/10.1644/08-MAMM-A-252.1

Whittington, J., St. Clair, C. C., & Mercer, G. (2005). Spatial responses of wolves to roads and trails in mountain valleys. *Ecological Applications, 15*(2), 543–553. https://doi.org/10.1890/03-5317

Wilmshurst, J. F., Fryxell, J. M., & Hudsonb, R. J. (1995). Forage quality and patch choice by wapiti (Cervus elaphus). *Behavioral Ecology, 6*(2), 209–217. https://doi.org/10.1093/beheco/6.2.209

Wright, E. F., Canham, C. D., & Coates, K. D. (2000). Effects of suppression and release on sapling growth for 11 tree species of northern, interior British Columbia. *Canadian Journal of Forest Research, 30*(10), 1571–1580. https://doi.org/10.1139/cjfr-30-10-1571

Chapter 12

Challenges and opportunities for yak herders in Bhutan

V. R. Monger and Susan C. Cork

Abstract

Yaks and yak hybrids constitute 12.2% of the livestock population in Bhutan, and yak rearing is one of the most important activities for highland communities. These communities lead a transhumant lifestyle with yaks as their main source of livelihood. The highland communities of Bhutan are known as Brokpas, Bjops and Lakhaps. The yak (*Bos grunniens*) is a multi-purpose animal, regarded as an iconic species of the Himalayan region adapted to the harsh climatic conditions in alpine and sub-alpine regions at altitudes ranging from 2500 to 6000 m above sea level (masl). In Bhutan, yak rearing is practised in ten highland districts and supports the livelihood of about 10% of the population. According to the Livestock Census 2020, the yak population of Bhutan was 40,897. However, if not supported, the population of yaks and yak-herding communities might further decline, as has occurred over past decades. There are various reasons for this, one being the urban migration of youth from rural villages in search of better economic opportunities, another being the harsh climatic conditions in yak-rearing areas. Other reasons include the lack of proper infrastructure, insufficient feed for yaks in winter, livestock diseases and challenges associated with a long-term flawed breeding programme. Hence, for the encouragement and survival of these marginal pastoral highlanders, the Royal Government of Bhutan is adopting a range of interventions that are being implemented through integrated interagency interventions, which we will examine in this chapter.

12.1 Introduction

Bhutan is surrounded by China to the north and India to the west, south and east. Yaks and yak hybrids constitute 12.2% of livestock population in the country (Department of Livestock, 2013). Yak rearing is one of the most important activities for highland communities, who lead a transhumant lifestyle, with yaks as their main source of livelihood. The highland communities of Bhutan are known such as Brokpas in the eastern region, Bjops in the western Bhutan

Susan C. Cork and Douglas P. Whiteside (eds) **Case Studies in EcoHealth**
DOI: 10.52517/9781789183368.012, © 5m Books Ltd, 2024

and Lakhaps in the west-central region, all of which basically mean 'pastoralists' (Gyamtsho, 2000). Yaks are the main source of livelihood of the highland people of India, Bhutan, Nepal and Tibet (Chettri, 2008; Dong et al., 2007; Karchung, 2011; Tshering, 2004). Yak herding is one of the most important activities for highland communities, who lead a transhumant lifestyle with yaks, along with some other livestock, as their main source of livelihood.

12.2 Yak-rearing practices in Bhutan

In Bhutan, yak rearing is practised in 10 highland districts (Figure 12.1b) and supports the livelihood of about 10% of the highland population, accounting for 12.2% of the total livestock population. The total population of yaks in 2019 was 41,918 but according to the Livestock Census 2020, the yak population in Bhutan was reduced to 40,897 yaks. Genetically there are two types of yak in Bhutan: western/central and eastern (Dorji, 2002). Genetically, western and central Bhutanese yaks are closely related but are less closely related to yaks from eastern Bhutan. In the western-central region, yaks reared are pure bred whereas in the east they are often hybridized with cattle (e.g. *Bos indicus*), involving up to 53% of the eastern yak population). Hybridization with cattle leads to a 200% increase in milk production.

The yak (*Bos grunniens*) is a medium-sized ruminant with long fur that hangs down below the belly and it is native to both the Himalayan range and the Qinghai-Tibetan Plateau; these animals are well adapted to high altitudes. It is considered an iconic Himalayan species that is exceptionally adapted to the harsh climatic conditions in alpine and sub-alpine regions at altitudes ranging from 2500 to 6000 m above sea level (masl). They are multi-purpose animals that provide herders with fibre, animal protein and fat – in the form of milk and meat – as well as draught power and manure. Milk is further processed into butter and cheese, dried or fermented for longer storage for months.

Figure 12.1a A prize domestic yak at the Royal Highland Show, Bhutan. Credit: Dr V. R. Monger.

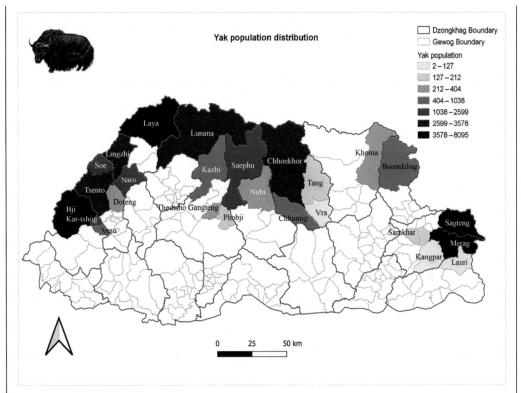

Figure 12.1b Yak distribution in Bhutan. Data sources: Department of Livestock (2019) and National Statistics Bureau (2020).

While wild yaks are generally dark blackish-brown in colouration, domestic yaks can be quite variable in colour, often having patches of rusty brown and cream. They have small ears and wide foreheads, with smooth horns that are generally dark in colour.

Yak herders in Bhutan live a transhumant (i.e. seasonally nomadic) life with their yaks. Transhumant pastoralism also maintains a socio-cultural landscape, which often has spiritual importance for the well-being of the community, and it is assumed to support and preserve biodiversity in the mountains. Yak-herding families have permanent homes near their winter pastures, with wooden fences or stone walls to provide some protection for livestock. Yak herders follow seasonal migration from higher elevations to lower elevations and vice versa to utilize the resources (Tsamdros) available during different times of the year (Dorji, 2013).

Winter pastures range from 3000 to 3800 masl and are generally used from November to April. During the summer, the herders migrate with their animals up to pastures at high altitudes ranging from 3500 to 5000 masl. At higher altitudes there are also wild animals such as snow leopards (*Panthera uncia*), blue sheep (*Pseudois nayaur*), red pandas (*Ailurus fulgens*), musk deer (*Moschus* sp.) and the national mammal, the takin (*Budorcas taxicolor*).

At the high altitudes where yak are found, the weather conditions are very cold and freezing with heavy snow in winter. The yak herders carry out some cultivation such as growing barley, wheat, mustard, potato, buckwheat, radish and turnip near their permanent settlements. The average size of a yak herd per household is 42–85 animals (Dorji, 2000).

In recent years yaks have increasingly been used for the purpose of tourism at high altitudes, for tourists and trekking groups. Yak dung is typically used as fuel along with wood for cooking by the nomads. Yaks also play special roles in the religious and cultural life of the herders. In fact, the yak is an important symbol for these nomadic societies and it has shaped the culture and tradition of these societies over a long period of time. In Bhutan, yak production is the main source of livelihood for highland people (Jamtsho, 1996). According to Dong et al. (2007), yak and yak hybrids provide milk, milk products and meat as important sources of

food, as well as hair and hides for textiles and leather.

Yak herders migrate to lower elevation pastures in the winter months in search of greener pastures, as well as to avoid the extreme cold and snowfall higher up. Yaks are not provided any shelter during the year except for protecting their calves during cold periods in the winter season. Typically, the milking yak cows are brought to the herders' homestead in the morning for milking after which the calves are allowed to suckle the remaining milk. The calf is removed from the yak cow in the evening and sheltered in a makeshift enclosure. The nutrition of the yak depends on grazing (see Figure 12.3) in alpine meadows during the summer months, with little or no feed supplement provided. In the winter months, milking cows are fed

Transhumance is the migratory practice of moving livestock from one grazing ground to another in a seasonal cycle, typically to lowlands in winter and to highlands in summer, depending on the seasonal availability of forage and pastures (see Figure 12.2a,b). The alpine summer pastures, at around 5000 masl, are grazed until late September or early October and the herds begin to descend to winter pastures on lower ground at an altitude of around 2500–3500 masl. All yak-herding families in Bhutan have permanent homes near their winter pastures.

(a) (b)

Figure 12.2a,b. Domestic yaks on migration route with villagers in Marek, North East Bhutan. Photo credits: Dr Susan Cork.

with hay and also dried turnip, oilcake and salt. While the nutrition of grazing yaks in the summer is considered adequate, they are generally underfed during the winter due to feed scarcity and can be in very poor physical condition, thus producing relatively less milk in winter.

12.3 Challenges for yak herders in Bhutan

In spite of the various challenges, highlanders have adopted suitable strategies to sustain their livelihoods due to the efficient utilization of grassland resources by yaks that have evolved in the Himalayan mountains (Mishra

et al., 2010). However, over the years there has been a marked decline in the number of yak herders in Bhutan. This is for various reasons, but mainly due to the neglect of yak-herding communities in mainstream international development plans and limited yak husbandry support. Yaks and yak herding are losing their importance as the main source of livelihood for transhumant pastoralists in high-altitude mountain areas, and one of the main drivers of change is socio-economic development (i.e. alternative sources of income such as foraging for *Cordicpes* spp.) and urbanization (Tulachan and Neupane, 1999; Roder et al., 2002; Wangchuk et al., 2014). Some of the common issues faced by yak herders will now be discussed.

Figure 12.3 Domestic yak grazing on pasture in Haa district, northwest Bhutan. Photo credit: Dr Susan Cork.

12.3.1 Lack of basic amenities

Yak-herding communities have gained minimal benefit from resources allocated for development compared with lowland communities. Due to their migratory life style, most of the mainstream development efforts, including animal health service delivery, were unable to effectively reach these migratory and scattered populations. Furthermore, the difficult terrain and harsh environment of the mountain area has encouraged the younger generation to abandon the traditional eco-friendly lifestyle of highland livestock rearing and triggered a rural–urban migration in search of better living conditions. This has also created rural labour shortages across Bhutan. Although there have been some development interventions to support enhancement of yak communities' livelihoods, including pasture development and supply of inputs such as fencing materials and milking and milk-processing equipment, this has had little or no impact.

12.3.2 Insufficient feed in winter

Rangelands, locally known as Tshamdro, are the main summer grazing grounds for yaks at elevations of 4000–5000 masl, but these are degrading due to lack of management interventions (Roder et al., 2001). The winter grazing grounds are located near villages in the lowlands and there is greater pressure on rangelands, which has led to overgrazing. In the lowlands, pastures are shared with a wide variety of other livestock (domestic ruminants such as cattle and sheep, and horses) as well as wild animals including takin, blue sheep, sambar and musk deer (Richard et al., 2000; Wangchuk et al., 2009), leading to overgrazing in these limited pasturelands and resulting in a threat to yak husbandry (Behnke, 2003). As a result of inadequate forage, poor nutrition leads to slow growth,

health-related problems such as intestinal parasites and reduced fertility, weight loss and even death. Heavy snowfall at high altitudes during the winter makes the situation even worse and can cause mass mortality up to 50% in compromised populations (Wu and Yan, 2002).

12.3.3 Alternative source of income

In areas where collection of the fungal genus *Cordyceps* (*Ophiocordyceps sinensis*) is prevalent, this foraged niche crop has become the dominant income source for highland communities. *Cordyceps sinensis* is a composite consisting of the stromata of a fungus that parasitizes the larvae of some species of insects (Family: Hepialidae). It is thought to have significant medicinal benefits. This dead caterpillar, with its fungal infestation, is distributed across the Tibetan Plateau and its surrounding regions at an altitude above 3,000 m of the Himalayas and occurs in China, Bhutan, India and Nepal.

In Bhutan, *Cordyceps* is found in the remotest northern parts of Bhutan, at an altitude of 3400–4100 masl, and it fetches herding households a good annual income of up to US$24,500 (Wangchuk et al., 2012). The income earned from *Cordyceps* is very good and the profits are often invested in lowlands and urban areas, further accelerating rural–urban migration and abandonment of the traditional transhumance lifestyle.

12.3.4 Modernization

Yak-herding practices and traditions in the highlands have been passed down from generation to generation. However, as modernization of the country has taken place rapidly, more and more children are enrolled in schools and families are more likely to consider an alternative, less difficult lifestyle. Yak-herding families prefer to send their children to school to ensure a better future, thus leaving only the adults and elderly to tend the yaks (Dervillé, 2010).

12.3.5   Poor marketing opportunities

The processing of yak milk and milk products is usually done using age-old traditional knowledge and equipment. This is not only laborious and time consuming, but also potentially unhygienic with poor packaging and difficulty in marketing yak dairy products in urban areas. Transporting these dairy products from the mountains to the nearest farmers' markets is a key constraint, with spoilage having occurred by the time they reach the towns. Challenges to the sale of yak products include lack of supporting infrastructure, reliance on inefficient traditional equipment for processing, food safety concerns and lack of a proper marketing strategy. Thus, yak herding is no longer attractive as a main source of income for herders.

12.3.6   Poor yak-breeding practices

Lack of facilities and technical capacity for yak breeding has also contributed to the decline in yak herding. Yak breeding has not received adequate attention over past decades due to the lack of a strategic breeding plan. The scattered nature of yak herds and a lack of quality selection for yak bulls are the major constraints to improving breeding quality. Yak breeding through artificial insemination, with semen imported from China, did achieve limited success (Tshering et al., 2000), but the technology remains to be tested in Bhutan on a larger scale.

12.3.7   Poor animal health care coverage

Gid, also known as coenurosis, is the most common disease, causing high mortality among

yaks and affecting mainly young animals under 1–2 years of age (Tenzin, 1979). The dog is the definitive host for the tapeworm responsible, *Taenia multiceps,* and the yak is the intermediate host, in which cysts can form in the brain (Figure 12.4). Poisoning by plants such as *Senecio* spp., which contain the highly toxic chemical pyrrolizidine alkaloid (Dahal, 2000), and by contaminated water are also reported to be a major cause of yak mortality. Yaks can also be infested with a range of internal parasites. Animal health care is available in Bhutan but it is hard to provide sufficient access in highland regions due to the difficult terrain.

12.3.8   Climate change

Climate change, land-use change and population dynamics are the main drivers of environmental change in high-altitude areas (Sharma, 2012). Climate change is known to have a significant impact on species distribution and diversity patterns, and can also have a negative impact on yak populations because of their lack of tolerance to heat, the reduction in suitable grazing areas and associated decline in yak survival and/ or reproduction (Haynes et al., 2014). Research also shows that climate change may increase the risk of occurrence of disease in yaks (Wangchuk et al., 2013). According to studies conducted in different parts of the Himalayan region, climate change is forcing yak-herding communities to migrate to higher elevations in search of productive grazing lands, with an early start in the upward migration due to shortening of the winter period. Climate change is known to have a synergistic effect on the already existing challenges of dwindling yak populations, yak husbandry, degradation of high-altitude pastures and shortage of feed and fodder, and even on changing social norms (Gyamtsho, 2000). Yaks are accustomed to very cold temperatures and can survive up to minus 40 degrees Celsius,

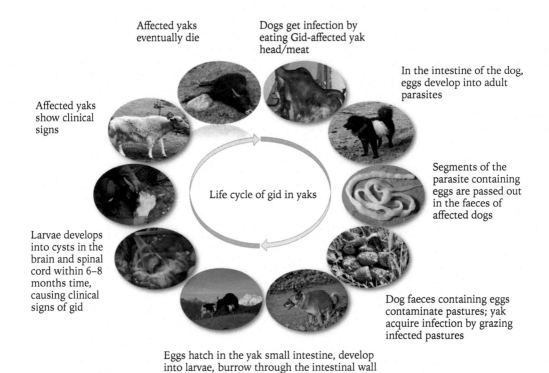

Affected yaks eventually die

Dogs get infection by eating Gid-affected yak head/meat

In the intestine of the dog, eggs develop into adult parasites

Affected yaks show clinical signs

Life cycle of gid in yaks

Segments of the parasite containing eggs are passed out in the faeces of affected dogs

Larvae develops into cysts in the brain and spinal cord within 6–8 months time, causing clinical signs of gid

Dog faeces containing eggs contaminate pastures; yak acquire infection by grazing infected pastures

Eggs hatch in the yak small intestine, develop into larvae, burrow through the intestinal wall and travel to the brain and spinal cord via the bloodstream

Figure 12.4 Life cycle of the tapeworm *Taenia multiceps*, the cause of gid disease in young yaks. The infected calf displays a circling movement, with the head tilting towards the location of the tapeworm cyst on the cerebral surface of the brain.

but they find is difficult when the temperature exceeds 13°C. Yaks can efficiently conserve their body heat during cold conditions but have minimal physiological mechanisms to dissipate heat, making them more susceptible to heat stress. Besides heat stress, high temperatures also affect growth and the availability of suitable fodder in alpine pasturelands. This, in turn, lowers the productivity of the animals. In addition to climate-related factors, there is a reduction in available grazing areas and also pasture lands have been degraded due to various urban developmental activities. The physiological and metabolic adjustment resulting from climate change may have negative consequences on yak production, and the occurrence of diseases has increased in the region (Sherpa and Kayastha,

2009). Some common livestock diseases, including foot-and-mouth disease, brucellosis, infectious bovine rhinotracheitis, haemorrhagic septicaemia, gid and tick-borne diseases also affect yak productivity in the Himalayan region (Gyamtsho, 2000).

12.4 Future actions to enhance the livelihoods of yak herders

All the challenges described above have led to a decline in yak-herding communities and their associated cultures and traditions in Bhutan's unique highland communities. There is a concern that these communities may vanish one day, with subsequent consequences to the highland ecosystem as well. Hence, for the encouragement and survival of these marginal pastoral highlanders, necessary interventions are being implemented in six broad areas:

- improving yak herder income and living conditions through product diversification
- supporting genetic improvement of the domestic yak
- gid elimination and improvement in yak health services
- rangeland restoration programmes
- pasture development support
- launch of a National Yak Federation.

12.5   Summary and conclusions

The highland communities of Bhutan, known as Brokpas, Bjops and Lakhaps, are the pastoralist groups that lead a transhumant lifestyle with yaks as their main source of livelihood. The Brokpas from the eastern region also maintain yak–cattle hybrids. Production of yak and yak hybrids is one of the most important activities for highland communities in Bhutan, and these animals are reared in ten highland districts supporting the livelihood of about 10% of the population. The domestic yak is a multipurpose animal and is regarded as an iconic species of the Himalayan region, because it is adapted to the harsh climatic conditions at altitudes ranging from 2500 to 6000 masl. With a population of 40,897 yaks, as per Livestock Census 2020, there has been a decline in yaks and yak herding in Bhutan for various reasons, including the rural–urban migration of youth in search of better economic opportunities and the harsh living conditions at high altitudes. Hence, to encourage the survival of these marginal pastoral highlanders, the Royal Government of Bhutan is adopting a range of interventions that are being implemented through integrated interagency interventions to support highland livestock herders and associated highland ecosystems.

References

Behnke, R. (2003). Reconfiguring property rights and land use. In C. Kerven (Ed.), *Prospects for pastoralism in Kazakhstan and Turkmenistan*. RoutledgeCurzon.

Chettri, N. (2008). *Local and indigenous practices on adaptation: An experience from herder's life of western Bhutan* [Mountain Forum bulletin], 8(1) (pp. 19–21).

Dahal, N. (2000). An investigation report on algae poisoning in yak. *Journal of Renewable Natural Resources of Bhutan*, 1(2), 65–77.

Dervillé, M. (2010). *Assessment of the opportunity to develop specific quality schemes for yak products in Bhutan*. Department of Livestock, Ministry of Agriculture and Forests (consultancy report).

Dong, S., Long, R., & Kang, M. (2007). Milking performance of China yak (*Bos grunniens*): A preliminary report. *African Journal of Agricultural Research*, 2(3), 52–57.

Dorji, K. (2013). *Rangeland tenure transfer; an analysis of policy and legal issues in Bhutan*. Centre for Bhutan Studies & GNH Studies.

Dorji, T. (2000). *Genotypic and phenotypic characterization of the yak (Bos grunniens) and yak farming systems in Bhutan* [MSc Thesis]. Institute of Land and Food Resources, the University of Melbourne.

Dorji, T. (2002). Yak production systems in Bhutan. In *Proceedings of the 5th TAPAFON Meeting*, Bajo. Rome, Bhutan, Italy. Food and Agriculture Organization of the United Nations.

Gyamtsho, P. (2000). Economy of yak herders. *Journal of Bhutan Studies*, 2(1), 90–135. https://lib.icimod.org/record/10412/files/6482.pdf

Haynes, M. A., Kung, K. S., Brandt, J. S., Yongping, Y., & Waller, D. M. (2014). Accelerated climate change and its potential impact on yak herding livelihoods in the eastern Tibetan Plateau. *Climatic Change*, 123(2), 147–160. https://doi.org/10.1007/s10584-013-1043-6

Jamtsho, P. (1996). Economy of herders. *Journal of Bhutan Studies*, 2(1/4).

Karchung, G. (2011). Diminishing cultures of Bhutan: Costume of Merag community. *SAARC Culture*, 2, 17–43.

Mishra, C., Bagchi, S., Namgail, T., & Bhatnagar, Y. V. (2010). Multiple use of trans-Himalayan rangelands: Reconciling human livelihoods with wildlife conservation. In J. T. du Toit, R. Kock & J. C. Deutsch (Eds.), *Wild rangelands: Conserving wildlife while maintaining livestock in semi-arid ecosystems* (pp. 291–311). Wiley-Blackwell.

National Statistics Bureau. (2020). Annual livestock statistics 2020. Renewable Natural Resources Statistics Division (RNR-SD) Directorate Services, Royal Government of Bhutan, MoAF. https://www.nsb.gov.bt/livestock-statistics-report/

Ning, W., Shaoliang, Y., Joshi, S., & Bisht, N. (Eds.) (2016). *Yak on the Move: Transboundary Challenges and Opportunities for Yak Raising in a Changing HKH Region.* International Centre for Integrated Mountain Development (ICIMOD). https://lib.icimod.org/record/31938/files/ICIMOD-Yak-2016.pdf?type=primary

Richard, C., Jainlin, H., Hanoote, O., McVeigh, C., & Rege, J. (2000). Yak production in central Asian highlands. In *Proceedings of the Third International Congresson Yak Held in Lhasa, P.R. China.* Nairobi, Kenya. International Livestock Research Institute.

Roder, W., Wangdi, K., Gyamtsho, P., & Dorji, K. (2001). *Feeding the herds: Improving fodder resources in Bhutan.* ICIMOD.

Roder, W., Gratzer, G., & Wangdi, K. (2002). Cattle grazing in the conifer forests of Bhutan. *Mountain Research and Development, 22*(4), 368–374. https://doi.org/10.1659/0276-4741(2002)022[0368:CGITCF]2.0.CO;2

Sharma, E. (2012). Climate change and its impacts in the Hindu Kush-Himalayas: An introduction. In A. Lamadrid, I. Kelman & R. Shaw (Eds.), *Climate change modelling for local adaptation in the Hindu Kush-Himalayan region* (5th ed) (pp. 17–32). Emerald Group Publishing Limited. https://doi.org/10.1108/S2040-7262(2012)0000011008

Sherpa, Y. D., & Kayastha, R. B. (2009). A study of livestock management patterns in Sagarmatha National Park, Khumbu region: Trends as affected by socio-economic factors and climate change. *Journal of Science, Engineering and Technology, 5*(11), 110–120.

Tenzin, D. (1979) *Studies of gid disease in yak with special reference to control measures Journal of Animal Husbandry of Bhutan pp. 1–4.*

Tshering, C. (2004). Rangeland of Bhutan. In *Proceedings of the International Congress on Yak.* IVIS Publication.

Tshering, L., Gurung, R., & Chungsila. (2000). Artificial insemination trial of yaks in Bhutan. In *Proceedings of the 3rd International Congress on Yak, Lhasa, Tibet.*

Tulachan, P. M., & Neupane, A. (1999). *Livestock in mixed farming systems of the Hindu Kush-Himalayas: Trends and sustainability.* Nepal International Centre for Integrated Mountain Development.

Wangchuk, T., Rai, M., Thinley, P., Nima, C., & Lhamu, Y. (2009). *Gamri watershed management plan.* SNV. Netherland Development Organization.

Wangchuk, T., Sangay, L., Norbu Nawang, A. & Sherub, T. (2012). *Impacts of Cordyceps collection on livelihoods and Alpine ecosystems in Bhutan as ascertained from questionnaire survey of Cordyceps collectors. Bumthang royal Government of Bhutan: UWICE press.*

Wangchuk, K., Wurzinger, M., Darabant, A., Gratzer, G., & Zollitsch, W. (2014). The changing face of cattle raising and forest grazing in the Bhutan Himalaya. *Mountain Research and Development, 34*(2), 131–138. https://doi.org/10.1659/MRD-JOURNAL-D-13-00021.1

Wangdi, J. (2016). The future of yak farming in Bhutan. Policy measures Government should adopt. *Rangeland Journal, 38*(4), 367–371. https://doi.org/10.1071/RJ15111

Wu, N., & Yan, Z. (2002). Climate variability and social vulnerability on the Tibetan Plateau. Dilemmas on the road to pastoral reform, Erdkunde. *Archive for Scientific Geography, 56*(1), 2–14.

Chapter 13

Human–wildlife conflict: An EcoHealth approach to addressing primate interactions with the public in Asia

Sangay Rinchen and Susan C. Cork

Abstract

As human populations expand into new ecosystems on a global scale, there is a higher likelihood of more frequent interactions between human communities and wildlife. The major causes of human–wildlife conflict include agricultural expansion, enlargement of human settlements, overgrazing by domestic livestock and deforestation. Although coexistence with wildlife is possible, fear of the unknown and competition for resources can result in conflict that can threaten wildlife populations. There are four main human–wildlife conflict types commonly recognized: (1) attack on humans, (2) livestock depredation, (3) crop raiding and (4) property damage. Other types of conflicts recognized are poaching of wildlife for meat and trade and retaliatory killing of wild animals. In this chapter we will discuss examples of human–wildlife conflict and consider what can be done to address these. We also present a case study from Bhutan examining public perspectives on human–monkey interaction and how government agencies are managing the interaction between wild macaques and humans along a major highway.

13.1 Introduction

There are numerous examples of human–wildlife conflict (HWC) recognized globally, these include elephant and monkey raids on crops in Asia, big cat predation of livestock in Africa and armadillo raids on honey hives in South America (Frank et al., 2019; Zimmerman, 2019). With the diversity of examples and the different cultural and varied socioeconomic contexts, it is clear that HWC requires an interdisciplinary approach, including a socioeconomic perspective, in order to be understood and effectively managed. There are a wide range of case studies available to illustrate this. Frank et al. (2019) examined a wide range of contemporary and historical case studies and found that some approaches to HWC failed because they were focused on the ecology and economics of wildlife-related losses but did not deal effectively

Susan C. Cork and Douglas P. Whiteside (eds) **Case Studies in EcoHealth**
DOI: 10.52517/9781789183368.013, © 5m Books Ltd, 2024

with the underlying causes of the conflict. They recognized that there needed to be more focus on the human dimension of HWC with a more thorough assessment of the reasons behind the conflict, as well as understanding the differing perspectives of local communities on the wild species with which they interact (Frank, 2016). There are now a number of studies describing the development and implementation of programmes to ensure community engagement in mitigating HWC and promoting tolerance of living with wildlife (Hill et al., 2017; Pooley et al., 2017; Woodroffe et al., 2005). There is also a growing recognition that many examples of what appears to be HWC might actually be a reflection of human–human conflicts, lack of understanding of animal behaviour and a lack of trust towards the agencies or authorities overseeing wildlife (Elbroch, 2020; Fraser-Celin et al., 2018).

Many effective programmes to mitigate HWC include a combination of economic incentives, technical solutions, legal protection and community involvement. However, in reality, lethal and non-lethal methods have been used to dissuade wildlife from consuming crops and attacking, or competing with livestock, while legal and economic instruments have generally been directed towards individuals and communities to prevent them from retaliating against wildlife. It is clear from the wider literature that every HWC situation is different and directly comparing one scenario with another can be misleading. HWC can become a significant challenge for wildlife conservation. However, it is clear that this can differ with the species involved, geographical region and cultural context. Frank et al. (2019) also suggest that it is apparent that no single approach has been successfully applied across a wide range of scenarios, although lessons learned can often lead to the development of novel new approaches to reduce HWC. From the various case studies considered, it is evident that the engagement

of social scientists and anthropologists, as well as considering the biology/ecology of the species and ecosystems involved, is important to ensure that sustainable and effective conservation plans can be developed. This also requires a shared responsibility and an understanding of the real causes of conflict between wildlife and communities.

Box 13.1   Examples of successful HWC mitigation programmes

Successful projects that have promoted human–wildlife coexistence include the use of beehive fences to protect crops from elephants and the 'lion guardians' project monitoring large carnivores to protect local communities in southern Kenya. Some other successful approaches have been applied elsewhere in similar contexts: for example, the 'Lion Defenders' in Tanzania (i.e. the Ruaha Carnivore Project). Both lion projects are supported by The Pride Alliance, which aims to conserve lions across Africa (http://pridelionalliance.org/).

It is clear that there are many factors to consider in the move from HWC to human–wildlife coexistence. This includes the species involved, the context and location in which the species is encountered and the culture of the local community as well as the sociocultural background, current conservation activities and enforcement practices and economic benefits (Frank, 2016). To fully appreciate this, a close understanding of the culture and experience base of the local communities engaged in the human–wildlife interaction is required (Fulton et al., 1996; Hrubes et al., 2001). Some approaches to human–monkey conflict are examined further below followed by a case study from Bhutan.

13.2   Human–monkey conflict in Asia

As mentioned above, it is well recognized that expansion of human activities into natural habitats frequently results in increased conflict between local wildlife and humans. In many parts of Asia this is well illustrated in studies on human conflict with monkeys. There are a number of reasons for this, including the destruction of natural food sources and replacement of these with more abundant food crops and waste produce from human activity. Monkeys and other wildlife that feed on such produce are then considered to be pests. In addition, the availability of such supplementary food can alter intergroup dynamics and may result in increased monkey populations. There is a lot of literature giving examples of monkeys damaging crops, stealing food and developing a dependence on human activity (Else, 1991). This change in human–monkey interaction can also result in aggressive encounters, which are especially seen around temple sites where monkeys expect to be fed by the visiting public.

The primate species most often cited as pests on the Indian subcontinent is the rhesus macaque (*Macaca mulatta*) of northern India and Nepal (Chaturvedi and Mishra, 2014; Reddy and Chander, 2016; Regmi et al., 2013; Srivastava and Begum, 2005) and the bonnet macaque (*M. radiata*) of southern India (Chakravarthy and Thyagaraj, 2005). In some regions the Hanuman or grey langur (*Semnopithecus* spp.) is also considered to be a pest (Chauhan and Pirta, 2010; Habiba et al., 2013), but these animals tend to be more wary of people and are less bold in their approach.

In natural, undisturbed forest habitats the availability of natural diet foods for primates, although variable with the season and year, is generally limited so that an individual monkey's access to food and water sets the limits to its physical growth and chances of reproduction and survival.

Recently there have also been observations that some monkey populations have become reliant on humans for food – for example, at temple and tourist sites – and that these have been significantly impacted as a result of fewer visitors during COVID-19 restrictions. In some locations this has resulted in more intergroup fighting and attacks on local villagers nearby (Anon., 2022)

https://india.mongabay.com/2020/05/absence-of-human-presence-impacting-monkey-behaviour-in-ayodhya/).

Survival rates under natural conditions are often low and are determined primarily by an interplay between environmental resource availability and social behaviours that influence the degree to which different individuals benefit from limited vital resources: i.e., to survive and reproduce (Dittus, 2004, 2013). It is considered that predation and disease usually have a secondary effect on mortality rates.

13.3   Dealing with monkey–human conflict

The global trend towards increasing wildlife–human conflict has resulted in more effort by wildlife managers and conservationists to develop ways to reduce it (Madden, 2004). A number of approaches have been developed, including designating natural protected areas and reserves for wildlife, preventing destructive human activity and promoting ways to tolerate and share habitat with wildlife. However, many nonhuman primates are a little harder to manage, especially those species that have developed a closer linkage with urban areas and are not afraid to interact with human populations. In addition, in many parts of Asia, monkeys have a significant cultural and religious significance (Nahallage and Huffman, 2013), which requires consideration of cultural differences in developing conservation strategies (Ferraro and Kiss, 2002; Fuentes and Wolfe, 2002; Fuentes et al., 2005; Lee and Priston, 2005). Dittus et al. (2019) suggest that this has given rise to the idea of ethnoprimatology, which is built on the premise that humans have perspectives on non-human primates that can contribute positively to their enduring survival (Lee, 2010).

Getting communities to coexist with monkeys takes careful consideration and requires active engagement of those directly impacted. Although it is widely appreciated that habitat loss leads to more human–monkey conflict, famers who are impacted by crop thefts tend to advocate for the removal of primate populations they consider to be a nuisance. In comparison, in the absence of crops, as in larger towns, the main reason that monkeys are attracted to human properties is the ready supply of food refuse and water (Diaz, 2000; Dittus, 2012). In the latter example, although there are more options to reduce access to these resources and to deter monkeys this requires a change in human habits, traditional ingrained ways of thinking and the expenditure of public and private resources, which is not easy.

Engagement of an interdisciplinary approach – i.e., to take an EcoHealth approach – is required in these situations. This needs to include social scientists to provide an anthropological perspective, as well as those with animal behaviour expertise to provide advice about how to achieve a positive and sustainable outcome. Examples of successful management of monkey–human conflict include those from Sri Lanka where villagers have learnt to deal successfully with this conflict. Although it has been found that some macaques are inveterate raiders, these villagers have developed methods to deter them. However, Sha et al. (2009) have found that although the intelligence of monkeys can favour

coexistence with humans (or not), the challenge lies in teaching not only humans but also the monkeys on how to share resources and habitat. In the next section we will consider an example of HWC in Bhutan.

13.4 Summary

A combination of approaches is often required to reduce human–monkey conflict, including stringently enforced food and refuse management that prevents monkeys' access, repelling monkeys from crops when necessary and, possibly, contraceptive interventions at localities such as tourist sites where there might be artificial overpopulation of monkeys. It is clear that there are many factors to consider in the move from HWC to human–wildlife coexistence. This includes the species involved, the context and location in which the species is encountered and the culture of the local community. To fully appreciate this, a close understanding of the culture, socioeconomic situation and experience base of the local communities engaged in human–wildlife interaction is required. These points are well illustrated in the following case study.

13.5 Case study: increasing macaque population along national highways in Bhutan

13.5.1 Context: addressing a potential public health and conservation problem

With more than 70% of the land under forest cover, Bhutan is known for its pristine and diverse flora and fauna. The constitution of the kingdom states that Bhutan shall maintain a minimum of 60% forest cover to 'conserve the country's natural resources and to prevent degradation of the ecosystem'. On the other hand, about 60% of its population (~750,000) live in rural areas and depend predominantly on agriculture and livestock farming for their livelihood. This unique setting, where people live in close proximity to nature, gives rise to an interface where humans frequently encounter wildlife. While conflict between humans and wildlife has been recognized for many years, it has intensified over recent years due to the rapid expansion of human settlements and agricultural activities into wildlife habitats in search of food and resources. HWC is a common occurrence in every part of the country. It is observed in the form of predation on livestock and crops by wild carnivores and omnivores/herbivores, respectively, mauling and trampling of people by bears and elephants and subsequent poaching and retaliatory killing of wild animals by people. Common predators for domestic animals in Bhutan are tigers (*Panthera tigris*), leopards (*Panthera pardus*), snow leopards (*Panthera uncia*), Asiatic wild dogs (*Cuon alpinus*) and Himalayan black bear (*Ursus thibettanus*) (Katel et al., 2014; NCD, 2008; Sangay and Vernes, 2008). Sambar deer (*Rusa unicolor*), barking deer (*Muntiacus muntjac*), musk deer (*Moschus moschiferus*), takin (*Budorcas taxicolor*), swamp deer (*Rucervus duvacelii*), hog deer (*Axis porcinus*), chital (*Axis axis*), gaur (*Bos gaurus*), water buffalo (*Bubalus bubalis*), Himalayan goral (*Naemorhedus goral*), Himalayan serow (*Capricornis thar*), wild boar (*Sus scrofa*), pygmy hog (*Porcula salvinia*), elephant (*Elephas maximus*) and primates are known to raid crops such as maize, wheat, buckwheat, paddy, potatoes, chilies, apples and oranges (NCD, 2008).

Aside from conflict with other wildlife species, conflicts with primates are rising in Bhutan (Tshering, 2020; Wangdi, 2015). There are reports of increasing macaque populations along the national highways, with reports of consequential interaction between humans and macaques (Tanden, 2016). Currently there are seven recorded primate species in Bhutan, namely the slow loris (*Nycticebus bengalensis*), Assamese macaque (*Macaca assemensis*), Rhesus macaque (*Macaca mulatta*), Hanuman langur (*Semnopithecus entellus*), golden langur (*Trachypithecus geei*), capped langur (*Trachypithecus pileatus*) and Arunachal macaque (*Macaca munzala*) (Choudhury, 2008). According to the IUCN red list of threatened species, the slow loris, golden langur and Arunachal macaques are endangered while Assamese macaque and capped langur are near threatened and the rhesus macaque and Hanuman langur fall within the category of least concern (IUCN, 2021).

13.5.2 Macaque populations and human interactions along a major highway

In the western parts of the country the Assamese macaque and the Hanuman langur are commonly sighted. Due to factors such as increasing developmental activities, provisioning of food by humans and improper management of waste, there is an increasing macaque population seen along the national highways of Bhutan (Figures 13.1, 13.2 and 13.3). Macaques are known as a menace for raiding crops, homes and vegetable and fruit stalls. In a study conducted in one of the districts of Bhutan, Wangdi et al. (2018) reported that, among wildlife species, after wild boars, macaques were responsible for causing the highest damage to crops. Further, given the similarity in genetical, physiological and behavioural characteristics, cross-transmission of pathogens is a concern when considering interactions between macaques and humans.

This case study reports the findings of a study that was carried out to understand the public perception about increasing macaque populations along the Thimphu–Phuentsholing

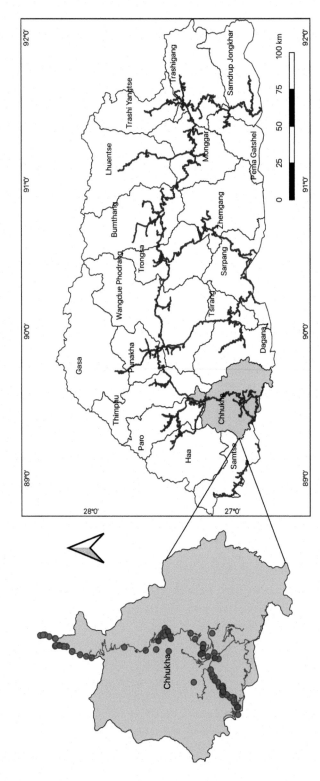

Figure 13.1 Map of Bhutan showing the networks of national highways and inset of Chhukha (study area) showing the locations where macaques were spotted and people were interviewed.

Figure 13.2 A troop of Assamese macaques by the side of a national highway (a), and road kill of a juvenile macaque (b).

Figure 13.3 A free-roaming dog feeding on a macaque carcass by the roadside (a) and Bhutanese travellers feeding oranges to macaques (b).

national highway, to design awareness education programmes and inform further study on the prevalence of zoonotic pathogens among the macaque population. The other objective of the study was to collect the geocoordinates of the locations along the western national highway where macaques were spotted, to inform placement of information and awareness signage. In the study, 129 respondents were interviewed of which 103 (80%) were females and 26 (20%) males.

In total, 125 (97%) participants have seen macaques near their residence. The majority of participants (71 (57%)) had seen macaques in their locality over 10 years and the remainder for up to 10 years. In assessing participants' perception of macaques and to understand their practice of provisioning food to them, 29 reported having a 'very high' level of tolerance to having macaques in their locality while 86 said they had a 'high' level of tolerance. Only 14 reported as having low tolerance. Among all participants, only 16 reported feeding macaques while 72% ($n = 93$) reported seeing other people feeding macaques along the highway. Participants reported that 98% ($n = 91$) of the people they saw feeding macaques were Bhutanese travellers while one was a tourist and the other a roadside vegetable vendor. The high level of tolerance for macaques among respondents and Bhutanese travellers mostly provisioning food to the macaques can be attributed to dominant religious beliefs and the mythological association of macaques with the Buddhist and Hindu religions. The majority of the Bhutanese population are Buddhist, followed by Hindu. Both Buddhists and Hindus believe that animals are sentient beings and need to be treated with love and kindness. In addition, in these religions, macaques are revered and worshipped as gods. Therefore, there is a need for awareness education among the travellers (both Bhutanese and tourists) regarding the potential consequences to the health of humans and macaques that would result from feeding the macaques.

One hundred and fourteen participants felt that the macaque population along the highway is increasing. However, eight participants reported that the population was not increasing while five were not sure about it. Eighty-nine (89%) respondents believed that people feeding macaques was the reason for their increasing population along the highway and four believed that deforestations was the cause, while the remaining participants were not sure. Among 127 participants (two missing), 120 felt that the growing macaque population along the highway was a problem while seven reported that it was not. Assamese macaques are omnivorous, with plants forming a major part of their diet (Koirala et al., 2017; Norbu et al., 2016; Zhou et al., 2011). However, humans feeding macaques can alter their feeding behaviour and activity budget (Koirala et al., 2017). The common foods provisioned by travellers are fruits, pack-processed foods such as chips, biscuits and breads and picnic leftovers. Because they do not have to move around in search for food when fed, they spend more time along the highways. While there are no past records to compare and assess the actual increase in macaque populations, frequent spotting of macaques along the highway could be the reason for the majority of participants reporting increases in macaque populations along the highway.

Seventeen respondents (14%) could recall some form of interactions (people killing macaques, macaques biting people) between humans and macaques while 45 (36%) had seen interaction between dogs and macaques. Fifty-four (42%) reported that humans can contract diseases from macaques while 51 (40%) were not sure. Twenty-four participants reported that humans cannot contract diseases from macaques. The information generated from this survey was shared with the Department of Forests and Park Services for development and implementation of further actions.

13.5.3   Risk of disease transmission

There are currently about 376 species of non-human primates globally (OIE, 2018). While there are numerous diseases that can be cross-transmitted between nonhuman primates and humans, for this chapter we list those that are relevant to macaques in Asia. Box 13.2 lists some of the important pathogens that can be transmitted from macaques to the human population.

Close interactions between humans and macaques can facilitate exchange of a wide range of pathogens and can have detrimental consequences on the health of both species (Koirala et al., 2017). Diseases can be cross-transmitted through different modes such as coming in direct contact with body fluids (e.g., simian type D virus, herpes B virus), through arthropod vectors (e.g., tickborne Kyasannur forest disease and mosquito-borne zoonotic malaria), aerosolized droplets (e.g., *Mycobacterium tuberculosis*, *Haemophilus influenzae*), contaminated feed and water (e.g., hepatitis A virus, salmonellosis, giardiasis) and saliva, bites and scratches (e.g., rabies) – see also Box 13.2. Furthermore, interactions between macaques and dogs can lead to the spread of rabies, which is a commonly reported notifiable zoonotic disease in free roaming dogs, in the macaque population. Although macaques are not a known reservoir for rabies, their social behaviour of grooming and dominance demonstration through fights can facilitate rabies transmission

Box 13.2   Zoonotic pathogens isolated from monkeys worldwide

- Viral: rabies, Herpes simiae, simian virus 40, simian type D retrovirus, simian foamy virus, Kyasanur forest disease virus, monkey pox virus, hepatitis A virus, measles and others.
- Bacterial: *Campylobacter jejuni*, *Salmonella* spp., *Shigella* spp., *Leptospira interrogans*, *Mycobacterium tuberculosis** and others. (Some organisms, such as *M. tuberculosis*, are primarily human-adapted pathogens and can be transmitted from humans to primates.)
- Fungal: *Trichophyton* spp. (ringworm)
- Parasitic (including protozoa): *Plasmodium* spp. (malaria), *Giardia* spp., *Trichuris* spp. *Ascaris* spp. and other helminths (worms).
- Typical transmission routes for viral and bacterial pathogens include fecal–oral, respiratory and via contaminated food and water.

within the macaque population, posing a potential conservation threat.

13.6   Conclusion

While urbanization and expansion of human settlement has led to widespread deforestation in many parts of the world, Bhutan still has about three-quarters of its land under forest coverage, providing home to a diverse wild flora and fauna. Human–wildlife conflict has remained an inherent problem associated with the development-driven expansion of human settlement. Livestock depredation by wild carnivores, crop raiding by macaques and other herbivores and damage to properties are not a new scenario in Bhutan. However, increases in the number of macaques

along the highways of Bhutan are largely due to the public feeding these animals. This is an emerging issue linked to a higher level of traffic and human activity on the roads. Humans provisioning food to macaques induces change in their foraging habits, thereby making them dependent on humans for food. Such dependence also compels macaques to move closer to human settlements, facilitating greater interaction with humans. While the relevant government agencies are implementing awareness programmes to discourage the feeding of wildlife along the highways, there is a need to strengthen such education programmes. Alongside awareness programmes, there is also a need to place signage prohibiting provisioning of food to macaques and ensuring proper management of waste along the highways, particularly in areas where macaques are commonly sighted.

In order to effectively address HWC it is important to engage an interdisciplinary team. An EcoHealth approach can be used, with the inclusion of social scientists and species biologists, to provide an anthropological perspective, as well as employing animal behaviour expertise to provide advice about how to achieve a positive and sustainable outcome. Managing HWC with primates can be especially challenging due to their intelligence and tenacity.

References

Anon. (2022). https://india.mongabay.com/2020/05/absence-of-human-presence-impacting-monkey-behaviour-in-ayodhya/Accessed Retrieved 11/11/2022

Brinton, M. A., Di, H., & Vatter, H. A. (2015). Simian hemorrhagic fever virus: Recent advances. *Virus Research, 202,* 112–119. https://doi.org/10.1016/j.virusres.2014.11.024

Cabral, S. J., Prasad, T., Deeyagoda, T. P., Weerakkody, S. N., Nadarajah, A., & Rudran, R. (2018). Investigating Sri Lanka's human-monkey conflict and developing a strategy to mitigate the problem. *Journal of Threatened Taxa, 10*(3), 11391–11398. https://doi.org/10.11609/jott.3657.10.3.11391-11398

Chakravarthy, A. K., & Thyagaraj, N. E. (2005). Coexistence of bonnet macaques (*Macaca radiata*) with planters in the cardamom (*Elettaria cardamomum* Maton) and coffee (*Coffea arabica* Linnaeus) plantations of Karnataka, South India: Hospitable or hostile? In J. D. Paterson & J. Wallis (Eds.), *Commensalism and conflict: The human-primate interface* (pp. 270–293). American Society of Primatologists.

Chapman, C. A., & Peres, C. A. (2001). Primate conservation in the new millennium: The role of scientists. *Evolutionary Anthropology, 10*(1), 16–33. https://doi.org/10.1002/1520-6505(2001)10:1<16::AID-EVAN1010>3.0.CO;2-O

Chaturvedi, S. K., & Mishra, M. K. (2014). Study of man-monkey conflict and its management in Chitrakoot, Madhya Pradesh, India. *International Journal of Global Science Research, 1,* 107–110.

Chauhan, A., & Pirta, R. S. (2010). Socio-ecology of two species of non-human primates, rhesus monkey (*Macaca mulatta*) and hanuman langur (*Semnopithecus entellus*), in Shimla, Himachal Pradesh. *Journal of Human Ecology, 30*(3), 171–177. https://doi.org/10.1080/09709274.2010.11906286

Choudhury, A. (2008). Primates of Bhutan and Observations of Hybrid Langurs. *Primate Conservation, 23*(1), 65–73. https://doi.org/10.1896/052.023.0107

Conly, J. M., & Johnston, B. L. (2008). The infectious diseases consequences of monkey business. *Canadian Journal of Infectious Diseases and Medical Microbiology, 19*(1), 12–14. https://doi.org/10.1155/2008/970372

Diaz, T. F. (2000). Cleaning up the mess: An everyday thing and expense. http://www.island.lk/2000/10/25/featur03.html

Dittus, W. P. J. (2004). Demography: A window to social evolution. In B. Thierry, M. Singh & W. Kaumanns (Eds.), *Macaque societies: A model for the study of social organization* (pp. 87–112). Cambridge University Press.

Dittus, W. P. J. (2012). Problems with pest monkeys: Myths and solutions. *Loris (journal of the wildlife and nature protection society of Sri Lanka)* 26: 18–23.

Dittus, W. P. J. (2013). Arboreal adaptations of body fat in wild toque macaques (*Macaca sinica*) and the evolution of adiposity in primates. *American Journal of Physical Anthropology, 152*(3), 333–344. https://doi.org/10.1002/ajpa.22351

Dittus, W. J., Gunathilake, S., & Felder, M. (2019). Assessing public perceptions and solutions to human-monkey conflict from 50 years in Sri Lanka. *Folia Primatologica, 90*(2), 89–108. https://doi.org/10.1159/000496025

Dixit, J., Zachariah, A., P K, S., Chandramohan, B., Shanmuganatham, V., & Karanth, K. P. (2018). Reinvestigating the status of malaria parasite (Plasmodium sp.) in Indian non-human primates. *PLOS Neglected Tropical Diseases, 12*(12), e0006801. https://doi.org/10.1371/journal.pntd.0006801

Elbroch, M. (2020). *The cougar conundrum: Sharing the world with a successful predator.* Island Press.

Else, J. G. (1991). Nonhuman primates as pests. In H. O. Box (Ed.), *Primate response to environmental change* (pp. 155–165). Chapman & Hall.

Engel, G. A., Jones-Engel, L. S., Michael, A., Suaryana, K. G., Putra, A., Fuentes, A., & Henkel, R. (2002). Human Exposure to herpesvirus B–seropositive macaques, Bali, Indonesia *Emerging Infectious Disease Journal, 8,* 7.

Faust, C., & Dobson, A. P. (2015). Primate malarias: Diversity, distribution and insights for zoonotic Plasmodium. *One Health, 1,* 66–75. https://doi.org/10.1016/j.onehlt.2015.10.001

Ferraro, P. J., & Kiss, A. (2002). Ecology. Direct payments to conserve biodiversity. *Science, 298*(5599), 1718–1719. https://doi.org/10.1126/science.1078104

Frank, B. (2016). Human-Wildlife conflicts and the need to include tolerance and coexistence: An introductory comment. *Society and Natural Resources, 29*(6), 738–743. https://doi.org/10.1080/089419 20.2015.1103388

Frank, B., Glikman, J. A., & Marchini, S. (2019). *Human-wildlife interactions: Turning conflict into coexistence.* Cambridge University Press.

Fraser-Celin, V. L., Hovorka, A. J., & Silver, J. J. (2018). Human conflict over wildlife: Exploring social constructions of African wild dogs (Lycaon pictus) in Botswana. *Human Dimensions of Wildlife, 23*(4), 341–358. https://doi.org/10.1080/108712 09.2018.1443528

Fuentes, A., & Wolfe, L. D. (2002). 2. *Cultural views of nonhuman primates.* In A. Fuentes, & L. D. Wolfe (Eds.) *Primates face to face* (pp. 61–62). Cambridge University Press.

Fuentes, A., Southern, M., & Suaryana, K. G. (2005). Monkey forests and human landscapes: Is extensive sympatry sustainable for *Homo sapiens* and *Macaca fascicularis* on Bali? In J. D. Paterson & J. Wallis (Eds.), *Commensalism and conflict: The human-primate interface* (pp. 168–196). American Society of Primatologists.

Fulton, D. C., Manfredo, M. J., & Lipscomb, J. (1996). Wildlife value orientations: A conceptual and measurement approach. *Human Dimensions of Wildlife, 1*(2), 24–47. https://doi.org/10.1080/ 10871209609359060

Habiba, U., Ahsan, F. M., & Røskaft, E. (2013). Local people's perceptions of crop damage by common langurs (*Semnopithecus entellus*) and human-langur conflict in Keshabpur of Bangladesh. *Environment and Natural Resources Research, 3*, 111–126.

Hill, C. M., Webber, A. D., & Priston, N. E. C. (2017). *Understanding conflicts about wildlife: A biological approach.* Berghahn.

Hrubes, D., Ajzen, I., & Daigle, J. (2001). Predicting hunting intentions and behavior: An application of the theory of planned behavior. *Leisure Sciences, 23*(3), 165–178. https://doi.org/10.1080/01490 4001316896855

Huff, J. L., & Barry, P. A. (2003). B-virus (Cercopithecine herpesvirus 1) infection in humans and macaques: Potential for zoonotic disease. *Emerging Infectious Diseases, 9*(2), 246–250. https://doi.org/10.3201/ eid0902.020272

International Union for Conservation of Nature and Natural Resources. (2021). The IUCN red list of threatened species. *Version 2021–3.* https://www.iucnredlist.org

Katel, O. N., Pradhan, S., & Schmidt-Vogt, D. (2014). A survey of livestock losses caused by Asiatic wild dogs, leopards and tigers, and of the impact of predation on the livelihood of farmers in Bhutan. *Wildlife Research, 41*(4), 300–310. https://doi.org/10.1071/WR14013

Koirala, S., Chalise, M. K., Katuwal, H. B., Gaire, R., Pandey, B., & Ogawa, H. (2017). Diet and activity of Macaca assamensis in wild and semi-provisioned groups in Shivapuri Nagarjun national park, Nepal. *Folia Primatologica; International Journal of Primatology, 88*(2), 57–74. https://doi.org/10.1159/000477581

Lee, P. C. (2010). Sharing space: Can ethnoprimatology contribute to the survival of nonhuman primates in human-dominated globalized landscapes? *American Journal of Primatology, 72*(10), 925–931. https://doi.org/10.1002/ajp.20789

Lee, P. C., & Priston, N. E. C. (2005). Human attitudes to primates: Perceptions of pests, conflict and consequencesmfor primate conservation. In J. D. Paterson & J. Wallis (Eds.), *Commensalism and conflict: The human-primate interface* (pp. 1–23). American Society of Primatologists.

Li, M., Zhao, B., Li, B., Wang, Q., Niu, L., Deng, J., Gu, X., Peng, X., Wang, T., & Yang, G. (2015). Prevalence of gastrointestinal parasites in captive non-human primates of twenty-four zoological gardens in China. *Journal of Medical Primatology, 44*(3), 168–173. https://doi.org/10.1111/jmp.12170

Madden, F. (2004). Creating coexistence between humans and wildlife: Global perspectives on local efforts to address human-wildlife conflict. *Human Dimensions of Wildlife, 9*(4), 247–257. https://doi.org/10.1080/10871200490505675

Maharajan, M. (2015). Endo-parasites of primates and their zoonotic importance. *National Zoonoses and Food Hygiene Research, 21*, 4.

Nahallage, C. A. D., & Huffman, M. A. (2013). Macaque-human interactions in past and present-day Sri Lanka. In S. Radhakrishna, M. A. Huffman & A. Sinha (Eds.), *The macaque connection: Cooperation, and conflict between humans and macaques* (pp. 135–148). Springer.

Nakayima, J., Hayashida, K., Nakao, R., Ishii, A., Ogawa, H., Ichiro, N., Moonga, L., Hang'ombe, B.

M., Mweene, A. S., Thomas, Y., Orba, Y., Sawa, H., & Sugimoto, C. (2014). Detection and characterization of zoonotic pathogens of free-ranging nonhuman primates from Zambia. *Parasites and Vectors, 7*, 490. https://doi.org/10.1186/s13071-014-0490-x

Nath, B. G., Chakraborty, A., & Rahman, T. (2012). Tuberculosis in non-human primates of Assam: Use of PrimaTB STAT-PAK Assay for detection of tuberculosis. Journal of Threatened Taxa, 4(4), 2541–2544. https://doi.org/10.11609/JoTT.o2860.2541-4.

National Council on Disability. (2008). *Bhutan national human -Wildlife ConfliCts management strategy*. Department of Forests.

Norbu, T., Wangda, P., Dorji, T., Chhetri, P. B., Rabgay, K., Dorji, R., Hamada, Y., Kawamoto, Y., Oi, T., & Chijiwa, A. (2016). *Ecology, Morphology and Genetic study of Macaque*.

OIE & WOAH. (2018). Terrestrial health code. Zoonoses Transmissible from nonhuman primates. https://www.woah.org/fileadmin/Home/eng/Health_standards/tahc/2018/en_chapitre_zoonoses_non_human_primate.htm

Pooley, S., Barua, M., Beinart, W., Dickman, A., Holmes, G., Lorimer, J., Loveridge, A. J., Macdonald, D. W., Marvin, G., Redpath, S., Sillero-Zubiri, C., Zimmermann, A., & Milner-Gulland, E. J. (2017). An interdisciplinary review of current and future approaches to improving human-predator relationships. *Conservation Biology, 31*(3), 513–523. https://doi.org/10.1111/cobi.12859

Reddy, A. R. M., & Chander, J. (2016). Human-monkey conflict in India: Some available solutions for conflict mitigation with special reference to Himachal Pradesh. *Indian Forester, 142*, 941–949.

Regmi, G. R., Nekaris, K. A. I., Kandel, K., & Nijman, V. (2013). Crop-raiding macaques: Predictions, patterns and perceptions from Langtang National Park, Nepal. *Endangered Species Research, 20*(3), 217–226. https://doi.org/10.3354/esr00502

Sangay, T., & Vernes, K. (2008). Human–wildlife conflict in the Kingdom of Bhutan: Patterns of livestock predation by large mammalian carnivores. *Biological Conservation, 141*(5), 1272–1282. https://doi.org/10.1016/j.biocon.2008.02.027

Sha, J. C. M., & Gumert, M. D., Lee BPY-H, Jones-Engel L, Chan S, Fuentes A (2009). Macaque-human interactions and the societal perceptions of macaques in Singapore. *American Journal of Primatology, 7*, 825–839.

Srivastava, A., & Begum, F. (2005). City monkeys (*Macaca mulatta*): A study of human attitudes. In J. D. Paterson & J. Wallis (Eds.), *Commensalism and conflict: The human-primate interface* (pp. 259–269). American Society of Primatologists.

Tabasshum, T., Mukutmoni, M., & Begum, A. (2018). Occurrence of gastrointestinal helminths in captive rhesus macaques (Macaca mulatta). *Bangladesh Journal of Zoology, 46*(2), 231–237. https://doi.org/10.3329/bjz.v46i2.39065

Tanden, Z. (2016). Speed on Thimphu-Phuentsholing highway leading to increasing number of wildlife kills. *Bhutanese, May 21*.

Tshering, S. (2020). Human-Wildlife Conflict, a never ending story in rural areas. *BBS, 15 December 2020*.

Wangdi, S., Norbu, N., Wangdi, N., Yoezer, D., Choden, K., & Wangchuk, J. (2018). *Demystifying the link between Rural Urban Migration and Human Wildlife Conflict: A case of Gangzur and Kengkhar, Eastern Bhutan*. UWICER press, Lamai Goempa, Bumthang.

Wangdi, T. (2015). Human wildlife conflict worsens in Khaptoe. *Kuensel*, 21 December 2015

Woodroffe, R., Thirgood, S., & Rabinowitz, A. (2005). *People and wildlife conflict or co-existence? Conservation biology, 9*. Cambridge University Press.

Zhou, Q., Wei, H., Huang, Z., & Huang, C. (2011). Diet of the Assamese macaque *Macaca assamensis* in lime-stone habitats of Nonggang, China. *Current Zoology, 57*, 8–17.

Zimmermann, A. (2019). Human-wildlife conflict, presentation by Alexandra Zimmerman. WildCRU Conservation Geopolitics Forum Plenary Talk Alexandra Zimmermann, IUCN and WildCRU, Zoology, University of Oxford, UK. https://www.youtube.com/watch?v=DA37BfSmG-A

Additional resources

https://www.hwctf.org/primates
http://www.indianforester.co.in/index.php/indianforester/article/view/104130
https://kuenselonline.com/feeding-monkeys-could-result-in-more-crop-damages/
https://kuenselonline.com/human-wildlife-conflict-worsens-in-khaptoe/
https://thebhutanese.bt/human-wildlife-conflict/
https://wwwnc.cdc.gov/eid/article/12/6/06–0030_article
https://india.mongabay.com/2021/09/quick-solutions-for-human-monkey-conflicts-could-lead-to-dangerous-results/

PART III

OCEANS, WETLANDS AND AQUACULTURE

Chapter 14

Aquaculture and aquatic biosecurity in a changing world: An insight from New Zealand

Aurelie Castinel and Henry Lane

Abstract

The spread and emergence of aquatic pests and diseases, such as infectious salmon anaemia virus in salmon and ostreid herpesvirus microvariant type 1 in Pacific oysters, in recent decades have become a major risk to aquaculture sustainability and growth globally. Factors contributing to this trend are wide-ranging – from the amplifying effect of denser populations inherent to aquaculture, to warmer temperatures and inadequate biosecurity practices. Marine and freshwater ecosystems present very different challenges and opportunities for farming and productivity, but also for conservation and restoration efforts of natural habitats. Biosecurity preparedness and risk management strategies can help sustain aquaculture production and its associated social benefits, and enhance ecosystem health as well as the indigenous values and practices depending on it. Through a few examples from Aotearoa New Zealand, we will explore how changes in the aquatic environment can impact animal health, ecosystem health and human health, and how biosecurity can help protect and sustain these important values.

14.1 Introduction

New Zealand, also known as Aotearoa, is an island nation of people who have been widely connected to the sea over centuries, and are reliant on its aquatic resources to thrive (Clapcott et al., 2018; McCarthy et al., 2014). The first people to reach its shores, around 1,500 years ago, are thought to have been sailors from West Polynesia (Anderson, 2002; Young, 2004). Once established, the indigenous people of Aotearoa, known as Māori, shifted their approach for sourcing food towards more sustainable practices, including food cultivation and stock relocation (Kirikiri and Nugent, 1995; Ross et al., 2018; Young, 2004). Significant additional stress on resources came with the arrival of European settlers in the 1840s, accelerating both biodiversity loss and remodelling of entire landscapes (Anderson, 2002). Nowadays the aquatic environment (broadly, lakes, rivers, wetlands and sea) and its resources (e.g. animal and plant proteins for food and pharmaceuticals, minerals and the water itself) are shared among a wide range of recreational,

Susan C. Cork and Douglas P. Whiteside (eds) **Case Studies in EcoHealth**
DOI: 10.52517/9781789183368.014, © 5m Books Ltd, 2024

commercial and customary users, occasionally creating tension between conservation and economic growth. This requires balanced policies to enable indigenous, public and industrial interests and rights to coexist (Le Heron et al., 2019). In addition to exploitation of aquatic resources, the introduction of non-indigenous pests and diseases via international and domestic pathways is also threatening aquatic ecosystems and associated environmental, cultural, social and economic values. The overall management of these risks is defined as biosecurity (Hewitt et al., 2004). Biosecurity risk management has become a worldwide focus over the past decades because poor practices, combined with increased movements of aquaculture stock, equipment and products, have led to a surge in trans-boundary diseases and opportunistic pathogens emerging as serious threats to aquatic animal health and biodiversity (Feist et al., 2019; Peeler and Taylor, 2011).

The growth and intensification of aquaculture have driven the emergence of aquatic diseases – for example, the emergence of white spot syndrome virus (WSSV) in prawns and infectious salmon anaemia virus (ISAV) in salmon. High-density, genetically homogenous populations often found in intensive farming practices support the emergence and rapid spread of diseases (Feist et al., 2019). Impacts from aquatic disease can be far reaching and extend beyond elevated levels of mortality on affected farms. For instance, outbreaks of disease that lead to a decrease in farm production can have a ripple effect through society via facility closures, increase in unemployment and human health impacts. Zoonotic diseases – i.e. infectious diseases that can transmit from animals to humans (e.g. highly pathogenic avian influenza (bird flu)) – are rarer in the aquatic environment compared with terrestrial animal diseases, but there is more to health than humans becoming physically ill. The emergence and spread of disease can spill over to wild populations, jeopardizing native biodiversity and compromising restoration efforts. For many people, in particular indigenous people, well-being is intrinsically linked to the natural world and its ecosystems, and cultural practices and identity are often interwoven with a healthy marine environment (Clapcott et al., 2018). Understanding what is at stake from the emergence and spread of aquatic diseases highlights the benefit of biosecurity in protecting cultural values, economic prosperity and aquatic environmental health.

Given the diversity of users and activities taking place in the aquatic environment, and the potential for disease impacts to be felt across different sections of society, biosecurity risk management cannot be expected to rely solely on the aquaculture industry and government officials (Figure 14.1). It also requires support and participation from the public, the indigenous community and other industry sectors (Le Heron et al., 2019). Strategies are developed to prevent or control risk at international, national, regional, local and even farm levels (Oidtmann et al., 2011). Surveillance allows for early detection of aquatic pathogens and remains a key mechanism in preventing the spread and establishment of diseases in both wild and farmed aquatic animals (Peeler and Taylor, 2011; Thrush et al., 2017). This is a particularly effective approach in the detection of emerging diseases in the presence of favourable host and environment conditions – for example, in relation to climate change and warmer sea temperatures or drought conditions in rivers (Lennox et al., 2019; Marcos-López et al., 2010; Peeler and Feist, 2011).

In this chapter we present aquaculture and aquatic biosecurity through the holistic lens of EcoHealth, highlighting the interconnectedness between animal health, human health and ecosystem health. We firstly introduce what a holistic view of aquatic health involves, before outlining the current status and trends of global aquaculture and documenting the direct and indirect impact of disease on aquaculture around the world. Lastly,

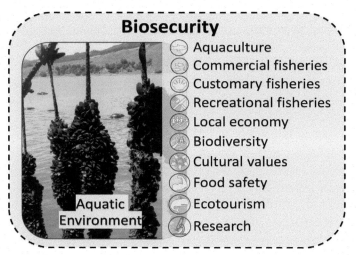

Figure 14.1 The emergence and spread of diseases impact across different sections of society, not just aquaculture. The success and health of values, activities and benefits associated with the aquatic environment are underpinned by biosecurity. Biosecurity is not the sole responsibility of one sector or community, but of everyone.

using three case studies from Aotearoa New Zealand, we demonstrate how aquatic diseases can impact ecosystem health and human health, and how biosecurity supports EcoHealth. The first case study relates how an investigation into a disease outbreak in native lampreys highlighted the importance of indigenous knowledge, as well as the network between wild and farmed fish species in rivers. Second, we present how biosecurity is critical to restoration of an important native taonga (treasure), the black-footed pāua. Third, we present and discuss offshore marine farming and its potential future role in biosecurity and maintaining EcoHealth. While this chapter has a predominant New Zealand focus on aquatic biosecurity and EcoHealth, much of what is presented is globally applicable.

14.2 A holistic approach to aquatic health

The concept of a healthy aquatic ecosystem is complex in itself and involves interlinked human, animal and environmental components that have evolved over centuries and remain highly dynamic (Stentiford et al., 2020). Despite the long history of indigenous peoples sustainably managing their fisheries, the inclusion of cultural perspectives in legislation and policies is relatively recent, and was only recently acknowledged through international policy frameworks on indigenous people's rights to access, restore and protect their natural heritage for future generations (Sherman, 2006; United Nations, 2007). This is a first step towards recognizing the deep connection between the natural elements and indigenous people, their traditional practices and beliefs, as well as their future. In addition, coastal communities, regardless of their cultural association, maintain close ties with rivers, catchments, wetlands and the ocean for their subsistence, well-being and identity (Hepburn et al., 2019; McCarthy et al., 2014). In contrast, the exploitation and extraction of freshwater, coastal and marine resources for commercial purposes have impacted ecosystems at a rapid pace. Aquaculture has developed very rapidly in some places, creating employment, food security and wealth expansion

(FAO, 2020); however, the profitable economic development seems to have happened at the expense of the wider environment, especially biodiversity and cultural values (Asche et al., 2018). The economic and animal health considerations will be addressed in the following section. Here we emphasize the aquatic network linkages and connections between indigenous people and the aquatic environment, as well as the overall implications for biosecurity and environmental health management.

14.2.1 Diversity of waterways and interconnectedness

Aquatic systems are interconnected, either physically through natural topography or economically or culturally (Peeler and Taylor, 2011; Rodgers, Mohan and Peeler, 2011; Ross et al., 2018). There is evidence that the health of lakes, rivers, wetlands and coastal ecosystems is interconnected, from the mountains to sea, or *Ki Uta Ki Tai* in Māori worldview. Thus negative impacts on one will be felt on the others – for example, when sediments and excess nutrients from forestry and farming activities or urban development upstream reduce estuarine water quality and cause habitat degradation (Collier et al., 2019; Ellis et al., 2015; Heggie and Savage, 2009). Modifications in salinity of the coastal waters will impact local biodiversity. Such physicochemical changes occur when drought reduces river flow and decreases water levels, leading to increased stress in fish, reduced feed availability and higher density of aquatic animals, together leading to changes in community dynamics (including emergence and establishment of pests and diseases) (Lennox et al., 2019).

Wetlands are another example of the fragile balance between freshwater and saline waters. They are a complex landscape where land meets streams, rivers, lakes and estuaries and where saltmarsh shrubs inland merge with mangrove

trees rooted in the seabed (Cometti and Morton, 1985). Mangroves are a unique and abundant ecological hub supporting thriving ecosystems, including nursery and breeding grounds for many fish species such as mullet, flounder, kahawai, snapper and trevally (Cometti and Morton, 1985). However, this thriving, oxygen-regulating ecosystem is under threat. In tropical regions the expansion of the profitable shrimp farming industry has resulted in the removal of extensive mangrove cover. Although work is under way to promote sustainable shrimp farming integrated within existing mangrove in a silvo-aquaculture system, there are still some challenges to overcome (Rahman et al., 2020). In this case, sustainability of aquaculture implies sourcing species that are resilient to environmental conditions as well as ensuring upskilling of farmers to adapt their farming practices (Rahman et al., 2020).

The strong ecological relationship between rivers and sea is embodied by marine fish species migrating up- and downstream to complete their life cycle. Aotearoa New Zealand has many native species of fish that depend on both healthy freshwater and marine environment to survive and flourish (McDowall, 2006). Among them, eels, lampreys and galaxiid species migrate seasonally up and down New Zealand rivers to reproduce and/or grow to maturity. Stress related to their natural metabolic needs, combined with man-made obstacles (such as dams), has made some species more vulnerable to opportunistic pathogens (Brosnahan et al., 2019; Egan et al., 2019; Kitson, 2012). Also, anthropogenic activities (such as recreational fishing, boating or fish farming) provide a physical link between waterways and coastal areas that would otherwise be geographically and hydrodynamically isolated and protected from the natural spread of pests and diseases (Peeler and Feist, 2011; Thrush et al., 2011). New Zealand aquatic systems are susceptible to introductions of new pests and diseases.

and pathway risk management is key to preventing their spread and establishment (Collier et al., 2019). The rate of introduction of nonindigenous aquatic species has significantly increased over the past decades and the extent is probably underestimated (Bailey et al., 2020). Most common risk pathways for pests include ballast water and biofouling on vessel hulls; for pathogens in particular, aquaculture operations provide numerous routes for transferring diseases, such as movement of stock and equipment (Bailey et al., 2020; Castinel et al., 2019; Hewitt et al., 2004; Rodgers et al., 2011). Risk pathways often overlap between types of activities, which makes domestic biosecurity challenging to regulate and enforce. For example, recreational fishing next to a commercial farm is a common sight in the Marlborough Sounds

(Figure 14.2). The same boat could be fishing in a different part of the Sounds the following day, unsuspectedly connecting farming areas and thus compromising the farms' biosecurity risk management.

14.2.2 Indigenous resource management

Aquatic ecosystems have always been central to food security and, in modern society, they have also become a key source of employment through fisheries and tourism (FAO, 2020). However, contemporary anthropogenic activities and poor resource management have had a significant impact on aquatic biodiversity which, in island nations like Aotearoa New Zealand,

Figure 14.2 Recreational fishing next to a commercial farm is a common sight in the Marlborough Sounds. The same boat could be fishing in a different part of the Sounds the following day, unsuspectedly connecting farming areas and thus compromising the farms' biosecurity risk management. Credit: R. Fraser.

had been effectively managed by indigenous guardians – *kaitiaki* for Māori – until the arrival of European settlers in the 19th century (Robinson et al., 2021). Exploitation of resources driven by non-indigenous parties has displaced traditional holistic practices, often at the expense of native biodiversity. In New Zealand, *Te Tiriti o Waitangi* (the Treaty of Waitangi), which is a constitutional document between Māori and the British Crown, guaranteed indigenous chiefs to retain *tino rangatiratanga* (broadly translated as sovereignty and autonomy in English) over their natural resources, which they consider *taonga* (treasure). The English version interpreted this term as shared authority with the Crown, which has been a prolific source of contentions ever since (Stenson, 2004). Today, the Treaty of Waitangi obligations are still not fulfilled and Māori's

customary authority and rights as *kaitiaki* have yet to be recognized in the law. Management of native aquatic resources has been largely delegated by the Crown to government agencies (e.g. the Department of Conservation, Ministry for Primary Industries). Restoring *tino rangatiratanga* over indigenous fauna and flora is the key driver behind the Wai 262 claim, lodged with the Waitangi Tribunal in 1991 (*Te Puni Kokiri*, 2021). The subsequent Waitangi Tribunal report has recommended changes to the Crown's laws, policies and practices relating to intellectual property, indigenous flora and fauna, resource management and conservation (Waitangi Tribunal, 2011); the Crown has decided to develop a whole-of-government approach, and work is still ongoing to consider and address these recommendations (*Te Puni Kokiri*, 2021).

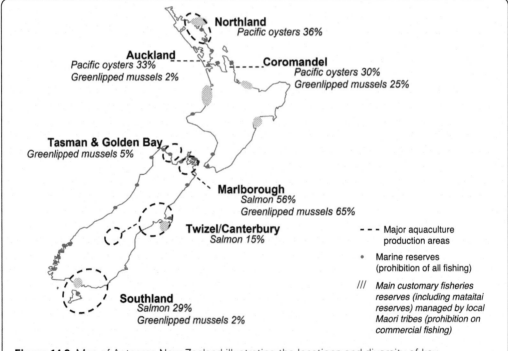

Figure 14.3 Map of Aotearoa New Zealand illustrating the locations and diversity of key aquaculture areas, marine reserves and customary fisheries. In 2020, the estimated total revenue was NZ$653 million for the three main farmed species – Chinook salmon (15,512 tonnes), greenlipped mussels (101,657 tonnes) and Pacific oysters (1,364 tonnes). Sources: Aquaculture NZ (2020); New Zealand Ministry for Primary Industries (n.d.).

Māori culture takes its strength from a unique combination of customary values, beliefs, principles (*kaupapa*) and practices (*tikanga*), based on kinship between Māori people and the elements of the natural world, including native fauna and flora (Marsden, 2003; Mead, 2016). Some of these values provide strong guidance for Māori to interact with aquatic ecosystems and to nurture the natural resources they rely upon for *kai* (food) and *hauora* (health).

Māori have long used customary practices for conservation and management of resources on the foreshore and in the ocean (Waitangi Tribunal, 1985). This includes measures to preserve habitats of fish and shellfish and to maintain stock numbers, such as *mātaitai* reserves, *taiāpure*, conservation *rāhui* and *kai-tiaki* of *mahinga kai* (places of food gathering) (Hepburn et al., 2019; Mead, 2003; Waitangi Tribunal, 1985). *Rāhui* is a mechanism in the Māori environmental management toolbox that temporarily prohibits access to an area to allow for a resource and its surrounding environment to recover (Best, 1904; Mead, 2003; Waitangi Tribunal, 1985). Any breach of this restriction would be regarded as a serious offence. Nowadays local Māori tribes may impose a conservation *rāhui* to close a fishery and enable fish stocks to recover, a procedure supported and respected by the Crown agencies.

The overall knowledge of a resource – for example, a fish species – and its interactions with its direct environment and other natural elements is key to informing its sustainable management for future generations (Young, 2004). This information, part of the indigenous knowledge system called *mātauranga*, is influenced by place and time and mostly orally transmitted within local tribes (Clapcott et al., 2018). Like all traditional indigenous knowledge, *mātauranga* is holistic and constantly adapting to a dynamic natural world (Moller, 1996). Monitoring of aquatic ecosystem health is often undertaken as a partnership between traditional

ecological knowledge (through *mātauranga*) and Western science methodologies, with support from local communities and schools (Clapcott et al., 2018; McCarthy et al., 2014). *Mātauranga* is being increasingly recognized as a source of information complementary to classical science disciplines in exploring local biosecurity issues (Moller, 2009); however, it has yet to be considered as a contributor to aquaculture biosecurity management, especially at the interface with wild stocks and climate challenges.

Aquatic species native to Aotearoa New Zealand and importance to Māori

Māori tribes have always had a close relationship with aquatic resources, not only through a spiritual kinship but also as a major source of food. Even though the arrival of European settlers in the 19th century marked the diversification of food sources for indigenous people, native species remain of great importance to Māori.

Indigenous management of aquatic environmental resources is guided by values and principles and informed by local knowledge called *mātauranga* Māori. Some species that once thrived in Aotearoa, such as eels (*tuna*) or lampreys (*piharau*) have to be carefully managed nowadays because stocks have declined. Others like freshwater crayfish (*kōura*) have become species of interest for farming.

The National Institute of Water & Atmospheric Research (NIWA) have produced some useful resources in collaboration with indigenous tribes to better understand native marine and freshwater species and support their sustainable management for future generations (see end of References sction).

14.3   Aquaculture growth: opportunities and challenges

Aquaculture is set to replace capture fisheries as the primary source of seafood proteins. Capture fisheries is the harvesting of aquatic species (e.g. fish, shellfish, prawns, crabs and lobster) from their naturally occurring wild populations across marine and freshwater environments (FAO, 2020). As fish stocks decrease globally, enabling the expansion of fish farming has become a priority for many governments around the world (FAO, 2020). Disease outbreaks constitute a very significant risk to sustainable aquaculture growth because they can suddenly reduce the volume and stability of production, thus threatening livelihoods and food supply (Stentiford et al., 2020). Farming intensification, transfer of stock and warming temperatures have led to the emergence and spread of aquatic animal diseases across both wild and farmed animal populations (Feist et al., 2019). Biosecurity measures, whether applied at international level through health standards or at farm level with biosecurity plans, contribute to preventing and reducing disease impacts (Oidtmann et al., 2011).

14.3.1   Current state of aquaculture and future trends

Seafood is a major component of future food security (Gephart et al., 2021). Globally, 156 million tonnes of seafood were produced for human consumption in 2018, with aquaculture contributing 52% of total production (FAO, 2020). Annual global fish consumption increased by 3.1% from 1961 to 2017, a rate twice that of global population growth and a rate higher than all other animal-sourced protein (i.e. dairy, meat) (FAO, 2020). The global increase in seafood consumption is mainly driven by growth in developing countries (2.4% annual increase

from 1961 to 2017), least-developed countries (1.3%) and low-income food-deficit countries (1.5%), while annual seafood consumption in developed countries peaked in 2007 at 26.4 kg per capita and sat steady at 24.4 kg per capita in 2017 (FAO, 2020).

Globally, aquaculture is replacing wild fisheries capture production as the primary source of seafood protein (FAO, 2020). Global aquaculture production has increased rapidly over the last two decades, whereas wild-capture fisheries production has more or less remained steady over the same period, mainly because there are few opportunities to either develop new sustainable fisheries or increase catches in existing ones. Consequently, aquaculture has had to diversify and intensify in order to meet demand for nutritious protein. Global aquaculture production is projected to continue to grow a further 32% between 2018 and 2030 and to remain the driving force behind global seafood production. Increasing production is expected to concomitantly boost employment. Global aquaculture employed 20.5 million people in 2018, of which 19% were women in production-related activities – i.e. farming operations – and approximately 50% were in post-harvesting activities (FAO, 2020).

Aquaculture feeds the human population and brings employment and economic activity to predominantly rural areas. This implies that any shock to production and, consequently, to its supply chain could jeopardize global food security. Disease is considered the biggest risk to the development and sustainable growth of aquaculture because of its ability to disrupt production, stop trade and impact wild stock and biodiversity. In fact, the average annual growth of global aquaculture is expected to slow in part due to an increase in disease outbreaks (FAO, 2020). The trans-boundary movement of pathogens has increasingly been linked to the emergence and impact of diseases (Peeler et al., 2011). Israngkura and Sae-Hae (2002) estimated that the economic impact of disease in aquaculture exceeded 40% of its overall production value. In Chile, production of Atlantic salmon declined by two-thirds, exports declined by around one-third and tens of thousands of jobs were lost following the introduction and spread of the pathogenic ISAV (Asche et al., 2009).

The increase in use of antimicrobials is a global emerging risk associated with the intensification of aquaculture. While vaccinations have been used to control many aquatic diseases of finfish and are seen as a proactive measure for managing disease (Brudeseth et al., 2013), administering antimicrobials remains a reactive measure to control a disease outbreak. The use of antimicrobials in aquatic food production is startlingly high, corresponding to 5.7% of global antimicrobial use, with some species such as catfish consuming antimicrobials at a rate per kilogram exceeding that in humans and terrestrial animal food production (Schar et al., 2020). The primary concern with excessive or consistent use of antimicrobials is the potential selection of pathogens with antimicrobial resistance, leading to treatment inefficacy. Such a scenario will not only compromise production but also enable the transmission of antimicrobial resistance genetic elements to humans or other animals via the consumption of farmed fish and associated products (Schar et al., 2020). The risk of antimicrobial resistance associated with aquatic food is an emerging area of research, and an important one. Seafood is one of the highest-traded commodities in the world, with total global exports in excess of US$164 billion, highlighting the threat of locally generated resistance easily becoming globally distributed. From available data, risks to production from antimicrobial resistance are likely to disproportionately affect low- and middle-income countries, further impacting food security (Schar et al., 2021).

The FAO has developed a Blue Growth policy framework to address the main challenges around sustainable use of marine resources and activities contributing to growing economic, environmental and social development (Bondad-Reantaso et al., 2018). This is in line with the 'Blue Economy', a concept used internationally to drive sustainable growth and exploitation of marine-based resources (Wenhai et al., 2019). It is argued that, to support the sustainable use of marine resources, social innovation is needed to change behaviours among the regulators, markets (consumers and producers) and public sectors (Soma et al., 2018). Fisheries and aquaculture are considered two key sectors in achieving the United Nations' 2030 Agenda Sustainable Development Goals, and the FAO published a strategic framework, the Blue Growth Initiative, to help countries transition into the Blue Economy (FAO, 2018). However, biosecurity and aquatic animal diseases are not directly emphasized in the framework, despite having risen as a serious challenge for the global aquaculture industry over the past three decades. The FAO (2020) acknowledges, through a diverse range of stakeholders, that greater effort is needed in biosecurity to protect and grow aquaculture production. Historically biosecurity has been reactive, being implemented following a disease event, compared with more effective and economical preventative biosecurity. A progressive management pathway for improving aquaculture biosecurity (PMP/AB) is an international initiative aimed at supporting the implementation of better biosecurity in aquaculture (Bondad-Reantaso et al., 2018). Outcomes and benefits from the PMP/AB include lower impact from disease, improved aquatic health at both farm and national level, minimized global spread of disease, greater social licence and maintenance of cultural literacy. These outcomes and benefits are all representative of the EcoHealth approach.

14.3.2 Emerging aquatic animal diseases and their spread

Emerging diseases include those diseases that are increasing in incidence or geographic range, or in severity (Feist et al., 2019). Also encompassed are diseases with a new pathology or those that have recently moved into a new host (Lane et al., 2016), are newly discovered, such as ostreid herpesvirus microvariant type 1 (OsHV-1) µvar in the Pacific oyster *Crassostrea gigas* (Segarra et al., 2010) or caused by a newly evolved pathogen, as exemplified by the phenotypic change of the oyster parasite *Perkinsus marinus* in farmed and wild American oyster *Crassostrea virginica* (Carnegie et al., 2021). A classic example of an emerging aquatic disease is WSSV, which causes white spot disease (WSD) in penaeid prawns. It emerged in the 1990s and rapidly spread throughout Asia via the translocations of live prawns for aquaculture; it then reached North, South and Central America, probably via a similar pathway (Rodgers et al., 2011). WSD remains a significant contributor to production losses in shrimp farming worldwide, estimated at around US$1 billion each year. Despite its publicity and reputation, WSSV continues to be introduced to areas previously free from the virus. In 2016 WSSV was detected in Australian prawn farms, at a significant financial cost to producers in the Logan River, Queensland (AU$43 million in production losses) and to the government as a result from having to respond to the outbreak (Scott-Orr et al., 2017). The most probable pathway for the virus entering Australia was found to be through imported infected consignments of frozen prawns for human food that were used as fish bait, that subsequently spread to naïve disease-free stock (Scott-Orr et al., 2017).

Viruses, bacteria, macro- and microparasites and fungi are all common components of the marine ecosystem, with many incapable of causing disease but instead providing critical ecosystem functions (Hewson, Chow and Fuhrman, 2010; Thomas et al., 1998). Pathogenic organisms can, however, emerge from non-pathogenic organisms through genetic mutations. Emerging diseases in aquaculture usually originate from pathogens in wild populations, but they will probably be first noticed in farmed populations following pathogen transmission from wild stock. Farmed populations provide ideal conditions for disease outbreaks because they generally consist of genetically homogenous populations with individuals in close proximity to one another, thus exacerbating disease transmission (Feist et al., 2019). For example, a non-pathogenic form of ISAV occurs in wild populations of the Atlantic salmon *Salmo salar*, from which a highly pathogenic strain evolved due to a mutation in a particular region of the virus genome, specifically deletions of amino acid-coding segments from the highly polymorphic region (HPR). The pathogenic strain is known as HPR deleted while the non-pathogenic strain is known as HPR0. Infectious salmon anaemia (ISA), the disease that occurs following infection by the pathogenic HPR-deleted ISAV, has emerged in farmed populations through one of the following two ways: (1) transmission of mutated pathogenic HPR-deleted ISAV from wild salmon to farmed salmon, or (2) infection of farmed salmon with the benign HPR0, which subsequently mutates into the pathogenic form. The emergence of the pathogenic HPR-deleted ISAV and its disease in farmed Atlantic salmon has had severe consequences for aquaculture production in Chile and parts of Europe – in particular, Norway and Scotland – and Canada (Asche et al., 2009, 2018). At the time of publication, the New Zealand salmon industry is free of major diseases and the introduction of ISAV would likely have significant impact on salmon production in that country (Georgiades et al., 2016). ISA is among the key diseases excluded during a biosecurity investigation of salmon mortalities (Norman et al., 2013).

The intensification and diversification of aquaculture are happening against the backdrop of climate change. Disease is governed by interaction between host, pathogen and the environment (Snieszko, 1974). As the environment changes so do the interactions between host and pathogen, potentially exacerbating disease-related issues. Warming waters and ocean acidification associated with climate change, pollution and increased sedimentation from human activities on land, as well as extraction pressures, can act as stressors that may compromise an organism's physiological performance and leave it more susceptible to infection and disease (Harvell, 2019). Opportunistic infections occur when a normally benign microorganism causes disease in a host with a compromised immunity. These are expected to be more common in the future as a consequence of ongoing environmental changes (Harvell, 2019).

Like diseases, the spread of marine pests can impact aquaculture and aquatic ecosystems. Marine pests can quickly foul aquaculture structure and shells of cultured molluscs, incurring an increase in production costs through treatments, reduced growth rates and marketability. Cumulative economic impacts of the invasive species clubbed tunicate *Styela clava* and the Mediterranean fanworm *Sabella spallanzanii* on farmed New Zealand greenlipped mussel, *Perna canaliculus*, were estimated at NZ\$23.9 million and NZ\$14 million, respectively, over a 24-year period (Soliman and Inglis, 2018), with societal impacts from these pests over the same period estimated at around NZ\$8–10 million (Soliman and Inglis, 2018). The trans-boundary spread of marine pests can also concomitantly spread marine pathogens. Costello et al. (2020) demonstrated that the haplosporidian parasite *Bonamia ostreae* could be isolated from *S. clava*. Furthermore, the oyster virus OsHV-1 µvar was isolated from the highly invasive European green crab *Carcinus maenas* (Bookelaar et al., 2018). While major disease outbreaks originating from marine pest reservoirs are rare, they remain a pathway in the trans-boundary spread of pathogens (Foster et al., 2021). While we acknowledge that pests can compromise global aquaculture production and harm natural ecosystems and the communities that depend on them, for the purposes of this chapter we will focus on the emergence and spread of aquatic diseases.

14.3.3 Impacts of aquatic diseases

Impacts from aquatic diseases are often measured in an economic sense because aquaculture is primarily a for-profit business (Carpenter, 2019); however, the impacts from disease can be much further reaching than an industry's perspective. Usually, the impacts of aquatic diseases are considered across three broad categories: economic, environmental and social (Groner et al., 2016). Impacts are rarely demarcated within each category, with much overlap between them. For instance, the emergence of disease in biogenic habitats such as corals not only impacts the ecosystem in which they occur through the loss of habitat and food provisions, but also impacts ecotourism and fisheries, and the jobs these sectors support (Aronson and Precht, 2001). Another category of impacts not widely recognized internationally is that of cultural relevance. This aspect has been introduced above and will be further discussed below through case studies.

Generally, the degree to which parasites and diseases of aquatic animals have been studied reflects the commercial importance of the host concerned and the severity of infection. For example, in New Zealand there is much information on the eel parasite *Anguilla australis* because it was identified as a potential aquaculture species in the mid-1970s (Hine, 1978). Diseases can cause direct impacts such as a reduction in flesh quality and marketability, reduced fecundity and increased mortality rates and, in the case of World Organisation for Animal Health (WOAH)-notifiable diseases, affect international trade (Lane et al., 2016; Segarra et al., 2010; Stentiford et al., 2010). The economic impacts from disease can be severe; for example, billions of US dollars in production were lost in the early 1990s as a result of the global emergence of WSD in farmed penaeid prawns (Rodgers et al., 2011). Often less obvious are the indirect impacts of disease on facility closures, job losses, human health risks and shocks to the food supply chain such as processing, distribution, storage and preparation. Even less obvious again are the cultural impacts most severely felt by indigenous communities that arise from a loss of wild populations through the spread of diseases.

The emergence of OsHV-1 μvar in the Pacific oyster *C. gigas* has impacted aquaculture production in a number of regions throughout the world. Mass mortality events associated with the emergence of this virus started in 2008 in France (Segarra et al., 2010), then appeared throughout Europe from the Mediterranean to Scandinavia. In 2010, the disease emerged as a global aquatic pandemic after its detection in New Zealand and Australia. The emergence and spread of OsHV-1 μvar has had profound economic and social consequences. In New Zealand the introduction of OsHV-1 μvar had immediate impacts on production, with up to 100% mortality of affected spat (Bingham et al., 2013) and up to 70% loss of production of market-sized oysters between 2010 and 2012 (Sim-Smith et al., 2016). Overall, production of Pacific oysters in New Zealand, because of reduced supply, actually raised market prices for that commodity – a trend that holds for markets of other affected countries including France and Ireland (Fuhrmann et al., 2019).

The severity of the OsHV-1 μvar outbreak in New Zealand led to the bankruptcy and abandonment of farms, and subsequent unemployment. Aquaculture often occurs in rural areas away from urban centres and that have regional economies linked to primary industries, leaving them vulnerable to impacts from disease outbreaks. Local communities in areas of northern New Zealand reliant on Pacific oyster farming and processing were impacted greatly from the emergence of OsHV-1 μvar. Before this oyster disease, Pacific oyster farming supported 336 full-time equivalent jobs and contributed NZ$19 million to Northland's regional economy. The closure of just one processing facility left 66 staff redundant (Fuhrmann et al., 2019). Smaller businesses suffered more greatly than larger companies, because the former had few cash reserves to move to unaffected areas or invest in other infrastructure, such as the construction of an oyster hatchery to produce disease-free stock. The impacts and responses from New Zealand communities to OsHV-1 μvar were mirrored in other regions of the world affected by the virus, including a massive decline in production following the detection of the virus, a raise in market prices through decreased supply, the consolidation of fewer, larger companies following the displacement of small farming enterprises and, with the largest impacts being felt in regional coastal communities, particularly through a loss of employment (Fuhrmann et al., 2019). It is clear from our experience responding to OsHV-1 μvar that consequences from disease can be similarly impactful irrespective of geographic location.

The overall global trends of aquaculture production do not necessarily reflect regional trends. For instance, a minor increase in production from capture fisheries in Asia has been offset by a large increase in aquaculture production. Conversely in Europe, the reduction in capture production is supplemented by only a slight increase in aquaculture production (FAO, 2020). Consequently, as seafood (capture fisheries and aquaculture) is one the highest-traded global commodities, any significant reduction in volume – for example, due to animal disease outbreaks – can impact all aspects of the supply chain from fish processing services, distribution, storage, sales and food preparation. All of these facets support jobs and income. The fragility of the seafood economy to supply shocks has recently been highlighted during the COVID-19 pandemic, when a real-time synthesis of news articles between January and September 2020 reported substantial declines in fresh seafood catches (–40%), imports (–37%) and exports (–43%) relative to the previous year (White et al., 2021).

The ISA epidemic in Chile demonstrates strikingly how production can quickly be halted by disease, no matter how large the industry (Asche et al., 2009). Before ISA, Chile was the fastest-growing salmon producer in the world and was projected to overtake Norway to become the largest global producer of salmon. In early 2007 the Chilean salmon industry enjoyed high market prices and rapid industry growth, supporting around 25,000 jobs directly related to farming activities and 20,000 jobs associated with downstream supply chain activities. In hindsight, there is consensus that Chilean salmon industry growth outpaced government regulations: in particular, there were few biosecurity regulations to control the introduction and spread of diseases, which ultimately provided the conditions for the emergence and rapid spread of ISA in Chile in 2007 (Alvial et al., 2012). The closure of several hatcheries and processing plants immediately after the outbreak increased unemployment from 5 to 13% in affected regions. However, a concerted effort to implement and enforce basic infectious disease control measures (i.e. biosecurity requirements) in response to the disease outbreak, such as managing stocking densities, surveillance testing, movement controls and single year class farming, led to the relatively rapid recovery of the Chilean salmon industry. Salmon production was back to near pre-epidemic levels within a few years and has

continued to increase since (FAO, 2020). It is conceivable that the ISA epidemic could have been avoided had these biosecurity measures been implemented in the first place and kept pace with the growth of the industry.

14.4 Aquatic animal health and biosecurity: examples from Aotearoa New Zealand

14.4.1 Overview of biosecurity risk management

Biosecurity covers the overall policies and measures aiming to prevent and control the diseases that threaten economic, social and ecological health. Biosecurity risks in the aquatic environment come from various sources: pests, parasites and pathogens, exotic or already present (e.g. expansion of distribution, spill-over into new species or emerging opportunistic pathogens). Traditionally, management of these risks can take place at different levels and at different phases of the biosecurity continuum. International standards to support the safe trade of aquatic animals and animal products are produced and kept current by the WOAH (2021). In addition, strategies will establish requirements specific to countries to safeguard their aquatic animal health status and facilitate import and export trade requirements. Within each country, additional biosecurity risk management can be organized at national and regional level by government authorities, at industry level and at farm level (Oidtmann et al., 2011).

In New Zealand, biosecurity interventions are considered pre-border with the aim to keep risk offshore (e.g. import health requirements), at the border (e.g. inspection of imported goods before clearance) and post border (e.g. surveillance to detect incursions of exotic pathogens). Regulations have consistently played a major role in preventing, responding to and containing incursions of animal diseases. In the aquatic environment, a largely open system, eradicating diseases following an incursion is extremely challenging and seldom successful, and most biosecurity activities focus on long-term management, including preventing the domestic spread of pathogens between regions or aquaculture production areas. Biosecurity planning at an aquaculture facility will aim to manage the introduction and spread of diseases onto, within and away from the farm. A lack of adequate biosecurity at aquaculture facilities can exert ecological risks on wild populations through the spread of diseases (Daszak et al., 2001). The subsequent impacts, beyond industry, as discussed above, have prompted collective discussions on biosecurity responsibilities and shared leadership in New Zealand (Figure 14.1). Examples of partnerships over the past years have included initiatives such as a government industry agreement with the main aquaculture industry body to share the cost of disease preparedness and responses, and *Kō Tātou* – This Is Us (https://www.thisisus.nz/), a community engagement movement emphasizing the

need to work together to protect Aotearoa's unique biodiversity.

14.4.2   Land-based aquaculture and freshwater ecosystems

Freshwater (often referred to as 'land-based') and marine aquaculture have fundamentally different operational requirements, but the two are often connected from a biosecurity perspective (Peeler and Feist, 2011). For example, land-based hatcheries and nurseries supply marine farms with young animals for on-growing to market size. Interactions also exist between fish farms and wildlife in lakes and rivers (Bouwmeester et al., 2021; Peeler and Feist, 2011). This interface between freshwater wild and farmed animals has raised concerns about potential pathogen transfers (Bouwmeester et al., 2021; Peeler and Feist, 2011; Thrush et al., 2011): this was brought to light in New Zealand with reports of mortality of lampreys in rivers connected with strong recreational fishery and salmonid farms (Brosnahan et al., 2019; Kitson, 2012).

 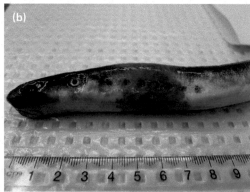

Figure 14.4 (a,b) New Zealand lamprey, *Geotria australis*, showed reddening lesions in association with significant mortalities, first reported in 2011 in rivers connected with fish farms. The biosecurity response prompted collective discussions on biosecurity risk management between aquaculture and conservation sectors. Lampreys have significant cultural value for Māori, the indigenous people of Aotearoa New Zealand. Credit: AHL (MPI).

Unusual mortality of lampreys, displaying reddening lesions (Figure 14.4), was first reported to the Ministry for Primary Industries in 2011. An initial investigation into the cause of death was conducted, primarily to exclude exotic diseases. Although the diagnostic picture was inconclusive, the bacterium *Aeromonas salmonicida* was isolated from a lamprey and from a captive trout in connected rivers (Brosnahan et al., 2019). The typical strain of *A. salmonicida* causes a severe disease called furunculosis, while atypical strains do not have such significant impacts. Until the bacterium could be cultured to ascertain the strain, trading partners had to be notified. During this waiting period, further mortality was occurring, prompting discussions on biosecurity practices between different users in the affected rivers. The Ministry's response team started engaging with various stakeholders, including land-based salmon industry, fish and game associations farming trout for stock enhancement, regional councils and *iwi* (local Māori tribes) (A. Castinel, Nelson, personal communication). Environmental information was also gathered, including traditional knowledge and observations from customary fishing shared by local *iwi*. It appeared that the reddening signs in lamprey had been seen for some years and in other rivers around New Zealand, albeit at much lower prevalence (Kitson, 2012).

This example illustrates a number of concepts that, if addressed, could significantly improve the biosecurity management of freshwater ecosystems and potentially land-based aquaculture in New Zealand, in line with an EcoHealth approach. First, working in partnership with local knowledge holders (indigenous and non-indigenous) could greatly enhance and expedite incursion investigations providing the information is shared in a culturally safe and respectful manner. Second, although the perception is generally that farmed fish impact wild fish in terms of disease transfer, the exchange dynamics can occur both ways and the

hypothetical mechanisms of disease spillover and spillback should be objectively considered (Bouwmeester et al., 2021). Third, this case study shows the importance of baseline data on the presence, distribution and prevalence of pathogens and clinical signs of disease in both farmed and wild populations. While the cost and complexity of pathogen surveillance in the aquatic environment is a major barrier (Bouwmeester et al., 2021; Peeler and Taylor, 2011), new technologies such as environmental DNA (e-DNA) in water samples could be helpful as an on-farm tool for the detection of pathogens and emerging diseases (Bass et al., 2015), but issues of validation and lack of regulatory framework should be addressed first.

14.4.3 Conservation, aquaculture and restoration

Ecological restoration and related restorative activities can have substantial benefits on human health. Ecosystem degradation has been shown to exacerbate communicable and non-communicable diseases, such as poor mental health (Thoma et al., 2021). Therefore, theoretically, if ecological degradation affects human well-being, then ecological restoration should reverse this interaction (Thoma et al., 2021). The holistic approach of EcoHealth recognizes indigenous knowledge and well-being. Although health impacts associated with ecological degradation and loss of biodiversity are difficult to quantify, these are expected to be felt most strongly in the indigenous communities because of their kinship links with the natural world, as outlined above. When these links are damaged or compromised, collective efforts to restore the ecological balance can be witnessed, as illustrated in New Zealand through the example of the native abalone health below.

The New Zealand abalone, *Haliotis iris*, locally referred to as *pāua*, supports commercial, recreational and customary fisheries. Concerns over the sustainability of catch limits led to a reduction in catch quotas in 1999. Enhancement of wild stocks by release of hatchery-reared seed is one method recognized as helping improve population numbers. Efforts to enhance wild stocks were underscored following the 7.8 Mw Kaikoura 2016 earthquake that caused up to 5 m of coastal uplift, stranding *pāua* and leading to an estimated loss of 21% of previously fished areas (Mccowan and Neubauer, 2018). Stock enhancement is not only attractive to commercial fishers, but also to indigenous *iwi* for whom *pāua* are *taonga* (treasures). However, the release of hatchery-reared stock into the wild is not without risks. For instance, breeding from a low number of broodstock may lead to inbreeding and emergence of deleterious traits in reseeded *pāua*. Detrimental effects could arise from the mixing of two genetically distinct lineages of seed and wild *pāua*. There is also the concern of transferring pathogens and disease, from the movement of wild broodstock into the hatchery and from the release of seed into the wild. All introductions and transfers of marine organisms carry risks associated with pathogens. The transfer of stock is arguably the largest contributor to the spread of pathogens, as demonstrated with *B. ostreae* decimating flat oysters throughout Europe after their introduction through infected stock shipped from the USA (Elston et al., 1986) or OsHV-1 µvar transferred across New Zealand through infected broodstock (Fuhrmann et al., 2019). In order to protect wild and hatchery-reared *pāua* from biosecurity threats, health assessments are regularly undertaken prior to any movement occurring. New Zealand is free from major diseases that have impacted abalone populations overseas (i.e. abalone viral ganglioneuritis or AVG, and withering syndrome); however, these assessments are usually limited to an absence of clinical signs or pathological findings on a small sample. This is significantly different to a proof of absence of specific pathogens. In New Zealand, surveillance of aquatic animal health relies almost exclusively on passive reporting of disease and mortality to the Ministry for Primary Industries by industry, the science community and the public. Occasionally pathogens may be inadvertently identified through pathological observations as part of research projects or routine screening.

The marine parasite *Perkinsus olseni* was first reported from farmed land-based and wild New Zealand *pāua* in 2013 (Muznebin et al., 2023). Despite an absence of clinical disease associated with this parasite in *pāua*, its emergence is curious because *P. olseni* has caused significant disease in Australian abalone populations (Goggin and Lester, 1995). As a warmer-water parasite, the emergence of *P. olseni* may be fuelled by climate change, signifying increased future disease risk that can be determined only by ongoing health surveillance. A growing interest in restoration and reports of new diseases underscore that any movement of individuals needs to be done with biosecurity in mind. For example, in Europe, restoration activities of the native flat oyster *Ostrea edulis* are accompanied by guidelines to assess and mitigate biosecurity risks associated with stock movements (zu Ermgassen et al., 2020). These guidelines clearly demonstrate that biosecurity needs to be considered in restoration activities to prevent the inadvertent spread of disease. As we act to reverse human-inflicted damage on wild populations through restoration, we must be mindful not to cause further harm and ensure an improvement in aquatic ecosystem health and human health through adequate biosecurity.

Land-based hatcheries present the advantage of providing a consistent supply of juveniles for on-growing aquaculture purposes or restocking for restoration. However, without adequate biosecurity protocols these can also negatively impact the same wild populations they are intended to support: for example, in Australia when an outbreak of AVG in abalone held in land-based facilities spilled back to wild populations (Hooper et al., 2007). AVG is caused by the abalone herpesvirus (AbHV). AbHV is relatively common in wild abalone and land-based facilities (Ellard et al., 2009), but the virus (and associated disease) can quickly amplify in land-based facilities, presumably as a result of stress from handling or overcrowding. AbHV spilled over into the wild populations via untreated effluent water from land-based facilities, causing outbreaks of AVG in wild abalone on reefs adjacent to the facility (Hooper et al., 2007). The disease subsequently spread to the wild abalone fishery along the coastline of southern Victoria, Australia, causing a 95% reduction in catch (Mayfield et al., 2015). Pathogen dispersal through the water is a primary pathway, and treatment of water into, within and away from shore-based facilities can significantly reduce disease transmission and is a major component of adequate on-farm biosecurity (Fernández-Boo et al., 2021; Whittington et al., 2015).

14.4.4 Marine aquaculture and coastal health

The vast majority of New Zealand's population lives near the coast. As of 2016, 75% of all New Zealanders lived within 10 km of the coast (Statistics New Zealand, 2021). The connectedness between urban human activities and the nearshore marine environment has never seemed so strong. Coastal sea, the interface between the land and open ocean, has been systemically degraded by resource extraction, urban development and pollution. Shellfish inhabiting the nearshore coastal zone bear the brunt of these impacts. The existence of healthy *kaimoana* (seafood) is intrinsically linked to the well-being of all communities that rely on

them (Hinemihi Rāwiri, 2018). For instance, a common sight at low tide on many New Zealand beaches is recreational gatherers of shellfish such as *pipi* or *tuatua*.

Open ocean aquaculture of both finfish and bivalves has emerged as a mechanism to not only intensify and maximize production from the aquaculture sector but also to reduce impacts from farming near the shore (Heasman et al., 2020). Open ocean (or, more accurately, offshore) greenlipped mussel farms in New Zealand are located approximately 5 km from the shoreline and are expected to bring many socio-economic benefits to the regional economy. Likewise, the expansion of offshore finfish aquaculture is forecast to produce 110,000 tonnes of salmon annually, equating to NZ$2 billion in sale (NZTE, 2020). This growth is expected to support hundreds of jobs when operating at full capacity. New Zealand Chinook salmon, by comparison with global Atlantic salmon industries, are relatively disease free. However, over the past 5 years new diseases have been reported from New Zealand salmon (Brosnahan et al., 2018), which is largely thought to be the result of warming water adding stress to fish under farming conditions. The development of offshore fish farming is expected not only to decouple impacts from land-based activities and the near shore but farming in deeper water will allow for cooler water temperatures with greater flow generally increasing fish health. Offshore aquaculture is also perceived as a more attractive alternative to land-based recirculating aquaculture farms from a social perspective, largely because of images of pristine open ocean farms compared with high-density land-based operations.

In coastal areas, increased sedimentation and contaminants from land, acidification and warming all act synergistically to negatively impact species physiology, leaving them more susceptible to disease and more prone to accumulate food-borne human pathogens like *Escherichia coli*.

Offshore shellfish aquaculture means that shellfish grown away from shores are less susceptible to the sanitation constraints that can close processing factories for periods at a time, which is another potential benefit advanced by the aquaculture sectors exploring this shift in farming operations (Heasman et al., 2020). Regardless of farming locations, the New Zealand aquaculture industry needs to continue to learn from overseas experience in regard to disease outbreaks and risk management. Biosecurity measures must be adapted to innovative farming practices: moving animal production away from shores does not mean that farms will be at a lesser risk from pest and disease transfers, especially with wild stocks.

14.5   Conclusion: future challenges for aquaculture

Wasterways are all interconnected in some way. As we look into the future, global linkages are likely to grow with increased maritime traffic, intensification of aquaculture, increased demand for food and urban development intruding into previously balanced ecosystems. From a biosecurity perspective this interconnectedness is a challenge, but also an opportunity. We have presented throughout this chapter the complexity in maintaining biosecurity to preserve EcoHealth and protecting values important to cultures and human and environmental health. The opportunity, however, is that biosecurity is not the responsibility of one sector or community, and that everyone can contribute to protect our EcoHealth. It takes all of us to safeguard our biological, cultural and economic heritage for future generations. This sentiment undoubtedly applies to global aquaculture. Certainly, the intensification and growth of aquaculture will bring employment, nutrition and well-being to much of the world's population. Its success, however, will be underpinned by biosecurity risk management. The aquaculture industry worldwide has been

slow to take up biosecurity, with many implementing reactive measures following disease outbreaks. A call to arms for proactive biosecurity has never been more pressing. This includes wider public education and surveillance to identify unusual events, strengthening on-farm biosecurity and coordinating efforts between farms and companies to achieve regional biosecurity. In addition to increased interconnectedness, climate change is another source of stress likely to significantly increase the risk of opportunistic diseases, and we can no longer treat it as a distant threat nor ignore its impacts.

References

Alvial, A. et al. (2012). *The recovery of the Chilean salmon industry. The ISA crisis and its consequences and lessons*. Puerto Montt, Chile.

Anderson, A. (2002). A fragile plenty: Pre-European Māori and the New Zealand environment. In E. Pawson & T. Brooking (Eds.), *Environmental histories of New Zealand* (pp. 19–34). Oxford University Press.

Aquaculture NZ. (2020). Aquaculture for New Zealand. A sector overview with key facts and statistics for 2020 (accessible from www.aquaculture.org.nz).

Aronson, R. B., & Precht, W. F. (2001). White-band disease and the changing face of Caribbean coral reefs. *Hydrobiologia*, *460*(1/3), 25–38. https://doi.org/10.1023/A:1013103928980

Asche, F., Hansen, H., Tveteras, R., & Tveterås, S. (2009). The salmon disease crisis in Chile. *Marine Resource Economics*, *24*(4), 405–411. https://doi.org/10.1086/mre.24.4.42629664

Asche, F., Cojocaru, A. L., & Sikveland, M. (2018). Market shocks in salmon aquaculture: The impact of the Chilean disease crisis. *Journal of Agricultural and Applied Economics*, *50*(2), 255–269. https://doi.org/10.1017/aae.2017.33

Bailey, S. A., Brown, L., Campbell, M. L., Canning-Clode, J., Carlton, J. T., Castro, N., Chinho, P., Chan, F. T., Creed, J. C., Curd, A., Darling, J., Fofonoff, P., Galil, B. S., Hewitt, C. L., Inglis, G. J., Keith, I., Mandrak, N. E., Marchini, A., McKenzie, C. H., . . . Zhan, A. (2020). Trends in the detection of aquatic nonindigenous species across global marine, estuarine and freshwater ecosystems. *Diversity and Distributions, 26*(12), 1780–1797. https://doi.org/10.1111/ddi.13167

Bass, D., Stentiford, G. D., Littlewood, D. T. J., & Hartikainen, H. (2015). Diverse applications of environmental DNA methods in parasitology. *Trends in Parasitology, 31*(10), 499–513. https://doi.org/10.1016/j.pt.2015.06.013

Best, E. (1904). Notes on the custom of "rahui". *Journal of the Polynesian Society, 13*(50), 83.

Bingham, P. et al. (2013). Marine and freshwater investigation into the first diagnosis of ostreid herpesvirus type 1 in Pacific oysters. *Surveillance (Wellington), 40,* 20–24.

Bondad-Reantaso, M. G. et al. (2018). Progressive management pathway to improve aquaculture biosecurity (PMP/AB) 1. *FAO Aquaculture Newsletter, 58,* 9–11.

Bookelaar, B. E., O'Reilly, A. J., Lynch, S. A., & Culloty, S. C. (2018). Role of the intertidal predatory shore crab Carcinus maenas in transmission dynamics of ostreid herpesvirus-1 microvariant. *Diseases of Aquatic Organisms, 130*(3), 221–233. https://doi.org/10.3354/dao03264

Bouwmeester, M. M., Goedknegt, M. A., Poulin, R., & Thieltges, D. W. (2021). Collateral diseases: Aquaculture impacts on wildlife infections. *Journal of Applied Ecology, 58*(3), 453–464. *https://doi.org/10.1111/1365-2664.13775*

Brosnahan, C. L., Munday, J. S., Ha, H. J., Preece, M., & Jones, J. B. (2019). New Zealand rickettsia-like organism (NZ-RLO) and Tenacibaculum maritimum: Distribution and phylogeny in farmed chinook salmon (Oncorhynchus tshawytscha). *Journal of Fish Diseases, 42*(1), 85–95. https://doi.org/10.1111/jfd.12909

Brosnahan, C. L., Pande, A., Keeling, S. E., van Andel, M., & Jones, J. B. (2019). Lamprey (Geotria australis; Agnatha) reddening syndrome in Southland rivers, New Zealand 2011–2013: Laboratory findings and epidemiology, including the incidental detection of an atypical Aeromonas salmonicida. *New Zealand Journal of Marine and Freshwater Research, 53*(3), 416–436. https://doi.org/10.1080/00288330.2018.1556167

Brudeseth, B. E., Wiulsrød, R., Fredriksen, B. N., Lindmo, K., Løkling, K. E., Bordevik, M., Steine, N., Klevan, A., & Gravningen, K. (2013). Status and future perspectives of vaccines for industrialised fin-fish farming. *Fish and Shellfish Immunology, 35*(6), 1759–1768. https://doi.org/10.1016/j.fsi.2013.05.029

Carnegie, R. B., Ford, S. E., Crockett, R. K., Kingsley-Smith, P. R., Bienlien, L. M., Safi, L. S. L., Whitefleet-Smith, L. A., & Burreson, E. M. (2021). A rapid phenotype change in the pathogen Perkinsus marinus was associated with a historically significant marine disease emergence in the eastern oyster. *Scientific Reports, 11*(1), 12872. https://doi.org/10.1038/s41598-021-92379-6

Carpenter, T. E. (2019). Measuring the impacts of aquatic animal diseases: The role of economic analysis. *Revue Scientifique et Technique (International Office of Epizootics), 38*(2), 511–522. https://doi.org/10.20506/rst.38.2.300

Castinel, A., Webb, S., Jones, J., Peeler, E., & Forrest, B. (2019). Disease threats to farmed green-lipped mussels Perna canaliculus in New Zealand: Review of challenges in risk assessment and pathway analysis. *Aquaculture Environment Interactions, 11,* 291–304. https://doi.org/10.3354/aei00314

Clapcott, J., Ataria, J., Hepburn, C., Hikuroa, D., Jackson, A., Kirikiri, R., & Williams, E. (2018). Mātauranga Māori: Shaping marine and freshwater futures. *New Zealand Journal of Marine and Freshwater Research, 52*(4), 457–466. https://doi.org/10.1080/00288330.2018.1539404

Collier, K. J., Pilditch, C. A., & Lundquist, C. J. (2019). Mountains-to-the-sea conservation: An island perspective. *Aquatic Conservation: Marine and Freshwater Ecosystems, 29*(9), 1383–1390. https://doi.org/10.1002/aqc.3197

Cometti, R., & Morton, J. (1985). Manawas of the north. In, *Margins of the sea, exploring New Zealand coastline* (pp. 34–37). Hoddes & Stoughton.

Costello, K. E. et al. (2020). The role of invasive tunicates as reservoirs of molluscan pathogens. *Biological Invasions, 23*(2), 641–655. https://doi.org/10.1007/s10530–020–02392–5

Daszak, P., Cunningham, A. A., & Hyatt, A. D. (2001). Anthropogenic environmental change and the emergence of infectious diseases in wildlife. *Acta Tropica, 78*(2), 103–116. https://doi.org/10.1016/s0001-706x(00)00179-0

Egan, E. M. C., Hickford, M. J. H., & Schiel, D. R. (2019). Understanding the life histories of amphidromous fish by integrating otolith-derived growth reconstructions, post-larval migrations and reproductive traits. *Aquatic Conservation: Marine and Freshwater Ecosystems, 29*(9), 1391–1402. https://doi.org/10.1002/aqc.3145

Ellard, K., Pyecroft, S., Handlinger, J., & Andrewartha, R. (2009). Findings of disease investigations following the recent detection of AVG in Tasmania. In M. Crane & J. Slater (Eds.) *Proceedings of the Fourth National FRDC Aquatic Animal Health Scientific Conference, Cairns, Australia.*

Ellis, J. I., Hewitt, J. E., Clark, D., Taiapa, C., Patterson, M., Sinner, J., Hardy, D., & Thrush, S. F. (2015). Assessing ecological community health in coastal estuarine systems impacted by multiple stressors. *Journal of Experimental Marine Biology and Ecology, 473*, 176–187. https://doi.org/10.1016/j.jembe.2015.09.003

Elston, R. A., Farley, C. A., & Kent, M. L. (1986). Occurence and significance of bonamiasis in European flat oysters Ostrea edulis in North America. *Diseases of Aquatic Organisms, 2*, 49–54. https://doi.org/10.3354/dao002049

Food and Agriculture Organization. (2018). *Achieving Blue Growth: Building vibrant fisheries and aquaculture communities.* https://www.fao.org/policy-support/tools-and-publications/resources-details/en/c/1234976/

Food and Agriculture Organization. (2020). Sustainability in action. Rome. *State of World Fisheries and Aquaculture.* https://doi.org/10.4060/ca9229en

Feist, S. W., Thrush, M. A., Dunn, P., Bateman, K., & Peeler, E. J. (2019). The aquatic animal pandemic crisis. *Revue Scientifique et Technique, 38*(2), 437–457. https://doi.org/10.20506/rst.38.2.2997

Fernández-Boo, S., Provot, C., Lecadet, C., Stavrakakis, C., Papin, M., Chollet, B., Auvray, J., & Arzul, I. (2021). Inactivation of marine bivalve parasites using UV-C irradiation: Examples of Perkinsus olseni and Bonamia ostreae. *Aquaculture Reports, 21*, 100859. https://doi.org/10.1016/j.aqrep.2021.100859

Foster, R., Peeler, E., Bojko, J., Clark, P. F., Morritt, D., Roy, H. E., Stebbing, P., Tidbury, H. J., Wood, L. E., & Bass, D. (2021). Pathogens co-transported with invasive non-native aquatic species: Implications for risk analysis and legislation. *NeoBiota, 69*, 79–102. https://doi.org/10.3897/neobiota..71358

Fuhrmann, M., Castinel, A., Cheslett, D., Furones Nozal, D., & Whittington, R. J. (2019). 'L'impact des microvariants du virus herpétique ostreid herpesvirus 1 sur la culture d'huîtres creuses dans les hémisphères Nord et Sud depuis 2008. *Revue Scientifique et Technique (International Office of Epizootics), 38*(2), 491–509. https://doi.org/10.20506/rst.38.2.3000

Georgiades, E., Fraser, R., & Jones, B. (2016). *Options to strengthen on-farm biosecurity management for commercial and noncommercial aquaculture.* Wellington.

Gephart, J. A., Henriksson, P. J. G., Parker, R. W. R., Shepon, A., Gorospe, K. D., Bergman, K., Eshel, G., Golden, C. D., Halpern, B. S., Hornborg, S., Jonell, M., Metian, M., Mifflin, K., Newton, R., Tyedmers, P., Zhang, W., Ziegler, F., & Troell, M. (2021). Environmental performance of blue foods. *Nature, 597*(7876), 360–365. https://doi.org/10.1038/s41586-021-03889-2

Goggin, C. L., & Lester, R. J. G. (1995) 'Perkinsus. Perkinsus, a protistan parasite of abalone in Australia: A review. *Marine and Freshwater Research, 46*(3), 639–646. https://doi.org/10.1071/MF9950639

Groner, M. L., Maynard, J., Breyta, R., Carnegie, R. B., Dobson, A., Friedman, C. S., Froelich, B., Garren, M., Gulland, F. M., Heron, S. F., Noble, R. T., Revie, C. W., Shields, J. D., Vanderstichel, R., Weil, E., Wyllie-Echeverria, S., & Harvell, C. D. (2016). Managing marine disease emergencies in an era of rapid change. *Philosophical Transactions of the Royal Society of London. Series B, Biological Sciences, 371*(1689), 20150364. https://doi.org/10.1098/rstb.2015.0364

Harvell, C. (2019). *Ocean outbreak: Confronting the rising tide of marine disease.* University of California Press. https://doi.org/10.2307/j.ctvfxvc83

Heasman, K. G., Scott, N., Ericson, J. A., Taylor, D. I., & Buck, B. H. (2020). Extending New Zealand's marine shellfish aquaculture into exposed environments – Adapting to modern anthropogenic challenges. *Frontiers in Marine Science, 7.* https://doi.org/10.3389/fmars.2020.565686

Heggie, K., & Savage, C. (2009). Nitrogen yields from New Zealand coastal catchments to receiving estuaries. *New Zealand Journal of Marine and Freshwater Research, 43*(5), 1039–1052. https://doi.org/10.1080/00288330.2009.9626527

Hepburn, C. D., Jackson, A., Pritchard, D. W., Scott, N., Vanderburg, P. H., & Flack, B. (2019). Challenges to traditional management of connected ecosystems within a fractured regulatory landscape: A case study from southern New Zealand. *Aquatic Conservation: Marine and Freshwater Ecosystems, 29*(9), 1535–1546. https://doi.org/10.1002/aqc.3152

Le Heron, E., Logie, J., Allen, W., Le Heron, R., Blackett, P., Davies, K., Greenaway, A., Glavovic, B., & Hikuroa, D. (2019). Diversity, contestation, participation in Aotearoa New Zealand's multi-use/user marine spaces. *Marine Policy, 106,* 103536. https://doi.org/10.1016/j.marpol.2019.103536

Hewitt, C. L., Willing, J., Bauckham, A., Cassidy, A. M., Cox, C. M. S., Jones, L., & Wotton, D. M. (2004). New Zealand marine biosecurity: Delivering outcomes in a fluid environment. *New Zealand Journal of Marine and Freshwater Research, 38*(3), 429–438. https://doi.org/10.1080/00288330.2004.9517250

Hewson, I., Chow, C.-E., & Fuhrman, J. (2010). Ecological role of viruses in aquatic ecosystems. In. https://doi.org/10.1002/9780470015902.a0022546

Hine, P. M. (1978). Distribution of some parasites of freshwater eels in New Zealand. *New Zealand Journal of Marine and Freshwater Research, 12*(2), 179–187. https://doi.org/10.1080/00288330.1978.9515739

Hinemihi Rāwiri, A. (2018). *Tahi Ki a Maru: Water, fishing and tikanga in Ngti Raukawa Ki Te Tonga.* Te wananga o Raukawa.

Hooper, C., Hardy-Smith, P., & Handlinger, J. (2007). Ganglioneuritis causing high mortalities in farmed Australian abalone (*Haliotis laevigata* and *Haliotis rubra*). *Australian Veterinary Journal, 85*(5), 188–193. https://doi.org/10.1111/j.1751-0813.2007.00155.x

Israngkura, A., & Sae-Hae, S. (2002). A review of the economic impacts of aquatic animal disease. In J. R. Arthur, M. J. Phillips, R. P. Subasinghe, M. B. Reantaso, & I. H. MacRae (Eds.) *Primary Aquatic Animal Health Care in Rural, Small-scale, Aquaculture Development.* FAO Fish. Tech. Pap. No. 406.

Kirikiri, R., & Nugent, G. (1995). Harvesting of New Zealand native birds by Māori. In G. Grigg, P. Hale & D. Lunney (Eds.), *Conservation through sustainable use of wildlife* (pp. 54–59). Centre for Conservation Biology, the University of Queensland.

Kitson, J. (2012) *Kanakana harvest mātauranga: Potential tools to monitor population trends on the Waikawa River, Southland/Murihiku.* Prepared for Ngā Pae o te Māramatanga. https://www.maramatanga.ac.nz/index.php/media/208/download?attachment

Lane, H. S., Webb, S. C., & Duncan, J. (2016). Bonamia ostreae in the New Zealand oyster Ostrea chilensis: A new host and geographic record for this haplosporidian parasite. *Diseases of Aquatic Organisms, 118*(1), 55–63. https://doi.org/10.3354/dao02960

Lennox, R. J., Crook, D. A., Moyle, P. B., Struthers, D. P., & Cooke, S. J. (2019). Toward a better understanding of freshwater fish responses to an increasingly drought-stricken world. *Reviews in Fish Biology and Fisheries, 29*(1), 71–92. https://doi.org/10.1007/s11160-018-09545-9

Marcos-López, M., Gale, P., Oidtmann, B. C., & Peeler, E. J. (2010). Assessing the impact of climate change on disease emergence in freshwater fish in the United Kingdom. *Transboundary and Emerging Diseases, 57*(5), 293–304. https://doi.org/10.1111/j.1865-1682.2010.01150.x

Marsden, M. (2003). The natural world and natural resources: Māori values systems and perspectives. In T. A. C. Royal (Ed.), *The woven universe: Selected writings of Rev. Māori Marsden* (pp. 24–53). Otaki: Estate of Rev. Maori Marsden.

Mayfield, S. et al. (2015) *Southern zone abalone (Haliotis rubra and H. laevigata) fishery.*

McCarthy, A., Hepburn, C., Scott, N., Schweikert, K., Turner, R., & Moller, H. (2014). Local people see and care most? Severe depletion of inshore fisheries and its consequences for Māori communities in New Zealand. *Aquatic Conservation: Marine and Freshwater Ecosystems, 24*(3), 369–390. https://doi.org/10.1002/aqc.2378

Mccowan, T., & Neubauer, P. (2018). *Paua biomass estimates and population monitoring in areas affected by the November 2016 Kaikoura earthquake.* New Zealand Fisheries Assessment Report.

McDowall, R. M. (2006). Fish, fish habitats and fisheries in New Zealand. *Aquatic Ecosystem Health and Management, 9*(4), 391–405. https://doi.org/10.1080/14634980601026352

Mead, H. (2003). Rahui, aukati: Ritual prohibitions. In *Tikanga Māori* (pp. 193–207). Huia Publishers.

Mead, H. (2016). Mana whenua, mana moana: Authority over land and the ocean. In T. Māori (Ed.) (rev. ed): *Living by Māori values.* Huia Publishers.

Moller, H. (1996). Customary use of indigenous wildlife – Towards a bicultural approach to conserving New Zealand's biodiversity. In *Science and Research Division, D. of C.* (Ed.) *Biodiversity: Papers from a Seminar Series on Biodiversity* (pp. 89–125). Wellington.

Moller, H. (2009). Matauranga Maori, science and seabirds in New Zealand. *New Zealand Journal of Zoology, 36*(3), 203–210. https://doi.org/10.1080/03014220909510151

Muznebin, F., Alfaro, A. C., & Webb, S. C. (2023). Occurrence of Perkinsus olseni and other parasites in New Zealand black-footed abalone (Haliotis iris). New Zealand Journal of Marine and

Freshwater Research, 57(2), 261–281. https://doi.org/10.1080/00288330.2021.1984950

New Zealand Ministry for Primary Industries. (n.d.). National Aquatic Biodiversity Information System. https://www.mpi.govt.nz/legal/legislation-standards-and-reviews/fisheries-legislation/maps-of-nz-fisheries/#NABIS

Norman, R. et al. (2013). *Salmon mortality investigation* [MPI technical paper]. Wellington.

NZTE. (2020). *Open ocean finfish aquaculture: Business case. Prepared by Envirostrat Ltd for New Zealand trade and enterprise.* Wellington.

Oidtmann, B. C., Thrush, M. A., Denham, K. L., & Peeler, E. J. (2011). International and national biosecurity strategies in aquatic animal health. *Aquaculture, 320*(1–2), 22–33. https://doi.org/10.1016/j.aquaculture.2011.07.032

Peeler, E. J., Oidtmann, B. C., Midtlyng, P. J., Miossec, L., & Gozlan, R. E. (2011). Non-native aquatic animals introductions have driven disease emergence in Europe. Biological Invasions, 13(6), 1291–1303. https://doi.org/10.1007/s10530-010-9890-9

Peeler, E. J., & Feist, S. W. (2011). Human intervention in freshwater ecosystems drives disease emergence: Human intervention and disease emergence. *Freshwater Biology, 56*(4), 705–716. https://doi.org/10.1111/j.1365-2427.2011.02572.x

Peeler, E. J., & Taylor, N. G. H. (2011). The application of epidemiology in aquatic animal health - opportunities and challenges. *Veterinary Research (Paris), 42*(1), 94. https://doi.org/10.1186/1297-9716-42-94. Te Kokiri, Puni. (2021). Wai. *Te Pae Tawhiti. Te puni Kokiri.* https://www.tpk.govt.nz/en/a-matou-kaupapa/wai-262-te-pae-tawhiti

Rahman, K. S., Islam, M. N., Ahmed, M. U., Bosma, R. H., Debrot, A. O., & Ahsan, M. N. (2020). Selection of mangrove species for shrimp based silvo-aquaculture in the coastal areas of Bangladesh. *Journal of Coastal Conservation, 24*(5). https://doi.org/10.1007/s11852-020-00770-8

Robinson, J. M., Gellie, N., MacCarthy, D., Mills, J. G., O'Donnell, K., & Redvers, N. (2021). Traditional ecological knowledge in restoration ecology: A call to listen deeply, to engage with, and respect Indigenous voices. *Restoration Ecology, 29*(4). https://doi.org/10.1111/rec.13381

Rodgers, C. J., Mohan, C. V., & Peeler, E. J. (2011). The spread of pathogens through trade in aquatic animals and their products, [Revue scientifique et technique (International Office of Epizootics)].

Revue *Scientifique et Technique, 30*(1), 241–256. https://doi.org/10.20506/rst.30.1.2034

Ross, P. M., Knox, M. A., Smith, S., Smith, H., Williams, J., & Hogg, I. D. (2018). Historical translocations by Māori may explain the distribution and genetic structure of a threatened surf clam in Aotearoa (New Zealand). *Scientific Reports, 8*(1), 17241. https://doi.org/10.1038/s41598-018-35564-4

Schar, D., Klein, E. Y., Laxminarayan, R., Gilbert, M., & Van Boeckel, T. P. (2020). Global trends in antimicrobial use in aquaculture. *Scientific Reports, 10*(1), 21878. https://doi.org/10.1038/s41598-020-78849-3

Schar, D., Zhao, C., Wang, Y., Larsson, D. G. J., Gilbert, M., & Van Boeckel, T. P. (2021). Twenty-year trends in antimicrobial resistance from aquaculture and fisheries in Asia. *Nature Communications, 12*(1), 5384. https://doi.org/10.1038/s41467-021-25655-8

Scott-Orr, H., Jones, J., & Bhatia, N. (2017). *Uncooked prawn imports: Effectiveness of biosecurity controls.* https://www.igb.gov.au/sites/default/files/documents/final-uncooked-prawn-imports_0.pdf

Segarra, A., Pépin, J. F., Arzul, I., Morga, B., Faury, N., & Renault, T. (2010). Detection and description of a particular Ostreid herpesvirus 1 genotype associated with massive mortality outbreaks of Pacific oysters, Crassostrea gigas, in France in 2008. *Virus Research, 153*(1), 92–99. https://doi.org/10.1016/j.virusres.2010.07.011

Sherman, D. J. (2006). Seizing the cultural and political moment and catching fish: Political development of Māori in New Zealand, the Sealord Fisheries Settlement, and social movement theory. *Social Science Journal, 43*(4), 513–527. https://doi.org/10.1016/j.soscij.2006.08.002

Sim-Smith, C., Faire, S., & Lees, A. (2016). *Managing biosecurity risk for business benefit.* Aquaculture Biosecurity Practices Research Report. Wellington.

Snieszko, S. F. (1974). The effects of environmental stress on outbreaks of infectious diseases of fishes. *Journal of Fish Biology, 6*(2), 197–208. https://doi.org/10.1111/j.1095-8649.1974.tb04537.x

Soliman, T., & Inglis, G. J. (2018). Forecasting the economic impacts of two biofouling invaders on aquaculture production of green-lipped mussels Perna canaliculus in New Zealand. *Aquaculture Environment Interactions, 10*, 1–12. https://doi.org/10.3354/aei00249

Soma, K., van den Burg, S. W. K., Hoefnagel, E. W. J., Stuiver, M., & van der Heide, C. M. (2018). Social innovation – A future pathway for Blue growth? *Marine Policy, 87*, 363–370. https://doi.org/10.1016/J.MARPOL.2017.10.008

Statistics New Zealand. (2021). Population statistics. https://www.stats.govt.nz/topics/population

Stenson, M. (2004) 'Chapter 1'. In M. Stenson (Ed.). *The Treaty: Every New Zealander's guide to the Treaty of Waitangi* (1st ed) (pp. 111–121). Random House New Zealand,.

Stentiford, G. D., Oidtmann, B., Scott, A., & Peeler, E. J. (2010). Crustacean diseases in European legislation: Implications for importing and exporting nations. *Aquaculture, 306*(1–4), 27–34. https://doi.org/10.1016/j.aquaculture.2010.06.004

Stentiford, G. D., Bateman, I. J., Hinchliffe, S. J., Bass, D., Hartnell, R., Santos, E. M., Devlin, M. J., Feist, S. W., Taylor, N. G. H., Verner-Jeffreys, D. W., van Aerle, R., Peeler, E. J., Higman, W. A., Smith, L., Baines, R., Behringer, D. C., Katsiadaki, I., Froehlich, H. E., & Tyler, C. R. (2020). Sustainable aquaculture through the One Health lens. *Nature Food, 1*(8), 468–474. https://doi.org/10.1038/s43016-020-0127-5

Thoma, M. V., Rohleder, N., & Rohner, S. L. (2021). Clinical ecopsychology: The mental health impacts and underlying pathways of the climate and environmental crisis. *Frontiers in Psychiatry, 12*, 675936. https://doi.org/10.3389/fpsyt.2021.675936

Thomas, F., Renaud, F., de Meeûs, T., & Poulin, R. (1998). Manipulation of host behaviour by parasites: Ecosystem engineering in the intertidal zone? *Proceedings of the Royal Society of London. Series B, 265*(1401), 1091–1096. *https://doi.org/10.1098/rspb.1998.0403*

Thrush, M. A., Murray, A. G., Brun, E., Wallace, S., & Peeler, E. J. (2011). The application of risk and disease modelling to emerging freshwater diseases in wild aquatic animals. *Freshwater Biology* [Manuscript], *56*(4), 658–675. https://doi.org/10.1111/j.1365-2427.2010.02549.x

Thrush, M. A., Pearce, F. M., Gubbins, M. J., Oidtmann, B. C., & Peeler, E. J. (2017). A simple model to rank shellfish farming areas based on the risk of disease introduction and spread. *Transboundary and Emerging Diseases, 64*(4), 1200–1209. https://doi.org/10.1111/tbed.12492

United Nations. (2007). United Nations declaration on the rights of indigenous peoples. https://www.un.org/development/desa/indigenouspeoples/wp-content/uploads/sites/19/2018/11/UNDRIP_E_web.pdf

Waitangi tribunal. (1985) 'The loss of the waters – a Māori perspective'. In *Report of the Waitangi Tribunal on the Manukau claim (WAI 8)* (pp. 37–43). Government printer.

Waitangi tribunal. (2011) *Ko Aotearoa tēnei: A report into claims concerning New Zealand law and policy affecting Māori culture and identity. Te Taumata tuarua.* Wellington.

Wenhai, L., Cusack, C., Baker, M., Tao, W., Mingbao, C., Paige, K., Xiaofan, Z., Levin, L., Escobar, E., Amon, D., Yue, Y., Reitz, A., Neves, A. A. S., O'Rourke, E., Mannarini, G., Pearlman, J., Tinker, J., Horsburgh, K. J., Lehodey, P., . . . Yufeng, Y. (2019). Successful blue economy examples with an emphasis on international perspectives. *Frontiers in Marine Science, 6.* https://doi.org/10.3389/fmars.2019.00261

White, E. R., Froehlich, H. E., Gephart, J. A., Cottrell, R. S., Branch, T. A., Agrawal Bejarano, R., & Baum, J. K. (2021). Early effects of COVID-19 on US fisheries and seafood consumption. *Fish and Fisheries, 22*(1), 232–239. *https://doi.org/10.1111/faf.12525*

Whittington, R. J., Hick, P. M., Evans, O., Rubio, A., Alford, B., Dhand, N., & Paul-Pont, I. (2015). Protection of Pacific oyster (Crassostrea gigas) spat from mortality due to ostreid herpesvirus 1 (OsHV-1 μVar) using simple treatments of incoming seawater in land-based upwellers. *Aquaculture, 437*, 10–20. https://doi.org/10.1016/j.aquaculture.2014.11.016

WOAH. (2021). Aquatic animal health code, 23rd edition, 2021. https://doc.woah.org/dyn/portal/index.xhtml?page=alo&aloId=41547

Young, D. (2004). *Our islands, our selves: A history of conservation in New Zealand.* University of Otago Press.

zu Ermgassen, P. et al. (2020). *European guidelines on biosecurity in native oyster restoration.* https://nativeoysternetwork.org/wp-content/uploads/sites/27/2020/11/ZSL00161%20Biosecurity%20Handbook%20ONLINE.pdf

For further reading:

McDowall, R. M. (2011). Ikawai: freshwater fishes in Māori culture and economy. Canterbury University Press, Christchurch, New Zealand 832 p.

NIWA. (n.d.). Taonga Species Series. https://niwa.co.nz/te-kuwaha/tools-and-resources/taonga-species-series

Roberts, C. D. et al. (2015). *The Fishes of New Zealand.* Te Papa Press, Wellington, New Zealand, 2000 pp.

Chapter 15

Aquaculture – the interdependency of food security and the natural environment

Jessica Wu and Sareeha Vasanthakumar

Abstract

Aquaculture plays a key role in securing global food security, as well as providing an important local source of protein and employment for billions of people. Due to environmental and anthropogenic factors, the industry is faced with numerous challenges. Globally, there is a growing need to ensure that aquaculture practices are sustainable in the years to come. In Asia, the significant reliance on the natural environment, especially in coastal zones and in the face of climate change and competing land use priorities, provides a challenge to aquaculture productivity and economic success. Using shrimp aquaculture in Sri Lanka as an example, this case study will examine the evolution of the aquaculture industry in the context of environmental needs, and the challenges of growing healthy and productive animals.

15.1 Aquaculture background

Global food consumption continues to see an increased focus on access to high-quality protein as the population increases, and the growing demand from a growing middle class (Blanchard et al., 2017). Consumption of fresh fish and seafood has grown steadily around the world, particularly in developing countries, and represents one of the largest sources of animal protein (FAO, 2020). As a result, the fisheries and aquaculture sector has more than tripled in the past 50 years, outpacing all other livestock sectors. Billions of people globally rely on fisheries and aquaculture for both protein and as a source of employment (FAO, 2020). Aquaculture itself contributed nearly half (46%) of total global fish harvest in 2018, with the remaining 54% from managed fisheries (FAO, 2020). This is a significant increase from the year 2000, when aquaculture contributed 25.7% of global fish production.

With this shift in food consumption patterns and decreasing wild stock, aquaculture has been considered a key contributor to food security and poverty alleviation. Aquaculture has subsequently been dubbed the "Blue Revolution" (Costa-Pierce, 2008; Golden et al., 2017). Particularly for developing countries, it is hoped that aquaculture can play an integral role

Susan C. Cork and Douglas P. Whiteside (eds) **Case Studies in EcoHealth**
DOI: 10.52517/9781789183368.015, © 5m Books Ltd, 2024

in meeting nutritional needs in a low-cost and sustainable manner (Bondad-Reantaso, 2012; Golden et al., 2017).

Globally, a number of studies have demonstrated the benefits of aquaculture although with significant differences between geographical regions, species cultured, local consumption patterns and sociopolitical context (Bondad-Reantaso, 2012). However, aquaculture has a mixed reputation in terms of its negative environmental impact: for example the use of wild fish products as feed, and pollution due to waste discharge versus its recognized benefits, such as reducing gender inequality and poverty alleviation. Aquaculture-associated pollution includes overuse of antibiotics and other chemicals required to manage pests and diseases in farmed species. The challenge for the future is developing aquaculture systems to ensure food security in a sustainable manner that balances productivity with both environmental and animal health.

Using shrimp aquaculture in Sri Lanka as an example, this case study will examine the interdependence of developing the aquaculture industry while respecting environmental health, and the challenges of growing healthy and productive captive animals.

15.2 Overview of aquaculture and shrimp farming in Sri Lanka

Many regions of the world have seen the growth of aquaculture in the past half century. There are approximately 450 species of aquatic organisms (fish, crustaceans, molluscs, amphibians, reptiles and invertebrates) cultivated globally (FAO, 2020; Naylor et al., 2021). Asia is the primary region for aquaculture growth in the world, contributing 69% of global aquaculture production with 35% from China alone in 2018. Nearly half of all global fish exports originated from low-income countries in Asia.

What is the Blue Revolution?

The term 'Blue Revolution' refers to the rapid increase in productivity of aquaculture, particularly farmed fish, for human consumption (Costa-Pierce, 2008). This began in Asia in the 1960s and subsequently spread to other parts of the world (Chua, 1997). In 2005, Dr. Modadugu V. Gupta was named the World Food Prize Laureate as one of the architects of the Blue Revolution in Asia, based on his early research on polyculture and use of farm waste as fish feed in India. Analogous to this term is the 'Green Revolution', which refers to the rapid increases in agricultural crop yield production that began in the 1960s (Evenson and Gollin, 2003). The scientific advances increasing food crop productivity were key to preventing famine predicted to occur due to population growth combined with decreasing land availability.

Island nations in particular rely heavily on consumption of aquatic species for animal protein. For example, 61% of animal protein consumed in Sri Lanka comes from fish or other fishery products (Jayasinghe et al., 2019). Within Sri Lanka, the fisheries and aquaculture sector provides livelihoods to 2.6 million individuals out of a population of approximately 21 million. The industry contributed 1.3% to the country's gross domestic product in 2019 (Jayasinghe et al., 2019). Fish production in the country is primarily from commercial marine fisheries (86%), with the remaining 14% from inland fisheries and aquaculture. Of this production, greater than 90% (by weight) was sold or consumed within Sri Lanka.

Although Sri Lanka is endowed with large fresh and brackish water resources, it does not have a tradition of aquaculture and only shrimp aquaculture and ornamental fish culture have been developed to any extent (FAO, 2016). Other species being cultured in the country include tilapia and carp (*Cyprinus* spp).

> The term 'tilapia' refers to several species of mostly freshwater fish species belonging to the cichlid family. Many wild tilapia are native to Africa, but the fish has been introduced throughout the world and is now farmed in over 135 countries. It is a suitable fish for farming because it can be kept at a high stocking rate and grows quickly at relatively low input cost.

Shrimp farming in the country started in the late 1970s in the Batticaloa district of the Eastern Province of Sri Lanka, established by a few large multinational companies and some medium-scale entrepreneurs (Munasinghe et al., 2010). Shrimp farming expansion was particularly attractive due to high export demand for shrimp, with the potential for significant profit when successfully executed (FAO, 2012). The total number of farms numbered close to 1,400, with over 70 hatcheries and a total area of 4,500 ha allocated for shrimp farming. The farms in this region were largely abandoned in the 1980s due to civil unrest and the industry spread to the opposite side of the country, Puttalam district in the North Western Province (NWP).

> Shrimp are free-swimming crustaceans with long antennae, slender legs and a laterally compressed, muscular abdomen. The latter is highly adapted for both forward swimming and a backward escape response.

The shrimp-farming industry boomed in NWP in the 1980s and 1990s, with continued growth of the international export market and investment incentives from the government, encouraging ongoing input from multinational companies and medium-scale entrepreneurs that owned a large proportion of the shrimp farms (Drengstig, 2013). By the end of 1994, the estimated employment generated through shrimp farming was over 4,000 jobs (Wijegoonawardena and Siriwardena, 1996). However, due to a lack of proper law enforcement and poor environmental planning, this rapid development culminated in close to 47% of all farms being illegal establishments operating without proper licenses. The lack of coastal zone management and infrastructure development led the industry into a 'self-pollution effect', with renewal of water resources being outpaced by effluent polluting common waterways (Cattermoul and Devendra, 2002; Wijegoonawardena and Siriwardena, 1996). Disease became a significant problem in cultivated shrimp with the introduction of *Penaeus monodon* nucloepolyhedrovirus virus, white spot syndrome virus (WSSV) and yellowhead disease, causing a production drop of 64% from 1988 to 1989 (Harkes et al., 2015; Munasinghe et al., 2012). Combined with challenges associated with the civil war, this led to the abandonment of many shrimp farms and, from 1996, the industry started to decline significantly with almost all farming activities restricted to a narrow coastal belt of approximately 120×10 km^2 in NWP (Jayasinghe and Niroshana, 2016; Munasinghe et al., 2010). Currently, shrimp farming is primarily carried out by local farmers with shrimp from small-holdings sold to local markets (Harkes et al., 2015). See Figures 15.1 and 15.2.

With significant water resources available in the country there is still great potential for increasing aquaculture and inland fisheries (Landesman et al., 2009). According to the National Aquaculture Development Authority in Sri Lanka, 8,500 hectares of land are suitable for aquaculture within Sri Lanka with only 25%

Figure 15.1 Shoreline aquaculture in southern Asia.

Figure 15.2 Aquaculture products for sale in local food markets in southern Asia.

of this area currently utilized (Drengstig, 2013). Access to coastal land previously restricted due to the civil conflict in the eastern and northern Provinces may allow opportunities for expansion of the aquaculture industry to these areas in the future. The history of a successful Sri Lankan shrimp export market indicates the potential to revive this industry (Huxham, 2015; Landesman et al., 2009). This could provide employment opportunities for the rural poor and improvements to food security. The Sri Lankan government recognizes this potential by their inclusion of aquaculture within the country's development plans. Despite this growth potential, there are a number of key issues that need to be addressed. While some of the required changes appear to be superficially simple, such as better management for disease prevention, many other actions are neither straightforward nor easy to implement and sustain.

15.3 The human dimension of aquaculture

15.3.1 Aquaculture in poverty alleviation and food security

With a growing global population, sustainable food production will be the key to reducing malnutrition and hunger (FAO, 2013). Smallholder farming is considered to be a key method in alleviating extreme poverty (Godfray et al., 2010; HLPE, 2013). Aquaculture can contribute to poverty alleviation and food security through facilitation of direct consumption of fish and aquaculture products, providing employment and opportunities for income generation and improving the status of women through aquaculture participation (Ahmed and Lorica, 2002; Kawarazuka and Béné, 2010). Fish and aquatic products are nutrient-dense foods that can be an important source of animal protein and micronutrients, particularly when consumed whole (Roos et al., 2007), and therefore there is a major impact in terms of food security and nutrition if consumed directly (FAO, 2020).

Food security is defined as 'when all people, at all times, have physical and economic access to sufficient, safe, and nutritious food to meet their dietary needs and food preferences for an active and healthy life' (World Food Summit, 1996). It is comprised of four key components: availability, accessibility, utilization and – more recently – stability (FAO, 2009).

Globally, the outcomes and impact of aquaculture development vary greatly depending on the local context such as the species being cultured (i.e., whether the household prefers that type of fish), the value of the species (i.e., whether the income from local or export sales is more desirable), traditional household diets and cultural dietary preferences (Barraclough and Finger-Stich, 1996; Stevenson and Irz, 2009). Although households involved in aquaculture will have greater access to fish and aquatic products, increased household consumption is not guaranteed. In Bangladesh, an intervention program that provided training for three consecutive years in freshwater fish rearing resulted in an increase in 6.6% of household fish consumption over the course of the project (Jahan et al., 2010). However, other work in other communities in Bangladesh revealed no change in consumption patterns between households involved in fish farming and nonproducing households (Kawarazuka, 2010). Overall interpreting the conflicting evidence around direct improved food security, particularly for women and infants, through fish farming, has been challenging due to the added economic incentives to sell the products (Béné et al., 2016).

Employment in aquaculture has been growing steadily in the past few decades, with a reported 18.9 million people worldwide employed as fish farmers in 2012 (FAO, 2014). Including secondary employment (such as in fish product processing), fisheries and aquaculture provide employment to 10–12% of the world population. This is in line with the role of shrimp farming in poverty alleviation, which largely stems from increased income and employment opportunities (Irz et al., 2007). Sri Lanka is no different, with shrimp being sold in the local market with a further 30–40% exported (Jayasinghe et al., 2019; Landesman et al., 2009). Reports from the early 2000s indicated that 50% of total fisheries export earnings in Sri Lanka came from farmed shrimp (FAO, 2012).

Food security data from shrimp-farming households in Sri Lanka have shown a high level of food security among these farmers, likely due to increased income (Wu et al., 2016). Compared with the general population, these households had a higher level of food security although differences in measurement tools may have affected results. At the same time, only a small proportion of the shrimp harvested was consumed within households – i.e., under 2% of harvests. It would appear that increased household shrimp consumption was likely not a significant direct factor in increased food security.

Smallholder shrimp farmers have faced significant income uncertainty due to unpredictable production yields as a result of disease outbreaks (Westers et al., 2017). Feed costs are also a large proportion of farm expenditure and, for some farms, this input comprises 20–40% of total production costs. Feed costs for shrimp quadrupled within a decade (Munasinghe, 2010) while the sale price of shrimp simultaneously declined. All shrimp farms in Sri Lanka import commercial feed from Thailand, which may require them taking out loans in order to cover production costs (Munasinghe et al., 2010); between 10 and 27% of farmers have taken out loans to maintain their farms (Westers et al., 2017). This instability in income presents many challenges for sustainable and predictable household food security. In order to mitigate this risk, households had sought out alternative income sources.

15.3.2 Gender gaps in aquaculture

Data on gendered employment in aquaculture is limited. Women's roles in aquaculture vary across Asia. Overall, 33% of women in rural China participate in aquaculture, 24% in India and 42% of rural aquaculture workers are women in Indonesia and 80% in Vietnam (Kusakabe and Kelkar, 2001; Nuruzzaman, 2012). These roles tend to be considered as "low-skill" and subsequently low pay; duties range from overseeing hatching and harvesting to repairing equipment. Some women in Thailand and China are primarily responsible

for all tasks in aquaculture due to male migration for better work in the cities (Ahmed et al., 2012; Weeratunge et al., 2010). Specifically, in the shrimp-farming industry, approximately 5% of the global workforce are women (Williams et al., 2010).

Within Sri Lanka, women's roles in fisheries and aquaculture are culturally influenced and are primarily in post-harvest activities (De Silva and Yamao, 2004). Women may have equal access to education, but not to employment opportunities (Asian Development Bank, 2008). More than 90% of employees in seafood-processing plants are women and, while gender discrimination is officially outlawed in Sri Lanka, there are significant differences in wages between men and women (De Silva and Yamao, 2004).

In fisheries, women in coastal communities in Sri Lanka are involved in sorting, gutting, salting, drying and marketing of fish (Dissanayaka and Wijeyaratne, 2010). High rates of illiteracy were noted in women involved in fisheries compared with those not involved in the industry, as well as limited access to electricity and piped water. Research by Rasanayagam (1999) studied women working in shrimp farming in contrast to those working in the handicrafts sector. The former were able to obtain higher and more steady incomes from shrimp farming compared with other livelihoods in fishing and agriculture (Rasanayagam, 1999); however, overall participation of women in shrimp aquaculture remains limited (Mahagamage and Jayakody, 2020).

15.4 Environmental impacts of aquaculture

15.4.1 The challenges of water

When considering water management in aquaculture, rearing systems may be open, closed or semi-closed. In an open system, water is extracted from a natural water source and effluents are discharged back to the water source without treatment (Jayasinghe et al., 2019). This is contrasted with a closed system where the aquaculture system is isolated from the natural source and water undergoes extensive treatment for recirculation.

The amount of water required for aquaculture systems is dependent on several factors such as the quality of the available water supply, the physiological tolerance of the cultured species and the type of aquaculture system (FAO, 1994). Land-based aquaculture requires water for maintaining pond levels, compensating for water loss through seepage, evaporation and intentional discharge. Water input may be from harvesting rainwater, or diversion and removal of water from rivers or canals.

Culture of shrimp presents its own challenge in that it requires brackish water – that is, water with salinity between that of freshwater and sea-water. Establishment of shrimp farms in highly saline areas can result in the exhaustion of the freshwater table, thus creating competition for water between aquaculture, agriculture, other industry and domestic use (Wijegoonawardena and Siriwardena, 1996). Alternatively, development of shrimp farms in low-saline areas can lead to salinization of drinking water sources. Salinization can also have adverse effects on freshwater flora and fauna and can result in a loss of biodiversity and aquatic animal migration (Lorenz, 2014).

In Sri Lanka, water management for aquaculture is primarily open and many of the shrimp farms located in NWP in the Puttalam region are connected to the environment via a canal, the Dutch Canal, linking water from three adjacent lagoons. However, because this method exceeds water access capacity, without any further improvements in water renewal and water treatment, water quality parameters of the Dutch Canal have changed to improve effluent discharge from shrimp farms. Without these improvements, higher levels of suspended solids and the depletion of dissolved oxygen in the water resulted in the death of aquatic wildlife living in the interconnected waterways (Wijegoonawardena and Siriwardena, 1996).

While aquaculture can be a source of water pollution, the effects of water pollution from other sources can also have significant effects on the industry: i.e., discharge of domestic, agricultural and industrial waste into coastal waters can affect aquaculture production and profitability (Chua et al., 1989). Globally, 80% of municipal wastewater is discharged into the water bodies, with millions of tonnes of heavy metals, solvents, toxic sludge and other waste entering the aquatic system each year (WWAP, 2017). Heavy metal contamination of water can then enter the food chain through cultured seafood, and this has serious consequences for human health (Sonone et al., 2020). Consequently, water pollution can pose a risk to ensuring safe and stable aquaculture systems (Beveridge et al., 1997; Wongbusarakum et al., 2019).

With Sri Lanka's location in the Indian Ocean, there are a number of important marine shipping routes in close proximity, including transport of oil from the Middle East (Arachchige et al., 2021). In May of 2021, a large container ship en route to Sri Lanka's capital, Colombo, caught fire approximately 9 nautical miles from shore along the coast near Puttalam. The ship was carrying nitric acid and other chemicals, as well as plastic micropellets and oil for fuel. The ship eventually sank, leading to significant environmental contamination. The release of these hazardous materials into the ecosystem will have direct impacts on marine life and also on the livelihoods of the coastal communities that rely on natural fisheries (Arachchige et al., 2021). Impact will also be felt in aquaculture because contaminated ocean water sources will require monitoring and potential treatment prior to use.

Climate change and other weather disturbances can also have significant effects on water quality. Increased temperatures, combined with high levels of organic matter in ponds, can quickly compromise water quality, generate algal blooms and lead to die-off (Jana and Sarkar, 2005). Flooding can also cause ponds to overflow, not only leading to loss of animal life but also changing the water chemistry parameters required for shrimp survival. Drought can similarly negatively affect water quality (Ataguba et al., 2013; Islam et al., 2016; Reid et al., 2019). The December 26, 2004 tsunami in the Indian Ocean had numerous effects on Sri Lanka. Significant damage to coastal regions led to displacement of people, loss of infrastructure and loss of vegetation (Kaplan et al., 2009). In addition, seawater infiltrated groundwater and subsequently impacted aquaculture productivity, resulting in lower income and reduced employment (Illangasekare et al., 2006; Kaplan et al., 2009).

15.4.2   Effects of land use for aquaculture

Aquaculture farms use land for infrastructure, including buildings, water storage, discharge canals, settling basins, buildings and roads (Boyd, 2002). This can have considerable impact on local land-use configurations and cause ecological perturbations. However, compared with land-based agriculture, land-use requirements for aquaculture are considerably less even with the anticipated future doubling or tripling of aquaculture (Boyd and McNevin, 2015).

Brackish water shrimp culture is one of the major reasons for massive destruction of mangrove ecosystems (UNEP, 2014). Valiela et al. (2001) stated that around 1.89 million ha of global mangrove forests have been lost due to coastal fish cultivation, mainly shrimp farming. Mangrove swamps are unique ecosystems that sustain a variety of flora and fauna and provide many ecosystem services such as coastal protection from storms, reduction of shoreline and riverbank erosion, stabilizing sediments and absorption of pollutants (Duke and Schmitt, 2015). Mangroves provide nutrients for marine microorganisms and refuge and nursery grounds for juvenile fish, crabs, shrimps and mollusks, as well as prime nesting and habitat sites for hundreds of migratory and resident bird species. Large proportions of mangrove ecosystems have been exploited for several purposes, including commercial development, aquaculture, agriculture, residential land use, tourism, mining and industrial growth (Romañach et al., 2018). More data around the benefits of mangroves, and the need for ecosystem rehabilitation, will be necessary to better understand how to balance environmental concerns and economic development (Friess et al., 2019).

15.5   Challenges of healthy shrimp aquaculture

15.5.1   Shrimp diseases

Infectious diseases are considered to be one of the most significant challenges to successful shrimp farming, whether parasitic, viral, bacterial or fungal. Diseases are considered to be significant barriers to the sustainability and long-term growth of aquaculture (Bondad-Reantaso et al., 2005; Stentiford et al., 2017). It is reported that annual losses due to aquatic animal disease amount to approximately US$6 billion globally. Within the shrimp-farming industry, infectious diseases cause losses of more than 40% of global capacity (Stentiford et al., 2020). A number of factors such as intensification, interactions between cultured and wild species and poor or absent biosecurity have been identified as contributors to the challenges of disease in aquaculture (Bondad-Reantaso et al., 2005).

White spot disease, caused by WSSV, is of particular concern for shrimp culture globally. It was first identified in cultured shrimp in Asia in the 1990s and had spread broadly through the Americas by the mid-2000s, and then into Australia in 2017 (Oakey et al., 2019). It was first identified in Sri Lanka in 1996 (Munasinghe et al., 2010; Wijegoonawardena and Siriwardena, 1996). Subsequently WSSV has also been isolated from wild shrimp (Bondad-Reantaso et al., 2005; Oakey et al., 2019). This disease spread was believed to be due to movement of infected live postlarvae and broodstock. There is no cure for this viral disease, and improving biosecurity and implementation of better management practises (BMPs) are considered the ideal solutions (Jayasinghe, 2019). Among these BMPs are better site selection, appropriate pond preparation, good water quality and sediment management and informed postlarval selection.

Adoption and uptake of BMPs can vary based on the farmers' perceptions of health management issues and associated financial constraints associated with implementation of BMPs: e.g., the cost of regular PCR (molecular) screening of shrimp for disease and recommended water treatment (Jayasinghe, 2019). In spite of the development of local shrimp BMPs in Sri Lanka, wide adoption of these practices has not been demonstrated (Mahagamage and Jayakody, 2020; Westers et al., 2017). The barriers to implementation warrant further investigation, with suggestions that economic constraints prevail (Munasinghe, 2012), along with knowledge gaps (Wu et al., 2016) and acceptance of the uncertainty of successful shrimp production (Panduwawala et al., 2018).

Some farmers may also prophylactically apply antimicrobials or chemicals to their ponds based on the advice of service providers without waiting for a proper disease diagnosis (Munasinghe, 2010). At the postlarval stage of growth, shrimp hatcheries in many areas of Sri Lanka have been found to have high levels of oxytetracycline in their effluent (Manage, 2018). With the increased use of antimicrobials, there is also the potential risk of development of antimicrobial resistance (Schar et al., 2020). The intensive use of antibiotics in aquaculture has environmental and health risks, including impacts on aquatic biodiversity toxicity and the emergence of multiantibacterial-resistant strains due to residue accumulation (Lulijwa et al., 2020; Miranda et al., 2018). The application of antimicrobials in aquaculture systems also has wider environmental effects, either from direct application to water or in feed, and residues that remain in the environment. Within the aquatic environment this has implications for terrestrial animal and human health from the development of a reservoir of antimicrobial-resistant bacteria (Schar et al., 2020). However, among the significant challenges is lack of documented monitoring and surveillance of antimicrobial use in many countries.

Increased use of preventive measures as part of BMPs may assist in reducing the use of antimicrobials. Increased diagnostic capacity is required to determine the causative agent of disease, to allow for the judicious selection of antimicrobials and other medications (Zrnčić, 2020). Determining key barriers, and subsequently developing sustainable and locally acceptable solutions to overcome these barriers to BMP implementation, will be necessary for future success of the industry.

15.5.2   Biodiversity

Aquaculture can also have negative impacts on ecosystems and biodiversity because some cultured species utilize wild-caught larvae, with large numbers of nontarget species caught and subsequently disrupting these other wildlife species (Ahmed and Troell, 2010; Naylor et al., 2000; Primavera, 2006). There are also concerns about escape of non-native fish that could result in competition for food with wild fish, potentially displacing native fish species. The introduction of exotic fish species in aquaculture, in the case of escape, can cause serious threats to ecosystems and biodiversity (De Silva and Davy, 2010; Naylor et al., 2001). Cyclones and flooding have caused significant harm to the aquaculture sector, resulting in overflowing shrimp ponds and allowing the mixing of cultured with wild stock and the potential for disease spread (Jayasinghe and Niroshana, 2016; Jayasinghe et al., 2019). The long-term effects of this mixing of farmed and wild species is unknown.

15.5.3 Feed efficiency

The feed conversion efficiency of farmed finfish is similar to that of poultry, but more efficient than that for beef (Waite et al., 2014). These feed conversion ratios could be further improved through better feeding and management practices (White, 2013). In intensive aquaculture systems there are considerable portions of organic waste produced due to uneaten feed and feces, in both particulate form and as soluble substances, which increase biochemical oxygen demand and elevate the concentrations of dissolved nitrates and phosphates (White, 2013). Overfeeding of captive species, due to improper feeding strategies, can therefore lead to excess nutrients polluting the water, subsequently impacting health. Shrimp often consume only 20–30% of feed, which leaves significant residues in the pond water (Avnimelech, 2006). The sludge that is removed from ponds, which may also be polluted with chemicals, poses an environmental risk (Harkes et al., 2015). However, as a result of this there has been a shift from culture of black tiger shrimp (*P. monodon*) to other species such as the whiteleg shrimp (*Litopenaeus vannamei*) that leave less waste (Naylor et al., 2021).

15.6 Conclusion

In the development of sustainable aquaculture systems that balance global needs for food security in an environmentally responsible and equitable way, we must acknowledge the interdependence of multiple factors. Learning from the lessons of the past and contextualization of potential solutions can provide a way forward, including application of metrics using an EcoHealth lens. In Sri Lanka, shrimp farming has resulted in environmental degradation through destruction of mangroves, water pollution, conflict over shared resources (e.g., salinization of drinking water) and lack of benefits to the local community by large shrimp-producing enterprises. Simultaneously, however, there is evidence that households participating in shrimp aquaculture have benefited from improved household food security, including increased income for women. While the costs and benefits must be carefully assessed, there is the belief that extensive or semi-intensive shrimp farming that is more environmentally and socially sustainable may benefit the rural poor. An integrated interdisciplinary team, using an EcoHealth approach, will need to include environmental experts, aquaculture specialists and social scientists to develop useful and appropriate plans for sustainability moving forwards.

References

Ahmed, M., & Lorica, M. H. (2002). Improving developing country food security through aquaculture development—Lessons from Asia. *Food Policy*, 27(2), 125–141. https://doi.org/10.1016/S0306-9192(02)00007-6

Ahmed, M. K., Halim, S., & Sultana, S. (2012). Participation of women in aquaculture in three coastal districts of Bangladesh: Approaches toward sustainable livelihood. *World Journal of Agricultural Sciences*, 8(3), 253–268.

Ahmed, N., & Troell, M. (2010). Fishing for prawn larvae in Bangladesh: An important coastal livelihood causing negative effects on the environment. *Ambio*, *39*(1), 20–29. https://doi.org/10.1007/s13280-009-0002-y

Arachchige, U. S., Sathsara, K. L. T., Preethika, P., Miyuranga, K. V., De Silva, S. J., & Thilakarathne, D. (2021). The impact of shipping on marine environment-A study of Sri Lankan water ways. *International Journal of Scientific Engineering and Science*, *5*(7), 30–38.

Asian Development Bank. (2008). *Country gender assessment*. Asian Development Bank. Phillipines.

Ataguba, G. A., Ogbe, F., G., Solomon, S. G., Ameh, M. I., & Okomoda, V. T. (2013). Synthesis of climate change effects on water sources for aquaculture. In *Proceedings of the 28th Annual Conference of Fisheries Society of Nigeria* (pp. 104–108). Fisheries Society of Nigeria.

Avnimelech, Y. (2006). Bio-filters: The need for a new comprehensive approach. *Aquacultural Engineering*, *34*(3), 172–178. https://doi.org/10.1016/j.aquaeng.2005.04.001

Barraclough, S., & Finger-Stich, A. (1996). *Some ecological and social implications of commercial shrimp farming in Asia*. United Nations Research Institute for Development.

Beveridge, M. C. M., Phillips, M. J., & Macintosh, D. J. (1997). Aquaculture and the environment: the supply of and demand for environmental goods and services by Asian aquaculture and the implications for sustainability. *Aquaculture research*, *28*(10), 797–807.

Beveridge, M. C. M., & Brummett, R. E. (2015). Aquaculture and the environment. In J. F. Craig (Ed.), *Freshwater fisheries ecology* (pp. 794–803). Wiley-Blackwell.

Blanchard, J. L., Watson, R. A., Fulton, E. A., Cottrell, R. S., Nash, K. L., Bryndum-Buchholz, A., Büchner, M., Carozza, D. A., Cheung, W. W. L., Elliott, J., Davidson, L. N. K., Dulvy, N. K., Dunne, J. P., Eddy, T. D., Galbraith, E., Lotze, H. K., Maury, O., Müller, C., Tittensor, D. P., & Jennings, S. (2017). Linked sustainability challenges and trade-offs among fisheries, aquaculture and agriculture. *Nature Ecology and Evolution*, *1*(9), 1240–1249. https://doi.org/10.1038/s41559-017-0258-8

Bondad-Reantaso, M. G., Subasinghe, R. P., Arthur, J. R., Ogawa, K., Chinabut, S., Adlard, R., Tan, Z., & Shariff, M. (2005). Disease and health management in Asian aquaculture. *Veterinary Parasitology*, *132*(3–4), 249–272. https://doi.org/10.1016/j.vetpar.2005.07.005

Bondad-Reantaso, M. G., Bueno, P., Demaine, H., & Pongthanapanich, T. (2009). Development of an indicator system for measuring the contribution of small-scale aquaculture to sustainable rural development. In M. G. Bondad-Reantaso & M. Prein (Eds.), *Measuring the contribution of small-scale aquaculture: An assessment. FAO* Fisheries *and* Aquaculture Technical *Paper* 534 (pp. 161–179).

Bondad-Reantaso, M. G., Subasinghe, R. P., Josupeit, H., Cai, J., & Zhou, X. (2012). The role of crustacean fisheries and aquaculture in global food security: Past, present and future. *Journal of Invertebrate Pathology*, *110*(2), 158–165. https://doi.org/10.1016/j.jip.2012.03.010

Boyd, C. E. (2002). Land and water use issues in aquaculture. Global aquaculture advocate. https://www.aquaculturealliance.org/advocate/land-and-water-use-issues-in-aquaculture/?headlessPrint=AAAAAPIA9c8r7gs82oWZBA Retrieved May 30, 2021

Boyd, C. E., & McNevin, A. A. (2015). Land use. In *Aquaculture, resource use, and the environment* (pp. 81–100). Wiley-Blackwell. https://doi.org/10.1002/9781118857915.ch5

Cattermoul, N., & Devendra, A. (2002). *The impacts of shrimp farming on communities in the Chilaw Lagoon area. Effective management for biodiversity conservation in Sri Lanka coastal wetlands*. Final report. University of Sri Jayewardenapura.

Chua T. E., Paw, J. N., & Guarin, F. Y. (1989). The environmental impact of aquaculture and the effects of pollution on coastal aquaculture development in Southeast Asia. Marine Pollution Bulletin, 20(7), 335–343. https://doi.org/10.1016/0025-326X(89)90157-4

Chua, T. E. (1997). Sustainable aquaculture and integrated coastal management. In J. E. Bardach (Ed.) *Sustainable aquaculture.*, (pp. 177–200). John Wiley & Sons.

Costa-Pierce, B. A. (Ed.). (2008). *Ecological aquaculture: The evolution of the blue revolution*. John Wiley & Sons.

De Silva, D., & Yamao, M. (2004). The involvement of female labor in seafood processing in Sri Lanka: Impact of organizational fairness and supervisor evaluation on employee commitment. Global Symposium on Gender and Fisheries: Seventh Asian Fisheries Forum.

De Silva, S. S., & Davy, F. B. (2010). Aquaculture successes in Asia: Contributing to sustained development and poverty alleviation. In S. S. De Silva & F.

B. Davy (Eds.), *Success stories in Asian aquaculture* (pp. 1–14). Springer.

Dissanayaka, D., & Wijeyaratne, M. J. S. (2010). Impact of women involvement in fisheries on socio-economics of fisher households in Negombo, Sri Lanka. *Sri Lanka Journal of Aquatic Sciences, 14,* 45–57. https://doi.org/10.4038/sljas.v14i0.2199

Drengstig, A. (2013). Aquaculture in Sri Lanka: History, current status, and future potential. *Aqua. Nor Exhibition, VisorAq.*

Duke, N. C., & Schmitt, K. (2015). Mangroves: Unusual forests at the seas edge. In L. Pancel & M. Kohl (Eds.), *Tropical forestry handbook* (pp. 1–24). Springer.

E-Jahan, K. M., Ahmed, M., & Belton, B. (2010). The impacts of aquaculture development on food security: Lessons from Bangladesh. *Aquaculture Research, 41*(4), 481–495. https://doi.org/10.11 11/j.1365-2109.2009.02337.x

Evenson, R. E., & Gollin, D. (2003). Assessing the impact of the Green Revolution, 1960 to 2000. *Science, 300*(5620), 758–762. https://doi.org/10.1 126/science.1078710

Food and Agriculture Organization. (2009). Gender equity in agriculture and rural development. In: Development, E.a.S. (Ed.) FAO, Rome.

Food and Agriculture Organization. (2011). *World aquaculture 2010.* Technical Paper. No. 500/1. FAO Fisheries and Aquaculture Department.

Food and Agriculture Organization. (2012). *The state of world fisheries and aquaculture 2012.* Food and Agriculture Organization.

Food and Agriculture Organization. (2012). *2004 – 2012. Fishery and Aquaculture Country profiles. Sri Lanka. Fishery and Aquaculture Country Profiles.* In: FAO Fisheries and Aquaculture Department. Food and Agriculture Organization.

Food and Agriculture Organization. (2013). *The state of food insecurity in the world 2013. The multiple dimensions of food security.* Food and Agriculture Organization.

Food and Agriculture Organization. (2014). *The state of world fisheries and aquaculture 2014.* Food and Agriculture Organization.

Food and Agriculture Organization. (2016). The *State of world fisheries and aquaculture, 2016.* Food and Agriculture Organization.

Food and Agriculture Organization. (2018). *The state of world fisheries and aquaculture 2018—Meeting the sustainable development goals.* Food and Agriculture Organization.

Food and Agriculture Organization. (2020). *The state of world fisheries and aquaculture 2020 – Sustainability in action.* Food and Agriculture Organization.

Friess, D. A., Rogers, K., Lovelock, C. E., Krauss, K. W., Hamilton, S. E., Lee, S. Y., Lucas, R., Primavera, J., Rajkaran, A., & Shi, S. (2019). The state of the world's mangrove forests: Past, present, and future. *Annual Review of Environment and Resources, 44*(1), 89–115. https://doi.org/10.1146/annurev-envi ron-101718-033302

Godfray, H. C. J., Beddington, J. R., Crute, I. R., Haddad, L., Lawrence, D., Muir, J. F., Pretty, J., Robinson, S., Thomas, S. M., & Toulmin, C. (2010). Food security: The challenge of feeding 9 billion people. *Science, 327*(5967), 812–818. https://doi.org/10.1126/science.1185383

Golden, C. D., Seto, K. L., Dey, M. M., Chen, O. L., Gephart, J. A., Myers, S. S., Smith, M., Vaitla, B., & Allison, E. H. (2017). Does aquaculture support the needs of nutritionally vulnerable nations? *Frontiers in Marine Science, 4,* 159. https://doi. org/10.3389/fmars.2017.00159

Harkes, I. H. T., Drengstig, A., Kumara, M. P., Jayasinghe, J. M. P. K., & Huxham, M. (2015). Shrimp aquaculture as a vehicle for climate compatible development in Sri Lanka. The case of Puttalam Lagoon. *Marine Policy, 61,* 273–283. https://doi.org/10.1016/j.marpol.2015.08.003

HLPE. (2013). *Investing in smallholder agriculture for food security.* A report by the High Level Panel of Experts on Food Security and Nutrition of the Committee on World Food Security. Food and Agriculture Organization.

Illangasekare, T., Tyler, S. W., Clement, T. P., Villholth, K. G., Perera, A. P. G. R. L., Obeysekera, J., Gunatilaka, A., Panabokke, C. R., Hyndman, D. W., Cunningham, K. J., Kaluarachchi, J. J., Yeh, W. W.-G., van Genuchten, M. T., & Jensen, K. (2006). Impacts of the 2004 tsunami on groundwater resources in Sri Lanka. *Water Resources Research, 42*(5). https://doi.org/10.1029/2006WR004876

Irz, X., Stevenson, J. R., Tanoy, A., Villarante, P., & Morissens, P. (2007). The equity and poverty impacts of aquaculture: Insights from the Philippines. *Development Policy Review, 25*(4), 495–516. https://doi.org/10.1111/j.1467-7679.20 07.00382.x

Islam, M. A., Islam, M. S., & Wahab, M. A. (2016). Impacts of climate change on shrimp farming in the South-West coastal region of Bangladesh. *Research in Agriculture Livestock and*

Fisheries, 3(1), 227–239. https://doi.org/10.3329/ralf.v3i1.27881

Jana, B. B., & Sarkar, D. (2005). Water quality in aquaculture-impact and management: A review. *Indian Journal of Animal Sciences, 75*(11), 1354–1361.

Jayasinghe, A. D., & Niroshana, K. H. H. (2016). Potential impacts of climate change on fisheries and aquaculture in Sri Lanka. *Journal of Fisheries Science and Technology, 3,* 11–16.

Jayasinghe, J. M. P. K. (1999). Shrimp culture in Sri Lanka: Key issues in sustainability and research. In P. T. Smith (Ed.), *Towards sustainable shrimp culture in Thailand and the region.* Proceedings of a workshop held at Hat Yai, Songkhla, Thailand, 28 October – 1 November 1996. ACIAR Proceedings No. 90 (pp. 48–51).

Jayasinghe, J. M. P. K., Gamage, D. G. N. D., & Jayasinghe, J. M. H. A. (2019). Combating climate change impacts for shrimp aquaculture through adaptations: Sri Lankan perspective. In A. Sarkar, S. R. Sensarma & G. W. van Loon (Eds.), *Sustainable solutions for food security combating climate change by adaptation* (pp. 287–308). Springer Nature.

Kaplan, M., Renaud, F. G., & Lüchters, G. (2009). Vulnerability assessment and protective effects of coastal vegetation during the 2004 Tsunami in Sri Lanka. *Natural Hazards and Earth System Sciences, 9*(4), 1479–1494. https://doi.org/10.5194/nhess-9-1479-2009

Kawarazuka, N. (2010). *The contribution of fish intake, aquaculture, and small-scale fisheries to improving food and nutrition security: A literature review.* WorldFish Center working paper.

Kawarazuka, N., & Béné, C. (2010). Linking small-scale fisheries and aquaculture to household nutritional security: An overview. *Food Security, 2*(4), 343–357. https://doi.org/10.1007/s12571-010-0079-y

Kusakabe, K., & Kelkar, G. (2001). *Gender concerns in aquaculture in Southeast Asia.* Asian Institute of Technology.

Landesman, L., Amandakoon, H. P., & Varley, J. W. (2009). *Sri Lanka connecting regional economies (USAID/CORE): Assessment of aquaculture and inland fisheries in Eastern Sri Lanka.* United States Agency for International Development.

Lorenz, J. J. (2014). A review of the effects of altered hydrology and salinity on vertebrate fauna and their habitats in northeastern Florida Bay. *Wetlands, 34*(S1), 189–200. https://doi.org/10.1007/s13157-013-0377-1

Lulijwa, R., Rupia, E. J., & Alfaro, A. C. (2020). Antibiotic use in aquaculture, policies and regulation, health and environmental risks: A review of the top 15 major producers. *Reviews in Aquaculture, 12*(2), 640–663. https://doi.org/10.1111/raq.12344

Mahagamage, M. G. Y. L., & Jayakody, S. (2020). Influence of current culture practices on disease outbreaks in shrimp farms located in North Western Province of Sri Lanka. *Egyptian Journal of Aquatic Biology and Fisheries, 24*(7), 453–469. https://doi.org/10.21608/ejabf.2020.126043

Manage, P. M. (2018). Heavy use of antibiotics in aquaculture; emerging human and animal health problems–a review. *Sri Lanka Journal of Aquatic Sciences, 23*(1), 13–27. https://doi.org/10.4038/sljas.v23i1.7543

Miranda, C. D., Godoy, F. A., & Lee, M. R. (2018). Current status of the use of antibiotics and the antimicrobial resistance in the Chilean salmon farms. *Frontiers in Microbiology, 9,* 1284. https://doi.org/10.3389/fmicb.2018.01284

Munasinghe, M. N., Stephen, C., Abeynayake, P., & Abeygunawardena, I. S. (2010). Shrimp farming practices in the Puttallam District of Sri Lanka: Implications for disease control, industry sustainability, and rural development. *Veterinary Medicine International, 2010,* 1–7. https://doi.org/10.4061/2010/679130

Munasinghe, M. N., Stephen, C., Robertson, C., & Abeynayake, P. (2012). Farm level and geographic predictors of antibiotic use in Sri Lankan shrimp farms. *Journal of Aquatic Animal Health, 24*(1), 22–29. https://doi.org/10.1080/08997659.2012.667049

Naylor, R. L., Goldburg, R. J., Primavera, J. H., Kautsky, N., Beveridge, M. C. M., Clay, J., Folke, C., Lubchenco, J., Mooney, H., & Troell, M. (2000). Effect of aquaculture on world fish supplies. *Nature, 405*(6790), 1017–1024. https://doi.org/10.1038/35016500

Naylor, R. L., Williams, S. L., & Strong, D. R. (2001). Ecology. Aquaculture—A gateway for exotic species. *Science, 294*(5547), 1655–1656. https://doi.org/10.1126/science.1064875

Naylor, R., Hindar, K., Fleming, I. A., Goldburg, R., Williams, S., Volpe, J., Whoriskey, F., Eagle, J., Kelso, D., & Mangel, M. (2005). Fugitive salmon: Assessing the risks of escaped fish from net-pen aquaculture. *BioScience, 55*(5), 427–437. https://doi.org/10.1641/0006-3568(2005)055[0427:FSATRO]2.0.CO;2

Naylor, R. L., Hardy, R. W., Buschmann, A. H., Bush, S. R., Cao, L., Klinger, D. H., Little, D. C., Lubchenco,

J., Shumway, S. E., & Troell, M. (2021). A 20-year retrospective review of global aquaculture. *Nature, 591*(7851), 551–563. https://doi.org/10.1038/s41586-021-03308-6

Nuruzzaman, M. D. (2012). Gender roles in development of small-scale shrimp farming and recent challenges in the coastal region of Bangladesh. *Asian Fisheries Science, 25s,* 187–197.

Oakey, J., Smith, C., Underwood, D., Afsharnasab, M., Alday-Sanz, V., Dhar, A., Sivakumar, S., Sahul Hameed, A. S., Beattie, K., & Crook, A. (2019). Global distribution of white spot syndrome virus genotypes determined using a novel genotyping assay. *Archives of Virology, 164*(8), 2061–2082. https://doi.org/10.1007/s00705-019-04265-2

Panduwawala, P. K. C. D., Premarathne, W., & V. D. Kithsiri. (2018). Impact of best management practices on farmed prawn production in Sri Lanka. *International Journal of Research and Analytical Reviews, 5*(4), 875–884.

Primavera, J. H. (2006). Overcoming the impacts of aquaculture on the coastal zone. *Ocean & Coastal Management, 49*(9–10), 531–545. https://doi.org/10.1016/j.ocecoaman.2006.06.018

Rasanayagam, Y. (1999). Women as users and victims of marine and coastal resources in the south and west of Sri Lanka. *GeoJournal, 48*(3), 231–236. https://doi.org/10.1023/A:1007040011136

Reid, G., Gurney-Smith, H., Marcogliese, D., Knowler, D., Benfey, T., Garber, A., Forster, I., Chopin, T., Brewer-Dalton, K., Moccia, R., Flaherty, M., Smith, C., & De Silva, S. (2019). Climate change and aquaculture: Considering biological response and resources. Aquaculture Environment Interactions, 11, 569–602. https://doi.org/10.3354/aei00332

Romañach, S. S., DeAngelis, D. L., Koh, H. L., Li, Y., Teh, S. Y., Raja Barizan, R. S. R., & Zhai, L. (2018). Conservation and restoration of mangroves: Global status, perspectives, and prognosis. *Ocean & Coastal Management, 154,* 72–82. https://doi.org/10.1016/j.ocecoaman.2018.01.009

Roos, N., Wahab, M. A., Chamnan, C., & Thilsted, S. H. (2007). The role of fish in food-based strategies to combat vitamin A and mineral deficiencies in developing countries. *Journal of Nutrition, 137*(4), 1106–1109. *https://doi.org/10.1093/jn/137.4.1106*

Schar, D., Klein, E. Y., Laxminarayan, R., Gilbert, M., & Van Boeckel, T. P. (2020). Global trends in antimicrobial use in aquaculture. *Scientific Reports, 10*(1), 21878. https://doi.org/10.1038/s41598-020-78849-3

Sonone, S. S., Jadhav, S., Sankhla, M. S., & Kumar, R. (2020). Water contamination by heavy metals and their toxic effect on aquaculture and human health through food chain. *Letters in Applied Nanobioscience, 10*(2), 2148–2166. https://doi.org/10.33263/LIANBS102.21482166

Stentiford, G. D., Sritunyalucksana, K., Flegel, T. W., Williams, B. A., Withyachumnarnkul, B., Itsathitphaisarn, O., & Bass, D. (2017). New paradigms to help solve the global aquaculture disease crisis. *PLOS Pathogens, 13*(2), e1006160. https://doi.org/10.1371/journal.ppat.1006160

Stentiford, G. D., Bateman, I. J., Hinchliffe, S. J., Bass, D., Hartnell, R., Santos, E. M., Devlin, M. J., Feist, S. W., Taylor, N. G. H., Verner-Jeffreys, D. W., van Aerle, R., Peeler, E. J., Higman, W. A., Smith, L., Baines, R., Behringer, D. C., Katsiadaki, I., Froehlich, H. E., & Tyler, C. R. (2020). Sustainable aquaculture through the One Health Lens. *Nature Food, 1*(8), 468–474. https://doi.org/10.1038/s43016-020-0127-5

Stevenson, J. R., & Irz, X. (2009). Is aquaculture development an effective tool for poverty alleviation? A review of theory and evidence. *Agriculture, 18*(2), 292–299. *https://doi.org/10.1684/agr.2009.0286*

UN Environmental Program. (2014). *The importance of mangroves to people: A call to action.* UN Environmental Program World Conservation Monitoring Centre.

Valiela, I., Bowen, J. L., & York, J. K. (2001). Mangrove forests: One of the world's threatened major tropical environments. *BioScience, 51*(10), 807–815. https://doi.org/10.1641/0006-3568(2001)051[0807:MFOOTW]2.0.CO;2

Waite, R., Beveridge, M., Brummett, R., Castine, S., Chaiyawannakarn, N., Kaushik, S., Mungkung, R., Nawapakpilai, S., & Phillips, M. (2014). *Improving productivity and environmental performance of aquaculture.* World Resources Institute.

Weeratunge, N., Snyder, K. A., & Sze, C. P. (2010). Gleaner, fisher, trader, processor: Understanding gendered employment in fisheries and aquaculture. *Fish and Fisheries, 11*(4), 405–420. https://doi.org/10.1111/j.1467-2979.2010.00368.x

Westers, T., Ribble, C., Daniel, S., Checkley, S., Wu, J. P., & Stephen, C. (2017). Assessing and comparing relative farm-level sustainability of smallholder shrimp farms in two Sri Lankan provinces using indices developed from two methodological frameworks. *Ecological Indicators,*

83, 346–355. https://doi.org/10.1016/j.ecolind. 2017.08.025

White, P. (2013). *Environmental consequences of poor feed quality and feed management*. FAO Fisheries and Aquaculture technical paper.

Wijegoonawardena, P. K. M., & Siriwardena, P. P. G. S. N. (1996). Shrimp farming in Sri Lanka: Health management and environmental considerations. In R. P. Subasinghe, J. R. Arthur & M. Shariff (Eds.), *Health management in Asian aquaculture. Proceedings of the Regional Expert Consultation on Aquaculture Health Management in Asia and the Pacific*. FAO fisheries technical Paper no. 360, Rome (pp. 127–139). Food and Agriculture Organization.

Williams, M., Agbayani, R., Bhujel, R., Bondad-Reantaso, M. G., Brugere, C., Choo, P. S., Dhont, J., Galmiche-Tejeda, A., Ghulam, K., Kusakabe, K., Little, D., Nandeesha, M. C., Sorgeloos, P., Weeratunge, N., Williams, S., & Xu, P. (2010). Sustaining aquaculture by developing human capacity and enhancing opportunities for women. *Farming the waters for people and food: Global conference on aquaculture, Phuket, Thailand*.

Wongbusarakum, S., De Jesús-Ayson, E. G., Weimin, M., & DeYoung, C. (2019). *Building climate-resilient fisheries and aquaculture in the Asia-Pacific region – FAO/APFIC regional consultative workshop*. Bangkok, 14–16 November 2017. Food and Agriculture Organization.

World Food Summit. (1996). *Rome declaration on world food security and world food summit plan of action*. Food and Agriculture Organization.

Wu, J. P., Burns, T., Kumara, K. P., Djager, T., Westers, T., Checkley, S., Ribble, C., Daniel, S., & Stephen, C. (2016). Linkages between social-learning networks and farm sustainability for smallholder shrimp farmers in Sri Lanka. *Asian Fisheries Science, 29*(2), 96–111. https://doi.org/10.33997/j.afs.2 016.29.2.003

Wu, J., Checkley, S., Westers, T., Burns, T., Ribble, C., & Stephen, C. (2016). Shrimp farming in Sri Lanka: A case study at the interface of human, shrimp and environmental health. In S. Cork, D. Hall & K. Liljebjelke (Eds.), *One health case studies: Addressing complex problems in a changing world*. 5m Publishing (pp. 271–281). Sheffield, UK.

WWAP (United Nations World Water Assessment Programme). (2017). Wastewater: The untapped resource. In *The United Nations world water development report 2017* (pp. 74–78). UNESCO.

Zrnčić, S. (2020). Correct diagnostics: Prerequisite for prudent and responsible antimicrobial administration. Technical Seminar on Aquaculture Biosecurity: Understanding Antimicrobial Resistance (AMR) in Aquaculture. Food and Agriculture Organization.

Additional resources

Ahmed, N., Thompson, S., & Glaser, M. (2019). Global aquaculture productivity, environmental sustainability, and climate change adaptability. *Environmental Management, 63*(2), 159–172. https://doi.org/10.1007/s00267-018-1117-3

Béné, C., Arthur, R., Norbury, H., Allison, E. H., Beveridge, M., Bush, S., Campling, L., Leschen, W., Little, D., Squires, D., Thilsted, S. H., Troell, M., & Williams, M. (2016). Contribution of fisheries and aquaculture to food security and poverty reduction: Assessing the current evidence. *World Development, 79*, 177–196. https://doi.org/10.1016/j.worlddev. 2015.11.007

Bergquist, D. A. (2007). Sustainability and local people's participation in coastal aquaculture: Regional differences and historical experiences in Sri Lanka and the Philippines. *Environmental Management, 40*(5), 787–802. https://doi.org/10.1007/s00267-006-0108-y

Boone Kauffman, J., Arifanti, V. B., Hernández Trejo, H., Carmen Jesús García, M., Norfolk, J., Cifuentes, M., Hadriyanto, D., & Murdiyarso, D. (2017). The jumbo carbon footprint of a shrimp: Carbon losses from mangrove deforestation. *Frontiers in Ecology and the Environment, 15*(4), 183–188. https://doi. org/10.1002/fee.1482

Bouwman, A. F., Beusen, A. H. W., Overbeek, C. C., Bureau, D. P., Pawlowski, M., & Glibert, P. M. (2013). Hindcasts and future projections of global inland and coastal nitrogen and phosphorus loads due to finfish aquaculture. *Reviews in Fisheries Science, 21*(2), 112–156. https://doi.org/10.1080/ 10641262.2013.790340

Brugere, C., & Williams, M. (2017). Profile: Women in aquaculture. https://genderaquafish.org/portfo lio/women-in-aquaculture/

Galappaththi, E. K. (2013). *Community-based shrimp aquaculture in Northwestern Sri Lanka*. Natural Resources Institute, University of Manitoba. Master of natural resources management.

Chapter 16

Amphibian health as an indicator of wetland health

Regina M. Krohn and Danna M. Schock

Abstract

Wetlands are integral components of hydrological cycles and provide vital ecosystem services, including water purification and providing wildlife habitat. However, degradation and destruction of wetlands is a global issue that is increasing in rate and spatial scale. Assessing and monitoring the health of wetlands is best accomplished by examining several lines of evidence. Because wetlands are critical habitat for thousands of species of amphibians, they are often studied to gather information about wetland health. However, not all amphibian species are likely to provide useful information on wetland ecosystem health. There are more than 7400 species of amphibians, and ecology and physiology differ widely across the taxonomic class. These differences are further compounded by the ecologically distinct life stages that typify many amphibians (e.g. egg, aquatic tadpole, terrestrial adult lifecycles of many anurans). Environmental contamination is a major threat to wetland ecosystems, including the amphibians that inhabit them. Herein we focus on strengths and limitations of using the health of amphibians to assess wetland health, focusing especially on contamination of wetlands. We highlight the need for caution and explicit consideration of ecology using, as an example, a field study involving wood frog (*Rana sylvatica* = *Lithobates sylvaticus*) tadpoles reared under conditions associated with a controlled diluted bitumen spill into freshwater environments.

16.1   Introduction

Wetlands are areas of land that are saturated with freshwater or saltwater frequently and/or long enough that the underlying soils show evidence of being altered by water (e.g. anaerobic conditions, processes such as the accumulation of peat) and the vegetation and other biota associated with the area are tolerant of, or adapted to, wet environments (adapted from Ramsar, 2018). This broad definition of wetlands reflects the wide diversity of wetlands globally, which range from vast ocean-side estuaries to peatlands in the circumpolar boreal forest, as well as human-made wetlands such as rice paddies and naturalized borrow pits in agricultural landscapes (Ramsar, 2018). Ecosystem services provided by wetlands are diverse and vital, including water

Susan C. Cork and Douglas P. Whiteside (eds) **Case Studies in EcoHealth**
DOI: 10.52517/9781789183368.016, © 5m Books Ltd, 2024

purification, carbon storage, wildlife habitat, and human food production (e.g. waterfowl that are hunted, naturally occurring plants that are gathered, production of agricultural crops such as rice). Unfortunately, both the quality and quantity of wetlands are declining (Albert et al., 2021; Ramsar, 2018). Cumulatively on a global scale, the area occupied by wetlands is greater than the area of Greenland but has declined substantially due to human activities. The true extent of the loss of wetlands due to human actions is difficult to quantify because human alterations of the landscape have been ongoing, and escalating in speed and geographic scale, for hundreds, if not thousands, of years (e.g. wetlands drained to create agricultural land). Where reliable data exist, more than one-third of the area of wetlands in some regions has been lost in less than 50 years (Ramsar, 2018). Declines in the quantity of wetlands are accompanied by declines in the quality of wetlands because of multipronged stressors such as aquatic invasive species, climate change, resource development, and pollution; these stressors often act on wetlands simultaneously.

Assessing and monitoring the health of wetlands is imperative to slowing, mitigating, and reversing the loss of wetlands and the services they provide. However, assessing the health of wetlands is complex because the ecology of wetlands is complex, and so too are the causes and consequences of declining wetland health. Consequently, multiple sources of evidence, each providing distinct information, are needed (e.g. Adams, 2003; Cairns et al., 1993). Here we discuss ways in which the health of amphibians can provide information about the health of wetlands – that is, the ways in which amphibian health can be used as an indicator of wetland health. Although there is vigorous and important debate in the literature about precise definitions of what '(bio)indicators' are and how they may or may not differ from 'sentinels' or 'biomarkers' (e.g. Burger, 2006; McCarty et al.,

2002; Sewell and Griffiths, 2009), we are using the term 'indicator' here in an inclusive way to facilitate discussion about facets of amphibian biology that allow for their use as indicators of wetland health at different levels of biological organization. We also include comment and caveats with respect to the inferences that can be made about wetland health from assessment of amphibian health. Our chapter largely focuses on amphibian health from the perspective of environmental contaminants in wetlands. However, concepts we discuss also apply to the use of amphibian health for monitoring other aspects of wetland health.

16.2 Amphibians are biologically diverse – a frog is not a frog is not a frog

A phrase that may be familiar to many is that 'amphibians are canaries in a coalmine' (e.g. *Prince Edward News*, 2019; *Washington Post*, 2021). The phrase refers to the idea that amphibians provide early warnings about the health of their environment in the same way that canaries in cages were once carried by coalminers into mines to act as an indicator that deadly gases had accumulated, because the canaries were more sensitive to the gases than humans. However, the vast oversimplification of amphibian biology wrapped up in this phrase makes it misleading and problematic because it is used as justification for the use of amphibians as 'indicators' simply by virtue of a species' membership in the clade Amphibia.

There are currently 7,400+ recognized species of amphibians (i.e. members of Class Amphibia), from across three taxonomic Orders (frogs – Anura; salamanders – Caudata; caecilians – Gymnophiona). Amphibians are a biologically diverse taxon. They are found on all continents except Antarctica, in diverse ecosystems, from deserts to boreal forests, from fast-moving alpine rivers to tropical cloud forests (IUCN Redlist, 2023; Vitt and Caldwell, 2013). Related to this diversity of habitats is a diversity of life history strategies. For example, modes of reproduction include the somewhat familiar 'biphasic lifecycle' whereby eggs are laid in water where the larvae grow, develop, and then go on to metamorphose into terrestrial animals. However, viviparity, ovoviviparity, and direct development (fully formed young that hatch from eggs, usually in species with entirely terrestrial lifecycles) also occur in all three taxonomic orders of amphibians (Vitt and Caldwell, 2013). These differences in reproduction have implications for exposure to, and depuration of, contaminants (e.g. maternal transfer of contaminants to offspring; Bergeron et al., 2010; Todd et al., 2011).

Globally more than 41% of amphibian species are threatened with extinction, including hundreds of species that are critically endangered or extinct in the wild; we are truly in the midst of a global amphibian decline crisis (IUCN Redlist, 2023; McCallum, 2007). However, more than 45% of amphibian species are not currently thought to be declining, and there are more than 1,800 species whose biology is too poorly understood at this time to gauge whether their populations are stable or not (IUCN Redlist, 2023). It is also worth noting that there are amphibian species that are formidable invasive species, wreaking havoc in their non-native ranges, including the decimation of local amphibian populations through competition, the spread of pathogens, and depredation (e.g. Garner et al., 2006; Kraus, 2015). A particularly notorious example is the American bullfrog (*Rana catesbiana = Lithobates catesbiana*), which is native to eastern North America but has been widely established (e.g. Jamaica, Brazil, South Korea, Japan) as a result of escapees from ranaculture facilities and from deliberate releases for well-intentioned but misguided biological control purposes.

Another aspect of amphibian biology that requires consideration is that the drivers of amphibian declines are diverse (Collins and Storfer, 2003) and have been ongoing for considerable periods of time – several decades or more in the case of environmental contamination, 1000+ years in the case of draining wetlands for agricultural and other human uses. It is possible, even probable, that some of the most sensitive species of amphibians were lost (regionally or globally) before issues were even recognized (Collins, 2010; Collins and Crump, 2009). Equally important and probable is that natural selection for resistance to stressors will have been taking place all the while, and complicating matters further is that amphibians may exhibit phenotypically plastic responses to stressors such as contaminants (e.g. Jones and Relyea, 2015). For these reasons, it is perhaps not surprising that Kerby et al. (2010) conducted a meta-analysis of toxicological studies and found that amphibians were not the most sensitive taxon for several categories of environmental contaminants, including various classes of pesticides. It should be noted, though, that the number of amphibian species used across the studies examined by Kerby et al. (2010) was limited to a handful.

As a final comment regarding the problematic nature of blanket statements about amphibians as 'canaries in a coalmine', it is instructive to consider that it is uncommon to encounter phrases indicating that 'all mammals are canaries in the coal mine' even though over a quarter of the ~6,000 recognized species of mammals are threatened with extinction (IUCN Redlist, 2023). The declines we are witnessing in mammals are alarming and are also part of the larger global biodiversity crisis (e.g. Albert et al., 2021). However, declarations are not routinely made about the utility of all mammals to serve as ecosystem health indicators. Indeed, there are vigorous debates about the utility of a single mammal species that has been extensively studied, the American mink (*Neovison vison*), to act as a useful indicator of environmental contamination (e.g. Basu et al., 2007; Bowman et al., 2012; Bowman and Schulte-Hostedde, 2009). Perhaps mammals are not routinely inappropriately clustered together as indicators because the diversity of ecologies among mammals (e.g. primates versus ungulates) is more widely appreciated than it is for amphibians. In any event, it is untenable from an ecological standpoint to argue (or assume) that all 7,400+ species of amphibians are equally informative with respect to environmental health such as the effects of contaminants. It is imperative to explicitly consider the biology (e.g. physiology, ecology) of the particular species proposed as an indicator of wetland health because of the variability, including intraspecific variability, that permeates Class Amphibia.

16.3 Demonstrating causality – easier said than done, but (some) amphibians can help

For a species to be useful as an indicator of environmental health, it must be possible to demonstrate causality between the environmental stressor(s) and the effect(s) observed in the indicator species. However, causality can be difficult to demonstrate because multiple stressors are always simultaneously operating on natural systems and there can be direct and indirect interactions among stressors that mask or exacerbate effects. For example, Relyea (2004) demonstrated that the organophosphate insecticide malathion was moderately lethal to tadpoles of six North American anurans but became twice as lethal to gray treefrog tadpoles (*Hyla versicolor*) when they were simultaneously exposed to malathion and the chemical cues of the presence of one of their natural predators, the eastern newt (*Notophthalmus viridescens*). Environmental contamination can also interact

with infectious diseases in intricate ways. For example, Bosch et al. (2021) exposed tadpoles of the common midwife toad (*Alytes obstetricans*) to different concentrations of microplastics and the highly pathogenic chytrid fungus *Batrachochytrium dendrobatidis* (Bd). Among their findings was that tadpoles infected with Bd had lower accumulations of microplastics in their gut, likely because Bd damages tadpole mouthparts which interferes with the tadpoles' ability to ingest food – and microplastics. These are just two studies among many that demonstrate the importance of interactions among stressors. We recommend that study designs should allow for the detection of interactions wherever possible and, where not possible, interpretation of findings should explicitly consider what role(s) interactions among stressors may play.

Recognizing the ecological realities that complicate attempts to demonstrate straightforward cause-and-effect relationships, Adams (2003) proposed seven criteria, or lines of evidence, that indicate causality between suspected contaminant stressors and observed effects in aquatic ecosystems. When taken in concert in an 'eco-epidemiological approach', or 'weight-of-evidence approach', data that satisfy these criteria indicate causality between a putative stressor and its effects (Adams, 2003; Johnson et al., 2021). We highlight how amphibians are particularly well suited for providing evidence of causality for three of the criteria proposed by Adams (2003): (1) experimental evidence of causality, (2) biological plausibility of mechanisms linking suspected causes and effects, and (3) evidence of biological gradients in natural systems.

Experimental evidence of causality derives from controlled laboratory or mesocosm studies, and that evidence can be used to predict effects or to corroborate suspected cause–effect relationships observed in natural systems. This line of evidence is perhaps the most intuitive and best-known type of evidence used to link causes and effects. The use of amphibians in experimental set-ups is facilitated by the relatively small size, egg-laying behavior, and tractable husbandry requirements of many species. Toxicological assays involve exposing amphibians (usually eggs or very young larvae of anurans) to different doses of chemicals of interest under standardized, highly controlled settings. These types of assay are a key source of evidence used by government agencies when establishing environmental protection regulations. The FETAX assay is one example of a widely used assay. This assay involves exposing the developing embryos of the African clawed frog (*Xenopus laevis*) to the substance of interest and then measuring effects such as growth, developmental, and mortality rates (e.g. Xu et al., 2019). Mesocosm studies allow for the introduction of more ecological realism and complexity into study designs while still controlling many of the variables that could confound results and frustrate interpretations. They also allow for experimental designs with sufficient sample sizes to meaningfully analyze complex data. The literature is replete with studies that have used amphibians in mesocosm set-ups to investigate interactions among environmental contaminants and other stressors (e.g. Boone and James, 2003; Sievers et al., 2018). The case study discussed later in our chapter, which is based on Krohn et al. (2021), is also a mesocosm study.

The second criterion outlined by Adams (2003), where (some) amphibians are well suited to provide evidence of causality, is with respect to the biological plausibility of proposed mechanisms that link cause and effect. One notable example is the effect of contaminants on amphibian metamorphosis, particularly anuran tadpole growth and development rates, which in turn determine time to, and size at, metamorphosis. The progression from fertilized egg to metamorphosed froglet follows predictable, well-described stages that can readily be measured and analyzed (e.g. Gosner, 1960; Nieuwkoop and Faber, 1994). Frog metamorphosis is regulated by the activity of the hypothalamus–pituitary–thyroid (HPT) axis, a neuroendocrine system that responds to stress and regulates metabolism. Consequently, contaminants that stimulate or downregulate any part of the HPT axis – that is, act as endocrine disruptors – are likely to affect metamorphosis in measurable ways (Carr and Patiño, 2011; Hersikorn and Smits, 2011). Melatonin, for example, accelerates metamorphosis in *X. laevis* tadpoles (Rose and Rose, 1998), while exposure to the fungicide triadimefon causes developmental delay (Li et al., 2016). Linking individual-level effects (e.g. size at metamorphosis) to population- or higher-level effects (e.g. decline or increase) is a challenge across broad disciplines of ecology, and one that has not yet been overcome with respect to disruptions of the HPT axis in amphibians (Carr and Patiño, 2011).

The third criterion outlined by Adams (2003) that we highlight in our chapter is with respect to demonstrating a biological gradient – that is, demonstrating that there is a spatial or temporal dose–response relationship between the stressor and effect. Studying animals with small home ranges greatly reduces ambiguity about where, when, and for how long the animals are exposed to a particular contaminated environment. In amphibians, home ranges and dispersal distances are usually much less than 1 km from their natal location, with many species spending their entire lives less than 100 m from their natal location (Smith and Green, 2005). Larvae of pond-breeding amphibians are even more fine-scaled, reliable indicators of local habitat conditions than their post-metamorphic counterparts. By comparing the health of amphibians across multiple locations that vary in distance from the putative source of contamination, it is possible to gather evidence of a causal relationship between stressors and effects (e.g.

Hayes et al., 2003). Demonstrating causality can be further strengthened by assessing effects at multiple levels of biological organization at each location in the study (e.g. molecular biomarkers of stress, bioaccumulation of contaminants, recruitment rates) because all effects can reasonably be attributed to local conditions. However, this is another situation where it is necessary to be aware of the biology of the species being proposed as an indicator, or at least to be explicit about the assumptions being made, because it is not rare for some individuals to make movements of up to ~10 km, particularly juvenile anurans in undisturbed habitat (Sinsch, 2014; Smith and Green, 2005).

The case study below makes use of frog tadpoles as indicators of contaminant bioaccumulation in the local habitat. It presents an example of how choosing to investigate contaminant accumulation in developing larvae may lead to very different conclusions if morphological changes during tadpole growth and metamorphosis are not considered.

16.4 Case study: changes in metal accumulation in anuran larvae during metamorphosis

Heavy metals and metalloids such as mercury, selenium, and arsenic are components of the Earth's crust and are naturally occurring in the environment. Although ecotoxic metal levels can be introduced into the environment through natural mobilization (e.g. weathering of metal-bearing rocks; Bradl, 2005), anthropogenic disturbances are making metals more biologically available, thereby creating threats to wildlife, ecosystem, and human health. For example, ore mining, the use of phosphate fertilizers, and metal-based pesticides all may lead to ecotoxic levels of heavy metals in aquatic environments. Endocrine-disrupting effects have been observed for some

metals. For example, Wang et al. (2016) and Sun et al. (2018) observed delayed metamorphosis in Chinese toad (*Bufo gargarizans*) tadpoles exposed to copper and cadmium at levels observed in polluted environments, respectively. The authors also reported a reduction in body size of the tadpoles. Similarly, Lanctôt et al. (2016) reported that environmentally relevant concentrations of metals disrupted tadpole development as well as behavior in striped marsh frog tadpoles (*Limnodynastes peronii*) exposed to coalmine wastewater. The exposed larvae also had high tissue levels of cobalt, selenium, and arsenic.

In general, the available literature suggests that most amphibian species examined to date bioaccumulate environmental contaminants of concern, including heavy metals and organochlorine pesticides (e.g. Sparling et al., 2010). However, the extent to which amphibians bioaccumulate contaminants varies across taxonomic orders (e.g. Smalling et al., 2021), among species within the same order (e.g. Boczulak et al., 2017), and across life stages of the same species (e.g. Krohn et al., 2021; Roe et al., 2005). Studying the exposure of anuran tadpoles to metals and other contaminants can serve as an indicator of threats to wetland health in the following ways.

- Anurans with biphasic lifecycles may function as vectors for metals from aquatic to terrestrial systems. → Some metals, such as selenium and arsenic, may be transferred to terrestrial ecosystems while others are depurated during metamorphosis (Rowe, 2014).
- Bioaccumulation of metals in frog tadpoles, which are an integral part of the food web, can potentially expose species of organisms at higher trophic levels to toxic levels of contaminants.
- Lethal, but also sublethal, metal exposure and mixtures of contaminants pose a threat to anuran populations due to behavioral and developmental effects. → Heavy metals

can reduce predator avoidance in tadpoles, making them more susceptible to predation; this may also lead to increased intake of metals by the predator (Lefcort et al., 1998).

Crude oils mainly consist of hydrocarbons (~80%) plus other organic compounds and small amounts of metals that may contribute to the adverse effects of oil spills (Overton et al., 2016). For example, elevated levels of copper and lead were detected in feathers of sea birds for 3 years after the Prestige oil spill off the Spanish northwest coast (Moreno et al., 2011). Crude oil spills into marine environments have been studied extensively, at least partly because of the large volumes of oil released. However, important knowledge gaps exist with respect to crude oil spills in freshwater ecosystems, including information on the efficacy of clean-up methods (Lee et al., 2015). To address these knowledge gaps, a multidisciplinary Canadian team commenced in 2018 with a series of experimental diluted bitumen (dilbit) spills into lake shoreline enclosures at the International Institute for Sustainable Development-Experimental Lakes Area (IISD-ELA) in Ontario, Canada. Bitumen is the form of petroleum extracted from oil sands deposits in northern Alberta, Canada, which is the third largest proven oil deposit in the world (Alberta Energy Regulator, 2023). Bitumen is viscous and must be diluted with lighter petroleum products (e.g. naphtha) in order for it to flow through pipelines (Government of Alberta, 2023). As part of this freshwater oil spill recovery study (FOReSt), effects of contaminants in wood frog (*R. sylvatica* = *L. sylvaticus*) tadpoles were investigated because contaminants related to oil and gas extraction can reduce survival of amphibians outright (e.g. Hersikorn et al., 2010), and lead to sublethal effects such as compromised immune function, endocrine disruption (e.g. Robert et al., 2019), and reduced growth (e.g. Pollet and Bendell-Young, 2000; Smalling et al., 2019).

A wood frog metamorphosis experiment was performed in an undisturbed shoreline of the experimental lake at the IISD-ELA prior to the oil spill experiment to acquire baseline data and validate the metamorphosis assay in the lake environment. Wood frogs were chosen as indicator organism because they are an abundant, nonthreatened species inhabiting the wetlands of the boreal forest. During this validation study, wood frog tadpoles were housed in wooden cages with mosquito net siding at a peat-organic shoreline of the experimental lake. Tadpoles were fed with frozen lettuce every other day until approximately 50% reached metamorphic climax, when front limbs emerge (Gosner stage 42 (G42)) (Gosner, 1960). At this time, several physiological endpoints were determined. Surprisingly, metal levels assessed in whole tadpoles showed a rapid decline in those that had reached metamorphic climax (within 24–48 h) (Krohn et al., 2021). With the exception of magnesium, selenium, zinc, and strontium, metals were several-fold lower in tadpoles at metamorphic climax. Differences were most pronounced for iron, lead, and manganese.

These observations indicated the importance of the developmental stage of frogs as indicator species when assessing contaminants, and led to further investigations of metal accumulation of tadpoles before and at metamorphic climax, and after exposure to dilbit in their habitat. For the dilbit spill experiment, shoreline enclosures (15 × 5 m) (Curry Industries, Winnipeg, Canada), were deployed in the experimental lake and sealed to the aquatic and terrestrial sediment/soil using a double row of sandbags. Additional booms were placed around the whole shoreline with enclosures. Seven kilograms of dilbit were weathered for 36 h in stainless steel evaporation pans over 25 cm of freshwater sourced from the experimental lake. After weathering, oil was retrieved from the surface of the water, immediately transported in glass jars to the site and applied to the water within

50 cm of the shoreline along the entire 5-m linear shoreline, estimated to result in a 1-mm oil film on the water. Two enclosures were left untreated as control systems. After 72 hours, sorbent pads were used to absorb the surface dilbit followed by micronutrient addition for enhanced natural recovery. After the clean-up, wood frog tadpoles were transferred to cages in the shoreline enclosures and fed with frozen kale until metamorphic climax (Figure 16.1). Some tadpoles were collected before metamorphic climax.

The digestive tract in premetamorphic anuran tadpoles can be tenfold the length of the tadpole (excluding the tail) and the so-called gut coil takes up most of the body volume. It is remodeled at metamorphosis to a shorter, more differentiated, organ comparable to a mammalian gastrointestinal tract (Ishizuya-Oka and Shi,

Figure 16.1 Cages with wood frog tadpoles in a shoreline enclosure at IISD-ELA, Ontario, Canada. Tadpole metal accumulation was determined in pre-and postmetamorphic tadpoles and premetamorphs without gut coil, in control and diluted bitumen spill sites. Photo: R. Krohn.

2005). *Xenopus laevis* digestive tracts shorten by about 75% within 8 days of the start of metamorphosis (Schreiber et al., 2005). Figure 16.2 shows photographs and microcomputer tomographic (mCT) scans of two boreal chorus frog tadpoles, one before metamorphic climax (G40) and one at metamorphic climax (G42). On the ventral mCT scans the gut coil in the premetamorphic tadpole is clearly visible while a much shorter, more differentiated, digestive tract is visible in the metamorphic tadpole. In boreal chorus frogs the gut coil volume was reduced by about 70% (unpublished observation) from stage G40–G42. While species-dependent differences exist, it is known that some metals concentrate in the tadpole gut coil. This is why in the dilbit spill experiment, premetamorphic tadpoles with and without the gut coil were analyzed for a 21-metal panel, as well as tadpoles at metamorphic climax including the digestive tract (G42) (Krohn et al., 2021). Interestingly, tadpoles reared in the control shoreline plots had lower metal levels than those from the validation experiment (undisturbed environment), which was likely due to the fact that, during that study, tadpole cages were positioned close to the sediment. Increased metal accumulation was likely due to ingestion of disturbed sediment during handling.

In both control and dilbit-exposed animals, whole premetamorphic tadpoles had higher metal levels than those tested without the gut coil, except for potassium, selenium, and zinc levels (strontium was not assessed in the spill study), which were not different, and magnesium, which was more concentrated when measured without the gut coil. These metals were also similar in tadpoles at metamorphic climax. Molybdenum, vanadium, and cadmium were significantly increased in dilbit-exposed tadpoles with gut coil when compared with control tadpoles. Molybdenum was also higher in dilbit-exposed premetamorphs without the gut coil as compared with control counterparts.

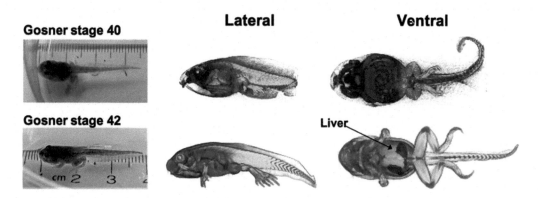

Figure 16.2 Gut coil remodeling and other morphological changes happen rapidly in boreal chorus frog (*Pseudacris maculata*) tadpoles after they reach metamorphic climax (tadpole shown at bottom). The digestive tract is highlighted in purple. At metamorphosis the liver moves to the ventral side. Photos: Ahmad Yaghi and Judit Smits, Dept. of Ecosystem and Public Health, Faculty of Veterinary Medicine, University of Calgary; mCT images of boreal chorus frogs: Jason Pardo and Jason Anderson, Department of Comparative Biology & Experimental Medicine, Faculty of Veterinary Medicine, University of Calgary.

Notably, mercury was the only element found to be significantly higher in dilbit-exposed metamorphs than in premetamorphs (with and without gut coil).

Another interesting finding was that tadpoles had much higher metal levels than those in the water they were reared in, and possibly accumulated more metals through the food source, which was frozen kale. Since higher levels of molybdenum, cadmium, and vanadium were observed in dilbit-exposed animals, it is conceivable that the kale adsorbed some of the metals from the water. Biosorption of metals into an aquatic plant has been reported previously (Keskinkan et al., 2004).

16.5 Conclusion

Wood frog tadpoles reared in shoreline plots for the experimental dilbit spill were housed in floating cages (see Figure 16.1), where they were not in contact with benthic plants or lake sediment. Tadpoles in their natural habitat ingest sediment and feed on plants and smaller invertebrates that may take up petroleum-related pollutants. As a result, their exposure to contaminants would likely be higher in nature than it was in the study. In addition, bitumen has the propensity to sink to the sediment within 1 week after a spill (Stoyanovich et al., 2019).

Metals accumulated mostly in the gut coil of premetamorphic wood frog tadpoles, with some exceptions of essential trace metals. In contrast, mercury was found to be highest in dilbit-exposed tadpoles at metamorphic climax, indicating higher tissue accumulation due to metabolic changes during metamorphosis. These findings highlight the importance of considering intraspecific differences in anurans generally, and the fact that studies measuring bioaccumulation must factor in the intestinal remodeling that occurs during metamorphosis. Speaking specifically with respect to wood frogs, these may serve as vectors of metals into terrestrial ecosystems for some metals, including mercury, zinc, and selenium.

References

Adams, S. M. (2003). Establishing causality between environmental stressors and effects on aquatic ecosystems. *Human and Ecological Risk Assessment, 9*(1), 17–35. https://doi.org/10.1080/713609850

Albert, J. S., Destouni, G., Duke-Sylvester, S. M., Magurran, A. E., Oberdorff, T., Reis, R. E., Winemiller, K. O., & Ripple, W. J. (2021). Scientists' warning to humanity on the freshwater biodiversity crisis. *Ambio, 50*(1), 85–94. https://doi.org/10.1007/s13280-020-01318-8

Alberta Energy (2023). Regulator. *Oil sands – Bitumen deposits in Alberta.* https://ags.aer.ca/research-initiatives/oil-sands

Basu, N., Scheuhammer, A. M., Bursian, S. J., Elliott, J., Rouvinen-Watt, K., & Chan, H. M. (2007). Mink as a sentinel species in environmental health. *Environmental Research, 103*(1), 130–144. https://doi.org/10.1016/j.envres.2006.04.005

Bergeron, C. M., Bodinof, C. M., Unrine, J. M., & Hopkins, W. A. (2010). Bioaccumulation and maternal transfer of mercury and selenium in amphibians. *Environmental Toxicology and Chemistry, 29*(4), 989–997. https://doi.org/10.1002/etc.125

Boczulak, S. A., Vanderwel, M. C., & Hall, B. D. (2017). Survey of mercury in boreal chorus frog (*Pseudacris maculata*) and wood frog (*Rana sylvatica*) tadpoles from wetland ponds in the Prairie Pothole Region of Canada. *FACETS, 2*(1), 315–329. https://doi.org/10.1139/facets-2016-0041

Boone, M. D., & James, S. M. (2003). Interactions of an insecticide, herbicide, and natural stressors in amphibian community mesocosms. *Ecological Applications, 13*(3), 829–841. https://doi.org/10.1890/1051-0761(2003)013[0829:IOAIHA]2.0.CO;2

Bosch, J., Thumsová, B., López-Rojo, N., Pérez, J., Alonso, A., Fisher, M. C., & Boyero, L. (2021). Microplastics increase susceptibility of amphibian larvae to the chytrid fungus Batrachochytrium dendrobatidis. *Scientific Reports, 11*(1), 22438. https://doi.org/10.1038/s41598-021-01973-1

Bowman, J., & Schulte-Hostedde, A. I. (2009). The mink is not a reliable sentinel species. *Environmental Research, 109*(7), 937–9; discussion 940. https://doi.org/10.1016/j.envres.2009.07.004

Bowman, J., Kidd, A. G., Martin, P. A., McDaniel, T. V., Nituch, L. A., & Schulte-Hostedde, A. I. (2012). Testing for bias in a sentinel species: Contaminants in free-ranging domestic, wild, and hybrid mink.

Environmental Research, 112, 77–82. https://doi.org/10.1016/j.envres.2011.11.004

Bradl, H. B. (2005). *Heavy metals in the environment: Origin, interaction and remediation.* Elsevier Science & Technology.

Burger, J. (2006). Bioindicators: Types, development, and use in ecological assessment and research. *Environmental Bioindicators, 1*(1), 22–39. https://doi.org/10.1080/15555270590966483

Cairns, J., McCormick, P. V., & Niederlehner, B. R. (1993). A proposed framework for developing indicators of ecosystem health. *Hydrobiologia, 263*(1), 1–44. https://doi.org/10.1007/BF00006084

Carr, J. A., & Patiño, R. (2011). The hypothalamus–pituitary–thyroid axis in teleosts and amphibians: Endocrine disruption and its consequences to natural populations. *General and Comparative Endocrinology, 170*(2), 299–312. https://doi.org/10.1016/j.ygcen.2010.06.001

Collins, J. P. (2010). Amphibian decline and extinction: What we know and what we need to learn. *Diseases of Aquatic Organisms, 92*(2–3), 93–99. https://doi.org/10.3354/dao02307

Collins, J. P., & Storfer, A. (2003). Global amphibian declines: Sorting the hypotheses. *Diversity Distributions, 9*(2), 89–98. https://doi.org/10.1046/j.1472-4642.2003.00012.x

Collins, J. P., & Crump, M. L. (2009). *Extinction in our time: Global amphibian decline.* Oxford University Press.

Garner, T. W., Perkins, M. W., Govindarajulu, P., Seglie, D., Walker, S., Cunningham, A. A., & Fisher, M. C. (2006). The emerging amphibian pathogen *Batrachochytrium dendrobatidis* globally infects introduced populations of the North American bullfrog, *Rana catesbeiana. Biology Letters, 2*(3), 455–459. https://doi.org/10.1098/rsbl.2006.0494

Gosner, K. L. (1960). A simplified table for staging anuran embryos and larvae with notes on identification. *Herpetologica, 16*, 183–190.

Government of Alberta. (2023). Oil. *S&S, 101.* https://www.alberta.ca/oil-sands-101.aspx Retrieved April 10, 2023.

Hayes, T., Haston, K., Tsui, M., Hoang, A., Haeffele, C., & Vonk, A. (2003). Atrazine-induced hermaphroditism at 0.1 ppb in American leopard frogs (*Rana pipiens*): Laboratory and field evidence. *Environmental Health Perspectives, 111*(4), 568–575. https://doi.org/10.1289/ehp.5932

Hersikorn, B. D., Ciborowski, J. J. C., & Smits, J. E. G. (2010). The effects of oil sands wetlands on wood frogs (*Rana sylvatica*). *Toxicological and Environmental Chemistry, 92*(8), 1513–1527. https://doi.org/10.1080/02772240903471245

Hersikorn, B. D., & Smits, J. E. G. (2011). Compromised metamorphosis and thyroid hormone changes in wood frogs (*Lithobates sylvaticus*) raised on reclaimed wetlands on the Athabasca oil sands. *Environmental Pollution, 159*(2), 596–601. https://doi.org/10.1016/j.envpol.2010.10.005

Ishizuya-Oka, A., & Shi, Y. B. (2005). Molecular mechanisms for thyroid hormone-induced remodeling in the amphibian digestive tract: A model for studying organ regeneration. *Development, Growth and Differentiation, 47*(9), 601–607. https://doi.org/10.1111/j.1440-169X.2005.00833.x

IUCN Redlist. (2023). Summary statistics. https://www.iucnredlist.org/resources/summary-statistics

Johnson, A. C., Sumpter, J. P., & Depledge, M. H. (2021). The weight-of-evidence approach and the need for greater international acceptance of its use in tackling questions of chemical harm to the environment. *Environmental Toxicology and Chemistry, 40*(11), 2968–2977. https://doi.org/10.1002/etc.5184

Jones, D. K., & Relyea, R. A. (2015). Here today, gone tomorrow: Short-term retention of pesticide-induced tolerance in amphibians. *Environmental Toxicology and Chemistry, 34*(10), 2295–2301. https://doi.org/10.1002/etc.3056

Kerby, J. L., Richards-Hrdlicka, K. L., Storfer, A., & Skelly, D. K. (2010). An examination of amphibian sensitivity to environmental contaminants: Are amphibians poor canaries? *Ecology Letters, 13*(1), 60–67. https://doi.org/10.1111/j.1461-0248.2009.01399.x

Keskinkan, O., Goksu, M. Z. L., Basibuyuk, M., & Forster, C. F. (2004). Heavy metal adsorption properties of a submerged aquatic plant (*Ceratophyllum demersum*). *Bioresource Technology, 92*(2), 197–200. https://doi.org/10.1016/j.biortech.2003.07.011

Kraus, F. (2015). Impacts from invasive reptiles and amphibians. *Annual Review of Ecology, Evolution, and Systematics, 46*(1), 75–97. https://doi.org/10.1146/annurev-ecolsys-112414-054450

Krohn, R. M., Palace, V., & Smits, J. E. G. (2021). Metal Changes in Pre- and Post-metamorphic Wood Frog (*Lithobates sylvaticus*) Tadpoles: Implications for Ecotoxicological Studies. Archives of Environmental Contamination and Toxicology, 80(4), 760–768. https://doi.org/10.1007/s00244-020-00735-w

Lanctôt, C., Bennett, W., Wilson, S., Fabbro, L., Leusch, F. D. L., & Melvin, S. D. (2016). Behaviour, development and metal accumulation in striped marsh frog tadpoles (*Limnodynastes peronii*) exposed to coal mine wastewater. *Aquatic Toxicology, 173*, 218–227. https://doi.org/10.1016/j.aquatox.2016.01.014

Lee, K., Boufadel, M., Chen, B., Foght, J., Hodson, P., Swanson, S. et al. (2015). *Expert panel report on the behaviour and environmental impacts of crude oil released into aqueous environments*. Royal Society of Canada.

Lefcort, H., Meguire, R. A., Wilson, L. H., & Ettinger, W. F. (1998). Heavy metals alter the survival, growth, metamorphosis, and antipredatory behavior of columbia spotted frog (*Rana luteiventris*) tadpoles. *Archives of Environmental Contamination and Toxicology, 35*(3), 447–456. https://doi.org/10.1007/s002449900401

Li, M., Li, S., Yao, T., Zhao, R., Wang, Q., & Zhu, G. (2016). Waterborne exposure to triadimefon causes thyroid endocrine disruption and developmental delay in *Xenopus laevis* tadpoles. *Aquatic Toxicology, 177*, 190–197. https://doi.org/10.1016/j.aquatox.2016.05.018

McCallum, M. L. (2007). Amphibian decline or extinction? Current declines dwarf background extinction rate. *Journal of Herpetology, 41*(3), 483–491. https://doi.org/10.1670/0022-1511(2007)41[483:ADOECD]2.0.CO;2

McCarty, L. S., Power, M., & Munkittrick, K. R. (2002). Bioindicators versus biomarkers in ecological risk assessment. *Human and Ecological Risk Assessment, 8*(1), 159–164. https://doi.org/10.1080/20028091056791

Moreno, R., Jover, L., Diez, C., & Sanpera, C. (2011). Seabird feathers as monitors of the levels and persistence of heavy metal pollution after the Prestige oil spill. *Environmental Pollution, 159*(10), 2454–2460. https://doi.org/10.1016/j.envpol.2011.06.033

Nieuwkoop, P. D., & Faber, J. (1994). Normal table of *Xenopus laevis* (Daudin). *Copeia.* Garland Publishing, Inc., 1958(1). https://doi.org/10.2307/1439568

Organization for Economic Co-operation and Development. (2004). *Detailed review paper on amphibian metamorphosis assay for the detection of thyroid active substances*. https://doi.org/10.1787/9789264079144-en

Overton, E. B., Wade, T. L., Radović, J. R., Meyer, B. M., Miles, M. S., & Larter, S. R. (2016).

Chemical composition of Macondo and other crude oils and compositional alterations during oil spills. *Oceanography*, *29*(3), 50–63. https://doi.org/10.5670/oceanog.2016.62

Pollet, I., & Bendell-Young, L. I. (2000). Amphibians as indicators of wetland quality in wetlands formed from oil sands effluent. Environmental Toxicology and Chemistry: An International Journal, 19(10), 2589-2597. https://doi.org/10.1002/etc.5620191027

Prince Edward Island News. (2019). Frog friend: P.E.I. advocate says amphibians are the new canary in the coal mine for water quality. https://www.saltwire.com/prince-edward-island/news/frog-friend-pei-advocate-says-amphibians-are-the-new-canary-in-the-coal-mine-for-water-quality-282625/

Ramsar Convention on Wetlands. (2018). Global wetland outlook: State of the World's wetlands and their services to people. Ramsar Convention Secretariat. https://ssrn.com/abstract=3261606

Relyea, R. A. (2004). Synergistic impacts of Malathion and predatory stress on six species of North American tadpoles. *Environmental Toxicology and Chemistry*, *23*(4), 1080–1084. https://doi.org/10.1897/03-259

Robert, J., McGuire, C. C., Nagel, S., Lawrence, B. P., & Andino, F. D. J. (2019). Developmental exposure to chemicals associated with unconventional oil and gas extraction alters immune homeostasis and viral immunity of the amphibian Xenopus. Science of the Total Environment, 671, 644-654. https://doi.org/10.1016/j.scitotenv.2019.03.395

Roe, J. H., Hopkins, W. A., & Jackson, B. P. (2005). Species- and stage-specific differences in trace element tissue concentrations in amphibians: Implications for the disposal of coal-combustion wastes. *Environmental Pollution*, *136*(2), 353–363. https://doi.org/10.1016/j.envpol.2004.11.019

Rose, M. F., & Rose, S. R. (1998). Melatonin accelerates metamorphosis in *Xenopus laevis*. *Journal of Pineal Research*, *24*(2), 90–95. https://doi.org/10.1111/j.1600-079x.1998.tb00372.x

Rowe, C. L. (2014). Bioaccumulation and effects of metals and trace elements from aquatic disposal of coal combustion residues: Recent advances and recommendations for further study. *Science of the Total Environment*, *485–486*, 490–496. https://doi.org/10.1016/j.scitotenv.2014.03.119

Schreiber, A. M., Cai, L., & Brown, D. D. (2005). Remodeling of the intestine during metamorphosis of *Xenopus laevis*. *Proceedings of the National Academy of Sciences of the United States of America*, *102*(10), 3720–3725. https://doi.org/10.1073/pnas.0409868102

Sewell, D., & Griffiths, R. A. (2009). Can a single amphibian species be a good biodiversity indicator? *Diversity*, *1*(2), 102–117. https://doi.org/10.3390/d1020102

Sievers, M., Parris, K. M., Swearer, S. E., & Hale, R. (2018). Stormwater wetlands can function as ecological traps for urban frogs. *Ecological Applications*, *28*(4), 1106–1115. https://doi.org/10.1002/eap.1714

Sinsch, U. (2014). Movement ecology of amphibians: From individual migratory behaviour to spatially structured populations in heterogeneous landscapes. *Canadian Journal of Zoology*, *92*(6), 491–502. https://doi.org/10.1139/cjz-2013-0028

Smalling, K. L., Anderson, C. W., Honeycutt, R. K., Cozzarelli, I. M., Preston, T., & Hossack, B. R. (2019). Associations between environmental pollutants and larval amphibians in wetlands contaminated by energy-related brines are potentially mediated by feeding traits. *Environmental Pollution*, *248*, 260–268. https://doi.org/10.1016/j.envpol.2019.02.033

Smalling, K. L., Oja, E. B., Cleveland, D. M., Davenport, J. M., Eagles-Smith, C., Campbell Grant, E. H., Kleeman, P. M., Halstead, B. J., Stemp, K. M., Tornabene, B. J., Bunnell, Z. J., & Hossack, B. R. (2021). Metal accumulation varies with life history, size, and development of larval amphibians. *Environmental Pollution*, *287*, 117638. https://doi.org/10.1016/j.envpol.2021.117638

Smith, M. A., & Green, D. M. (2005). Dispersal and the metapopulation paradigm in amphibian ecology and conservation: Are all amphibian populations metapopulations? *Ecography*, *28*(1), 110–128. https://doi.org/10.1111/j.0906-7590.2005.04042.x

Sparling, D. W., Linder, G., Bishop, C. A., & Krest, S. K. (2010). *Ecotoxicology of amphibians and reptiles* (2nd ed). CRC Press.

Stoyanovich, S. S., Yang, Z., Hanson, M., Hollebone, B. P., Orihel, D. M., Palace, V., Rodriguez-Gil, J. L., Faragher, R., Mirnaghi, F. S., Shah, K., & Blais, J. M. (2019). Simulating a spill of diluted bitumen: Environmental weathering and submergence in a model freshwater system. *Environmental Toxicology and Chemistry*, *38*(12), 2621–2628. https://doi.org/10.1002/etc.4600

Sun, N., Wang, H., Ju, Z., & Zhao, H. (2018). Effects of chronic cadmium exposure on metamorphosis, skeletal development, and thyroid endocrine disruption in Chinese toad *Bufo gargarizans* tadpoles. *Environmental Toxicology and Chemistry*, 37(1), 213–223. https://doi.org/10.1002/etc.3947

Todd, B. D., Bergeron, C. M., Hepner, M. J., & Hopkins, W. A. (2011). Aquatic and terrestrial stressors in amphibians: A test of the double jeopardy hypothesis based on maternally and trophically derived contaminants. *Environmental Toxicology and Chemistry*, 30(10), 2277–2284. https://doi.org/10.1002/etc.617

Vitt, L. J., & Caldwell, J. P. (2013). *Herpetology: An introductory biology of amphibians and reptiles* (4th ed.). Academic Press.

Wang, C., Liang, G., Chai, L., & Wang, H. (2016). Effects of copper on growth, metamorphosis and endocrine disruption of *Bufo gargarizans* larvae. *Aquatic Toxicology, 170*, 24–30. https://doi.org/10.1016/j.aquatox.2015.10.023

Washington Post. (2021). Frogs and toads: What's the difference? https://www.washingtonpost.com/lifestyle/kidspost/frogs-and-toads-whats-the-difference/2021/05/03/09223166-a9dd-11eb-bca5-048b2759a489_story.html

Xu, Y., Park, S. J., & Gye, M. C. (2019). Effects of nonylphenols on embryonic development and metamorphosis of *Xenopus laevis*: FETAX and amphibian metamorphosis toxicity test (OECD TG231). *Environmental Research, 174*, 14–23. https://doi.org/10.1016/j.envres.2019.04.010

Zhu, W., Chang, L., Zhao, T., Wang, B., & Jiang, J. (2020). Remarkable metabolic reorganization and altered metabolic requirements in frog metamorphic climax. *Frontiers in Zoology, 17*, 30. https://doi.org/10.1186/s12983-020-00378-6

Chapter 17

Marine mammals and plastics in the ocean

Cara Field and Pádraig Duignan

Abstract

The global mass production and use of plastics over the past 70 years has resulted in profound effects on human, animal and ecosystem health. The impact of plastic waste on human and terrestrial animal health and welfare is more commonly noted and described while marine environment impacts are less well studied. Plastics are found in waters throughout the world but, due to the vast expanse and depth of oceans and other bodies of water with relatively minimal exploration, the effects are neither well documented nor understood. Marine mammals inhabit all oceans as well as many freshwater systems, and are often referred to as sentinel species for human and other animal health given their trophic level in these environments. In this chapter we dive into the different forms of plastic documented in the marine environment, the reported effects of plastic on marine mammals as a model for numerous other aquatic species and the marine environment and how the One Health approach to human and animal health helps us better understand and assess past, current and future impacts on both individual- and population-level health.

17.1 Introduction

The abundance of plastic is likely to be the most important geologic signature of the Anthropocene, a proposed epoch characterized by the profound influence that humans have had on Earth's climate and ecosystems (Crutzen and Soermer, 2002). Since the beginning of mass production of plastic in the 1950s its growth has outpaced any other manufactured material, with an estimated total of 8.3 billion metric tons produced by 2015 and 368 million metric tons produced in 2019 alone (Geyer et al., 2017;

Jambeck et al., 2015). The increase in plastic production over time stems particularly from its extremely low cost production, durability, light weight and stability in the presence of chemicals, light and microbiota (Andrady, 2011). Due to these highly useful and versatile properties, large-scale plastic production has facilitated extensive innovation in medical applications, food packaging, textiles, transportation, drinking water infrastructure and building materials among many other areas. However, its remarkable resistance to degradation has resulted in the global accumulation of plastic debris in landfills

Susan C. Cork and Douglas P. Whiteside (eds) **Case Studies in EcoHealth**
DOI: 10.52517/9781789183368.017, © 5m Books Ltd, 2024

and ecosystems, and has led to a new proposed age in Earth's history that began with the proliferation of plastics in the 1950s: the Plasticene (Haram et al., 2020; Rangel-Buitrago et al., 2022).

Plastics are a group of materials made of polymers, which are chains of joined single units (monomers). The process of linking these units together is called polymerization, during which products can be shaped when soft then hardened to retain their shape. Many polymers, such as tar and cellulose, occur naturally and may be comprised simply of carbon and hydrogen (hydrocarbons) while others contain elements such as nitrogen, phosphorus and silicon. Although most plastics currently in use are derived from fossil fuels, synthetic polymers such as polypropylene, polystyrene, polyvinyl chloride, acrylic, nylon and many more forms are created through chemical processes. While recycling of some plastic is possible, the thousands of different types of plastics with their variety of different chemical additives and colorants often cannot be recycled together. Furthermore, plastic can adsorb chemicals and these often present toxic additives further limiting recycling options in addition to the high cost associated with plastic collection, sorting, transporting and reprocessing. Thus, the vast majority of plastic is discarded.

In 1972, the first report of plastic in the ocean highlighted widespread surface contamination of the Sargasso Sea with plastic fragments in the micro-size range (Carpenter and Smith, 1972). It was not until the early 2000s that research interest in microplastics accelerated, with over 90% of articles published since 2009 (Sorensen and Jovanović, 2021). In the decades that followed the first publications, the majority of plastic pollution reports documenting effects on animals focused on entanglement of marine mammals and numerous other species, likely due to readily observable injury

and mortality associated with entanglement in comparison with the less obvious sublethal and lethal toxic effects of plastic ingestion (Andrady, 2015; Gregory, 2009; Laist, 1997). Gall and Thompson (2015) reviewed 340 original publications that recorded encounters of 693 species with marine debris: 92% of these encounters were with plastic. Marine mammals inhabit all oceans as well as many freshwater bodies, and many dive as deep as several kilometers below the water surface. Many species, mostly seals and sea lions, also frequently congregate on shorelines to rest, and for breeding and pupping, expanding their exposure to plastics that remain on beaches. The purpose of this chapter is to provide the reader with an overview of plastic in its different forms in the ocean environment, how these forms of plastic are described and evaluated in marine mammals and how different forms of plastic affect marine mammal health, as a model of One Health.

17.2 Plastic pollution across Earth's ecosystems

17.2.1 Ubiquitous distribution

Although plastic accumulation in the world's oceans has drawn considerable media and research attention, plastic pollution is found ubiquitously in terrestrial, freshwater, estuarine and even atmospheric environments (Billard and Boucher, 2019). In oceans, plastic is found at every level: floating at the sea surface, suspended in the water column, settled onto the seafloor, trapped in sea ice and accumulated on beaches. Current estimates of the amount of annual plastic leakage into the environment vary between 4.8 and 12.7 million ton per year (Jambeck et al., 2015). Ocean plastic originates from many sources, including direct illegal dumping and mis-management of waste directly into the ocean, outflow of rivers and other waterways – especially with stormwater runoff through wastewater treatment effluent – and loss of fishing gear and from shipping activities (Billard and Boucher, 2019; Cózar et al., 2014). Winds and ocean currents can result in global transport of plastics thousands of miles beyond their origin. Beaches represent the interface between ocean and land, with high human activity, where rivers release acquired trash, as well as where ocean debris often washes up. The accumulation of plastic on beaches is one of the most heavily researched ecosystems due likely to ease of access and visibility.

Depending on formulation and environmental conditions, plastic can take centuries or longer to chemically degrade (Chamas et al., 2020). Over time, plastics that are exposed to ultraviolet light, thermal stress, atmospheric ozone and mechanical abrasion physically deteriorate into smaller and smaller pieces (Singh and Sharma, 2008). Macroplastics are generally defined as solid polymeric materials to which chemical additives or other substances may have been added, and are generally considered to have a diameter larger than approximately 5 millimeters (mm) (Arthur et al., 2009; GESAMP, 2015); however, this size distinction is not internationally standardized. The terms macro litter, plastic or marine litter, marine plastic and plastic debris are also used to describe macroplastics. Microplastics are particles having three dimensions all between 1 micrometer (1 µm) and 5 mm (Arthur et al., 2009). Researchers often differentiate nanoplastics – polymeric particles less than 1 micron in size – from microplastics due to their unique physical and toxicological behavior (Koelmans et al., 2015; Mattsson et al., 2018). Polymers that are derived in nature and that have not been chemically modified (other than by hydrolysis) are excluded (Koelmans et al., 2015). Due especially to their small size, micro-

and nanoplastics are capable of being transported thousands of miles via wind and water currents, resulting in their global distribution.

Studies of microplastics in sediment cores from around the world show a marked increase in the presence of microplastics toward the sediment surface. Evidence of microplastic contamination begins in the 1950s, with concentrations at each subsequent layer consistent with the history of plastic production (Barrett et al., 2020; Matsuguma et al., 2017). Carpenter and Smith (1972) detected concentrations of 3500 items and 290 grams per square kilometer from the water surface of the Sargasso Sea in 1972, which represented accumulation over approximately 20 years. With a world population increase from approximately 2.5 billion in 1950 to 3.7 billion in 1970, compared with current population estimates of 8 billion people in 2022 and our concomitant increase in plastic production and use, plastic demand and consumption have skyrocketed accordingly. A recent global estimate based on robust surface trawling data and multiple ocean circulation models reported 15–51 trillion microplastic particles in 2014, weighing 93–236 thousand metric tons (van Sebille et al., 2015). However, this is equivalent to only approximately 1% of global plastic waste that was estimated as having entered the ocean in 2010, highlighting the massive discrepancies in our understanding of the fate and location of ocean plastics (Jambeck et al., 2015; van Sebille et al., 2015). Dramatic surges in the use of single-use plastics associated with the Covid-19 pandemic are expected to further drastically exacerbate plastic pollution worldwide, due to the widespread use and mismanagement of disposable personal protective equipment such as gloves and masks, increased prepackaging and takeout container usage and biomedical waste among other factors, further threatening the health of animals, ecosystems and humans worldwide (Adyel, 2020; Benson et al., 2021).

17.2.2 Plastic distribution and degradation in the ocean

Plastics are distributed both horizontally and vertically through different transport routes, including passive movement by water currents and wind. Plastic location in the water column is affected by its size and surface area and, in addition, by the density of both the plastic and the water itself. Surface winds, waves and the Earth's rotation also affect plastic transportation (Lechthaler, 2020). Large subtropical oceanic gyres that represent large systems of ocean currents occur due to Ekman drift with the influence of wind and the Coriolis force. These gyres form in marine basins, can be millions of kilometers in area and are well known to accumulate all types of plastic (Lechthaler, 2020; Moore et al., 2001). Oceanic circulation models suggest potential accumulation regions in all five subtropical ocean gyres by conveyance of floating plastic debris released from continents into central convergence zones (Cózar et al., 2014; Lebreton et al., 2012). One widely known example is the Great Pacific Garbage Patch in the North Pacific between Hawaii and California. The term 'patch' is somewhat misleading because plastic debris is not only spread across the water surface, but extends all the way to the ocean floor (Egger et al., 2020) and, while microplastics are heavily concentrated within these gyres, larger marine debris such as discarded and lost fishing gear is also prevalent (NOAA-a). The amount of time it takes for plastic debris to reach these gyres varies, but one study found that plastic surface particles can move by approximately 1000 kilometers within 60 days, from eastern North America into the North Atlantic Subtropical Gyre (Law et al., 2014). These gyres are predicted to continue to rapidly accumulate plastic as more debris enters our oceans annually (Lebreton et al., 2018; van Sebille et al., 2020).

Plastic fragmentation and degradation in the ocean is driven primarily by photodegradation through exposure to radiation (UV-B), physical stress such as wave action, and microbial action (Cózar et al., 2014; SAPEA, 2019). Plastic object density is affected and changed by weathering and fragmentation as well as biofouling, which is the adhesion of organisms such as bacteria, algae, barnacles and more to other material on submerged surfaces (Gregory, 2009). Larger buoyant items and certain types of more buoyant plastics, such as polyethylenes and polypropylene with specific gravity less than 1, may remain on the surface longer while those with specific gravity greater than 1 (i.e., polystyrene and polyvinyl chloride) sink near point sources. However, biofouling increases the density of the plastic and these changes in density in turn cause the plastic to sink, thus impacting where microplastic will settle in the water column or sediment (Amaral-Zettler et al., 2021). The biofilm that forms is also dependent on sunlight, so plastics do not all settle all the way to the ocean floor, at least initially (Rech et al., 2018). Some researchers have estimated that up to 99% of plastic introduced since 1950 has indeed settled on the ocean floor, including in submarine canyons reported as 'hotspots' for plastic accumulation (Chiba et al., 2018, Kane et al., 2020; Pham et al., 2014; Pohl et al., 2020). Furthermore, macroplastics degrade very slowly on the seafloor because this is generally beyond light-flooded ocean zones (Andrady et al., 2015). However, other studies suggest that a high percentage of macroplastics was ashore in coastal regions (Figure 17.1) within about a year of their release into the environment due to wind, waves, tides, coastline morphology and water depth (Browne et al., 2011; Ho and Not, 2019; Lebreton et al., 2018; Ryan, 2020; Van Sebille et al., 2020). A further detrimental effect of plastic degradation is the release of hydrocarbon gases such as methane and ethylene from polyethylene during plastic degradation in both air and water (Royer et al., 2018). Intriguingly, despite the increase in production and disposal, surface plastic concentration in fixed ocean regions has not significantly increased since the 1980s (Cózar et al., 2014; Law et al., 2014; Thompson et al., 2004), indicating an alternate final destination for buoyant plastic debris. Biofouling, as well as shore deposition, nanofragmentation and ingestion have been proposed as mechanisms for removal of microplastics from the surface.

17.3 Plastics and marine mammals

17.3.1 Macroplastics and marine mammals

Macroplastics impact marine mammals worldwide, primarily through external entanglement and internally through ingestion, with lethal effects from all types of interactions. Commonly reported macroplastics affecting marine mammals include lost and abandoned fishing nets, also known as ghost nets, as well as nylon lines and ropes, packing straps, various fishing pot and trap gear and plastic bags and packaging of all sizes, among many other items (Laist, 1997; Lechthaler et al., 2020; Li et al., 2016). Approximately 640,000 tons of fishing gear are discharged annually, including 5.7% of all nets and 29% of all lines from commercial fisheries (Richardson et al., 2019), which accounts for approximately 10% of all marine debris (Good et al., 2010). Dumping of plastic waste at sea is internationally banned by the International Convention for the Prevention of Pollution from Ships (MARPOL Annex V, 1988), which has led to decreased ship-generated waste, but waste is still illegally dumped at sea.

Macroplastic entanglements cause a wide array of traumatic wounds that vary broadly in effect and appearance. Some interactions with macroplastics may result in superficial abrasions of the skin that may not appear to result in significant damage. However, because macroplastic degradation is so prolonged, the constant abrasion of soft tissue, muscle and bone results in extensive suffering associated with deep tissue trauma, infection, amputation and death with severe chronic exposure. A common presentation for marine mammals, as well as for other marine animals, is entanglement in mono- or multifilament gill nets. Gill nets are walls of netting with variable size mesh that hangs in both the marine and freshwater water columns, entrapping fish that can get their head and gills through the netting but cannot then back out of it, thus becoming caught. Mesh hole opening sizes vary widely among different fisheries but are commonly 10–150 millimeters in width or length. Nets are either set or anchored in substrate, or are drift nets kept afloat in the water column with buoys and weights (NOAA-b). Marine mammals, including odontocetes that use sonar to detect objects in the water, are often unable to detect this type of monofilament. Similar to fish, once a marine mammal encounters a net and becomes entangled – whether through a single mesh hole or multiple loops – the animal may not be able to back out of the net (Figure 17.2). Rapid drowning occurs because the animal often cannot overcome the size, strength and weight of the netting to reach the surface to breathe. In cases where the victim can successfully break away or is entangled in a smaller amount of material that allows it to reach the surface to breathe, the material may restrict its movement as well as cut into tissue, leading to strangulation, limited movement, reduced ability to forage, exhaustion, starvation and subsequent death. This incidental capture and mortality of nontarget species is referred to as bycatch.

The impact of external entanglement on marine mammals is often highly visible and well documented on individual animals that either wash ashore or strand, with obvious trauma and/or are viewed in the wild. Plastic entanglements have been widely reported for sea lions and seals, where materials such as packing-strap loops may attract curious young animals in particular (Arnould and Croxall, 1995; Hanni and Pyle, 2000; Lawson et al., 2015; Page et al., 2004). The smaller size of these animals easily results in loops passing over their heads and limbs, especially with larger mesh sizes, but the loops do not easily slip off again, particularly against the direction of their fur (Derraik, 2002; Raum-Suryan et al., 2009). Entangled juvenile fur seals (Figure 17.3) were recorded as expending higher amounts of energy when entangled in as little as 100 grams of netting and, despite spending more time resting and less time swimming when entangled, with higher relative swimming costs. The increased drag experienced by entangled animals increases exponentially with greater net size, resulting in concomitantly increased metabolic rates (Feldkamp, 1985). In addition to this increased need for food energy, entangled animals (particularly young, growing animals) experience long-term pain and debilitation through laceration, reduced fitness (including reduced ability to evade predators), secondary infection of exposed tissue, suffocation, tracheal perforation, sepsis and drowning (Derraik, 2002; Gall and Thompson, 2015; Laist, 1997; Raum-Suryan et al., 2009).

In addition to individual animal suffering, devastating impacts occur at marine mammal population levels. Global marine mammal bycatch from fisheries interactions is the greatest source of human-caused marine mammal mortality (Avila et al., 2018; Lewison et al., 2004; Read et al., 2005). Each year hundreds of thousands, to potentially millions, of marine mammals are killed as bycatch (Read et al., 2006), mostly trapped in macroplastic-based nets and ropes. Despite some international, regional and national regulatory body efforts over the past few decades to reduce global fisheries impacts on nontarget species, a recent review of marine mammal bycatch in gillnet and other entangling net fisheries found that at least 75% of toothed whale species (odontocetes, 90 species), 64% of baleen whale species (mysticetes, 14 species), 66% of pinnipeds (seals, sealions, fur seals and walrus, 34 species) and all sirenians (manatees and dugong, four species) and marine mustelids (sea otters, three subspecies) have been recorded as gillnet bycatch (Reeves et al., 2013). Entanglement-associated mortality has been associated with population-level impacts in northern fur seals (*Callorhinus ursinus*), Cape fur seals (*Arctocephalus pusillus*), Hawaiian monk seals (*Neomonachus schauinslandi*) and West Indian manatees (*Trichechus manatus*), with high concern also for other species (Fowler, 1987; Gregory, 2009; Harting et al., 2020; Laist, 1997; Reinert et al., 2017). A powerful example of the severe impact of bycatch on a marine mammal population is evidenced by the vaquita (*Phocoena sinus*), a small porpoise that only inhabits the northern Gulf of California. Vaquita are similar in size to a targeted species of fish, the totoaba, which is also endangered. Despite a ban on totoaba fishing since 1975, the vaquita population has declined precipitously secondary to bycatch in gillnets (D'Agrosa et al., 2000; Gulland et al., 2020; Jaramillo-Legorreta et al., 2019) and vaquita now faces extinction with fewer than 15 individuals remaining (Rojas-Bracho et al., 2022). Other marine mammals – particularly large baleen whales such as the critically endangered North Atlantic right whale (*Eubalaena glacialis*) – often become entangled in large fixed fisheries buoy lines (aka endlines) and other fisheries gear (Moore et al., 2005; Sharp et al., 2019). Some drown rapidly, but often these powerful animals continue swimming for months or even years while entangled, slowly becoming exhausted, malnourished and often starving to death while dragging ropes, buoys and traps (Cassoff et al., 2011; Moore et al., 2012; Sharp et al., 2019). Severity of injury associated with these debilitating entanglements worsened with the introduction and broad adoption of stronger ropes, particularly those introduced in the mid-1990s (Knowlton et al., 2016). These robust synthetic fiber-based ropes were developed using copolymers that allow blending of different plastic resins to create stronger and more abrasion-resistant floating ropes, such as Polysteel, made from polypropylene and polyethylene (Knowlton et al., 2016), and remain widely in use with even more deadly effects on nontargeted species.

Efforts to reduce marine mammal bycatch and entanglement mostly involve attempts to prevent marine mammals from becoming entangled in fishing lines and gear through deterrents and gear modification, rather than through reduction of plastics use. Mitigation measures that have been implemented with varied, but usually minimal, success include visual or acoustic deterrent devices (i.e., pingers) attached to gillnets, exclusion devices and pot and trap guard designs that reduce entrapment of marine mammals (Hamilton and Baker, 2019; Pace et al., 2014). For example, currently efforts are underway to assess the viability and effectiveness of 'ropeless' fishing gear – gear that does not use vertical buoy lines prior to gear retrieval – to reduce North Atlantic right whale entanglement (Meyers et al., 2019). However, the use and presence of macroplastics in fisheries and other industries remains high and newer sources of macroplastics such as those used in the rapidly expanding aquaculture industry (Skirtun et al., 2022) continue to magnify the problem.

In addition to external entrapment and entanglement, ingestion of plastics has been widely reported in marine mammals primarily through post-mortem examination. Plastic consumption by marine mammals under human care has been well documented and can result in severe gastrointestinal ulceration, impaction and death (Brown et al., 1960; Caldwell et al., 1965; Walker and Coe, 1989) The full impacts of plastic ingestion on free-ranging marine mammals are not truly known, however, because dead stranded marine mammals that undergo a full post-mortem examination with complete documentation and reporting of gastrointestinal content represent only a small fraction of the total number of potentially affected animals (Baulch and Perry, 2014; Roman et al., 2021). Marine mammals ingest plastic directly or indirectly, likely via prey that has previously consumed plastic, and reported ingestion of macroplastics varies widely among species.

Plastic ingestion is best described in odontocete cetaceans. A study of by-caught Franciscana dolphins (*Pontoporia blainvillei*) in Argentina found that 28% (of 106 individuals) had ingested plastic debris consisting mostly of packaging debris (cellophane, bags and bands) and fishing gear fragments (monofilament lines, ropes and nets) (Denuncio et al., 2011). There were no reported detrimental obstructions or digestive tract ulcerations reported in this particular group of dolphins, and plastic ingestion correlated with age-related changes in feeding strategy, being higher in juvenile, recently weaned individuals. However, numerous other studies have reported both lethal and sublethal effects of macroplastic ingestion in marine mammals (Figure 17.4a&b), particularly teuthophagous (cephalopod-consuming) species such as sperm whales (*Physeter macrocephalus*, *Kogia* sp.) and various beaked whales (Baulch and Perry, 2014; Fernández et al., 2009; Jacobson et al., 2010; Lusher et al., 2015; Walker and Coe, 1989). Benthic-feeding cetaceans in particular are thought to accidentally ingest plastic when they are feeding at or near the ocean floor, because it is not uncommon to find rocks and gravel in their stomachs. A similar hypothesis has been suggested for harbor seals with various plastic particles in their digestive tracts (Bravo Rebolledo et al., 2013). Alternatively, plastic objects and fragments may resemble food and are mistaken for prey, or are ingested during play or curious investigation. Another possibility is that some marine mammals ingest plastic due to concurrent disease (Walker and Coe, 1989).

Macroplastics can cause a wide range of problems for marine wildlife. Examples are shown in Figures 17.1, 17.2, 17.3, 17.4a and 17.4b.

Figure 17.1 Extensive plastic debris litters the beach of an uninhabited island (Kauo/Laysan Island) in the Northwestern Hawaiian Islands in the Pacific. Photo: Cara Field.

Figure 17.2 California sea lion (*Zalophus californianus*) entangled in plastic gillnet, with associated deep lacerations around the entire head. Photo: Cara Field, The Marine Mammal Center; NOAA permit no. 18786.

Figure 17.3 Guadalupe fur seal (*Arctocephalus townsendi*) pup severely entangled in a synthetic plastic fiber fishing net. Photo: Cara Field, The Marine Mammal Center; NOAA permit no. 18786.

Figure 17.4b Post-mortem examination of a dead stranded pygmy sperm whale revealed that the stomach was filled with ten large plastic garbage bags, which completely obstructed the stomach resulting in the death of the whale. Photo: Cara Field.

Figure 17.4a Dead adult female pygmy sperm (*Kogia breviceps*) whale washed ashore in Connecticut, USA. Photo: Cara Field.

While some accumulation of plastics may not have obvious detrimental impacts on an individual animal, ingestion and accumulation of large amounts of macroplastics can result in a full stomach without any actual nutrient gain, and subsequent obstruction of the gastrointestinal tract, ulceration, gastric rupture, starvation and ultimately death, such as noted in multiple sperm whales (de Stephanis et al., 2013; Jacobson et al., 2010) and West Indian manatees (*Trichechus manatus*) (Beck and Barros, 1991).

17.3.2   Microplastics in marine mammals

Microplastics are ubiquitous worldwide across both marine and terrestrial environments, and their direct effects on animal and human health are of increasing concern. Microplastics, including synthetic fibers, textiles and tire debris, as well as manufactured microbeads and degraded debris from larger plastics, have been accumulating in the oceans and on beaches for decades (Thompson et al., 2004; Thompson, 2015) and are readily ingested from the marine environment by a wide variety of marine animals, due especially to their small size and variable buoyancy (Miller et al., 2020; Nelms et al., 2019). Microplastic presence and its effects are present throughout the food web, although their effects on smaller marine species such as zooplankton and filter feeders are increasingly described (Cole et al., 2013; Karlsson et al., 2017). Furthermore, microplastics colonized by potentially dangerous organisms – including algae that produce harmful biotoxins (Masó et al., 2003; Zettler et al., 2013), pathogenic microbes (Bowley et al., 2021; Radisic et al., 2020; Rogers et al., 2020) and protozoal parasites (Zhang et al., 2022) – can be vectors for these additional disease agents. As high-trophic-level, long-lived predators, marine mammals are not only susceptible to these diseases but, as sentinels for marine pollution, they represent important models to understand trophic transfer and the health effects of microplastics.

Marine mammals ingest microplastics directly through water, filter feeding (i.e., baleen whales) or drinking, and indirectly through trophic transfer with prey consumption. A recent systematic literature review found that microplastics were documented in marine mammals in 29 of 30 studies (Zantis et al., 2021), although there was considerable variability in methods and reporting. All studies focused on detection of microplastics within the gastrointestinal tract, and digestive tracts were examined from stranded, by-caught or hunted marine mammals in 12 studies (Zantis et al., 2021). In some cases the entire digestive tract was assessed and the total number of microplastic particles per animal reported (Lusher et al., 2015; Nelms et al., 2019), while others subsampled parts of the digestive tract (Moore et al., 2020) and nine studies analyzed marine mammal feces. A number of studies have recorded microplastics within the gastrointestinal tracts of wild-caught fish and shellfish (Lusher et al., 2013; Neves et al., 2015; Rummel et al., 2016; Tanaka and Takada, 2016; Wieczorek et al., 2018), and Nelms et al. (2018) confirmed trophic transfer by fully evaluating the gastrointestinal tracts of wild-caught fish fed to captive gray seals (*Halichoerus grypus*) followed by examination of the gray seals' feces. Because methods of evaluation of microplastic presence in marine mammals vary widely, comparison of results across studies remains problematic (Zantis et al., 2021). Standardization of methods, including collection, extraction, identification and prevention of contamination during all of these steps, will improve our ability to assess species differences and understand microplastic exposure and health risks. However, even the most thorough post-mortem examination methods provide only a small window of insight on individual stranded and by-caught marine mammals as opposed to comprehensive population effects from microplastics. Population exposure risk can be better assessed through studies that link marine mammal biology and foraging ecology with habitat and prey distribution (Burkhardt-Holm and N'Guyen, 2019), but risk assessment further requires determination of the probability and severity of adverse effects on health.

17.4 Health implications of microplastics in the marine environment: a One Health perspective

The potential health implications of microplastics in the environment are determined by numerous variables including route of exposure, which is not limited to ingestion and magnification up the food chain from filter feeding invertebrates to fish, marine mammals and humans but also includes inhalation of airborne particles and uptake through transcutaneous migration. Further variables are the physical and chemical characteristics of the particles. The former includes particle size and hydrophobicity while the latter includes additives incorporated during the manufacture of plastic products, to chemicals added in microplastic production (detergents, abrasives) to contaminants (polycyclic aromatic hydrocarbons, heavy metals etc.) adsorbed to particles in the environment. Exposure of marine mammals and humans to nano- and microplastics through the consumption of food is well established. Considering food of aquatic origin, the consumption of plastics varies considerably depending on the nature of the food item and by dietary preferences or patterns from country to country. Collating data on finfish fillet in European and American diets using data from the European Food Safety Authority and the National Oceanographic & Atmospheric Administration (NOAA), per capita consumption varies between from as low as 9600 g/year

in Brazil to 57,000 g/year in Portugal (Barboza et al., 2020). Based on the total mean quantity of microplastics in fish muscle as determined from three commercial North Atlantic species (0.054 microplastic (MP) items/g tissue), the intake of these particles by human consumers can vary from 112 MP items/year in infants to 842 MP items/year in the general adult population. However, when national differences are factored in, the quantity of microplastics ingested through consumption of fish varies from 518 MP items/year in Brazil to 3078 MP items/year in Portugal (Barboza et al., 2020). When an entire organism is consumed, as with many shellfish, the potential for microplastic ingestion is greatly increased. Thus, a consumer in Europe eating mussels is likely to ingest as many as 11,000 MP particles annually (Van Cauwenberghe and Janssen, 2014). The health implications for microplastic consumption in humans are still largely unknown but, using the One Health approach, it is possible to extrapolate from research on other species and *in vitro* studies using cell lines from a variety of vertebrates including human cell lines.

In recent years the numbers of empirical laboratory studies of the effects of microplastic toxicity in laboratory animals and *in vitro* studies on cell lines have increased substantially, but studies on free-ranging animals and the general human population remain limited (Barboza et al., 2020; Jeong and Choi, 2019; Kannan and Vimalkumar, 2021). When microplastics of size up to approximately 20 µm are ingested, they can translocate across the intestine by endocytotic uptake by M cells in the ileal mucosa overlying Payer's patches, or by dendritic cells and, from there, enter the lymphatic or blood stream (Smith et al., 2018). Depending on particle size and hydrophobicity, they may subsequently cross the blood–brain barrier or the placenta. A similar mode of entry is likely for particles inhaled into the respiratory tract. Despite this uptake, the majority of ingested microplastics are eliminated by normal excretory routes and it is estimated that >90% of ingested plastics are eliminated in feces (Wright and Kelly, 2017). The severity of adverse effects from absorbed microplastics is likely highly variable and will depend on the physical characteristics of the particle including size, shape and polymer type, to the chemical composition and individual susceptibility. Based on *in vivo* studies, once microplastics enter tissues the potential effects include oxidative stress, induction of apoptosis (cell death), inflammation and altered cell growth regulation or immune responses (Qiao et al., 2019). The physical presence of an inorganic substance in tissues may induce a prolonged chronic inflammatory response akin to 'frustrated phagocytosis', where macrophages fail to completely remove and destroy a particle but continue to produce and release cytokines in the immediate area, promoting a chronic inflammatory response. A similar pathogenesis has been documented for asbestos fibrils and wear debris from surgical implants in human patients (Kzhyshkowska et al., 2015). When particles are inhaled and absorbed through the respiratory epithelia, local granulomatous inflammation may ensue as was previously observed in the lungs of textile factory workers exposed to aerosolized fibrils (Wright and Kelly, 2017). *In vivo* studies using various genetic strains of laboratory mice have produced conflicting results confounded by differing protocols, methodology and endpoint evaluation (Braeuning, 2019; Deng et al., 2017; Stock et al., 2019; van Raamsdonk et al., 2020). In these studies, effects included hepatic inflammation and metabolomic changes that could indicate effects on energy balance and oxidative stress. Similar work with a different mouse strain in another laboratory found no pathologic changes in multiple organs and no indicators of oxidative stress, while studies in a third laboratory documented intestinal dysbiosis and alterations in hepatic and lipid metabolism (Lu et al., 2018; Stock et al., 2019). In a

study of laboratory zebrafish in which fish were exposed to low and high doses of microplastics in the water, the fish showed uptake of particles into the intestinal tissue, higher inflammatory response in the high exposure group and alterations in gut microbiota (Qiao et al., 2019). Rainieri et al. (2018) conducted a similar study but also compared the effect of exposure to microplastics alone and with adsorbed chemicals, including polychlorinated biphenyls, brominated flame retardants, perfluoroalkyl substances and methylmercury. In those groups exposed to microplastics alone there were no significant findings, while microplastics laced with chemicals induced hepatic lipidosis and induction of enzymes involved in detoxification. In addition to these effects on the gastrointestinal system and liver, neurotoxic effects – as measured by brain acetylcholinesterase activity – have been demonstrated in laboratory fish exposed to microplastics (Barboza et al., 2018; Oliveira et al., 2013). Expanding these laboratory findings to commercial finfish caught off the coast of Portugal, Barboza et al. (2020) found that wild-caught fish with microplastics in their tissues had significantly higher levels of lipid peroxidation in the brain, gills and dorsal muscle. Given the level of contamination of the dorsal muscle, and calculated exposure of humans to this source of microplastics as determined by published fish consumption statistics, it was suggested that this could be a risk factor for humans. Clearly, for predators such as marine birds and mammals, consumption of the whole fish would result in an even greater exposure to microplastics and their associated chemicals.

Further insights on the effect of microplastics on cells have been obtained using cell cultures from humans or other animal sources and measuring endpoints such as cell viability, intracellular localization, oxidative stress, membrane integrity and expression of immune response genes. In these studies the microplastic components tested included polystyrene, polyethylene and polypropylene polymers. As with *in vivo* rodent models, there was considerable variability in endpoints reported depending on the methodology, with some reporting decreased cell viability (Dong et al., 2020; Hesler et al., 2019; Hwang et al., 2019; Stock et al., 2019) while others saw no effect (Schirinzi et al., 2017; Wu et al., 2019). However, particle size mattered and smaller particles of polystyrene (1.00, 1.75 or 4.00 µm) reduced cell viability after 24 h, and even further at 48 h, exposure in a human colorectal adenocarcinoma (CACO-2) cell line (Stock et al., 2019). Chemical modification of the particles also had an effect with COOH-modified polystyrene, causing cytotoxicity to placental, embryonic and intestinal cells (Hesler et al., 2019). Wu and others (2019) showed that smaller polystyrene particles exerted their effect on CACO-2 cells by inhibiting cellular ABC transporters. Furthermore, when cells were coexposed to another toxin (arsenic) and polystyrene particles, uptake of the toxin was higher than in the absence of the microplastic.

Oxidative stress leading to loss of membrane integrity was an endpoint of several *in vitro* studies (Dong et al., 2020; Hwang et al., 2019; Schirinzi et al., 2017; Wu et al., 2019). Using different particle sizes for polystyrene and varying exposure time, Dong and others (2020) demonstrated elevated levels of reactive oxygen species and heme oxygenase-1 (HO-1) in exposed cells, with subsequent compromise of cell and mitochondrial membrane integrity (Dong et al., 2020; Wu et al., 2019). Hwang et al. (2019) exposed several cell lines, of human, sheep and rodent origin, to polypropylene microplastics of varying size (20–200 µm) and measured endpoints around oxidative stress and immune gene expression. They found minimal effect on cell viability in general but, when smaller particles were used at high concentration, there were increased levels of cytokines (IL-6 and TNF-α) and histamines in

some cell lines, suggesting that smaller particles can stimulate an immune response.

Endpoints such as genotoxicity, embryotoxicity and hemolysis have also been investigated to a limited extent. Hesler et al. (2019) found no evidence for genotoxicity using COOH-modified polystyrene particles in a p53 reporter assay in two human cell lines after 24 h exposure, but 0.5 µm polystyrene was weakly embryotoxic over a similar exposure timeframe. Sheep red blood cells were hemolyzed on exposure to polypropylene particles (Hwang et al., 2019). Regulation of gene expression in Caco-2 cells was demonstrated by 24 or 48 h exposure to 5 µm polystyrene particles (Wu et al., 2019). RNA sequencing analysis after 24 h showed upregulation of 80 differentially expressed genes and 94 downregulated genes. As the concentration of particles in the medium was increased, gene expression shifted from a metabolism profile to a cancer profile after 48 h exposure. Quantitative PCR with reverse transcription on cells exposed for 24 h to polystyrene showed downregulation of five proliferation-related genes (*Ras, ERK, MER, CDK4, Cyclin 1D*) and upregulation for four inflammatory genes (*TRPV1, iNOS, IL-1ß, IL-8*). In conjunction with data from *in vivo* studies, these *in vitro* results suggest that, depending on particle size, dose rate and exposure time, microplastics can alter gene expression and affect cell proliferation, viability and inflammatory responses. As a consequence, in free-living marine mammals or in the general human population with expected levels of seafood consumption, microplastics and their additives could contribute to cancer rates, morbidity from infectious pathogens and parasites and decreased survivorship. Furthermore, plasticizers such as bisphenol A (BPA) and phthalates, used in the manufacture of plastics and detected in marine mammals among other species (Fossi et al., 2016; Montoto-Martínez et al., 2021; Page-Karjian et al., 2020), have been linked in epidemiologic studies to endocrine disruption (reproductive disorders), obesity, cardiovascular disease, breast cancer and pre and postnatal development.

17.5   Future directions for plastic management and ecosystem health

Marine mammals are only one group of animals heavily impacted by the detrimental effects of plastic pollution. Reduction of the impacts of plastic pollution must take place at all levels of plastic production, utilization and impact including documentation of health effects, scientific investigation into mechanisms of disease, education to raise public awareness of health concerns and, ultimately, changes in policy, consumer behavior and plastic production practices. This chapter has provided background into the breadth of the problem, with a focus on health effects. Mitigation of ecosystem effects include policies that target plastic production before it becomes waste, as well as abatement of waste entering the environment. Methods include broad consumer education through public awareness campaigns, schools, universities, zoological institutions, recreational and commercial fishing industries, government agencies, international political agendas and more. The combination of policies, practices and behavior change around social activities has been the most successful in reducing environmental plastic, and investment in campaigns has led to greater reductions of environmental waste than investment in policies (Willis et al., 2018). Actions to reduce plastic pollution may be straightforward and targeted, such as the provision of tackle bins at popular wharfs and piers to promote appropriate fishing tackle disposal, beach clean-up programs and 'Lose the Loop' or 'Go Bandless' campaigns that urge consumers to decrease animal entanglement risk by cutting any encircling loop associated with fishing and packing practices (ADFG, 2022; Raum-Suryan et al., 2009). Arnould and Croxall (1995) reported a reduction in packing-strap entanglement (originating from longline fishing bait boxes) of Antarctic fur seals at South Georgia island following the success of an aggressive educational campaign and regulatory efforts through the Convention on the Conservation of Antarctic Marine Living Resources. Even more profound change includes switching from plastic packaging to nonplastic alternatives, such as paper strapping as an alternative to plastic straps (https://packagingguruji.com/paper-strapping/). Measures to reduce loss of fisheries gear loss can include port repair and disposal options, including incentives for these options, expanded use of biodegradable materials, restriction of fishing activity in conditions or locations where gear loss is likely and enforcement of penalties associated with failure to retrieve lost items. Actions to mitigate plastic production prior to waste include industry practice changes, such as the use of materials with high biodegradation rate rather than nondegradable material, require currently manufactured plastics to be increasingly produced from recycled material, especially with financial incentives, cap production of new 'virgin' plastic with concomitant recycling and upgrading of existing plastic, and cut demand for plastic products, especially single-use products. Numerous bans of plastic products have been enacted, such as microbeads in cosmetics and personal care products, plastic bags (especially single-use), plastic straws and Styrofoam in many different cities and countries worldwide. Many bans are still relatively new and data on outcomes often unfounded, but some reports found that plastic ban effectiveness is positively associated with reuse of old bags, higher pricing of plastic carrier bags reduces their purchase and discounts or subsidies for 'bringing your own bag' help reduce new plastic bag use (Wang and Li, 2021). It is clear that plastic pollution will remain a significant factor in ecosystem health for decades if not centuries, to come, and multiple global approaches to production, consumption and disposal must occur to mitigate this threat.

17.6 Conclusion

Marine mammals inhabit all oceans of the world and, as highly visible charismatic megafauna at the peak of the trophic level, are powerful sentinels of ocean health. Extensive studies over the past 50 years have clearly demonstrated the deadly impact of plastic pollution on animal, human and ecosystem health, yet plastic production and use continue to grow. Many detrimental effects of plastic exposure, particularly to micro- and nanoplastics, are only now being documented and much remains to be learned. The plastic crisis requires global leadership and broad interdisciplinary participation to develop policy and action plans to address this expanding health threat.

References

ADFG. Alaska Department of Fish and Game website, Entanglement in Marine Debris. https://www.adfg.alaska.gov/index.cfm?adfg=marinemammalprogram.entanglements. Retrieved December 2022

Adyel, T. M. (2020). Accumulation of plastic waste during COVID-19. *Science, 369*(6509), 1314–1315. https://doi.org/10.1126/science.abd9925

Amaral-Zettler, L. A., Zettler, E. R., Mincer, T. J., Klaassen, M. A., & Gallager, S. M. (2021). Biofouling impacts on polyethylene density and sinking in coastal waters: A macro/micro tipping point? *Water Research, 201*, 117289. https://doi.org/10.1016/j.watres.2021.117289

Andrady, A. L. (2011). Microplastics in the marine environment. *Marine Pollution Bulletin, 62*(8), 1596–1605. https://doi.org/10.1016/j.marpolbul.2011.05.030

Andrady, A. (2015). Persistence of plastic litter in the oceans. In B. Litter, M., L. Gutow & M. Klages (Eds.), *Marine anthropogenic* (pp. 57–72). Springer International Publishing.

Arnould, J. P. Y., & Croxall, J. P. (1995). Trends in entanglement of Antarctic fur seals (*Arctocephalus gazella*) in man-made debris at South Georgia. *Marine Pollution Bulletin, 30*(11), 707–712. https://doi.org/10.1016/0025-326X(95)00054-Q

Arthur, C., Baker, J., & Bamford, H. (Eds.). (2009). Sept 9–11, 2008. NOAA technical memorandum NOS-OR&R-30. In *Proceedings of the International Research Workshop on the Occurrence, Effects and Rate of Microplastic Marine Debris*. National Oceanic and Atmospheric Administration.

Avila, I. C., Kaschner, K., & Dormann, C. F. (2018). Current global risks to marine mammals: Taking stock of the threats. *Biological Conservation, 221*, 44–58. https://doi.org/10.1016/j.biocon.2018.02.021

Baini, M., Martellini, T., Cincinelli, A., Campani, T., Minutoli, R., Panti, C., Finoia, M. G., & Fossi, M. C. (2017). First detection of seven phthalate esters (PAEs) as plastic tracers in superficial neustonic/planktonic samples and cetacean blubber. *Analytical Methods, 9*(9), 1512–1520. https://doi.org/10.1039/C6AY02674E

Barboza, L. G. A., Vieira, L. R., Branco, V., Figueiredo, N., Carvalho, F., Carvalho, C., & Guilhermino, L. (2018). Microplastics cause neurotoxicity, oxidative damage and energy-related changes and interact with the bioaccumulation of mercury in the European seabass, *Dicentrarchus labrax* (Linnaeus, 1758). *Aquatic Toxicology, 195*, 49–57. https://doi.org/10.1016/j.aquatox.2017.12.008

Barboza, L. G. A., Lopes, C., Oliveira, P., Bessa, F., Otero, V., Henriques, B., Raimundo, J., Caetano, M., Vale, C., & Guilhermino, L. (2020). Microplastics in wild fish from North East Atlantic Ocean and its potential for causing neurotoxic effects, lipid oxidative damage, and human health risks associated with ingestion exposure. *Science of the Total Environment, 717*, 134625. https://doi.org/10.1016/j.scitotenv.2019.134625

Barrett, J., Chase, Z., Zhang, J., Holl, M. M. B., Willis, K., Williams, A., Hardesty, B. D., & Wilcox, C. (2020). Microplastic pollution in deep-sea sediments from the great Australian bight. *Frontiers in Marine Science, 7*. https://doi.org/10.3389/fmars.2020.576170

Baulch, S., & Perry, C. (2014). Evaluating the impacts of marine debris on cetaceans. *Marine Pollution Bulletin, 80*(1–2), 210–221. https://doi.org/10.1016/j.marpolbul.2013.12.050

Beck, C. A., & Barros, N. B. (1991). The impact of debris on the Florida manatee. *Marine Pollution Bulletin, 22*(10), 508–510. https://doi.org/10.1016/0025-326X(91)90406-I

Benson, N. U., Fred-Ahmadu, O. H., Bassey, D. E., & Atayero, A. A. (2021). COVID-19 pandemic and emerging plastic-based personal protective equipment waste pollution and management in Africa. *Journal of Environmental Chemical Engineering, 9*(3), 105222. https://doi.org/10.1016/j.jece.2021.105222

Besseling, E., Wegner, A., Foekema, E. M., van den Heuvel-Greve, M. J., & Koelmans, A. A. (2013). Effects of microplastic on fitness and PCB bioaccumulation by the lugworm Arenicola marina (L.). *Environmental Science and Technology, 47*(1), 593–600. https://doi.org/10.1021/es302763x

Billard, G., & Boucher, J. (2019). The challenges of measuring plastic pollution. *The Journal of Field Actions, 19*, 68–75. http://journals.openedition.org/factsreports/5319

Boren, L. J., Morrissey, M., Muller, C. G., & Gemmell, N. J. (2006). Entanglement of New Zealand fur seals in man-made debris at Kaikoura, New Zealand. *Marine Pollution Bulletin, 52*(4), 442–446. https://doi.org/10.1016/j.marpolbul.2005.12.003

Bowley, J., Baker-Austin, C., Porter, A., Hartnell, R., & Lewis, C. (2021). Oceanic Hitchhikers – Assessing Pathogen Risks from Marine Microplastic. *Trends in Microbiology, 29*(2), 107–116. https://doi.org/10.1016/j.tim.2020.06.011

Braeuning, A. (2019). Uptake of microplastics and related health effects: A critical discussion of Deng et al. *Scientific Reports* 7: 46687, 2017. Arch Toxicol, *93*(1), 219–220. https://doi.org/10.1007/s00204-018-2367-9

Bravo Rebolledo, E. L., Van Franeker, J. A., Jansen, O. E., & Brasseur, S. M. J. M. (2013). Plastic ingestion by harbor seals (*Phoca vitulina*) in the Netherlands. *Marine Pollution Bulletin, 67*(1–2), 200–202. https://doi.org/10.1016/j.marpolbul.2012.11.035

Brown, D. H., McIntyre, R. W., Delli Quadri, C. A., & Schroeder, R. J. (1960). Health problems of captive dolphins and seals. *Journal of the American Veterinary Medical Association, 137*, 534.

Browne, M. A., Crump, P., Niven, S. J., Teuten, E., Tonkin, A., Galloway, T., & Thompson, R. (2011). Accumulation of microplastic on shorelines woldwide: Sources and sinks. *Environmental Science and Technology, 45*(21), 9175–9179. https://doi.org/10.1021/es201811s

Burkhardt-Holm, P., & N'Guyen, A. (2019). Ingestion of microplastics by fish and other prey organisms of cetaceans, exemplified for two large baleen whale species. *Marine Pollution Bulletin, 144*, 224–234. https://doi.org/10.1016/j.marpolbul.2019.04.068

Caldwell, M. C., Caldwell, D. K., & Siebenaler, J. B. (1965). Observations on captive and wild Atlantic bottlenosed dolphins, *Tursiops truncatus*, in the northeastern Gulf of Mexico. *Contributions in Science, 91*, 1–10. https://doi.org/10.5962/p.241081

Carpenter, E. J., & Smith, K. L. (1972). Plastics on the Sargasso sea surface. *Science, 175*(4027), 1240–1241. https://doi.org/10.1126/science.175.4027.1240

Cassoff, R. M., Moore, K. M., McLellan, W. A., Barco, S. G., Rotsteins, D. S., & Moore, M. J. (2011). Lethal entanglement in baleen whales. *Diseases of Aquatic Organisms, 96*(3), 175–185. https://doi.org/10.3354/dao02385

Chamas, A., Moon, H., Zheng, J., Qiu, Y., Tabassum, T., Jang, J. H., Abu-Omar, M., Scott, S. L., & Suh, S. (2020). Degradation rates of plastics in the environment. *ACS Sustainable Chemistry and Engineering, 8*(9), 3494–3511. https://doi.org/10.1021/acssuschemeng.9b06635

Chiba, S., Saito, H., Fletcher, R., Yogi, T., Kayo, M., Miyagi, S., Ogido, M., & Fujikura, K. (2018). Human footprint in the abyss: 30 year records of deep-sea plastic debris. *Marine Policy, 96*, 204–212. https://doi.org/10.1016/j.marpol.2018.03.022

Cole, M., Lindeque, P., Fileman, E., Halsband, C., Goodhead, R., Moger, J., & Galloway, T. S. (2013). Microplastic ingestion by zooplankton. *Environmental Science and Technology, 47*(12), 6646–6655. https://doi.org/10.1021/es400663f

Cózar, A., Echevarría, F., González-Gordillo, J. I., Irigoien, X., Ubeda, B., Hernández-León, S., Palma, A. T., Navarro, S., García-de-Lomas, J., Ruiz, A., Fernández-de-Puelles, M. L., & Duarte, C. M. (2014). Plastic debris in the open ocean. *Proceedings of the National Academy of Sciences of the United States of America, 111*(28), 10239–10244. https://doi.org/10.1073/pnas.1314705111

Crutzen, P., & Soermer, E. (2002). The Anthropocene. *Glob. Change News, 41*, 17–18.

D'Agrosa, C., Lennert-Cody, C. E., & Vidal, O. (2000). Vaquita bycatch in Mexico's artisanal gillnet fisheries: Driving a small population to extinction. *Conservation Biology, 14*(4), 1110–1119. https://doi.org/10.1046/j.1523-1739.2000.98191.x

Deng, Y., Zhang, Y., Lemos, B., & Ren, H. (2017). Tissue accumulation of microplastics in mice and biomarker responses suggest widespread health risks of exposure. *Scientific Reports*, 7(1), 46687. https://doi.org/10.1038/srep46687

Denuncio, P., Bastida, R., Dassis, M., Giardino, G., Gerpe, M., & Rodríguez, D. (2011). Plastic ingestion in franciscana dolphins, *Pontoporia blainvillei* (Gervais and d'Orbigny, 1844), from Argentina. *Marine Pollution Bulletin*, 62(8), 1836–1841. https://doi.org/10.1016/j.marpolbul.2011.05.003

Derraik, J. G. B. (2002). The pollution of the marine environment by plastic debris: A review. *Marine Pollution Bulletin*, 44(9), 842–852. https://doi.org/10.1016/s0025-326x(02)00220-5

de Stephanis, R., Giménez, J., Carpinelli, E., Gutierrez-Exposito, C., & Cañadas, A. (2013). As main meal for sperm whales: Plastics debris. *Marine Pollution Bulletin*, 69(1–2), 206–214. https://doi.org/10.1016/j.marpolbul.2013.01.033

Dong, C. D., Chen, C. W., Chen, Y. C., Chen, H. H., Lee, J. S., & Lin, C. H. (2020). Polystyrene microplastic particles: *In vitro* pulmonary toxicity assessment. *Journal of Hazardous Materials*, 385, 121575. https://doi.org/10.1016/j.jhazmat.2019.121575

Egger, M., Sulu-Gambari, F., & Lebreton, L. (2020). First evidence of plastic fallout from the North Pacific Garbage Patch. Nature Sci. Rep. 10. *Scientific Reports*, 10(1), 7495. https://doi.org/10.1038/s41598-020-64465-8

Farrell, P., & Nelson, K. (2013). Trophic level transfer of microplastic: *Mytilus edulis* (L.) to *Carcinus maenas* (L.). *Environmental Pollution*, 177, 1–3. https://doi.org/10.1016/j.envpol.2013.01.046

Feldkamp, S. D. (1985). The effects of net entanglement on the drag and power output of a California sea lion, *Zalophus californianus*. *Fishery Bulletin*, 83, 692–695.

Feldkamp, S., Costa, D., & DeKrey, G. K. (1989). Energetics and behavioural effects of net entanglement on juvenile northern fur seals *Callorhinus ursinus*. *Fishery Bulletin*, 87, 85–94.

Fernández, R., Santos, M. B., Carrillo, M., Tejedor, M., & Pierce, G. J. (2009). Stomach contents of cetaceans stranded in the Canary Islands 1996–2006. *Journal of the Marine Biological Association of the United Kingdom*, 89(5), 873–883. https://doi.org/10.1017/S0025315409000290

Food and Agriculture Organization. (2009). *Abandoned, lost or otherwise discarded fishing gear*. Fisheries Ed Technical Paper Fisheries and Aquaculture, Rome, Italy, vol. 523.

Food and Agriculture Organization. (2016). *Abandoned, lost or otherwise discarded gill nets and trammel nets: Methods to estimate ghost fishing mortality, and the status of regional monitoring and management*. Technical paper fisheries and aquaculture, Rome, Italy, 600. Food and Agriculture Organization of the United Nations.

Fossi, M. C., Coppola, D., Baini, M., Giannetti, M., Guerranti, C., Marsili, L., Panti, C., de Sabata, E., & Clò, S. (2014). Large filter feeding marine organisms as indicators of microplastic in the pelagic environment: The case studies of the Mediterranean basking shark (*Cetorhinus maximus*) and fin whale (*Balaenoptera physalus*). *Marine Environmental Research*, 100, 17–24. https://doi.org/10.1016/j.marenvres.2014.02.002

Fossi, M. C., Marsili, L., Baini, M., Giannetti, M., Coppola, D., Guerranti, C., Caliani, I., Minutoli, R., Lauriano, G., Finoia, M. G., Rubegni, F., Panigada, S., Bérubé, M., Urbán Ramírez, J., & Panti, C. (2016). Fin whales and microplastics: The Mediterranean Sea and the Sea of Cortez scenarios. *Environmental Pollution*, 209, 68–78. https://doi.org/10.1016/j.envpol.2015.11.022

Fossi, M. C., Baini, M., Panti, C., & Baulch, S. (2018a). Impacts of marine litter on cetaceans: A focus on plastic pollution. In M. C. Fossi & C. Panti (Eds.), *Marine mammal ecotoxicology* (pp. 147–184). Academic Press, Chp 6. https://doi.org/10.1016/B978-0-12-812144-3.00006-1

Fossi, M. C., Pedà, C., Compa, M., Tsangaris, C., Alomar, C., Claro, F., Ioakeimidis, C., Galgani, F., Hema, T., Deudero, S., Romeo, T., Battaglia, P., Andaloro, F., Caliani, I., Casini, S., Panti, C., & Baini, M. (2018b). Bioindicators for monitoring marine litter ingestion and its impacts on Mediterranean biodiversity. *Environmental Pollution*, 237, 1023–1040. https://doi.org/10.1016/j.envpol.2017.11.019

Fossi, M. C., Romeo, T., Baini, M., Panti, C., Marsili, L., Campani, T., Canese, S., Galgani, F., Druon, J., Airoldi, S., Taddei, S., Fattorini, M., Brandini, C., & Lapucci, C. (2017). Plastic debris occurrence, convergence areas and fin whales feeding ground in the Mediterranean marine protected area pelagos sanctuary: A modeling approach. *Frontiers in Marine Science*, 4, 1–15. https://doi.org/10.3389/fmars.2017.00167

Fowler, C. W. (1987). Marine debris and northern fur seals: A case study. *Marine Pollution Bulletin*,

18(6), 326–335. https://doi.org/10.1016/S0025-326X(87)80020-6

Galgani, F., Hanke, G., & Maes, T. (2015). Global distribution, composition and abundance of marine litter. In B. Litter, M., L. Gutow & M. Klages (Eds.), *Marine anthropogenic* (pp. 29–56). Springer International.

Galgani, F., Leaute, J. P., Moguedet, P., Souplet, A., Verin, Y., Carpentier, A., Goraguer, H., Latrouite, D., Andral, B., Cadiou, Y., Mahe, J. C., Poulard, J. C., & Nerisson, P. (2000). Litter on the sea floor along European coasts. *Marine Pollution Bulletin, 40*(6), 516–527. https://doi.org/10.1016/S0025-326X(99)00234-9

Gall, S. C., & Thompson, R. C. (2015). The impact of debris on marine life. *Marine Pollution Bulletin, 92*(1–2), 170–179. https://doi.org/10.1016/j.marpolbul.2014.12.041

GESAMP. (2015). Sources, fate, and effects of microplastics in the marine environment: A global assessment. Joint group of experts on the scientific aspects of marine P. J. Kershaw (Ed.). *Environmental Protection* IMO/FAO/UNESCO-IOC/UNIDO/ WMO/IAEA/UN/UNEP/UNDP. Rep. Stud. GESAMP No. 90. London, UK.

Geyer, R., Jambeck, J. R., & Law, K. L. (2017). Production, use, and fate of all plastics ever made. *Science Advances, 3*(7), e1700782. https://doi.org/10.1126/sciadv.1700782

Good, T. P., June, J. A., Etnier, M. A., & Broadhurst, G. (2010). Derelict fishing nets in Puget Sound and the Northwest Straits: Patterns and threats to marine fauna. *Marine Pollution Bulletin, 60*(1), 39–50. https://doi.org/10.1016/j.marpolbul.2009.09.005

Gregory, M. R. (2009). Environmental implications of plastic debris in marine settings–entanglement, ingestion, smothering, hangers-on, hitch-hiking and alien invasions. *Philosophical Transactions of the Royal Society of London. Series B, Biological Sciences, 364*(1526), 2013–2025. https://doi.org/10.1098/rstb.2008.0265

Gulland, F., Danil, K., Bolton, J., Ylitalo, G., Okrucky, R. S., Rebolledo, F., Alexander-Beloch, C., Brownell, R. L., Mesnick, S., Lefebvre, K., Smith, C. R., Thomas, P. O., & Rojas-Bracho, L. (2020). Vaquitas (*Phocoena sinus*) continue to die from bycatch not pollutants. *Veterinary Record, 187*(7), e51. https://doi.org/10.1136/vr.105949

Hamilton, S., & Baker, G. B. (2019). Technical mitigation to reduce marine mammal bycatch and entanglement in commercial fishing gear: Lessons learnt and future directions. *Reviews in Fish Biology and Fisheries, 29*(2), 223–247. https://doi.org/10.1007/s11160-019-09550-6

Hanni, K. D., & Pyle, P. (2000). Entanglement of pinnipeds in synthetic materials at South-East Farallon Island, California, 1976–1998. *Marine Pollution Bulletin, 40*(12), 1076–1081. https://doi.org/10.1016/S0025-326X(00)00050-3

Haram, L. E., Carlton, J. T., Ruiz, G. M., & Maximenko, N. A. (2020). A plasticene lexicon. *Marine Pollution Bulletin, 150*, 110714. https://doi.org/10.1016/j.marpolbul.2019.110714

Harting, A. L., Barbieri, M. M., Baker, J. D., Mercer, T. A., Johanos, T. C., Robinson, S. J., Littnan, C. L., Colegrove, K. M., & Rotstein, D. S. (2021). Population-level impacts of natural and anthropogenic causes-of-death for Hawaiian monk seals in the main Hawaiian Islands. *Marine Mammal Science, 37*(1), 235–250. https://doi.org/10.1111/mms.12742

HESIS: Health Evaluation System and Information Service. (2022). *Understanding toxic substances. An introduction to chemical hazards in the workplace.* Occupational Health Branch California Department of Public Health. https://www.cdph.ca.gov/Programs/CCDPHP/DEODC/OHB/HESIS/CDPH%20Document%20Library/introtoxsubstances.pdfRetrieved November 2, 2022. State of California Department of Public Health Department of Industrial Relations.

Hesler, M., Aengenheister, L., Ellinger, B., Drexel, R., Straskraba, S., Jost, C., Wagner, S., Meier, F., von Briesen, H., Büchel, C., Wick, P., Buerki-Thurnherr, T., & Kohl, Y. (2019). Multi-endpoint toxicological assessment of polystyrene nano- and microparticles in different biological models in vitro. Toxicology in Vitro, 61, 104610. https://doi.org/10.1016/j.tiv.2019.104610.

Ho, N. H. E., & Not, C. (2019). Selective accumulation of plastic debris at the breaking wave area of coastal waters. *Environmental Pollution, 245*, 702–710. https://doi.org/10.1016/j.envpol.2018.11.041

Hwang, J., Choi, D., Han, S., Choi, J., & Hong, J. (2019). An assessment of the toxicity of polypropylene microplastics in human derived cells. *Science of the Total Environment, 684*, 657–669. https://doi.org/10.1016/j.scitotenv.2019.05.071

IMO. (1987). Resolution MEPC.36(28) Marine Environment Protection Committee Adoption of Amendments to the Annex of the Protocol of Relating to the International Convention for the Prevention of Pollution from Ships (Amendment to Annex V of MARPOL 73/78) p. 1989. IMO.

Jacobsen, J. K., Massey, L., & Gulland, F. (2010). Fatal ingestion of floating net debris by two sperm whales (*Physeter macrocephalus*). *Marine Pollution Bulletin*, *60*(5), 765–767. https://doi.org/10.1016/j.marpolbul.2010.03.008

Jambeck, J. R., Geyer, R., Wilcox, C., Siegler, T. R., Perryman, M., Andrady, A., Narayan, R., & Law, K. L. (2015). Marine pollution. Plastic waste inputs from land into the ocean. *Science*, *347*(6223), 768–771. https://doi.org/10.1126/science.1260352

Jaramillo-Legorreta, A. M., Cardenas-Hinojosa, G., Nieto-Garcia, E., Rojas-Bracho, L., Thomas, L., Ver Hoef, J. M., Moore, J., Taylor, B., Barlow, J., & Tregenza, N. (2019). Decline towards extinction of Mexico's vaquita porpoise (*Phocoena sinus*). *Royal Society Open Science*, *6*(7), 190598. https://doi.org/10.1098/rsos.190598

Jeong, J., & Choi, J. (2019). Adverse outcome pathways potentially related to hazard identification of microplastics based on toxicity mechanisms. *Chemosphere*, *231*, 249–255. https://doi.org/10.1016/j.chemosphere.2019.05.003

Kane, I. A., Clare, M. A., Miramontes, E., Wogelius, R., Rothwell, J. J., Garreau, P., & Pohl, F. (2020). Seafloor microplastic hotspots controlled by deep-sea circulation. *Science*, *368*(6495), 1140–1145. https://doi.org/10.1126/science.aba5899

Kannan, K., & Vimalkumar, K. (2021). A review of human exposure to microplastics and insights into microplastics as obesogens. *Frontiers in Endocrinology*, *12*, 724989. https://doi.org/10.3389/fendo.2021.724989

Karlsson, T. M., Vethaak, A. D., Almroth, B. C., Ariese, F., van Velzen, M., Hasselöv, M., & Leslie, H. A. (2017). Screening for microplastics in sediment, water, marine invertebrates and fish: Method development and microplastic accumulation. *Marine Pollution Bulletin*, *122*(1–2), 403–408. https://doi.org/10.1016/j.marpolbul.2017.06.081

Knowlton, A. R., Robbins, J., Landry, S., McKenna, H. A., Kraus, S. D., & Werner, T. B. (2016). Effects of fishing rope strength on the severity of large whale entanglements. *Conservation Biology*, *30*(2), 318–328. https://doi.org/10.1111/cobi.12590

Koelmans, A. A., Besseling, E., & Shim, W. J. (2015). Nanoplastics in the aquatic environment. Critical review. In B. Litter, M., L. Gutow & M. Klages (Eds.), *Marine anthropogenic litter* (pp. 325–340). Springer. https://doi.org/10.1007/978-3-319-16510-3_12

Kühn, S., Bravo Rebolledo, E. L., & van Franeker, J. A. (2015). Deleterious effects of litter on marine life. In B. Litter, M., L. Gutow & M. Klages (Eds.), *Marine anthropogenic* (pp. 75–116). Springer. https://doi.org/10.1007/978-3-319-16510-3_4

Kzhyshkowska, J., Gudima, A., Riabov, V., Dollinger, C., Lavalle, P., & Vrana, N. E. (2015). Macrophage responses to implants: Prospects for personalized medicine. *Journal of Leukocyte Biology*, *98*(6), 953–962. https://doi.org/10.1189/jlb.5VMR0415-166R

Laist, D. W. (1997). Impacts of marine debris: Entanglement of marine life in marine debris including a comprehensive list of species with entanglement and ingestion records. In J. M. Coe & D. B. Rogers (Eds.), *Marine debris* (pp. 99–139). Springer. https://doi.org/10.1007/978-1-4613-8486-1_10

Law, K. L., Morét-Ferguson, S. E., Goodwin, D. S., Zettler, E. R., Deforce, E., Kukulka, T., & Proskurowski, G. (2014). Distribution of surface plastic debris in the eastern pacific ocean from an 11-year data set. *Environmental Science and Technology*, *48*(9), 4732–4738. https://doi.org/10.1021/es4053076

Lawson, T. J., Wilcox, C., Johns, K., Dann, P., & Hardesty, B. D. (2015). Characteristics of marine debris that entangle Australian fur seals (Arctocephalus pusillus doriferus) in southern Australia. Marine Pollution Bulletin, 98(1–2), 354–357. https://doi.org/10.1016/j.marpolbul.2015.05.053

Lebreton, L. C. M., Greer, S. D., & Borrero, J. C. (2012). Numerical modelling of floating debris in the world's oceans. *Marine Pollution Bulletin*, *64*(3), 653–661. https://doi.org/10.1016/j.marpolbul.2011.10.027

Lebreton, L., Slat, B., Ferrari, F., Sainte-Rose, B., Aitken, J., Marthouse, R., Hajbane, S., Cunsolo, S., Schwarz, A., Levivier, A., Noble, K., Debeljak, P., Maral, H., Schoeneich-Argent, R., Brambini, R., & Reisser, J. (2018). Evidence that the Great Pacific Garbage Patch is rapidly accumulating plastic. *Scientific Reports*, *8*(1), 4666. https://doi.org/10.1038/s41598-018-22939-w

Lechthaler, S., Waldschläger, K., Stauch, G., & Schüttrumpf, H. (2020). The way of macroplastic through the environment. *Environments, 7*(10), 73. https://doi.org/10.3390/environments7100073

Lewison, R. L., Crowder, L. B., Read, A. J., & Freeman, S. A. (2004). Understanding impacts of fisheries bycatch on marine megafauna. *Trends in Ecology and Evolution, 19*(11), 598–604. https://doi.org/10.1016/j.tree.2004.09.004

Li, W. C., Tse, H. F., & Fok, L. (2016). Plastic waste in the marine environment: A review of sources, occurrence and effects. *Science of the Total Environment, 566–567*, 333–349. https://doi.org/10.1016/j.scitotenv.2016.05.084

Lu, L., Wan, Z., Luo, T., Fu,s Z., & Jin, Y. (2018). Polystyrene microplastics induce gut microbiota dysbiosis and hepatic lipid metabolism disorder in mice. *Science of the Total Environment, 631–632*, 449–458. https://doi.org/10.1016/j.scitotenv.2018.03.051

Lusher, A. L., McHugh, M., & Thompson, R. C. (2013). Occurrence of microplastics in the gastrointestinal tract of pelagic and demersal fish from the English Channel. *Marine Pollution Bulletin, 67*(1–2), 94–99. https://doi.org/10.1016/j.marpolbul.2012.11.028

Lusher, A. L., Hernandez-Milian, G., O'Brien, J., Berrow, S., O'Connor, I., & Officer, R. (2015). Microplastic and macroplastic ingestion by a deep diving, oceanic cetacean: The True's beaked whale *Mesoplodon mirus. Environmental Pollution, 199*, 185–191. https://doi.org/10.1016/j.envpol.2015.01.023

MARPOL. (1973) Retrieved December 31, 1988. https://www.imo.org/en/about/Conventions/Pages/International-Convention-for-the-Prevention-of-Pollution-from-Ships-(MARPOL).aspx: International Convention for the Prevention of Pollution from Ships, Convention; Annex V modification: Prevention of Pollution by Garbage from Ships

Masó, M., Garcés, E., Pagés, F., & Camp, J. (2003). Drifting plastic debris as a potential vector for dispersing Harmful Algal Bloom (HAB) species. *Scientia Marina, 67*(1), 107–111. https://doi.org/10.3989/scimar.2003.67n1107

Mato, Y., Isobe, T., Takada, H., Kanehiro, H., Ohtake, C., & Kaminuma, T. (2001). Plastic resin pellets as a transport medium for toxic chemicals in the marine environment. *Environmental Science and Technology, 35*(2), 318–324. https://doi.org/10.1021/es0010498

Matsuguma, Y., Takada, H., Kumata, H., Kanke, H., Sakurai, S., Suzuki, T., Itoh, M., Okazaki, Y., Boonyatumanond, R., Zakaria, M. P., Weerts, S., & Newman, B. (2017). Microplastics in sediment

cores from Asia and Africa as indicators of temporal trends in plastic pollution. *Archives of Environmental Contamination and Toxicology, 73*(2), 230–239. https://doi.org/10.1007/s00244-017-0414-9

Mattsson, K., Jocic, S., Doverbratt, I., & Hansson, L. (2018). Nanoplastics in the aquatic environment. In Microplastic Contamination in Aquatic Environments, (pp. 379–399).

Miller, M. E., Hamann, M., & Kroon, F. J. (2020). Bioaccumulation and biomagnification of microplastics in marine organisms: A review and meta-analysis of current data. *PLOS ONE, 15*(10), e0240792. https://doi.org/10.1371/journal.pone.0240792

Montoto-Martínez, T., De la Fuente, J., Puig-Lozano, R., Marques, N., Arbelo, M., Hernández-Brito, J. J., Fernández, A., & Gelado-Caballero, M. D. (2021). Microplastics, bisphenols, phthalates and pesticides in odontocetes species in the Macronesian Region (eastern North Atlantic). *Marine Pollution Bulletin, 173*, 113105. https://doi.org/10.1016/j.marpolbul.2021.113105

Moore, C. J., Moore, S. L., Leecaster, M. K., & Weisberg, S. B. (2001). A comparison of plastic and plankton in the North Pacific central gyre. *Marine Pollution Bulletin, 42*(12), 1297–1300. https://doi.org/10.1016/S0025-326X(01)00114-X

Moore, C. J. (2008). Synthetic polymers in the marine environment: A rapidly increasing, long-term threat. *Environmental Research, 108*(2), 131–139. https://doi.org/10.1016/j.envres.2008.07.025

Moore, M. J., Knowlton, A. R., Kraus, S. D., McLellan, W. A., & Bonde, R. K. (2005). Morphometry, gross morphology and available histopathology in North Atlantic right whale (*Eubalaena glacialis*) mortalities (1970–2002). *J. Cetacean Res. Manage, 6*(3), 199–214. https://doi.org/10.47536/jcrm.v6i3.762

Moore, M. J., & van der Hoop, J. M. (2012). The painful side of trap and fixed net fisheries: Chronic entanglement of large whales. *Journal of Marine Biology, 2012*, article ID 230653. https://doi.org/10.1155/2012/230653

Moore, R. C., Loseto, L., Noel, M., Etemadifar, A., Brewster, J. D., MacPhee, S., Bendell, L., & Ross, P. S. (2020). Microplastics in beluga whales (*Delphinapterus leucas*) from the eastern Beaufort Sea. *Marine Pollution Bulletin, 150*, 110723. https://doi.org/10.1016/j.marpolbul.2019.110723

Myers, H. J., Moore, M. J., Baumgartner, M. F., Brillant, S. W., Katona, S. K., Knowlton, A. R., Morissette, L., Pettis, H. M., Shester, G., & Werner, T. B. (2019). Ropeless fishing to prevent large whale entanglements: Ropeless consortium report. *Marine Policy, 107*, 103587. https://doi.org/10.1016/j.marpol.2019.103587

NOAA (National Oceanic and Atmospheric Administration)-a. https://marinedebris.noaa.gov/info/patch.html. United States Department of Commerce, National Oceanic and Atmospheric Administration, Marine Debris Program.

NOAA (National Oceanic and Atmospheric Administration)-b). United States Department of Commerce, National Oceanic and Atmospheric Administration, fishing gear: Bycatch. https://www.fisheries.noaa.gov/national/bycatch/fishing-gear-gillnets

Nelms, S. E., Galloway, T. S., Godley, B. J., Jarvis, D. S., & Lindeque, P. K. (2018). Investigating microplastic trophic transfer in marine top predators. *Environmental Pollution, 238*, 999–1007. https://doi.org/10.1016/j.envpol.2018.02.016

Nelms, S. E., Barnett, J., Brownlow, A., Davison, N. J., Deaville, R., Galloway, T. S., Lindeque, P. K., Santillo, D., & Godley, B. J. (2019). Microplastics in marine mammals stranded around the British coast: Ubiquitous but transitory?. *Scientific Reports, 9*(1), 1075. https://doi.org/10.1038/s41598-018-37428-3

Neves, D., Sobral, P., Ferreira, J. L., & Pereira, T. (2015). Ingestion of microplastics by commercial fish off the Portuguese coast. *Marine Pollution Bulletin, 101*(1), 119–126. https://doi.org/10.1016/j.marpolbul.2015.11.008

Oliveira, M., Ribeiro, A., Hylland, K., & Guilhermino, L. (2013). Single and combined effects of microplastics and pyrene on juveniles (0+ group) of the common goby *Pomatoschistus microps* (Teleostei, Gobiidae). *Ecological Indicators, 34*, 641–647. https://doi.org/10.1016/j.ecolind.2013.06.019

Pace, R. M. I. I. I., Cole, T. V. N., & Henry, A. G. (2014). Incremental fishing gear modifications fail to significantly reduce large whale serious injury rates. *Endangered Species Research, 26*(2), 115–126. https://doi.org/10.3354/esr00635

Page, B., McKenzie, J., McIntosh, R., Baylis, A., Morrissey, A., Calvert, N., Haase, T., Berris, M., Dowie, D., Shaughnessy, P. D., & Goldsworthy, S. D. (2004). Entanglement of Australian sea lions and New Zealand fur seals in lost fishing gear and other marine debris before and after Government and industry attempts to reduce the problem. *Marine Pollution Bulletin, 49*(1–2), 33–42. https://doi.org/10.1016/j.marpolbul.2004.01.006

Page-Karjian, A., Lo, C. F., Ritchie, B., Harms, C. A., Rotstein, D. S., Han, S., Hassan, S. M., Lehner, A. F., Buchweitz, J. P., Thayer, V. G., Sullivan, J. M., Christiansen, E. F., & Perrault, J. R. (2020). Anthropogenic contaminants and histopathological findings in stranded cetaceans in the southeastern United States, 2012–2018. *Frontiers in Marine Science, 7*. https://doi.org/10.3389/fmars.2020.00630

Pham, C. K., Ramirez-Llodra, E., Alt, C. H. S., Amaro, T., Bergmann, M., Canals, M., Company, J. B., Davies, J., Duineveld, G., Galgani, F., Howell, K. L., Huvenne, V. A., Isidro, E., Jones, D. O., Lastras, G., Morato, T., Gomes-Pereira, J. N., Purser, A., Stewart, H., . . . Tyler, P. A. (2014). Marine litter distribution and density in European seas, from the shelves to deep basins. *PLOS ONE, 9*(4), e95839. https://doi.org/10.1371/journal.pone.0095839

Philipp, C., Unger, B., Ehlers, S. M., Koop, J. H. E., & Siebert, U. (2021). First evidence of retrospective findings of microplastics in harbour porpoises (*Phocoena phocoena*) from German waters. *Frontiers in Marine Science, 8*, 682532. https://doi.org/10.3389/fmars.2021.682532

Pohl, F., Eggenhuisen, J. T., Kane, I. A., & Clare, M. A. (2020). Transport and burial of microplastics in deep-marine sediments by turbidity currents. *Environmental Science and Technology, 54*(7), 4180–4189. https://doi.org/10.1021/acs.est.9b07527

Qiao, R., Sheng, C., Lu, Y., Zhang, Y., Ren, H., & Lemos, B. (2019). Microplastics induce intestinal inflammation, oxidative stress, and disorders of metabolome and microbiome in zebrafish. *Science of the Total Environment, 662*, 246–253. https://doi.org/10.1016/j.scitotenv.2019.01.245

Radisic, V., Nimje, P. S., Bienfait, A. M., & Marathe, N. P. (2020). Marine plastics from Norwegian west coast carry potentially virulent fish pathogens and opportunistic human pathogens harboring new variants of antibiotic resistance genes. *Microorganisms, 8*(8), 1200. https://doi.org/10.3390/microorganisms8081200

Rainieri, S., Conlledo, N., Larsen, B. K., Granby, K., & Barranco, A. (2018). Combined effects of microplastics and chemical contaminants on the organ toxicity of zebrafish (*Danio rerio*). *Environmental Research, 162*, 135–143. https://doi.org/10.1016/j.envres.2017.12.019

Rangel-Buitrago, N., Neal, W., & Williams, A. (2022). The plasticene: Time and rocks. *Marine*

Pollution Bulletin, 185(B), 114358. https://doi.org/10.1016/j.marpolbul.2022.114358

Raum-Suryan, K. L., Jemison, L. A., & Pitcher, K. W. (2009). Entanglement of Steller sea lions (*Eumetopias jubatus*) in marine debris: Identifying causes and finding solutions. *Marine Pollution Bulletin, 58*(10), 1487–1495. https://doi.org/10.1016/j.marpolbul.2009.06.004

Read, A. J., Reynolds, J. E., Perrin, W. F., Reeves, R. R., Montgomery, S., & Ragen, T. J. (2005). Bycatch and depredation. In T. J. Ragen, J. E. Reynolds, W. F. Perrin, R. R. Reeves & S. Montgomery (Eds.), *Marine mammal research: Conservation beyond crisis* (pp. 5–17). The John Hopkins University Press.

Read, A. J., Drinker, P., & Northridge, S. (2006). Bycatch of marine mammals in US and global fisheries. *Conservation Biology, 20*(1), 163–169. https://doi.org/10.1111/j.1523-1739.2006.00338.x

Read, A. J. (2008). The looming crisis: Interactions between marine mammals and fisheries. *Journal of Mammalogy, 89*(3), 541–548. https://doi.org/10.1644/07-MAMM-S-315R1.1

Rech, S., Thiel, M., Borrell Pichs, Y. J., & García-Vazquez, E. (2018). Travelling light: Fouling biota on macroplastics arriving on beaches of remote Rapa Nui (Easter Island) in the South Pacific Subtropical gyre. *Marine Pollution Bulletin, 137*, 119–128. https://doi.org/10.1016/j.marpolbul.2018.10.015

Reeves, R. R., McClellan, K., & Werner, T. B. (2013). Marine mammal bycatch in gillnet and other entangling net fisheries, 1990 to 2011. *Endangered Species Research, 20*(1), 71–97. https://doi.org/10.3354/esr00481

Reinert, T. R., Spellman, A. C., & Bassett, B. L. (2017). Entanglement in and ingestion of fishing gear and other marine debris by Florida manatees, 1993 to 2012. *Endangered Species Research, 32*, 415–427. https://doi.org/10.3354/esr00816

Richardson, K., Hardesty, B. D., & Wilcox, C. (2019). Estimates of fishing gear loss rates at a global scale: A literature review and meta-analysis. *Fish and Fisheries, 20*(6), 1218–1231. https://doi.org/10.1111/faf.12407

Rogers, K. L., Carreres-Calabuig, J. A., Gorokhova, E., & Posth, N. R. (2020). Micro-by-micro interactions: How microorganisms influence the fate of marine microplastics. *Limnology and Oceanography Letters, 5*(1), 18–36. https://doi.org/10.1002/lol2.10136

Rojas-Bracho, L., Taylor, B., Booth, C., Thomas, L., Jaramillo-Legorreta, A., Nieto-García, E., Cárdenas Hinojosa, G., Barlow, J., Mesnick, S., Gerrodette, T., Olson, P., Henry, A., Rizo, H., Hidalgo-Pla, E., & Bonilla-Garzón, A.. (2022). More vaquita porpoises survive than expected. *Endangered Species Research, 48*, 225–234. https://doi.org/10.3354/esr01197

Roman, L., Schuyler, Q., Wilcox, C., & Hardesty, B. D. (2021). Plastic pollution is killing marine megafauna, but how do we prioritize policies to reduce mortality? *Conservation Letters, 14*(2), e12781. https://doi.org/10.1111/conl.12781

Royer, S. J., Ferrón, S., Wilson, S. T., & Karl, D. M. (2018). Production of methane and ethylene from plastic in the environment. *PLOS ONE, 13*(8), e0200574. https://doi.org/10.1371/journal.pone.0200574

Rummel, C. D., Loder, M.G.J., Fricke. NF., Lang, T., Griebeler, E.M., Janke, M., Gerdts, G., 2016. Plastic ingestion by pelagic and demersal fish from the North Sea and Baltic Sea. Mar Poll Bull 102, 134e141.

Ryan, P. G. (2020). The transport and fate of marine plastics in South Africa and adjacent oceans. *South African Journal of Science, 116*(5/6). https://doi.org/10.17159/sajs.2020/7677

Ryberg, M., Laurent, A., & Hauschild, M. (2018). *Mapping of global plastic value chain and plastics losses to the environment, with a particular focus on marine environment.* United Nations Environment Programme (UN Environmental Program).

Ryberg, M. W., Hauschild, M. Z., Wang, F., Averous-Monnery, S., & Laurent, A. (2019). Global environmental losses of plastics across their value chains. *Resources, Conservation and Recycling, 151*, 104459. https://doi.org/10.1016/j.resconrec.2019.104459

SAPEA: Science Advice for Policy by Europen Academies. (2019). A scientific perspective on microplastics in nature and society. https://sapea.info/topic/microplastics/

Schirinzi, G. F., Pérez-Pomeda, I., Sanchís, J., Rossini, C., Farré, M., & Barceló, D. (2017). Cytotoxic effects of commonly used nanomaterials and microplastics on cerebral and epithelial human cells. *Environmental Research, 159*, 579–587. https://doi.org/10.1016/j.envres.2017.08.043

Sharp, S. M., McLellan, W. A., Rotstein, D. S., Costidis, A. M., Barco, S. G., Durham, K. et al. (2019). Gross and histopathologic diagnoses from North Atlantic right whale *Eubalaena glacialis* mortalities between 2003 to 2018. Dis aqua Org, *135*, 1–31.

Singh, B., & Sharma, N. (2008). Mechanistic implications of plastic degradation. *Polymer Degradation and Stability*, *93*(3), 561–584. https://doi.org/10.1016/j.polymdegradstab.2007.11.008

Skirtun, M., Sandra, M., Strietman, W. J., van den Burg, S. W. K., De Raedemaecker, F., & Devriese, L. I. (2022) Plastic pollution pathways from marine aquaculture practices and potential solutions for the North-East Atlantic region. *Marine Pollution Bulletin*, *174*, 113178. https://doi.org/10.1016/j.marpolbul.2021.113178

Smith, M., Love, D. C., Rochman, C. M., & Neff, R. A. (2018). microplastics in seafood and the implications for human health. *Current Environmental Health Reports*, *5*(3), 375–386. https://doi.org/10.1007/s40572-018-0206-z

Sorensen, R. M., & Jovanović, B. (2021). From nanoplastic to microplastic: A bibliometric analysis on the presence of plastic particles in the environment. *Marine Pollution Bulletin*, *163*, 111926. https://doi.org/10.1016/j.marpolbul.2020.111926

Stamper, M. A., Whitaker, B. R., & Schofield, T. D. (2006). Case study: Morbidity in a pygmy sperm whale, *Kogia breviceps*, due to ocean-bourne plastic. Mar mam sci. *Marine Mammal Science*, *22*(3), 719–722. https://doi.org/10.1111/j.1748-7692.2006.00062.x

Stock, V., Böhmert, L., Lisicki, E., Block, R., Cara-Carmona, J., Pack, L. K., Selb, R., Lichtenstein, D., Voss, L., Henderson, C. J., Zabinsky, E., Sieg, H., Braeuning, A., & Lampen, A. (2019). Uptake and effects of orally ingested polystyrene microplastic particles in vitro and in vivo. *Archives of Toxicology*, *93*(7), 1817–1833. https://doi.org/10.1007/s002 04-019-02478-7

Tanaka, K., & Takada, H. (2016). Microplastic fragments and microbeads in digestive tracts of planktivorous fish from urban coastal waters. Nature Sci. Rep. 6. *Scientific Reports*, *6*, 34351. https://doi.org/10.1038/srep34351

Teuten, E. L., Saquing, J. M., Knappe, D. R. U., Barlaz, M. A., Jonsson, S., Björn, A., Rowland, S. J., Thompson, R. C., Galloway, T. S., Yamashita, R., Ochi, D., Watanuki, Y., Moore, C., Viet, P. H., Tana, T. S., Prudente, M., Boonyatumanond, R., Zakaria, M. P., Akkhavong, K., . . . Takada, H. (2009). Transport and release of chemicals from plastics to the environment and to wildlife. *Philosophical Transactions of the Royal Society of London. Series B,*

Biological Sciences, *364*(1526), 2027–2045. https://doi.org/10.1098/rstb.2008.0284

Thompson, R. C., Olsen, Y., Mitchell, R. P., Davis, A., Rowland, S. J., John, A. W. G., McGonigle, D., & Russell, A. E. (2004). Lost at sea: Where is all the plastic? *Science*, *304*(5672), 838. https://doi.org/10.1126/science.1094559

Thompson, R. C. (2015). Microplastics in the marine environment: Sources, consequences and solutions. In M. Bergman, L. Gutow & M. Klages (Eds.), *Marine anthropogenic litter* (pp. 185–200). Springer.

Van Cauwenberghe, L., & Janssen, C. R. (2014). Microplastics in bivalves cultured for human consumption. *Environmental Pollution*, *193*, 65–70. https://doi.org/10.1016/j.envpol.2014.06.010.

van Franeker, J. A., Bravo Rebolledo, E. L., & Hesse, E., IJsseldijk, L. L., Kühn, S., Leopold, M., et al. 2018. Plastic ingestion by harbour porpoises *Phocoena phocoena* in the Netherlands: Establishing a standardized method. *Ambio*, *47*, 387–397. https://doi.org/10.1007/s13280-017-1002-y

van Raamsdonk, L. W. D., van der Zande, M., Koelmans, A. A., Hoogenboom, R. L. A. P., Peters, R. J. B., Groot, M. J., Peijnenburg, A. A. C. M., & Weesepoel, Y. J. A. (2020). Current insights into monitoring, bioaccumulation, and potential health effects of microplastics present in the food chain. *Foods*, *9*(1), 9010072. https://doi.org/10.3390/foods9010072

van Sebille, E., Wilcox, C., Lebreton, L., Maximenko, N., Hardesty, B. D., Van Franeker, J. A., Eriksen, M., Siegel, D., Galgani, F., & Law, K. L. (2015). A global inventory of small floating plastic debris. *Environmental Research Letters*, *10*(12), 124006. https://doi.org/10.1088/1748-9326/10/12/124006

van Sebille, E., Aliani, S., Law, K. L., Maximenko, N., Alsina, J. M., Bagaev, A., Bergmann, M., Chapron, B., Chubarenko, I., Cózar, A., Delandmeter, P., Egger, M., Fox-Kemper, B., Garaba, S. P., Goddijn-Murphy, L., Hardesty, B. D., Hoffman, M. J., Isobe, A., Jongedijk, C. E., . . . Wichmann, D. (2020). The physical oceanography of the transport of floating marine debris. *Environmental Research Letters*, *15*(2), 023003. https://doi.org/10.1088/1748-9326/ab6d7d

Walker, W. A., & Coe, J. M. (1989). Survey of marine debris ingestion by odontocete cetaceans. In R. S. Shomura & M. L. Godfrey (Eds.). *Proceedings of the 2nd International Conference on Marine Debris, 1* (pp. 747–774). *United States Department of Commerce.*

Wang, B., & Li, Y. (2021). Plastic bag usage and the policies: A case study of China. *Waste Management, 126*(8), 163–169. https://doi.org/10.1016/j.wasman.2021.03.010

Waring, G. T., Josephson, E., Maze-Foley, K., & Rosel, P. E. (2016). US Atlantic and Gulf of Mexico. Marine mammal stock assessments, *2015*. NOAA Technical Memorandum NMFS-NE-238.

Wieczorek, A. M., Morrison, L., Croot, P. L., Allcock, A. L., MacLoughlin, E., Savard, O., Brownlow, H., & Doyle, T. K. (2018). Frequency of microplastics in mesopelagic fishes from the northwest Atlantic. *Frontiers in Marine Science, 5.* https://doi.org/10.3389/fmars.2018.00039

Willis, K., Maureaud, C., Wilcox, C., & Hardesty, B. D. (2018). How successful are waste abatement campaigns and government policies at reducing plastic waste into the marine environment? *Marine Policy, 96,* 243–249. https://doi.org/10.1016/j.marpol.2017.11.037

Wright, S. L., & Kelly, F. J. (2017). Plastic and human health: A micro issue? *Environmental Science and Technology, 51*(12), 6634–6647. https://doi.org/10.1021/acs.est.7b00423

Wu, B., Wu, X., Liu, S., Wang, Z., & Chen, L. (2019). Size-dependent effects of polystyrene microplastics on cytotoxicity and efflux pump inhibition in human Caco-2 cells. *Chemosphere, 221,* 333–341. https://doi.org/10.1016/j.chemosphere.2019.01.056

Zantis, L. J., Carroll, E. L., Nelms, S. E., & Bosker, T. (2021). Marine mammals and microplastics: A systemic review and call for standardization. *Environmental Policy and Law, 269,* 116142. https://doi.org/10.1016/j.envpol.2020.116142

Zettler, E. R., Mincer, T. J., & Amaral-Zettler, L. A. (2013). Life in the "Plastisphere": Microbial communities on plastic marine debris. *Environmental Science and Technology, 47*(13), 7137–7146. https://doi.org/10.1021/es401288x

Zhang, E., Kim, M., Rueda, L., Rochman, C., VanWormer, E., Moore, J., & Shapiro, K. (2022). Association of zoonotic protozoan parasites with microplastics in seawater and implications for human and wildlife health. Nature Sci. Rep. 12. *Scientific Reports, 12*(1), 6532. https://doi.org/10.1038/s41598-022-10485-5

Ziccardi, L. M., Edgington, A., Hentz, K., Kulacki, K. J., & Kane Driscoll, S. (2016). Microplastics as vectors for bioaccumulation of hydrophobic organic chemicals in the marine environment: A state-of-the-science review. *Environmental Toxicology and Chemistry, 35*(7), 1667–1676. https://doi.org/10.1002/etc.3461

Chapter 18

Beaver reintroductions in Scotland: Disease risk assessment, welfare and population growth

Simon J. Girling and Roisin Campbell-Palmer

Abstract

Beavers are well known as environmental engineers, modifying their habitat with subsequent benefits for biodiversity and sustainability of wetland areas. With growing concern about habitat protection and a global biodiversity crisis, there has been more support for the reintroduction of the Eurasian beaver (*Castor fiber*) across Europe and its former native range. At the same time, there have been concerns about the potential introduction of disease and threats to other species which are endemic. In this chapter we consider the reintroduction of the beaver into Scotland and discuss disease risks, welfare and population growth associated with the process. The importance of robust assessments of the health and welfare of beavers reintroduced has been vital to the success of the project, which has made this the first native/ extinct mammalian species to be successfully reintroduced to Britain.

18.1 Introduction

Beavers can significantly modify freshwater ecosystems to suit their needs and, by doing so, increase wetland habitat and its associated biodiversity. The Eurasian beaver (*Castor fiber*) is recognized as a significant positive ecological engineer increasing plant, invertebrate and aquatic diversity in particular (Brazier et al., 2021; Law et al., 2016; Rosell et al., 2005; Stringer and Gaywood, 2016). As such, it is listed as a species that member states of the European Union should consider its reintroduction where evidence indicates its previous existence under the

'Habitats Directive' (Council of Europe Directive 92/43/EEC on the Conservation of Natural Habitats and of Wild Fauna and Flora (Annex II and Annex IV(a)) and the Bern Convention, Annex III (the Convention on the Conservation of Flora, Wildlife and Natural Habitats). As a consequence, the Eurasian beaver population has largely recovered across much of its former continental European range through various conservation measures, including legal protection as a European Protected Species and planned official and unofficial releases (Nolet and Rosell, 1998). Britain represents the furthest western range of the Eurasian beaver, where they were

Susan C. Cork and Douglas P. Whiteside (eds) **Case Studies in EcoHealth**
DOI: 10.52517/9781789183368.018, © 5m Books Ltd, 2024

hunted to extinction by the mid-16th century (Coles, 2006).

Beaver translocations within Europe date from 1922, when beavers were reintroduced to Sweden from Norway; since then at least 203 translocations to distinct sites have been recorded outside the former Soviet Union (Halley et al., 2021). However, despite these numerous Eurasian beaver reintroduction programmes in Europe outside of Britain, surprisingly little has been published on the health surveillance, disease, welfare and mortality associated with reintroductions, and no formal disease risk assessment for release has been published (Campbell-Palmer et al., 2021; Goodman et al., 2012a; Nolet et al., 1997).

Figure 18.1 Eurasian beaver (*C. fiber*) eating bracken. Photo: Elliot McCandless.

Figure 18.2 Eurasian beaver (*C. fiber*) swimming at Argaty, Doune, near Stirling, Scotland. Photo: Elliot McCandless.

> Beavers (*Castor* spp.) are large, semi-aquatic rodents from the Northern Hemisphere. There are two species: the Eurasian beaver (*C. fiber*) (Figure 18.1) and the North American beaver (*C. canadensis*). Beavers have stout bodies with large heads, long incisors, typically brown fur (can range from blond to black), dextrous front feet, webbed hind feet and a characteristically flattened tail. Eurasian beavers swim well (Figure 18.2) and can be found in a number of freshwater habitats including rivers, streams, ponds and lakes. They are herbivorous, consuming tree cambium, grasses, aquatic plants and sedges. Beavers typically build dams and lodges using tree branches and vegetation. They can fell sizeable trees as well as saplings with their teeth to make these structures. Beavers are considered 'keystone' species because their activities create and maintain the wetland habitats needed to support many other species.

Scotland represents the first devolved British nation to seriously investigate beaver restoration. In October 2007 a public consultation was carried out, which responded positively to a widespread reintroduction of the species. However, concerns raised from certain stakeholders meant eventually that a scientific trial release was chosen initially to demonstrate, among other things, whether a wider reintroduction could be undertaken successfully without risk to the environment, wildlife, domestic animals and humans. This project was called the Scottish Beaver Trial (SBT) and, sanctioned by the Scottish Government, was carried out under licence granted by NatureScot (then Scottish Natural Heritage), by the Royal Zoological Society of Scotland (RZSS) and the Scottish Wildlife Trust. With permission from

Forestry and Land Scotland, Knapdale Forest, in mid-Argyll, was selected as an unenclosed trial site with five families of beavers released in May 2009 marking the start of a 5-year trial. Wild beavers were imported from Norway by the RZSS and underwent a strict quarantine and health-screening process. Beavers were sourced from Norway in order to follow International Union for Conservation of Nature (IUCN) best-practice guidelines because it was believed, at that time, that those found historically in Scotland more closely resembled the current Scandinavian Eurasian wild beaver and would share climatic adaptations (Kitchener and Lynch, 2000; IUCN/SSC, 2013). The source country of Norway, specifically the Telemark region, was also chosen because this was officially free of two zoonotic parasites, *Echinococccus multilocularis* and *Giardia lamblia*, as well as the salmon fluke *Gyrodactylus salsii*, making it less likely that these significant pathogens could be inadvertently introduced to the release site in Scotland.

Although there was much hype around SBT, it became increasingly evident earlier that unauthorized wild beaver populations were identified living around the River Tay catchment in eastern Scotland. One of the first surveys identified 38 beaver groups (Campbell et al., 2012), whereas figures 10 years on suggest that a minimum of 251 active beaver territories now exist (Campbell-Palmer et al., 2021). These wild populations, of unknown source and health status, posed a potential disease threat to native wildlife, domestic animals and humans. These threats have been largely resolved at the time of writing through subsequent health testing and genetic screening. This became particularly relevant in light of an imported captive Eurasian beaver in England being diagnosed at post-mortem in 2010 with *E. multilocularis*, a significant human zoonotic pathogen and legally notifiable disease and one from which Britain is currently free (Barlow

et al., 2011). In addition, because the Tayside population was not part of any monitoring programme, there was no assessment as to whether the released animals were coping with the environment into which they had been released.

The official trial licensed by the Scottish Government (SBT) and the assessment of the Tayside population, a now significant wild population across eastern Scotland, provided the dataset required to assess the disease risk, health, welfare and overall ecological impact of the reintroduction of the Eurasian beaver to Scotland.

18.2 Pre-release health-screening Scottish Beaver Trial

The source population for the SBT came from the Telemark region, Norway. Twenty-seven wild beavers were trapped under licence by the Norwegian Government and then underwent 30 days' quarantine in Norway. This was then followed by a further 6 months' quarantine after importation to the UK as dictated at the time by national legislation (Rabies Importation Order 1974). During this protracted quarantine period the beavers underwent health screening. This comprised a full physical examination by a veterinary surgeon, including assessment of pelage, body score, weight, dentition and cardiovascular and respiratory systems; full biochemical and haematological analysis; and specific pathogen testing as outlined in Table 18.1 below.

Giardia spp., *Franciscella tularensis* and *Cryptosporidium parvum* are zoonotic diseases and have been associated with many rodent species, including *Castor* spp., and are easily transmitted via faecal contamination of water (Ahlen, 2001; Bajer et al., 1997, 2008). *Leptospira* spp. and *Yersinia* spp. have both been associated with beaver mortalities in previous

Table 18.1 Diagnostic tests used to screen for pathogens while the beavers were in quarantine for the SBT

Pathogens tested for during SBT quarantine	Method
Cryptosporidium parvum	Faecal acid-fast staining
Giardia spp.	Faecal SNAP® *Giardia* immunoassay kit (IDDEX Labs., Wetherby, UK)
Nematodes Cestodes Trematodes	Faecal saturated salt solution with flotation for nematodes/cestodes and sedimentation for trematodes
Franciscella tularensis	ELISA serological testing PCR testing on blood samples
Leptospira spp.	Serological microscopic agglutination test for six pools of *Leptospira* spp.
Salmonella spp.	Faecal culture with *Salmonella* spp. on selective media
Campylobacter spp.	Faecal culture with *Campylobacter* spp. on selective media
Yersinia spp.	Faecal culture with *Yersinia* spp. on selective media. Serological agglutination test for *Y. pseudotuberculosis* and *Y. enterocolitica*

reintroduction projects and wild beavers, and so the screening programme paid particular interest to these pathogens (Marreros et al., 2018; Nolet et al., 1997).

The screening outlined above failed to demonstrate evidence of any zoonotic parasites or bacteria with the exception of exposure to *Leptospira* spp., although this was not associated with mortalities. The lengthy quarantine period also showed that Eurasian beavers responded poorly to prolonged captivity post importation, required under the legislation, resulting in a 20% UK quarantine mortality rate (Goodman et al., 2012a). The six beavers that died during quarantine (of 27 beavers imported) had no specific pathogen identified, and this led to a recommendation to reduce the length of quarantine to the IUCN guidelines of 35 days after the author (Girling) produced a risk assessment accepted by Scottish Government (Goodman et al., 2012a).

The health assessment of beavers during the initial stages of the trial identified previously

There are a number of endemic strains of *Leptospira* spp. that can cause leptospirosis in a range of domestic animals, as well as humans. Not all strains cause disease. Some wild animals, especially rodents, may act as reservoirs of *Leptospira* spp. and carry the organism without showing any clinical signs of disease.

Echinococcosis due to *E. multilocularis* is endemic to many areas of Europe. It is not currently found in the UK and is a significant zoonosis and, as such, a notifiable disease under UK and EU legislation. Beavers may act as an occasional intermediate host.

Giardia spp. is a zoonotic parasite that may infect the small intestine of most mammals. Because it is easily disseminated via water, beavers may become infected and also pass on infection to other species.

unreported issues such as audible flow murmurs during the cardiac contraction cycle. These were subsequently determined as 'normal' for the species due to the subsequent testing of wild beavers in Norway by the team involved in the SBT (Devine et al., 2011). In addition, to give context to those findings of the quarantined SBT beavers, wild Norwegian beavers were also assessed in situ and showed no evidence of significant zoonotic diseases such as haemoparasites, *Leptospira* spp. or Hantavirus (Cross et al., 2012; Girling et al., 2019a,b).

18.3 Post-release health-screening Scottish Beaver Trial

The first 16 imported beavers were released into the trial site in 2009/2010. The trial lasted 5 years, during which each individual was monitored as closely as possible via regular observations and remote camera traps to confirm both presence and general body condition. Annual trapping to check physical condition, microchip any new offspring, collect faecal and blood samples, along with post-mortem examinations when relevant, were all undertaken on as many animals as possible to monitor the health and welfare status of the population. During such routine animal management and health screening the host-specific beaver beetle, *Platypsyllus castoris*, an obligate commensal, was discovered on one of the nine kits born into the trial by 2012 (Duff et al., 2013). Subsequently, after 5 years of various scientific monitoring programmes and much deliberation, the Scottish Government permitted the beavers to remain. After several years, an augmentation phase to bolster genetic diversity of the nine remaining beavers was undertaken. New individuals, including wild Scottish and captive bred individuals – both presumed to be British born – were trapped, health screened as per the previous assessments and transferred to the trial site in Knapdale, Argyll

(Dowse et al., 2020). All beavers previously tested remained disease and parasite free, with the exception of the parasite *Stichorchis subtriquetrus*, a beaver-specific intestinal trematode that is not considered pathogenic and was now obviously completing its life cycle in the wild (Campbell-Palmer et al., 2021; Dowse et al., 2020). Those beavers translocated for the augmentation phase of this project demonstrated no significant pathogens, including an absence of *Salmonella*, *Giardia*, *Cryptosporidium* and *Leptospira* spp. (Campbell-Palmer et al., 2021; Dowse et al., 2020).

18.4 Health screening of wild Tayside beavers

As the presence of unauthorized beavers continued to expand and become more evident, a beaver stakeholder forum was established chaired by NatureScot including representatives from the land management, fisheries and conservation sectors. One identified concern from this forum was the presence of beavers from unknown origins with any potential health concerns they might represent. Following experience gained through SBT, including animal trapping and specific beaver health screening, the RZSS tested a subset of this wild population for evidence of a range of diseases previously reported in beavers elsewhere in Europe (*Salmonella*, *Yersinia*, *Campylobacter*, *Leptospira*, *Mycobacterium*, *Giardia* and *Cryptosporidium* spp., *F. tularensis* and *E. multilocularis*) – and to make an assessment of their adjustment and ability to fully function in their new environment.

Post-mortems of beavers found dead (usually having been shot or involved in road traffic collisions) in the Tayside area between 2012 and 2020 were undertaken to increase the data on this population.

Overall the results of these tests showed an absence of significant disease for beavers,

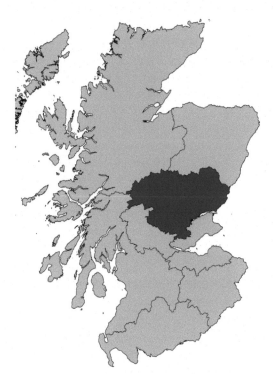

Figure 18.3 Tayside district in Scotland. Source: Wikipedia

and of any pathogens significant to other wildlife, domestic animals or humans (Campbell-Palmer et al., 2015a,c, 2021; Dowse et al., 2020; Girling et al., 2019a,b). There was evidence of a host-specific beaver parasite, the intestinal trematode *Stichorchis subtriquetrus*; because these had not previously been reported in other species and had not been associated with primary pathology in Eurasian beavers, they were not considered significant (Campbell-Palmer et al., 2013).

18.5 Genetics of Eurasian beavers in Scotland

There are two extant beaver species: *C. canadensis*, the North American beaver, evolutionarily diverged from the Eurasian beaver (*C. fiber*) somewhere around 7.5 million years ago (Horn

et al., 2011). Both species have undergone significant reductions in genetic diversity due to overhunting, bringing them to the brink of extinction. Both are very similar in appearance, ecology and biology, although differentiation between the two species is genetically straightforward because the North American beaver has 48 and the Eurasian beaver only 40 chromosomes, with no hybridization (Lavrov and Orlov, 1973). However, only fairly recently was this genetic distinction made and previously they were treated as the same species, resulting in large populations of North American beavers in parts of Finland and Russia as part of sanctioned government programmes (Dewas et al., 2012; Halley et al., 2021). Therefore, concerns were raised that beavers appearing from unknown origins should not be as a result of North American introductions to Britain.

An important aspect of the Tayside beavers was a genetic assessment to speciate individual animals. All beavers sampled in Tayside, whether by skeletal muscle (post-mortem) or blood analysis, were confirmed as Eurasian in origin (Campbell-Palmer et al., 2015a,d, 2020c). This was essential, because both EU law and Scottish Government licensing allowed the release of only Eurasian beavers in Scotland. The process of determining species was based on differences at two single-nucleotide polymorphism (SNPs) loci of the 16s rRNA (McEwing et al., 2015). An assessment of the accuracy of data on any beavers culled – including radiography and gross post-mortem – was performed to further define the humane methods of dispatch that formed advice to the Scottish Government.

The origin of the SBT beavers, as previously mentioned, was Norway, and their genetic analysis further confirmed this (Senn et al., 2014). Combined with a genetic assessment of those from Tayside translocated to the trial site and identified as being of central European origin (Bavaria, Germany), McEwing et al. (2015) confirmed that the Tayside beavers had significantly improved the genetic outbreeding potential of the population (Campbell-Palmer et al., 2020c).

Another aspect of genetic assessment to consider for the long-term survival of the reintroduced animals was the level of relatedness of existing beavers in Scotland. Norwegian wild beavers (as per all relict beaver populations) have low genetic diversity as a consequence of being hunted to almost extinction in the late 19th century (Collett, 1897). Beavers released in the original SBT were deliberately sourced from Norway because these were believed to be closest in form and type to the historical population in Britain as assessed by fossil and museum records (Kitchener and Conroy, 1997; Kitchener and Lynch, 2000; Halley et al., 2011). Tayside beavers comprised wild animals of unknown lineage and therefore assessment of their relatedness was another vital component of the health and welfare assessment of the released populations to inform future decisions on releases (Campbell-Palmer et al., 2015d, 2020c).

In 2017, an unpublished survey by the RZSS showed that only nine beavers of the original SBT cohort were still alive. In addition, two of these were a father–daughter pairing. This, combined with the closed catchment nature of the Knapdale site – meaning that the series of lochs all drained directly into the sea – made it important to increase the genetic diversity of the original trial population following permission by the Scottish Government for them to remain. Genetic analysis of the beavers originally released, and their subsequent offspring, was reviewed and, where gaps occurred, animals were trapped and resampled under licence prior to a planned augmentation with translocated wild beavers from Tayside.

Beavers translocated from Tayside were individually tested before release into Knapdale, with analysis taking approximately 24 hours using the species identification method described above (McEwing et al., 2015). Previous work identifying the geographic origins of the Tayside population (McEwing et al., 2015) as Bavarian confirmed that any augmentation of the SBT population with Tayside individuals would importantly increase the likelihood of outbreeding, and increase genetic diversity of the population in the hope of making it less vulnerable to extinction. Additionally, and for the first time, evidence of Polish/Lithuanian origins in a Tayside individual was shown (Campbell-Palmer et al., 2020c). A sample of 22 Tayside beavers from which sufficient genetic material of sufficiently good quality was obtained were assessed using 275 SNPs for molecular relatedness. The Tayside population in general was admixed and outbred within the context of the species, although it was demonstrated that these 22 beavers were fairly closely related and could have originated from as few as three family groups (Campbell-Palmer et al., 2020c). However, more recent genetic analysis of a now wider sampling of the Tayside beavers with 1938 SNPs has indicated they have genetic diversity comparable to other European populations and show ancestry from three of the known European fur trade refugia, with lower levels of relatedness than previously reported (Ritchie-Parker et al., 2021). Hence their use as founders in carefully managed translocations to other parts of Britain is supported, assuming that inbreeding can be avoided in the new populations and negative effects on the Tayside population are mitigated.

18.6 Disease screening results in the context and development of a disease risk analysis

Overall, the information gathered in this significant reintroduction project helped to define the first published disease risk analysis for the reintroduction of the Eurasian beaver (Girling et al., 2019c). This information has helped to clarify the significance of diseases such as leptospirosis, which has been viewed as a significant pathogen in Eurasian beavers and blamed for previous reintroduction failures (Marreros et al., 2018; Nolet et al., 1997). Girling et al. (2019a), observing over 150 samples, showed through testing of SBT and Tayside beavers that infection with *Leptospira* spp. was uncommon in Britain, Germany and Norway but that where it did occur it could do so without the presence of clinical disease. Geographical variations in the occurrence of *Leptospira* spp. infection/exposure showed its absence in Scotland (Argyll and Tayside, specifically) and Norway (Telemark region, specifically), with a low evidence in Devon (the area of post-import quarantine) and Germany (Bavaria). The same publication highlighted that there was no evidence of persistent infection with *Leptospira* spp. Testing for *Leptospira* spp. was therefore considered an essential part of the pre- and post-release monitoring programme (Goodman et al., 2012a,b, 2017). In addition, there was no evidence of persistent infection with *Leptospira* spp., which helped to allay concerns about the species acting as a reservoir and risk to human and animal health.

Negative findings for other zoonoses, including *Giardia* spp., *Salmonella* spp. *Yersinia enterocolitica, Y. pseudotuberculosis, F. tularensis* and Hantavirus were equally important (Campbell-Palmer et al. 2021; Girling et al., 2019b). Many diseases of concern are believed to be influenced by proximity to those domestic animals, wild rodents and humans that are more commonly associated with those diseases, a situation referred to as 'synurbization' (Andersen-Ranberg et al., 2016; Girling et al., 2019a). This may be due to the often physical separation, particularly in Scotland, from human conurbations and intensive animal farming, thereby removing exposure to the pathogens previously reported in this species, and may help to explain the low incidence in beavers in Scotland of any of the four pathogens (*Giardia* spp., *Leptospira* spp., *Salmonella* spp. and *Yersinia* spp.) known to be found in other species.

This raises a potentially important point: many of the deaths associated with beaver reintroductions and translocations reported in Scotland and elsewhere are frequently associated with opportunistic infections. It may be that the more significant aspect is the degree of physiological stress that the animal is undergoing during translocation, particularly if associated with prolonged quarantine or holding periods while test results are returned, as experienced initially in the SBT (Campbell-Palmer et al., 2021; Dowse et al., 2020; Girling et al., 2011; Goodman et al., 2012a,b, 2017;). Physiological stress and elevated glucocorticoids associated with this condition can result in immunosuppression and lifespan reduction, and has been demonstrated in a wide range of captive wild and domestic animals (Capitanio et al., 1998; Carlstead et al., 1999; Dreschel, 2010; Swaisgood et al., 2006; Weiss et al., 2011).

The summation of the series of publications regarding health screening of beavers in Scotland and the wider European source populations enabled the development of disease risk assessments of Eurasian beaver reintroduction to Britain as a whole (Girling et al., 2019c). This enabled the identification of potentially high-risk pathogens because of evidenced presence in Eurasian beavers and/or their impact on humans, wildlife and domestic animals (Campbell-Palmer et al., 2020a,b; Girling et al., 2019c). These included pathogens such as *Chrysosporium parvum* (*Emmonsia parva*), *Cryptosporidium parvum, E. multilocularis, Eimeria* spp., *Escherichia coli, Fasciola hepatica, F. tularensis, Giardia* spp., *Mycobacterium avium, Salmonella* spp., *Trichinella britovi, Yersinia* spp. and terrestrial rabies virus. The process of disease risk meta-analysis following modified IUCN guidelines allowed the risks associated with most of these to be mitigated by sourcing beavers from Britain. Where this is not possible, disease risks could still be mitigated by pre-release testing procedures that are already well established and have high sensitivity and specificity testing protocols. Girling et al. (2019c) thus showed that the risk of introducing significant disease by releasing Eurasian beavers in Britain was low, assuming that the beavers released were either captive bred in Britain or a translocated wild beaver from Scotland, using the suggested pre-release health screening techniques developed in the processes created by the SBT and its further augmentation over the past 12 years.

18.7 Ecological impacts

Beaver reintroductions have often been used to improve habitat quality, increase species biodiversity and generate ecosystem services. A meta-analysis of published ecological beaver studies demonstrated that beavers have an 'overwhelmingly positive influence' on biodiversity; this is expected for Scotland once beavers are more widely restored (Stringer and Gaywood, 2016). The potential for utilization of beavers as agents of habitat restoration, especially in association with agriculturally degraded land, has been demonstrated in a Scottish context, offering a working demonstration of how their activities can modify landscapes if we permit (Law et al., 2017). For example, after 12 years of beaver presence, plant species richness increased by 46% while cumulative species number increased by 148% through their grazing and waterlogging activities compared with un-grazed enclosures (Law et al., 2017). Beaver damming activities in some degraded environments where they have been tolerated have also demonstrated improvements in physical features such as flood attenuation, sediment and nutrient retention, and increased macro-invertebrates at a landscape scale (Law et al., 2016, Brazier et al., 2020). As beaver presence in other parts of Britain becomes evident, and the findings from numerous enclosed projects are beginning to be published, the picture

of beavers having multiple, measurable positive impacts on British landscapes is emerging fast (Brazier et al., 2020). This is offering an exciting potential picture of how this species could function throughout our freshwater ecosystems if we are prepared to co-exist and develop practical management and mitigation strategies in areas where their activities are challenging.

18.8 Population growth and management

Since the SBT concluded, beaver restoration has continued across Britain, through both officially recognised projects and unofficial releases and escapes. England saw its first licensed trial, the River Otter Beaver Project, conclude, with unofficially released animals permitted to stay after being retrospectively licensed following strong public support, landowner engagement and scientific monitoring (the findings from which are summarized in Brazier et al., 2020). Release of beavers into the wild is still not permitted in Britain outside of Scotland, though a growing number of enclosed licensed projects are currently in progress, predominantly to promote beaver restoration and demonstrate their ecological benefits. Additionally, growing wild populations exist in Devon, Wiltshire, Kent, Scotland, and in parts of Wales. Eurasian beavers became formally protected as a European Protected Species in May 2019 in Scotland and in October 2022 in England. The wider restoration of this species and the positive ecological services they bring have generated much interest and excitement in the conservation and wider public community (Brazier et al., 2020; Law et al., 2016, 2017; Stringer and Gaywood, 2016). However, certain land management sectors and interest groups have strong concerns because beavers' modifying abilities and activities can be challenging and their management can require resource investment in modern, heavily modified landscapes. After much political lobbying and alongside gained protected status, lethal control licences have been issued by NatureScot in Scotland under specific circumstances including demonstration of significant impacts to agriculture. In the first year of protection, 39 lethal control licences were issued and a reported 87 animals dispatched, and a further 115 in the second year (NatureScot, 2021). An extensive population survey was undertaken in 2020/2021 demonstrating that beavers in Tayside were increasing in numbers and distribution, with a minimum of 251 active territories representing a more than doubling since 2017/2018 (Campbell-Palmer et al., 2021). Therefore it can be concluded that, even with licensed lethal control, the Eurasian beaver is re-establishing. However, it cannot be taken for granted that prolonged lethal control will not have an impact, and therefore continued monitoring and investment in mitigation promoting co-existence is key in order to secure this species as a successfully returned native.

18.9 Conclusion

In this chapter we have illustrated the importance of taking a broad interdisciplinary approach when considering the benefits and risks associated with the reintroduction of a keystone species such as the Eurasian beaver to a habitat in which it was previously rendered absent. The reintroduction programme in Scotland began officially in 2009 and was accepted by the Scottish Government as a success in 2019. Despite potential setbacks to the successful reintroduction of beavers to Scotland, such as the finding of unofficially released beavers on the Rivers Tay and Earn catchments, the development of appropriate disease risk assessments and their revision during the process as more evidence became available ensured that this wider population was also granted official

permission, in 2019, to remain. The reintroduction outlined here demonstrates the importance of robust disease screening prior to and after release, not only to satisfy legislative and human health requirements but, importantly, to ensure the health and welfare of the animals being reintroduced and viable populations going forward.

References

Ahlen, P. A. (2001). Umea: 3 Department of Animal Ecology. *The parasitic and commensal fauna of the European beaver (Castor fiber) in Sweden* [Honors Thesis]. Saint Louis University p. 21.

Andersen-Ranberg, E. U., Pipper, C., & Jensen, P. M. (2016). Global patterns of *Leptospira* prevalence in vertebrate reservoir hosts. *Journal of Wildlife Diseases, 52*(3), 468–477. https://doi.org/10.7589/2014-10-245

Bajer, A., Benarska, M., & Siński, E. (1997). Wildlife rodents from different habitats as a reservoir for *Cryptosporidium parvum. Acta Parasitologica, 42*, 192–194.

Bajer, A., Bednarska, M., Paziewska, A., Romanowski, J., & Siński, E. (2008). Semi-aquatic animals as a source of water contamination with *Cryptosporidium* and *Giardia. Wiadomosci Parazytologiczne, 54*(4), 315–318.

Barlow, A. M., Gottstein, B., & Mueller, N. (2011). *Echinococcus multilocularis* in an imported captive European beaver (*Castor fiber*) in Great Britain. *Veterinary Record, 169*(13), 339. https://doi.org/10.1136/vr.d4673

Bennett, E., Clement, J., Sansom, P., Hall, I., Leach, S., & Medlock, J. M. (2010). Environmental and ecological potential for enzootic cycles of Puumala hantavirus in Great Britain. *Epidemiology and Infection, 138*(1), 91–98. https://doi.org/10.1017/S095026880999029X

Brazier, R. E., Puttock, A., Graham, H. A., Auster, R. E., Davies, K. H., & Brown, C. M. L. (2021). Beaver: Nature's ecosystem engineers. *WIREs. Water, 8*(1), e1494. https://doi.org/10.1002/wat2.1494

Campbell, R. D., Harrington, A., Ross, A., & Harrington, L. (2012). *Distribution, population assessment and activities of beavers in Tayside*. Scottish Natural Heritage Commissioned Report No. 540.

Campbell-Palmer, R., Girling, S., Rosell, F., Paulsen, P., & Goodman, G. (2012). Echinococcus risk from imported beavers. *Veterinary Record, 170*(9), 235. https://doi.org/10.1136/vr.e1508

Campbell-Palmer, R., & Rosell, F. (2013). *Captive management guidelines for Eurasian beavers (Castor fiber).* Royal Zoological Society of Scotland.

Campbell-Palmer, R., Girling, S., Pizzi, R., Hamnes, I. S., Øines, Ø., & Del-Pozo, J. (2013). *Stichorchis subtriquetrus* in a free-living beaver in Scotland. *Veterinary Record, 173*(3), 72. https://doi.org/10.1136/vr.101591

Campbell-Palmer, R., Pizzi, R., Dickinson, H., & Girling, S. (2015a). *Trapping and health screening of free-living beavers within the River Tay catchment, east Scotland.* Scottish Natural Heritage Commissioned Report No. 681. Scottish Natural Heritage, Battleby, Perthshire.

Campbell-Palmer, R., Schwab, G., Girling, S., Lisle, S., & Gow, D. (2015b). *Managing wild Eurasian beavers: A review of European management practices with consideration for Scottish application.* Scottish Natural Heritage commissioned report, 812 Scottish Natural Heritage, Battleby. Perthshire.

Campbell-Palmer, R., Gottstein, B., Del-Pozo, J., Girling, S., Cracknell, J., Schwab, G., Rosell, F., & Pizzi, R. (2015c). Combination of laparoscopy and abdominal ultrasound for *Echinococcus multilocularis* detection in Eurasian beavers (*Castor fiber*) *PLOSONE* Jul, *13*(10), e0130842. https://doi.org/10.1371/journal.pone.0130842. .

Campbell-Palmer, R., Girling, S., Senn, H., & Pizzi, R. (2015d). *Health and genetic screening report for wild beavers on the River Otter, Devon.* River Otter Beaver Trial: Science and Evidence Report. http://www.exeter.ac.uk/creww/research/beavertrial/appendix5/.

Campbell-Palmer, R., Gow, D., Campbell, R., Dickinson, H., Girling, S., Gurnell, J., Halley, D., Jones, S., Lisle, S., Parker, H., Schwab, G., & Rosell, F. (2016). *The Eurasian beaver handbook, ecology and management of Castor fiber.* Pelagic Publishing.

Campbell-Palmer, R., & Girling, S. (2019). *Final beaver trapping and health screening report river otter beaver trial.* River otter beaver trial: Science and evidence report available via University of Exeter. http://www.exeter.ac.uk/creww/research/beavertrial/appendix5/.

Campbell-Palmer, R., Naylor, A., & Girling, S. (2020a). *Veterinary Risk Assessment – Wild release of Eurasian beaver (Castor fiber).* Report submitted to the Welsh government on behalf of the Welsh Beaver Project.

Campbell-Palmer, R., Naylor, A., & Girling, S. (2020b). *Veterinary Risk Assessment – Captive enclosure release of Eurasian beaver (Castor fiber).* Report submitted to the Welsh government on behalf of the Welsh Beaver Project.

Campbell-Palmer, R., Senn, H., Girling, S., Pizzi, R., Elliott, M., Gaywood, M., & Rosell, F. (2020c). Beaver genetic surveillance in Britain. *Global Ecology and Conservation, 24*(4), e01275. https://doi.org/10.1016/j.gecco.2020.e01275

Campbell-Palmer, R., Puttock, A., Needham, R. N., Wilson, K., Graham, H., & Brazier, R. E. (2021). *Survey of the Tayside area beaver population 2020–2021.* NatureScot Research Report, 1274.

Campbell-Palmer, R., Rosell, F., Naylor, A., Cole, G., Mota, S., Brown, D., Fraser, M., Pizzi, R., Elliott, M., Wilson, K., Gaywood, M., & Girling, S. (2021). Eurasian beaver (*Castor fiber*) health surveillance in Britain: Assessing a disjunctive reintroduced population. *Veterinary Record, 188*(8), e84. https://doi.org/10.1002/vetr.84

Capitanio, J. P., Mendoza, S. P., Lerche, N. W., & Mason, W. A. (1998). Social stress results in altered glucocorticoid regulation and shorter survival in simian acquired immune deficiency syndrome. *Proceedings of the National Academy of Sciences of the United States of America, 95*(8), 4714–4719. https://doi.org/10.1073/pnas.95.8.4714

Carlstead, K., Fraser, J., Bennett, C., & Kleiman, D. G. (1999). Black rhinoceros (*Diceros bicornis*) in U.S. zoos: II. behaviour, breeding success, and mortality in relation to housing facilities. *Zoo Biology, 18*(1), 35–52. https://doi.org/10.1002/(SICI)1098-2361(1999)18:1<35::AID-ZOO5>3.0.CO;2-L

Carroll, C. L., & Huntington, P. J. (1988). Body condition scoring and weight estimation of horses. *Equine Veterinary Journal, 20*(1), 41–45. https://doi.org/10.1111/j.2042-3306.1988.tb01451.x

Ćirović, D., Pavlović, I., Kulišić, Z., Ivetić, V., Penezić, A., & Ćosić, N. (2012). *Echinococcus multilocularis* in the European beaver (*Castor fiber* L.) from Serbia [First report]. *Veterinary Record, 171*(4), 100. https://doi.org/10.1136/vr.100879

Coles, B. (2006). *Beavers in Britain's past.* Oxford Books.

Collen, P., & Gibson, R. J. (2000). The general ecology of beavers (*Castor* spp.), as related to their influence on stream ecosystems and riparian habitats, and the subsequent effects on fish – A review. *Reviews in Fish Biology and Fisheries, 10*(4), 439–461. https://doi.org/10.1023/A:1012262217012

Collett, R. (1897). Bæveren I Norge, dens Utbredelsen og Levemaade. *Bergens Museums Aarbog, 1*, 1–139.

Cronstedt-Fell, A., Stalder, G. L., & Kübber-Heiss, A. (2010). Echinococcosis in a European beaver (*Castor fiber*) in Austria. Poster presentation 36, 9th EWDA Conference, Vieland, September.

Cross, H. B., Campbell-Palmer, R., Girling, S., & Rosell, F. (2012). The Eurasian beaver (*Castor fiber*) is apparently not a host to blood parasites in Norway. *Veterinary Parasitology, 190*(1–2), 246–248. ISSN 0304-4017. https://doi.org/10.1016/j.vet par.2012.06.008

Cunningham, A. A. (1996). Disease risks of wildlife translocations. *Conservation Biology, 10*(2), 349–353. https://doi.org/10.1046/j.1523-1739.1996.10020349.x

Devine, C., Girling, S., Pizzi, R., Martinez-Pereira, Y., Campbell-Palmer, R., & Rosell, F. (2011). Physiological ('flow') murmurs are a common clinical finding in isoflurane anaesthetized European beavers. In *Proceedings of the European Association of Zoo and Wildlife Veterinarians, Lisbon, Portugal*.

Dewas, M., Herr, J., Schley, L., Angst, C., Manet, B., Landry, P., & Catusse, M. (2012). Recovery and status of native and introduced beavers Castor fiber and Castor canadensis in France and neighbouring countries. *Mammal Review, 42*(2), 144–165. https://doi.org/10.1111/j.1365-2907.2011.00196.x

Dowse, G., Taylor, H. R., Girling, S., Costanzi, J.-M., Robinson, S., & Senn, H. (2020). *Beavers in Knapdale: Final report from the Scottish Beavers Reinforcement Project*. Scottish Beavers. https://issuu.com/rzss/docs/scottish_beavers_reinforcement_report

Dracz, R. M., Ribeiro, V. M., Pereira, C. A., & Wdos, L. (2016). Occurrence of *Fasciola hepatica* (Linnaeus, 1758) in capybara (*Hydrochoerus hydrochaeris*) (Linnaeus, 1766) in Minas Gerais, Brazil. *Revista Brasileira de Parasitologia Veterinaria* Jul-Sep, 25(3), 364–367. https://doi.org/10.1590/S1984-29612016021

Dreschel, N. A. (2010). The effects of fear and anxiety on health and lifespan in pet dogs. *Applied Animal Behaviour Science, 125*(3–4), 157–162. https://doi.org/10.1016/j.applanim.2010.04.003

Dróżdż, J., Demiaszkiewicz, A. W., & Lachowicz, J. (2004). Endoparasites of the beaver *Castor fiber* (L.) in northeast Poland. *Helminthologia, 41*, 99–101.

Duff, A. G., Campbell-Palmer, R., & Needham, R. (2013). The beaver beetle *Platysyllus castoris* Ristema (Leiodidae: Platypsyllinae) apparently established on reintroduced beaver in Scotland, new to Britain. *Coleopterist, 22*, 9–19.

Durka, W., Babik, W., Ducroz, J. F., Heidecke, D., Rosell, F., Samjaa, R., Saveljev, A. P., Stubbe, A., Ulevicius, A., & Stubbe, M. (2005). Mitochondrial phylogeography of the Eurasian beaver *Castor fiber* L. *Molecular Biology, 14*(12):384303856, 3843–3856. https://doi.org/10.1111/j.1365-294X.2005.02704.x

Eckert, J. (1997). Epidemiology of *Echinococcus multilocularis* and *E. granulosus* in central Europe. *Parasitologia, 39*, 337–344.

Eckert, J., Conraths, F. J., & Tackmann, K. (2000). Echinococcosis: An emerging or re-emerging zoonosis? *International Journal for Parasitology, 30*(12–13), 1283–1294. https://doi.org/10.1016/s0020-7519(00)00130-2, PubMed: 11113255

Eckert, J., Gemmell, M. A., Meslin, F.-X., & Pawlowski, Z. S. (2001). *WHO/OIE Manual on Echinococcus in humans and animals: A public health problem of global concern*. https://apps.who.int/iris/handle/10665/42427

Erlandsen, S. L., Sherlock, L. A., Bemrick, W. J., Ghobrial, H., & Jakubowski, W. (1990). Prevalence of *Giardia* spp. in beaver and muskrat populations in Northeastern States and Minnesota: Detection of intestinal tropozoites at necropsy provides greater sensitivity than detection of cysts in faecal samples. *Applied and Environmental Microbiology, 56*(1), 31–36. https://doi.org/10.1128/aem.56.1.31-36.1990

Ewen, J. G., Sainsbury, A. W., Jackson, B., & Canessa, S. (2015). Disease risk management in reintroduction. In D. Armstrong, M. Hayward, D. Moro & P. Seddon (Eds.), *Advances in reintroduction biology of Australian and New Zealand fauna* (pp. 43–57). CSIRO Publishing.

Freeman, A. R., & Levine, S. A. (1933). The clinical significance of the systolic murmur. *Annals of Internal Medicine, 6*(11), 1371–1385. https://doi.org/10.7326/0003-4819-6-11-1371

Gaydos, J. K., Zabek, E., & Raverty, S. (2009). *Yersinia pseudotuberculosis* septicemia in a beaver from Washington State. *Journal of Wildlife Diseases, 45*(4), 1182–1186. https://doi.org/10.7589/0090-3558-45.4.1182

Gaywood, M. J. (2015). *Beavers in Scotland: A report to the Scottish government*. Scottish Natural Heritage. Inverness, UK.

Gelling, M., Zochowski, W., MacDonald, D. W., Johnson, A., Palmer, M., & Mathews, F. (2015). Leptospirosis acquisition following reintroduction of wildlife. *Veterinary Record, 177*(17), 440. https://doi.org/10.1136/vr.103160

Girling, S., Pizzi, R., Devine, C., Martinez-Pereira, Y., Campbell-Palmer, R., Goodman, G., & Rosell, F. (2011). Aspects of veterinary health screening during trial reintroduction of the Eurasian beaver (*Castor fiber*) to the wild in Scotland. In *Proceedings of the European Association of Zoo and Wildlife Veterinarians, Lisbon, Portugal.*

Girling, S. (2013) Chapter 7. Health and veterinary care. In R. Campbell-Palmer & F. Rosell (Eds.), *Captive management guidelines for Eurasian beavers (Castor fiber)* (pp. 56–77). Royal Zoological Society of Scotland.

Girling, S., Fraser, M., Arnemo, J., Pizzi, R., Campbell-Palmer, R., Cracknell, J., & Rosell, F. (2015). Haematology and serum biochemistry parameters and variations in the Eurasian beaver (*Castor fiber*). *PLOSONE Jun 12*; 10, 6, e0128775. https://doi.org/10.1371/journal.pone.0128775

Girling, S., & Campbell-Palmer, R. (2016). *Veterinary report Eurasian beavers – Prerelease health check.* River otter beaver trial: Science and evidence report, University of Exeter. http://www.exeter.ac.uk/creww/research/beavertrial/appendix5

Girling, S., & Campbell-Palmer, R. (2017). *Report on trapped Eurasian Beavers on the River Otter, spring 2017.* River otter beaver trial: Science and evidence report, University of Exeter. http://www.exeter.ac.uk/creww/research/beavertrial/appendix5

Girling, S. J., Goodman, G., Burr, P., Pizzi, R., Naylor, A., Cole, G., Brown, D., Fraser, M., Rosell, F. N., Schwab, G., Elliott, M., & Campbell-Palmer, R. (2019a). Evidence of *Leptospira* spp. and their significance during reintroduction of Eurasian beavers (*Castor fiber*) to Britain. *Veterinary Record, 185*(15), 482. http://doi.org/10.1136/vr.105429

Girling, S. J., McElhinney, L. M., Fraser, M. A., Gow, D., Pizzi, R., Naylor, A., Cole, G., Brown, D., Rosell, F., Schwab, G., & Campbell-Palmer, R. (2019b). Absence of Hantavirus in water-voles and Eurasian beavers in Britain. *Veterinary Record, 184*(8), 253. https://doi.org/10.1136/vr.105246

Girling, S. J., Naylor, A. N., Fraser, M. A., & Campbell-Palmer, R. (2019c). Reintroducing beavers to Britain: A disease risk analysis. *Mammal Review, 49*(4), 300–323. https://doi.org/10.1111/mam.12163

Girling, S. (2020). Health screening and post mortems-Section 9. In *Beavers in Knapdale: Final report from the Scottish Beavers Reinforcement Project* (pp. 28–30). Scottish Beavers.

Goodman, G., Girling, S., Pizzi, R., Meredith, A., Rosell, F., & Campbell-Palmer, R. (2012a). Establishment of a health surveillance programme for the reintroduction of the Eurasian beaver (*Castor fiber*) into Scotland. *Journal of Wildlife Diseases, 48*(4), 971–978. https://doi.org/10.7589/2011-06-153

Goodman, G., Girling, S., Pizzi, R., Rosell, F., & Campbell-Palmer, R. (2012b). Animal health-screening protocol for the trial reintroduction of beavers to Scotland. *Proceedings of the 6th International Beaver Symposium, City of Ivanic-Grad, Croatia, 17–20 September.*

Goodman, G., Meredith, A., Girling, S., Rosell, F., & Campbell-Palmer, R. (2017). Outcomes of a 'One Health' Monitoring Approach to a Five-Year Beaver (Castor fiber) Reintroduction Trial in Scotland. *EcoHealth, 14*(1) Suppl. 1, 139–143. https://doi.org/10.1007/s10393-016-1168-y

Gurnell, J., Demeritt, D., Lurz, P. W. W., Shirley, M. D. F., Rushton, S. P., Faulkes, C. G., Nobert, S., & Hare, E. J. (2009) *The feasibility and acceptability of reintroducing the European beaver to England.* Report prepared for Natural England and the People's Trust for Endangered Species. https://publications.naturalengland.org.uk/file/50080

Hady, P. J., Domecq, J. J., & Kaneene, J. B. (1994). Frequency and precision of body condition scoring in dairy cattle. *Journal of Dairy Science, 77*(6), 1543–1547. https://doi.org/10.3168/jds.S0022-0302(94)77095-8

Halley, D. J. (2011). Sourcing Eurasian beaver *Castor fiber* stock for reintroductions in Great Britain and Western Europe. *Mammal Review, 41*(1), 40–53. https://doi.org/10.1111/j.1365-2907.2010.00167.x

Halley, D. J., Saveljev, A. P., & Rosell, F. (2021). Population and distribution of beavers *Castor fiber* and *Castor canadensis* in Eurasia. *Mammal Review, 51*(1), 1–24. https://doi.org/10.1111/mam.12216

Heidecke, D. (1986). Erste Ergebnisse der Biberumsiedlungen in der DDR. *Zoologische Abhandlungen, 41,* 137–142.

Horn, S., Durka, W., Wolf, R., Ermala, A., Stubbe, A., Stubbe, M., & Hofreiter, M. (2011). Mitochondrial genomes reveal slow rates of molecular evolution and the timing of speciation in beavers (*Castor*), one of the largest rodent species. *PLOS ONE, 6*(1), e14622. https://doi.org/10.1371/journal.pone.0014622

International Union for Conservation of Nature and Natural Resources, & S. S. C. (2013). *Guidelines for* reintroductions and other conservation translocations version 1.0. International Union for Conservation of Nature and Natural Resources Species Survival Commission. https://portals.iucn.org/library/sites/library/files/documents/2013-009.pdfRetrieved 15/7/2021. (2014). Guidelines for disease risk analysis World Organization for Animal Health OIE Paris, France. http://www.iucn-whsg.org/DRA Retrieved 15/7/2021. *IUCN/SSC.*

Jamieson, I. G., & Lacy, R. C. (2012). Managing genetic issues in reintroduction biology. In J. G. Ewen, D. P. Armstrong, K. A. Parker & P. J. Seddon (Eds.), *Reintroduction biology: Integrating science and management* (pp. 441–475). Wiley-Blackwell.

Janovsky, M., Bacciarini, L., Sager, H., Gröne, A., & Gottstein, B. (2002). *Echinococcus multilocularis* in a European beaver from Switzerland. *Journal of Wildlife Diseases, 38*(3), 618–620. https://doi.org/10.7589/0090-3558-38.3.618

Kitchener, A. C., & Conroy, J. W. H. (1997). The history of the Eurasian beaver *Castor fiber* in Scotland. *Mammal Review, 27*(2), 95–108. https://doi.org/10.1111/j.1365-2907.1997.tb00374.x

Kitchener, A. C., & Lynch, J. M. (2000). A morphometric comparison of the skulls of fossil British and extant European beavers, *Castor fiber. Scottish Natural Heritage Review No. 127.* https://media.nature.scot/record/~1e3063235d

Knudsen, G. J. (1962). *Relationship of beavers to forest, trout and wildlife in Wisconsin.* Technical Bulletin 25. Wisconsin Conservation Department.

Kock, R. A., Woodford, M. H., & Rossiter, P. B. (2010). Disease risks associated with the translocation of wildlife. *Revue Scientifique et Technique, 29*(2), 329–350. https://doi.org/10.20506/rst.29.2.1980

Lavrov, L. S., & Orlov, V. N. (1973). Karyotypes and taxonomy of modern beavers (*Castor, Castor idae,* Mammalia). *Zoologische Zhurnal, 52,* 734–742. (in Russian with English summary.)

Law, A., Gaywood, M. J., Jones, K. C., Ramsay, P., & Willby, N. J. (2017). Using ecosystem engineers as tools in habitat restoration and rewilding: Beavers and wetlands. *Science of the Total Environment, 605–606,* 1021–1030. https://doi.org/10.1016/j.scitotenv.2017.06.173

Law, A., McLean, F., & Willby, N. J. (2016). Habitat engineering by beaver benefits aquatic biodiversity and ecosystem processes in agricultural streams. *Freshwater Biology, 61*(4), 486–499. https://doi.org/10.1111/fwb.12721

Lawrence, W. H., Fay, L. D., & Graham, S. A. (1956). A report on the beaver die-off in Michigan. *Journal of Wildlife Management, 20*(2), 184–187. https://doi.org/10.2307/3797425

Leighton, F. A. (2002). Health risk assessment of the translocation of wild animals. *Revue scientifique et technique de l'office internationale des epizooties, 21,* 187–195. https://doi.org/10.20506/rst.21.1.1324

López-Pérez, A. M., Carreón-Arroyo, G., Atilano, D., Vigueras-Galván, A. L., Valdez, C., Toyos, D., Mendizabal, D., López-Islas, J., & Suzán, G. (2017). Presence of antibodies to *Leptospira* spp. in black-tailed prairie dogs (*Cynomys ludovicianus*) and beavers (*Caster canadensis*) in Northwestern Mexico. *Journal of Wildlife Diseases, 53*(4), 880–884. https://doi.org/10.7589/2016-11-240

Máca, O., Pavlásek, I., & Vorel, A. (2015). *Stichorchis subtriquetrus* (Digenea: Paramphistomatidae) from Eurasian beaver (*Castor fiber*) in the Czech Republic. *Parasitology Research, 114*(8), 2933–2939. https://doi.org/10.1007/s00436-015-4495-y

Macdonald, D. W., Tattersall, F. H., Brown, E. D., & Balharry, D. (1995). Reintroducing the European Beaver to Britain: Nostalgic meddling or restoring biodiversity? *Mammal Review, 25*(4), 161–200. https://doi.org/10.1111/j.1365-2907.1995.tb00443.x

Mackie, P. (2014). *Scottish Beaver Trial independent public health monitoring 2009–2014 report and recommendations. Battleby, Perthshire UK* http://www.snh.gov.uk/docs/A1450214.pdf.

Marreros, N., Zürcher-Giovannini, S., Origgi, F. C., Djelouadji, Z., Wimmershoff, J., Pewsner, M., Akdesir, E., Batista Linhares, M., Kodjo, A., & Ryser-Degiorgis, M. P. (2018). Fatal leptospirosis in free-ranging Eurasian beavers (*Castor fiber* L.), Switzerland. *Transboundary and Emerging Diseases, 65*(5), 1297–1306. https://doi.org/10.1111/tbed.12879

Marti, R., Zhang, Y., Tien, Y. C., Lapen, D. R., & Topp, E. (2013). Assessment of a new *Bacteroidales* marker targeting North American beaver (*Castor canadensis*) fecal pollution by real-time PCR. *Journal of Microbiological Methods, 95*(2), 201–206. https://doi.org/10.1016/j.mimet.2013.08.016

McElhinney, L., Fooks, A. R., Featherstone, C., Smith, R., & Morgan, D. (2016). Hantavirus (Seoul virus) in pet rats: A zoonotic threat. *Veterinary Record, 178*(7), 171–172. https://doi.org/10.1136/vr.i817

McEwing, R., Senn, H., & Campbell-Palmer, R. (2015). *Genetic assessment of free-living beavers in and around the River Tay catchment, east Scotland.*

Scottish Natural Heritage Commissioned Report No. 682.

Miller, A. L., Olsson, G. E., Walburg, M. R., Sollenberg, S., Skarin, M., Ley, C., Wahlström, H., & Höglund, J. (2016). First identification of *Echinococcus multilocularis* in rodent intermediate hosts in Sweden. *International Journal for Parasitology. Parasites and Wildlife*, 5(1), 56–63. https://doi.org/10.1016/j.ijppaw.2016.03.001

Mol, J. P. S., Carvalho, T. F., Fonseca, A. A., Sales, E. B., Issa, M. A., Rezende, L. C., Hodon, M. A., Tinoco, H. P., Malta, M. C. C., Pessanha, A. T., Pierezan, F., Mota, P. M. P. C., Paixão, T. A., & Santos, R. L. (2016). Tuberculosis caused by *Mycobacterium bovis* in a capybara (*Hydrochoerus hydrochaeris*). *Journal of Comparative Pathology*, 155(2–3), 254–258. https://doi.org/10.1016/j.jcpa.2016.05.014

Mörner, T., Avenäs, A., & Mattsson, R. (1999). Adiaspiromycosis in a European beaver from Sweden. *Journal of Wildlife Diseases*, 35(2), 367–370. https://doi.org/10.7589/0090-3558-35.2.367

Mörner, T., Sandström, G., & Mattsson, R. (1988). Comparison of serum and lung extracts for surveys of wild animals for antibodies to *Franciscella tularensis* biovar *palaearctica*. *Journal of Wildlife Diseases*, 24(1), 10–14. https://doi.org/10.7589/0090-3558-24.1.10

Mörner, T. (1992). The ecology of tularaemia. *Revue Scientifique et Technique*, 11(4), 1123–1130. https://doi.org/10.20506/rst.11.4.657

Napolitani, M., Troiano, G., Bedogni, C., Messina, G., & Nante, N. (2019). *Kocuria kristinae*: An emerging pathogen in medical practice. *Journal of Medical Microbiology*, 68(11), 1596–1603. https://doi.org/10.1099/jmm.0.001023

NatureScot. (2021) *Beaver management report for 2020 beaver management report for 2020* | NatureScot

Nolet, B. A., Broekhuizen, S., Dorrestein, G. M., & Rienks, K. M. (1997). Infectious diseases as main causes of mortality to beavers *Caster fiber* after translocation to the Netherlands. *Journal of Zoology*, 241(1), 35–42. https://doi.org/10.1111/j.1469-7998.1997.tb05497.x

Nolet, B. A., & Rosell, F. (1998). Comeback of the beaver *Castor fiber*: An overview of old and new conservation problems. *Biological Conservation*, 83(2), 165–173. https://doi.org/10.1016/S0006-3207(97)00066-9

OIE. (2012). Terrestrial animal health code. http://www.oie.int/doc/ged/D12376.PDF

Oliver, M. K., Telfer, S., & Piertney, S. B. (2009). Major histocompatibility complex (MHC) heterozygote superiority to natural multi-parasite infections in the water vole (*Arvicola terrestris*). *Proceedings. Biological Sciences*, 276(1659), 1119–1128. https://doi.org/10.1098/rspb.2008.1525

Parker, H., Nummi, P., Hartman, G., & Rosell, F. (2012). Invasive North American beaver *Castor canadensis* in Eurasia: A review of potential consequences and strategy for eradication. *Wildlife Biology*, 18(4), 354–365. https://doi.org/10.2981/12-007

Pizzi, R., Campbell-Palmer, R., Girling, S., & Cracknell, J. (2012a). Antemortem screening of Eurasian beavers (*Castor fiber*) for *Echinococcus multilocularis*. Oral presentation. Proceedings of the British Veterinary Zoological Society Autumn Meeting.

Pizzi, R., Girling, S., Cracknell, J., Rosell, F., Robstad, C., Needham, R., Brown, D., & Campbell-Palmer, R. (2012b). Finding a viable nonlethal method of screening Eurasian beavers (*Castor fiber*) for *Echinococcus multilocularis*. Proceedings 6th International Beaver Symposium, City of Ivanic-Grad, Croatia, 17–20 September.

Platt-Samoraj, A., Syczyło, K., Bancerz-Kisiel, A., Szczerba-Turek, A., Giżejewska, A., & Szweda, W. (2015). *Yersinia enterocolitica* strains isolated from beavers (*Castor fiber*). *Polish Journal of Veterinary Sciences*, 18(2), 449–451. https://doi.org/10.1515/pjvs-2015-0058

Ritchie-Parker, H., Ball, A., Campbell-Palmer, R., Taylor, H., & Senn, H. (2021). *Genetic diversity analysis of beavers (Castor fiber) in England*. Natural England commissioned report. https://publications.naturalengland.org.uk/file/5494588969320448

Romašov, V. A. (1969) *Results of zoogeographic helminth fauna characteristics for river beavers in USSR. [Black earth, Voroņež [In Russian: Ромашов ВА (1969) Результаты зоогеографических исследований гельминто фауны речных бобров СССР. Чернозем. кн. изд. Воронеж].*

Romašov, B. V. (1992). Krankheiten der biber [beaver diseases], *Semiaquatische Säugetiere, wiss. Beitr.* University Halle.

Rosell, F., Rosef, O., & Parker, H. (2001). Investigations of waterborne pathogens in Eurasian beaver (*Castor fiber*) from Telemark County, Southeast Norway. *Acta Veterinaria Scandinavica*, 42(4), 479–482. https://doi.org/10.1186/1751-0147-42-479

Rosell, F., Bozsér, O., Collen, P., & Parker, H. (2005). Ecological impact of beavers *Castor fiber* and *Castor canadensis*and their ability to modify ecosystems. *Mammal Review*, 35(3–4), 248–276. https://doi.org/10.1111/j.1365-2907.2005.00067.x

Roy, A. (2005). An annotated list of the nonmarine molluscs of Britain and Ireland. *Journal of Conchology, 38*, 607–637.

Saëz, H. (1976). Cutaneous candidiosis in a European beaver, *Castor fiber*. Epidimiological aspect and parasitic form of *Candida albicans*. *Acta Zoologica et Pathologica Antverpiensia*, (66), 101–110.

Schiffman, C., Clauss, M., Hoby, S., & Hatt, J.-M. (2017). Visual body condition scoring in zoo animals-composite, algorithm and overview approaches. *Journal of Zoo and Aquarium Research, 5*(1), 1–10.

Segliņa, Z., Bakasejevs, E., Deksne, G., Spuņģis, V., & Kurjušina, M. (2015). New finding of *Trichinella britovi* in a European beaver (*Castor fiber*) in Latvia. *Parasitology Research, 114*(8), 3171–3173. https://doi.org/10.1007/s00436-015-4557-1

Shimalov, V. V., & Shimalov, V. T. (2000). Findings of *Fasciola hepatica* Linnaeus, 1758 In wild animals in Belorussian Polesie. *Parasitology Research, 86*(6), 527. https://doi.org/10.1007/s004360050707

Sjöberg, G. (1998). *Ecosystem engineering in forest streams, invertebrate fauna in beaver ponds* p. 158. European–American Mammal Congress. Universidad de Spain.

Stringer, A. P., & Gaywood, M. J. (2016). The impacts of beavers on biodiversity and the ecological basis for reintroduction to Scotland. *Mammal Review, 46*(4), 270–283. https://doi.org/10.1111/mam.12068

Swaisgood, R., Owen, M. A., Czelka, N. M., Mauroo, N., Hawk, K., & Tang, J. C. (2006). *Evaluating stress and well-being in the giant panda: A system for monitoring. Giant pandas, biology, veterinary medicine and management* (pp. 299–314). Cambridge University Press.

Torgerson, P. R., & Craig, P. S. (2009). Risk assessment of importation of dogs infected with *Echinococcus multilocularis*into the UK. *Veterinary Record, 165*(13), 366–368. https://doi.org/10.1136/vr.165.13.366

Ullman-Culleré, M. H., & Foltz, C. J. (1999). Body condition scoring: A rapid and accurate method for assessing health status in mice. *Laboratory Animal Science, 49*(3), 319–323.

Viggers, K. L., Lindenmayer, D. B., & Spratt, D. M. (1993). The importance of disease in reintroduction programmes. *Wildlife Research, 20*(5), 687–698. https://doi.org/10.1071/WR9930687

Weiss, A., Adams, M. J., & King, J. E. (2011). Happy orang-utans live longer lives. *Biology Letters, 7*(6), 872–874. https://doi.org/10.1098/rsbl.2011.0543

Wimmershoff, J., Robert, N., Mavrot, F., Hoby, S., Boujon, P., Frey, C., Weber, M., Café-Marcal, V., Hussy, D., Mattsson, R., Pilo, P., Nimmervoll, H., Marreros, N., Pospischil, A., Angst, C., & Ryser-Degiorgis, M.-P. (2012). Causes of mortality and diseases in the reintroduced European beaver population in Switzerland from 1989 to 2009. In Proceedings of the Joint WDA/EWDA Conference, Lyon, July 22–27 p. 37.

Woodford, M. H. (2001). *Quarantine and health screening protocols for wildlife prior to translocations and release to the wild*. Office international de Epizootics (OIE) Veterinary Specialist Group/Species Survival Commission of the World Conservation Union (IUCN), Care for the Wild International, and the European Association of Zoo and Wildlife Veterinarians.

Zamora, J., & Enriquez, R. (1987). *Yersinia enterocolitica, Yersinia frederiksenii* and *Yersinia intermedia* in *Cyprinus carpio* (Linneo 1758). *Zentralblatt Fur Veterinarmedizin. Reihe B. Journal of Veterinary Medicine. Series B, 34*(2), 155–159. https://doi.org/10.1111/j.1439-0450.1987.tb00381.x

Żurowski, W., & Kasperczyk, B. (1988). Effects of reintroduction of European beaver in the lowlands in the Vistula Basin. *Acta Theriologica, 33*, 325–338. https://doi.org/10.4098/AT.arch.88-26

PART IV

RESEARCH, OUTREACH AND EDUCATION

Chapter 19

Ticks and tick-borne diseases in Eastern Bhutan

Jamyang Namgyal, Tenzin Tenzin, Ratna B. Gurung and Susan C. Cork

Abstract

Ticks and tick-borne diseases have the potential to pose challenges to both public and animal health. Besides causing skin irritation and blood loss, some species of ticks can serve as reservoirs and vectors for a variety of pathogens including bacteria, viruses and protozoa. Most tick species spend much of their life cycle in the environment, away from their vertebrate hosts, resulting in a complex life cycle that is significantly influenced by environmental conditions. Understanding the complexities of the tick life cycle is important in the development and adoption of effective tick management and control strategies. In Eastern Bhutan, recent studies have shown that there are at least five genera of ixodid ticks present on livestock but that the most common species is *Rhipicephalus microplus*. The latter is a one-host tick that develops on the host before the adult female drops off to lay eggs in the environment. Although cattle and other bovids are its primary hosts, it can be found on other domestic animals and wildlife. This tick species is known to transmit a number of diseases to cattle. In our case study, we illustrate how an interdisciplinary team worked with communities and policy makers to develop prevention and control plans to deal with common ticks and tick-borne diseases in Eastern Bhutan. It is clear that farmer education plays an important role in the implementation of such disease prevention and control plans and that community engagement is key to ensuring ongoing compliance with tick control guidelines.

19.1 Introduction

Nestled in the Himalayas and located between China to the north and India to the south, east and west, Bhutan is a small landlocked country covering an area of 38,394 km². Geographically it is characterized by high mountains, dense forests and fast-flowing rivers that form narrow valleys before flowing out onto the vast north Indian plains. As a result of a pronounced south–north elevation gradient and an inverse north–south precipitation gradient, the country experiences diverse climatic conditions; these are further challenged by the impacts of climate change, ranging from wet subtropical ecozones in the southern foothills to temperate and alpine ecozones in the northern mountainous areas (MoAF, 2019). About 62% of Bhutan's population live

Susan C. Cork and Douglas P. Whiteside (eds) **Case Studies in EcoHealth**
DOI: 10.52517/9781789183368.019, © 5m Books Ltd, 2024

in rural areas and depend on agriculture and livestock farming for their livelihoods (National Statistics Bureau, 2018). However, Bhutanese farmers face a unique set of challenges in ensuring efficient food production. With just 2.75% of the country's land available for cultivation (MoAF, 2019), individual landholdings are limited in their ability to achieve any significant crop or livestock productivity. This is further challenged by climate change (Wang et al., 2017), erratic rainfall and other biophysical conditions such as steep terrain and poor soil fertility. Anthropogenic factors such as economic pressures, urban encroachment and changing demographics are also impacting agricultural productivity, as well as pests and diseases. The last of these impact livestock productivity, with tick infestations being a common animal health challenge reported by cattle farmers.

19.2   Livestock farming in Bhutan

Livestock farming, especially cattle rearing, is one of the most important sources of income for the vast majority of Bhutan's rural population. Besides generating direct income through the sale of dairy products, cattle also provide manure to enrich the soil, draught power for challenging terrain and transportation in some parts of the country. Traditionally Siri cattle (*Bos taurus indicus* L.), a local indigenous breed, was the most predominant breed of cattle reared in Bhutan. However, from 1985 onwards the Royal Government of Bhutan launched a livestock improvement programme that aimed to intensify small-scale subsistence-oriented farms. This national cattle breeding policy encouraged the crossbreeding of local indigenous breeds with brown Swiss cattle (*B. taurus taurus* L.) in high-altitude areas and Jersey cattle (*B. taurus taurus* L.) in other regions (Samdup et al., 2010). As a result, in just one decade (1989–1999) the population of crossbred cattle (i.e., those

crossed with European breeds) increased by over 100% whereas the population of indigenous cattle decreased by 9% (Phanchung et al., 2007). From 1998 onwards, the cattle breeding policy was changed to provide artificial insemination services and breeding bulls of any breed to all districts based on farmers' preference and demand (Samdup et al., 2010). Historically, the traditional cattle-rearing system in Bhutan was predominantly a transhumant system in high-altitude areas involving cattle migration and a sedentary system in other areas involving crop–cattle integration (Phanchung et al., 2007). Currently, with the aim of encouraging self-sufficiency in dairy products, the government has been widely promoting the use of a 'stall-fed system' for cattle rearing with better feeding and management practices. Many farmers across Bhutan have adopted the stall-fed system largely because of government subsidies for the purchase of improved dairy cows, cattle shed construction and dairy farm management training sessions. This coincided with growing economic opportunities such as income through the sale of milk and milk products via farmers' groups and cooperatives. As a result, the number of farmers importing European breeds of cattle such as Jersey and Holstein Friesian from India has increased considerably. Despite all these efforts, 53.7% of the cattle population in the country (see Figure 19.1) is still comprised of indigenous cattle (DoL, 2018) and the cattle rearing system is still dominated by small-scale subsistence-oriented farms (i.e., with a herd size of five or six cattle) with low productivity.

19.3   Livestock health

Although there have been some good developments in livestock management, cattle farmers in Bhutan continue to face challenges due to various health constraints, among which infectious diseases such as foot-and-mouth disease,

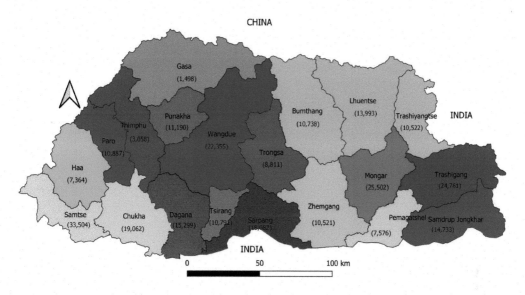

Figure 19.1 Map of Bhutan showing cattle population by district (derived from 2021 census figures).

haemorrhagic septicaemia, black quarter, anthrax, rabies, brucellosis, and parasitic diseases are ranked high (NCAH, 2019). Among parasite-related diseases in cattle in Bhutan, tick infestation is the most commonly reported (NCAH, 2019). For instance, in 2019, 89% of the reported parasite-related cases in cattle were due to tick infestation (NCAH, 2019). Tick-borne diseases (TBDs) like anaplasmosis, babesiosis, and theileriosis are also present in cattle in Bhutan, especially in the southern subtropical districts (NCAH, 2019; Phanchung et al., 2007). Globally, ticks are considered to be the most important ectoparasites of livestock. As blood feeding parasites, they can transmit a variety of pathogenic organisms including viruses, bacteria, and protozoa to animals and humans, many of which can cause significant health challenges (Fivaz et al., 2012). In animals, tick infestation by itself can cause production losses due to skin irritation (due to attachment), blood loss, and sometimes complications due to self-trauma, fly strike, and secondary bacterial infections (Minjauw and McLeod, 2003). In heavy infestations, ticks can result in anaemia and significant

weight loss, thereby causing animal welfare concerns and significant losses in productivity (de Castro, 1997). Some ticks can also produce toxins leading to toxicosis and subsequent tick paralysis (Estrada-Pena and Mans, 2013). The most commonly reported diseases transmitted by ticks to livestock are anaplasmosis, babesiosis, and theileriosis (Fivaz et al., 2012). In spite of the presence of these TBDs in Bhutan, there are currently no comprehensive data available on the actual number of clinical or subclinical cases due to the limited use of confirmatory diagnostic tests, limited surveillance capacity, and discrepancies in recorded data (Namgyal et al., 2021).

19.4 Case study: ticks and tick control in Eastern Bhutan

The current tick control practice in Bhutan largely depends on the use of chemical acaricides, applied directly by hand, to tick-infected areas on host animals. There have been no national guidelines or strategies developed for the control of ticks and TBDs, although a number of animal health teams have developed local guidelines. Nonetheless, the importance of ticks and TBDs as an animal health issue is increasingly becoming recognized among veterinary professionals across the country. In this section we will provide a general overview of ticks and TBDs in Bhutan and outline a cross-sectoral interdisciplinary (EcoHealth) approach to developing policy options regarding the prevention and control of ticks and TBDs in Eastern Bhutan.

Ticks are classified in the order Ixodida and are parasitic arachnids. They are external parasites that depend on finding a vertebrate host to take a blood meal. There are two major families, the Ixodidae (or hard ticks) and the Argasidae (or soft ticks). Ticks have four stages to their life cycle, namely the egg, larva, nymph, and adult. The nymphs and adults have four pairs of legs whereas the larvae have only three and can sometimes be mistaken for lice. Although all active stages can be seen feeding on a vertebrate host the adults are much larger, especially when engorged with blood, and so these are the stage most often noticed.

Hard ticks undergo either a one-, two-, or three-host lifecycle. In between hosts they spend time in the environment to moult before finding a new host. Ticks locate potential hosts by sensing body heat, moisture, and/or vibrations in the environment. Tick life cycles can take up to a year or may take several years depending on the species of tick and the local environmental conditions. In addition to having a hard shield on their dorsal surface (scutum), hard ticks have a beak-like structure at the front containing the mouthparts. These features are helpful when trying to identify ticks to genus and species level.

19.5 Tick species identified in Bhutan

In Bhutan, formal reports of scientifically identified ticks date back to the 1990s as outlined in one unpublished study conducted over a year in cattle in Eastern Bhutan (Dr Susan C. Cork, personal communication, 2018, Thimphu, Bhutan). The study reported *R. microplus* (Canestrini) as the most predominant tick species, with *Haemaphysalis* spp. and *Hyalomma* spp. identified to the genus level. More than two decades later, the Regional Livestock Development Center (RLDC) of the western region conducted a similar study on cattle from September 2018 to June 2019 in western Bhutan (RLDC Wangdue, 2019). They reported the genera *Rhipicephalus* (*Boophilus*) spp., *Rhipicephalus* spp., *Haemaphysalis* spp., *Ixodes* spp., and *Amblyomma* spp. but no information on species was available. According to current knowledge there are at least five genera and 13 species (including unidentified species of *Hyalomma* spp. and *Ixodes* spp.) of ticks in Eastern Bhutan (Table 19.1).

The diversity of tick species in Eastern Bhutan was found to be similar to the tick fauna of the Eastern Himalayan range (Namgyal et al., 2021). Like anywhere else in the world, tick infestation in animals in Bhutan was found to be generally influenced by both host and environmental factors (Namgyal et al., 2021). Host factors such as age and breed were found to influence the susceptibility of cattle to tick infestation in Bhutan (Namgyal et al., 2021). For instance, younger cattle that are deprived of maternal grooming and with free access to pastures and forests were found to be more susceptible to tick infestation. Asian or Zebu cattle (*B. taurus indicus*) are generally thought to be more resistant to ticks than European or Taurine cattle (*B. taurus taurus*) (Utech et al., 1978). However, in our study in Eastern Bhutan (see Figure 19.2), indigenous breeds were found to have a heavier tick infestation than the European breeds, but this can likely be attributed to the preferred management system

Table 19.1 Record of formally identified tick species in Bhutan

Tick species	Host species	Source
Amblyomma integrum Karsch	Cattle	Pem et al., 2021
Amblyomma testudinarium Koch	Cattle	Namgyal et al., 2021
Haemaphysalis aponommoides Warburton		Pem et al., 2021
Haemaphysalis bispinosa Neumann	Cattle	Namgyal et al., 2021; Pem et al., 2021
Haemaphysalis shimoga Trapido & Hoogstraal	Cattle	Jamyang Namgyal, unpublished data
Haemaphysalis spinigera Neumann	Cattle	Namgyal et al., 2021; Pem et al., 2021
Hyalomma spp.	Cattle	RLDC Wangdue, 2019
Ixodes apronophorus Schulze	Human	Jamyang Namgyal, unpublished data
Ixodes ovatus Neumann	Cattle	Pem et al., 2021
Ixodes spp.	Cattle	Namgyal et al., 2021
Rhipicephalus haemaphysaloides Supino	Cattle	Namgyal et al., 2021; Pem et al., 2021
Rhipicephalus microplus Canestrini	Cattle	Namgyal et al., 2021; Pem et al., 2021
Rhipicephalus sanguineus Latreille	Dog	Dr. Meena D. Samal, personal communication, 2021

used for these two breeds. European breeds, due to their greater economic value, are mostly reared in a stall-fed system with limited access to forests while the indigenous breeds are mostly reared in a free-grazing system with easy access to forests, thereby increasing the risk of exposure to ticks in these animals. Regardless, indigenous breeds of cattle in Bhutan are generally believed to be relatively resistant to many pests and diseases, and good husbandry and tick control practices like grooming, brushing, and acaricide application are rarely practised by Bhutanese farmers rearing these indigenous animals.

19.6 Tick biology

The typical life cycle of ixodid ticks consists of four stages—eggs, motile larvae, nymphs, and adults (male and female) (Apanaskevich and Oliver, 2013). Generally, temperature and moisture are is considered to be the key determinant of the development progress and activity of ticks (Estrada-Peña, 2015). For instance, questing larvae of the common cattle tick (*R. microplus*) depend on temperature and humidity for questing behaviour and survival (Leal et al., 2018; Sutherst and Bourne, 2006). Other factors, such as microclimate and host availability, are also known to influence the development and activity of this species and other ticks (Estrada-Peña, 2015). A wide range of potential tick hosts is available across Bhutan and so tick distribution was significantly influenced by regional environmental factors, including climatic variables such as temperature and rainfall.

Habitat suitability modelling has indicated that, besides temperature and rainfall, land cover classes also influence tick distribution in this country (Namgyal et al., 2021). For example, land cover classes such as Kamzhing (dry rain-fed land), meadows, Chuzhing (paddy fields), and shrubs were found to be the most suitable for tick species modelled in the east of the country. However, in a geographically diverse country like Bhutan, the high mountains and narrow valleys cause extreme variation in climate and the temperature is mainly affected by altitude and rainfall by latitude. Therefore, tick distribution might vary even within a short distance because of microclimate influencing tick development and activity (see Eastern study area in Figure 19.2).

> Habitat suitability modelling is a process that combines species occurrence data with environmental variables to predict species distribution across landscapes. Namgyal et al. (2021) used the presence data of four tick species in Eastern Bhutan and conducted habitat suitability modelling using MaxEnt (Maximum Entropy; Phillips et al., 2006), a machine-learning modelling approach that estimates probability distribution for a specific species.

19.7 Tick-borne diseases in cattle in Bhutan

In general, TBDs affect 80% of the world's cattle population and are widely distributed in the tropics and subtropics (de Castro, 1997). In Bhutan, TBDs such as anaplasmosis, babesiosis, and theileriosis are reported in cattle, especially in the southern subtropical districts of the country (Phanchung et al., 2007). Babesiosis is the most frequently reported TBD in cattle in Bhutan. The disease is also prevalent in both Assam (Kakati et al., 2015) and West Bengal (Debbarma et al., 2018), the two Indian states neighbouring the southern regions of Bhutan.

19.7.1 Babesiosis

Babesia bigemina is the most frequently reported tick-borne pathogen in the west-central region of Bhutan, and the prevalence of babesiosis in cattle was found to be positively correlated with the distribution of *R. microplus* (Pem et al., 2021). *Rhipicephalus microplus* is implicated as the principal vector of *B. bigemina* throughout the world (Bock et al., 2004). In the neighbouring Indian state of Assam, *R. microplus* is also considered to be the main vector for *B. bigemina* (Mushahary et al., 2019). While the positive correlation between the prevalence of *B. bigemina* and *R. microplus* distribution in Bhutan might suggest that this one-host tick is the main vector, further research is required to ascertain this (Namgyal et al., 2021; Pem et al., 2021). In west-central Bhutan, cattle age and management system were found to influence the prevalence of *B. bigemina*; for example, older cattle were found to be more susceptible compared with younger cattle and prevalence was more pronounced in free-grazing cattle, which are more likely to be exposed to ticks (Pem et al., 2021).

There were 72 cases of bovine babesiosis reported from 13 of the 20 districts in Bhutan in 2020 but, because babesiosis can be clinically confused with bovine enzootic haematuria associated with chronic ingestion of bracken fern (*Pteridium* spp.) (Hidano et al., 2017), these figures need to be considered with caution. In 2020 there were 1974 cases of bovine enzootic haematuria reported in the country (Sonam Deki, personal communication, 2021). In most parts of the country, suspected cases of babesiosis are rarely diagnosed using laboratory examinations but are largely based on clinical signs (i.e., fever and haemoglobinuria). In recent years babesiosis has been also detected in quarantine stations in cattle imported into Bhutan from the neighbouring Indian states of Assam and West Bengal (Namgyal, 2020).

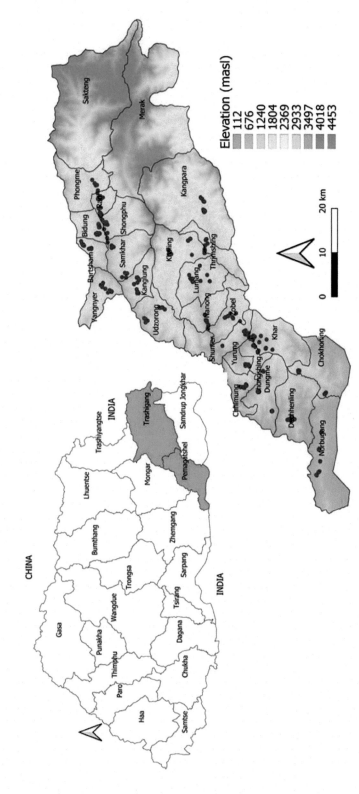

Figure 19.2 Topographic map of our study area in Eastern Bhutan showing low-altitude areas in green and high-altitude areas in red.

Tick-borne diseases

- Babesiosis is a disease caused by haemo-protozoan parasites of the genus *Babesia* (OIE, 2014). Diminazene aceturate (trade name Berenil) is used for treatment of suspected cases (NCAH, 2013).
- Theileriosis, a disease caused by an obligate intracellular parasite of the genus *Theileria*, affects both wild and domestic animals, especially bovines (OIE, 2018). Buparvaquone is used for treatment of suspected cases (NCAH, 2013).
- Anaplasmosis is mainly caused by the intracellular rickettsial pathogen *Anaplasma marginale*. A second species, *A. centrale*, causes benign infections with a limited degree of anaemia but clinical outbreaks in the field are extremely rare (OIE, 2015). Species *A. phagocytophilum* and *A. bovis* have been reported in cattle, but they do not cause clinical disease (OIE, 2015).
- There has been serological evidence of exposure to Crimean Congo Haemorrhagic Fever virus in goats in Bhutan. This is a TBD that can cause serious illness in humans exposed to infected animals (Wangchuk et al., (2016).

19.7.2   Theileriosis

Theileriosis is present in Bhutan but there is currently no information on which species of *Theileria* are involved. However, *T. orientalis* and *T. annulata* have been reported from Assam and West Bengal, respectively (Kakati et al., 2015) and, in both states, *R. microplus* is the vector. In Bhutan, diagnosis is rarely confirmed by laboratory tests but is based on clinical signs (i.e., fever and lymph node swelling). The veterinary information system in 2020 did not record any cases

(Sonam Deki, personal communication, 2021, Thimphu, Bhutan), but this may indicate a gap in case reporting due to a lack of diagnostic confirmation.

19.7.3   Anaplasmosis

Anaplasmosis caused by *A. marginale* mostly occurs in tropical and subtropical regions of the world, and it can have a significant economic impact on livestock productivity (Felsheim et al., 2010). The disease is present in Bhutan and there were only five cases reported in cattle in 2020 (Sonam Deki, personal communication, 2021, Thimphu, Bhutan), but this may indicate a gap in case reporting due to a lack of diagnostic confirmation. On the other hand, records from the National Veterinary Hospital in Thimphu showed 90 cases of anaplasmosis in dogs and 11 in cats (Dr. Meena D Samal, personal communication, 2021, Thimphu, Bhutan).

19.8   Diagnostics

Diagnosis of diseases caused by *Babesia*, *Theileria*, and *Anaplasma* spp. is often done by simple methods such as the examination of Giemsa-stained blood smears (OIE, 2014). Further, for *Theileria* spp. and *Anaplasma* spp., lymph node biopsy smears and organ impressions from dead animals can also be used (OIE, 2015, 2018). Additional serological tests, such as enzyme-linked immunosorbent assay, can be used as well as molecular testing of samples from infected animals. However, in Bhutan (as in many other parts of the world), although all veterinary laboratories in the districts are equipped with basic diagnostic equipment to examine blood smears, there are not many routine tests done to confirm suspected TBDs. As a result, the prevalence of these diseases in different regions of the country is not currently known.

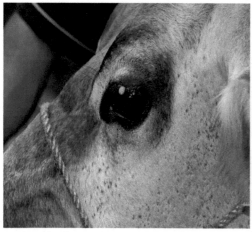

Figure 19.3 Tick-infested cattle (left) and dogs (right) in Eastern Bhutan.

19.9 Tick-borne diseases in humans

In the human health setting there is limited evidence of TBDs in Bhutan. However, there is a high incidence of acute undifferentiated febrile illness reported in the country (Tshokey et al., 2017) and rickettsioses are thought to be responsible for at least 15% of these cases (Tshokey et al., 2018). A study in 1044 patients with undifferentiated febrile illness in Bhutan reported that 47% had bush/forest contact and 45% had animal contact (Tshokey et al., 2018). In Bhutan it is not uncommon for animals to be infested with ticks (see Figure 19.3) and, considering that many rickettsiae are maintained and transmitted by ticks (Brouqui et al., 2004), they could be potential vectors for rickettsial organisms in Bhutan (Namgyal et al., 2021). A serological study in 294 animals found that 136 were seropositive to one or more rickettsiae (Tshokey et al., 2019). Among the animal species tested, dogs were found to show the highest seropositivity against spotted fever group rickettsia, horses against scrub typhus group rickettsia, and goats for *Coxiella* (Tshokey et al., 2019). Currently there are no clinical guidelines to guide the diagnosis and treatment of rickettsial diseases in humans in Bhutan (Tshokey et al., 2018). Since there is a huge gap in understanding the causes of rickettsial diseases in Bhutan (Tshokey et al., 2018), there is a need to increase collaboration between human and animal health teams in rickettsial disease research (Tshokey et al., 2019).

19.9.1 Rickettsiae in humans

Rickettsia is a genus of Gram-negative, nonmotile, highly pleomorphic bacteria that may occur in the form of cocci, bacilli, or threads. Rickettsiae are transmitted to humans by the bite of infected ticks and mites and by the feces of infected fleas and lice. They can enter via the skin and spread through the bloodstream to cause systemic disease affecting a wide range of organs including the vascular endothelium in the skin, brain, heart, lungs, kidneys, liver, and gastrointestinal tract.

Spotted fever group rickettsioses (spotted fevers) are a group of diseases caused by closely related bacteria (https://www.cdc.gov/otherspottedfever/index.html).

19.10 Current and traditional tick control practices in Bhutan

The prevention and control of tick infestation is important for reducing the direct and indirect effects of ticks on livestock productivity (Jongejan and Uilenberg, 1994). Many tick control approaches, such as the use of chemical acaricides, anti-tick vaccines, and selection of genetically resistant cattle breeds, are available but there is no single approach to tick control (Willadsen, 2006). Globally, the most widely used method for controlling ticks on livestock is the direct topical application of chemical acaricides to host animals (Minjauw and McLeod, 2003). Chemical acaricides, when used properly, are efficient and cost effective; however, their well-known disadvantages are the potential development of resistance by ticks to acaricides and concerns about environmental pollution, residues in food products – such as meat and milk, and natural toxicity (de Castro, 1997; Willadsen, 2006).

The wide use of chemical acaricides in Bhutan started in the 1990s; currently, chemical acaricide application is the primary method used for the control of ticks in cattle. The National Centre for Animal Health (NCAH) in the Department of Livestock (DoL) is responsible for the selection, procurement, and supply of acaricides to all districts in Bhutan. The most commonly used acaricides in Bhutan are liquid formulations of pyrethroid compounds (i.e., cypermethrin, deltamethrin, and flumethrin) and amidines (i.e., amitraz) imported from India (NCAH, 2013). These chemicals are provided free of charge to farmers by the government. The NCAH (funded through the government) spends approximately 2–3 million Bhutanese Ngultrum (1 USD = Nu.70) annually to provide free acaricides to farmers. Because most of the farms in the country are small-scale farms scattered over a large area with only five or six cattle per farm on average (DoL, 2018; National Statistics Bureau, 2018), community dip tanks and spray races are not practically feasible in Bhutan. Therefore farmers typically follow the hand-dressing method, which involves applying diluted acaricides to common tick attachment sites on the host animals with a sponge or cloth fabric. This method is considered ideal in some situations, such as for cattle reared in small and zero-grazed production systems (Minjauw and McLeod, 2003).

- Acaricides are chemical substances that are poisonous and primarily used for controlling ticks and mites. Some of these chemicals are irritants and can damage the environment and therefore must be handled and disposed of carefully.
- The main methods of applying acaricides to cattle are:
  - Dipping – cattle are made to move through dip tanks sufficiently deep to allow each animal to be completely wetted with acaricide and designed in such a way that excess acaricide running off from dipped animals is returned to the tank.
  - Spray race – cattle are made to pass through a corridor where acaricides are sprayed by pump at low pressure (but high volume) through nozzles placed along the corridor.
  - Hand spray – cattle are tethered to posts and acaricides are sprayed through a nozzle held close to their body.
  - Pour-on acaricide solutions or suspensions are poured along the back line of cattle.
  - Hand dressing – acaricide solutions are applied to the preferred tick attachment sites on the host, using a brush, sponge, or piece of cloth.
- In Bhutan, hand dressing is the most common method of applying acaricides.

Acaricide alone is not a sustainable, environmentally friendly, and cost-effective approach, and the development of resistance to these chemicals has become a serious threat to the sustainability of this approach.

Therefore, good tick control practice involves a systematic combination of two or more methods (de Castro, 1997). The timing of treatments based on tick species and their life cycles over an annual seasonal cycle is more important than an ad hoc frequency of application. A strategy adhering to well-timed, tick-specific treatment application should also be integrated with other control methods such as the use of resistant cattle breeds and pasture management (e.g., rotational grazing and controlled burning).

In Bhutan, besides manual removal of ticks – which is frequently practised by farmers rearing improved cattle breeds, there are also other indigenous methods used for controlling ticks. For example, *Zanthoxylum armatum* DC, commonly known as 'Thi-ngye' in Bhutan, is an important medicinal plant widely available in subtropical and temperate valleys of the Himalayas, including Bhutan, and it has several ethnopharmacological uses (Phuyal et al., 2019). Although not documented anywhere, farmers in Bhutan are known to have used *Zanthoxylum* spp. as both acaricide and insect repellant for a long time. The common practice, while using *Zanthoxylum* spp. as an acaricide, is that the seeds are soaked in water overnight and the resultant solution is then applied on tick attachment sites of the host animals (Namgyal et al., 2021). While the ethnoveterinary use of *Zanthoxylum* spp. as an acaricide is documented in studies in Pakistan (Sindhu et al., 2010) and Brazil (Nogueira et al., 2014), further studies are required to examine the efficacy of this culturally adopted approach

in Bhutan. Highly concentrated common salt solution is also applied by farmers to control ticks in Bhutan, a practice also documented from Punjab province in Pakistan (Muhammad et al., 2008). The most popular indigenous method of tick control in Bhutan is the use of a mixed solution of *Zanthoxylum* spp. seed and common salt. However, the effectiveness of this approach is also not well known.

19.11 Farmers' perspectives on ticks and TBDs in Eastern Bhutan

As indicated above, almost half of the Bhutanese population is comprised of subsistence farmers and their lives are intricately linked with animals, pastures, and forests. Understanding farmers' indigenous knowledge, management practices, and their perception of ticks and TBDs is crucial for the design and implementation of effective tick control and prevention strategies. To date, only one study (Namgyal et al., 2021) has been conducted in Eastern Bhutan to understand farmer perspectives on ticks and TBDs. *Rhipicephalus microplus*, the most predominant tick in Bhutan (Namgyal, 2020; Pem et al., 2021), is considered to be the most economically important tick globally because it is an important vector for major TBDs in cattle (Jongejan and Uilenberg, 2004). For example, every year, in Brazil and Mexico, annual losses from infestation of *R. microplus* were estimated at US\$3.24 billion (Grisi et al., 2014) and US\$ 573.61 million per annum (Rodríguez-Vivas et al., 2017). However, in Bhutan, most farmers might not even be able to recognize this tick, let alone understand the level of loss it can cause. The general trend among farmers in Bhutan is that, unless there is a large mortality of animals on their farms, they do not seek assistance from veterinary authorities for advice on suitable interventions. Veterinary officials too, in the absence of a major disease outbreak, often have limited resources to examine health and production issues in depth. Bhutanese farmers are generally more concerned about the lack of access to fodder and pasture for their cattle (Hidano et al., 2017) than anything else. This is also because livestock rearing in Bhutan is focused, to a large extent, around milk production. Tick-related direct injuries like hide damage caused are not of great concern and there is a lack of understanding about the direct and indirect impact of tick infestation on growth and productivity in cattle. In addition, in Eastern Bhutan, of the 246 farmers interviewed, 242 (98.4%) had never heard of any specific TBDs in cattle (Namgyal et al., 2021), indicating a very low level of awareness. Similarly, of the 148 farmers who had experienced tick bites at some point in their lives, 90 had incorrectly believed that these could not transmit disease to humans (Namgyal et al., 2021).

In Bhutan, Buddhism plays an important part in blending cultural principles with technology. While these principles have helped setting up the boundaries of moral and ethical responsibilities to manage forest cover, grazing land, and water bodies for sustainable ecological balance, there are also some problems associated with it. For instance, *Saga dawa* – the auspicious month according to the lunar calender (Kuensel, 2016) – is one example in the domain of animal health. *Saga dawa* has considerable influence on tick control practice in Bhutan. Despite animals being highly infested by ticks, most farmers avoid using acaricides during this month because, at this particular time, most Bhutanese farmers refrain from nonvirtuous acts and killing ticks is considered one of them (Namgyal et al., 2021). Coincidentally, this month sometimes falls in spring, the peak tick season in Bhutan. Therefore, it is important to embrace Buddhist cultural principles in the system of tick management in Bhutan.

Community engagement, involving efforts to promote dynamic relationships and mutual exchange of information, ideas, and resources between community members and public health/animal health policy makers, is crucial in achieving successful prevention and control of diseases (Morgan and Lifshay, 2006). Community engagement is not a new strategy in Bhutan – it has played an important role over the years in the prevention and control of many diseases. For instance, rabies control in Bhutan is a success story largely due to tremendous efforts from the government in engaging communities for discussions to determine suitable strategies to achieve a disease prevention goal (Rinchen et al., 2021). In the domain of ticks and TBDs, farmers were directly involved in the collection of ticks (Namgyal, 2020) and also as respondents in knowledge, attitude, and practice surveys (Namgyal et al., 2021), and these have helped in raising awareness about ticks and TBDs in Bhutan, at least in those communities involved. However, to effectively achieve this, community engagement and education are crucial. Unless farmers are made aware of the negative impacts of ticks on cattle and their economic gains, they are less likely to support animal health programmes such as tick collection or tick submission. Therefore, concerted efforts are required to further engage communities and to educate them about ticks and TBDs to achieve better prevention and control plans.

19.12 An interdisciplinary approach to monitoring and controlling ticks and tick-borne diseases in Eastern Bhutan

An interdisciplinary approach, engaging a wide range of stakeholders, is required to develop and implement effective programmes to control ticks and TBDs in Bhutan.

Currently information about ticks on cattle in Bhutan is limited to two regions, the Eastern

Veterinary services in Bhutan are provided free of charge to farmers. The following are the services provided by the state:

- All vaccines against notifiable animal diseases in Bhutan are free of charge.
- All veterinary drugs are free of charge.
- All clinical veterinary services including major surgeries are free of charge.
- All veterinary laboratory diagnostic services are free of charge.
- All surveillance and monitoring of animal diseases are free of charge.
- All disease outbreak response costs are borne by the state.

region (Namgyal, 2020) and central-western region (Pem et al., 2021). Although the presence of ticks might vary depending on their life cycle, season, environmental conditions, and host availability, there is still a potential to collect ticks from other regions. Although initiating active surveillance will be very costly, the government can leverage on a wide network of veterinarians or veterinary laboratory technicians at the District Veterinary Hospitals (DVHs) and paraveterinarians at subdistrict Livestock Extension Centres (LECs) to submit ticks from domestic livestock and companion animals (Figure 19.4). Similarly, the Bhutan Food and Drug Authority (BFDA) has officials in all animal quarantine centres in Bhutan and therefore there is an opportunity to collaborate with BFDA for submission and testing of ticks from imported animals (Figure 19.4). Likewise, there is also a potential in initiating passive surveillance in wildlife in collaboration with the Department of Forests and Park Services, in regard to especially tick collection from wild animals and birds during rescue and relocation (Figure 19.4). Hospitals in the Ministry of Health can also submit ticks from humans

(Figure 19.4). Such collaboration in passive surveillance can help in establishing a comprehensive database for tick species in Bhutan. As shown in Figure 19.4, there is a tremendous potential in Bhutan to collaborate on surveillance and research related to ticks and TBDs. Community engagement is ongoing, and farmer participation in the development and implementation of livestock disease prevention and control policies is fundamental. Additionally, there is already an ongoing linkage with experts from Canada and Japan.

19.13 Conclusion

Because climate change is affecting tick distribution and disease transmission globally, TBDs are increasingly becoming a health burden throughout the world (Gray et al., 2009). Bhutan, by virtue of its location in the Himalayas, is considered to be one of the most vulnerable places on earth that will be impacted by climate change (Hoy et al., 2016). In 2021, for the first time in Lunana, Bhutan at 3500 m above sea level, mosquitoes were sighted. Similarly, mountain regions of Nepal have already reported 900 cases of malaria (Sureis, 2019). These examples indicate that mountain regions are becoming ever warmer. In this context, tropical and subtropical arthropods, including ticks …. and TBDs might also relocate to higher latitudes in Bhutan as a result of climate change, bringing potential animal and public health threats to new areas. Further, anthropological activities, demographic shifts, urban encroachment onto agricultural areas, and other factors will also play a role in tick distribution. In our case study we have illustrated how an interdisciplinary team worked with communities and policy makers to develop prevention and control plans to deal with common ticks and TBDs in Eastern Bhutan. Currently in Bhutan, as ticks have a complex life cycle influenced by changing environmental

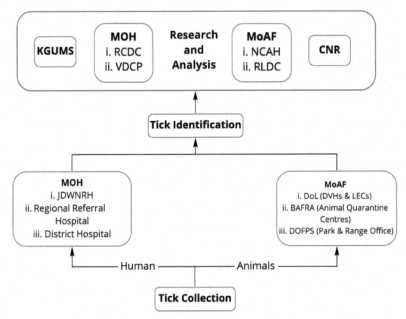

Figure 19.4 Suggested One Health collaboration for ticks and TBDs in Bhutan.

Note: KGUMS, Khesar Gyelpo University of Medical Sciences; RCDC, Royal Centre for Disease Control; VDCP, Vector Borne Disease Control Program; MoAF, Ministry of Agriculture and Forests; JDWNRH, Jigme Dorji Wangchuck National Referral Hospital; CNR, College of Natural Resources.

conditions, there is a need to consider components such as season, host availability, and local climatic conditions while designing suitable tick prevention and control plans. It is clear that farmer education plays an important role in the implementation of such disease prevention and control plans and that community engagement is key to ensuring ongoing compliance with tick control guidelines.

References

Apanaskevich, D. A., & Oliver, J. H. (2013). Life cycles and natural history of ticks. In *Biology of ticks*, 1 (pp. 59–73). Oxford University Press.

Bock, R., Jackson, L., de Vos, A., & Jorgensen, W. (2004). Babesiosis of cattle. *Parasitology*, *129*(Suppl.), S247–S269. https://doi.org/10.1017/S0031182004005190

Brouqui, P., Bacellar, F., Baranton, G., Birtles, R. J., Bjoёrsdorff, A., Blanco, J. R., Caruso, G., Cinco, M.,

Fournier, P. E., Francavilla, E., Jensenius, M., Kazar, J., Laferl, H., Lakos, A., Lotric Furlan, S., Maurin, M., Oteo, J. A., Parola, P., Perez-Eid, C., . . . European Network for Surveillance of Tick-Borne Diseases. (2004). Guidelines for the diagnosis of tick-borne bacterial diseases in Europe. *Clinical Microbiology and Infection*, *10*(12), 1108–1132. https://doi.org/10.1111/j.1469-0691.2004.01019.x

de Castro, J. J. (1997). Sustainable tick and tick-borne disease control in livestock improvement in developing countries. *Veterinary Parasitology*, *71*(2–3), 77–97. https://doi.org/10.1016/s0304-4017(97)00033-2

Debbarma, A., Pandit, S., Jas, R., Baidya, S., Mandal, S. C., & Jana, P. S. (2018). Prevalence of hard tick infestations in cattle of West Bengal, India. *Biological Rhythm Research*, *49*(5), 655–662. https://doi.org/10.1080/09291016.2017.1395527

DoL. (2018). *Livestock Statistics*. http://www.dol.gov.bt/?page_id=394

Estrada-Peña, A. (2015). Ticks as vectors: Taxonomy, biology and ecology. *Revue Scientifique et Technique*, *34*(1), 53–65. https://doi.org/10.20506/rst.34.1.2345

Estrada-Pena, A., & Mans, B. J. (2013). Tick-induced paralysis and toxicoses. In *Biology of ticks volume 2.* illustrated (rev. ed). Oxford University Press.

Felsheim, R. F., Chávez, A. S., Palmer, G. H., Crosby, L., Barbet, A. F., Kurtti, T. J., & Munderloh, U. G. (2010). Transformation of Anaplasma marginale. *Veterinary Parasitology, 167*(2–4), 167–174. https://doi.org/10.1016/j.vetpar.2009.09.018

Fivaz, B., Petney, T., & Horak, I. (2012). *Tick vector biology: Medical and veterinary aspects.* Springer Science+Business Media.

Gray, J. S., Dautel, H., Estrada-Peña, A., Kahl, O., & Lindgren, E. (2009). Effects of climate change on ticks and tick-borne diseases in Europe. *Interdisciplinary Perspectives on Infectious Diseases, 2009*, 593232. https://doi.org/10.1155/2009/593232

Grisi, L., Leite, R. C., Martins, J. R., Barros, A. T., Andreotti, R., Cançado, P. H., León, A. A., Pereira, J. B., & Villela, H. S. (2014). Reassessment of the potential economic impact of cattle parasites in Brazil. *Revista Brasileira de Parasitologia Veterinária, 23*(2), 150–156. https://doi.org/10.1590/s1984-29612014042

Hidano, A., Sharma, B., Rinzin, K., Dahal, N., Dukpa, K., & Stevenson, M. A. (2017). Revisiting an old disease? Risk factors for bovine enzootic haematuria in the Kingdom of Bhutan. *Preventive Veterinary Medicine, 140*, 10–18. https://doi.org/10.1016/j.prevetmed.2017.02.011

Hoy, A., Katel, O., Thapa, P., Dendup, N., & Matschullat, J. (2016). Climatic changes and their impact on socio-economic sectors in the Bhutan Himalayas: An implementation strategy. *Regional Environmental Change, 16*(5), 1401–1415. https://doi.org/10.1007/s10113-015-0868-0

Jongejan, F., & Uilenberg, G. (1994). Ticks and control methods. *Revue Scientifique et Technique de l'OIE, 13*(4), 1201–1226. https://doi.org/10.20506/rst.13.4.818

Jongejan, F., & Uilenberg, G. (2004). The global importance of ticks. *Parasitology, 129*(Suppl.), S3–S14. https://doi.org/10.1017/S0031182004005967

Kakati, P., Sarmah, P. C., Ray, D., Bhattacharjee, K., Sharma, R. K., Barkalita, L. M., Sarma, D. K., Baishya, B. C., Borah, P., & Stanley, B. (2015). Emergence of oriental theileriosis in cattle and its transmission through Rhipicephalus (Boophilus) microplus in Assam, India. *Veterinary World, 8*(9), 1099–1104. https://doi.org/10.14202/vetworld.2015.1099-1104

Kuensel. (2016). Saga Dawa, the auspicious month. https://kuenselonline.com/saga-dawa-the-auspicious-month/

Leal, B., Thomas, D. B., & Dearth, R. K. (2018). Population dynamics of off-host Rhipicephalus (Boophilus) microplus (acari: Ixodidae) larvae in response to habitat and seasonality in South Texas. *Veterinary Sciences, 5*(2). https://doi.org/10.3390/vetsci5020033

Minjauw, B., & McLeod, A. (2003). Tick-borne diseases and poverty. The impact of ticks and tick-borne diseases on the livelihood of small-scale and marginal livestock owners in India and Eastern and southern Africa. http://agris.fao.org/agris-search/search.do?recordID=GB2012100456

MoAF. (2019). Bhutan RNR statistics – 2018. Ministry of Agriculture and Forests. http://www.moaf.gov.bt/bhutan-rnr-statistics-2018/

Morgan, M. A., & Lifshay, J. (2006). Community engagement in public health. California Endowment under the sponsorship of Contra Costa Health Services (CCHS), pp. 1–8.

Muhammad, G. et al. (2008). Tick control strategies in dairy production medicine. *Pakistan Veterinary Journal, 28*(1), 43.

Mushahary, D., Bhattacharjee, K., Sarmah, P. C., Kr. Deka, D., Upadhyaya, T. N., & Saikia, M. (2019). Prevalence of ixodid ticks on local and crossbred cattle in Indo-Bhutan border districts of Assam, India. *International Journal of Current Microbiology and Applied Sciences, 8*(5), 2168–2183. https://doi.org/10.20546/ijcmas.2019.805.256

Namgyal, J. (2020). Identification and distribution of tick species in cattle in Eastern Bhutan. https://prism.ucalgary.ca/handle/1880/112725

Namgyal, J., Tenzin, T., Checkley, S., Lysyk, T. J., Rinchen, S., Gurung, R. B., Dorjee, S., Couloigner, I., & Cork, S. C. (2021). A knowledge, attitudes, and practices study on ticks and tick-borne diseases in cattle among farmers in a selected area of Eastern Bhutan. *PLOS ONE, 16*(2), e0247302. https://doi.org/10.1371/journal.pone.0247302

Namgyal, J., Lysyk, T. J., Couloigner, I., Checkley, S., Gurung, R. B., Tenzin, T., Dorjee, S., & Cork, S. C. (2021). Identification, distribution, and habitat suitability models of ixodid tick species in cattle in Eastern Bhutan. *Tropical Medicine and Infectious Disease, 6*(1), 27. https://doi.org/10.3390/tropicalmed6010027

National Statistics Bureau. (2018). 2017 population and housing census of Bhutan: National report

['Brug gi mi rlobs dang khyim gyi grangs rtsis 2017]. http://www.nsb.gov.bt/publication/files/PHCB2017_national.pdf

NCAH. (2013). External parasiticides. https://www.ncah.gov.bt/Downloads/File_3.pdf. In *National veterinary drug formulary* (2nd ed) p. 35. Department of Livestock, Royal Government of Bhutan.

NCAH. (2019). Veterinary information system. http://vis.ncah.gov.bt/php/index.php Retrieved February 15, 2020

Nieto, N. C., Porter, W. T., Wachara, J. C., Lowrey, T. J., Martin, L., Motyka, P. J., & Salkeld, D. J. (2018). Using citizen science to describe the prevalence and distribution of tick bite and exposure to tickborne diseases in the United States. *PLOS ONE, 13*(7), e0199644. https://doi.org/10.1371/journal.pone.0199644

Nogueira, J., Vinturelle, R., Mattos, C., Tietbohl, L. A., Santos, M. G., Junior, I. S., Mourão, S. C., Rocha, L., & Folly, E. (2014). Acaricidal Properties of the Essential Oil from Zanthoxylum caribaeum against Rhipicephalus microplus. *Journal of Medical Entomology, 51*(5), 971–975. https://doi.org/10.1603/ME13236

OIE. (2014). Manual of diagnostic tests and vaccines for terrestrial animals 2019. https://www.oie.int/en/standard-setting/terrestrial-manual/access-online/

OIE. (2018). Manual of diagnostic tests and vaccines for terrestrial animals 2019. https://www.oie.int/fr/normes/manuel-terrestre/acces-en-ligne/

Pem, R. et al. (2021). Seroprevalence of babesia bigemina in cattle in West-central region of Bhutan. *Bhutan Journal of Animal Science, 5*(1), 82–88.

Phanchung, P. et al. (2007). Smallholder dairy farming in Bhutan: Characteristics, constraints, and development opportunities. http://lib.icimod.org/record/21365/files/c_attachment_79_552.pdf. In.

Phillips, S. J., Anderson, R. P., & Schapire, R. E. (2006). Maximum entropy modeling of species geographic distributions. *Ecological Modelling, 190*(3–4), 231–259. https://doi.org/10.1016/j.ecolmodel.2005.03.026

Phuyal, N., Jha, P. K., Prasad Raturi, P., & Rajbhandary, S. (2019). Zanthoxylum armatum DC.: Current knowledge, gaps and opportunities in Nepal. *Journal of Ethnopharmacology, 229,* 326–341. https://doi.org/10.1016/j.jep.2018.08.010

Rinchen, S. et al. (2021). One health in policy development: Options to prevent rabies in cattle in Bhutan. In J. Zinsstag et al. (Eds.), *One Health:*

The theory and practice of integrated health approaches (2nd ed) (pp. 382–393). CABI Publishing. https://doi.org/10.1079/9781789242577.0382

RLDC Wangdue (2019). *9th Annual Progress Report of RLDC Wangdue (2018–19).* http://www.dol.gov.bt/?wpdmpro=rldc-wangdue-annual-progress-report-fy-2018–19

Rodríguez-Vivas, R. I., Grisi, L., Pérez de León, A. A., Silva Villela, H., Torres-Acosta, J. Fd. J., Fragoso Sánchez, H., Romero Salas, D., Rosario Cruz, R., Saldierna, F., & García Carrasco, D. (2017). Potential economic impact assessment for cattle parasites in Mexico. Review [Review]. *Revista Mexicana de Ciencias Pecuarias, 8*(1), 61. https://doi.org/10.22319/rmcp.v8i1.4305

Samdup, T., Udo, H. M. J., Eilers, C. H. A. M., Ibrahim, M. N. M., & van der Zijpp, A. J. (2010). Crossbreeding and intensification of smallholder crop-cattle farming systems in Bhutan. *Livestock Science, 132*(1–3), 126–134. https://doi.org/10.1016/j.livsci.2010.05.014

Sindhu, Z.-D. et al. (2010). Documentation of ethnoveterinary practices used for treatment of different ailments in a selected hilly area of Pakistan. *International Journal of Agriculture and Biology, 12*(3), 353–358.

Sureis. (2019). Malaria in hilly, mountainous regions. *The Himalayan Times.* https://thehimalayantimes.com/nepal/malaria-in-hilly-mountainous-regions

Sutherst, R. W., & Bourne, A. S. (2006). The effect of desiccation and low temperature on the viability of eggs and emerging larvae of the tick, Rhipicephalus (Boophilus) microplus (Canestrini) (Ixodidae). *International Journal for Parasitology, 36*(2), 193–200. https://doi.org/10.1016/j.ijpara.2005.09.007

Tshokey, T., Stenos, J., Durrheim, D. N., Eastwood, K., Nguyen, C., & Graves, S. R. (2017). Seroprevalence of rickettsial infections and Q fever in Bhutan. *PLOS Neglected Tropical Diseases, 11*(11), e0006107. https://doi.org/10.1371/journal.pntd.0006107

Tshokey, T., Stenos, J., Durrheim, D. N., Eastwood, K., Nguyen, C., Vincent, G., & Graves, S. R. (2018). Rickettsial infections and Q fever among febrile patients in Bhutan. *Tropical Medicine and Infectious Disease, 3*(1), 12. https://doi.org/10.3390/tropicalmed3010012

Tshokey, T., Stenos, J., Tenzin, T., Drukpa, K., Gurung, R. B., & Graves, S. R. (2019). Serological evidence of Rickettsia, Orientia, and Coxiella in domestic animals from Bhutan: Preliminary findings.

Vector Borne and Zoonotic Diseases, 19(2), 95–101. https://doi.org/10.1089/vbz.2018.2336

Utech, K. B. W., Wharton, R. H., & Kerr, J. D. (1978). Resistance to Boophilus microplus (Canestrini) in different breeds of cattle. *Australian Journal of Agricultural Research, 29*(4), 885–895. https://doi.org/10.1071/AR9780885

Wang, S. W., Lee, W.-K., & Son, Y. (2017). An assessment of climate change impacts and adaptation in South Asian agriculture. *International Journal of Climate Change Strategies and Management, 9*(4), 517–534. https://doi.org/10.1108/IJCCSM-05-2016-0069

Wangchuk, S., Pelden, S., Dorji, T., Tenzin, S., Thapa, B., Zangmo, S., Gurung, R., Dukpa, K., & Tenzin, T.

(2016). Crimean-Congo hemorrhagic fever virus IgG in goats, Bhutan. *Emerging Infectious Diseases, 22*(5), 919–920. https://doi.org/10.3201/eid2205.151777

Willadsen, P. (2006). Tick control: Thoughts on a research agenda. *Veterinary Parasitology, 138*(1–2), 161–168. https://doi.org/10.1016/j.vetpar.2006.01.050

Zinsstag, J., Mackenzie, J. S., Jeggo, M., Heymann, D. L., Patz, J. A., & Daszak, P. (2012). Mainstreaming one health. *EcoHealth, 9*(2), 107–110. https://doi.org/10.1007/s10393-012-0772-8

Chapter 20

Citizen science and conserving pollinators in Utah

Jacqualine Grant, Rachel Bolus, Ashlee Hardin, Matt Ogburn, Jacob Olvera, Julie Pynn, Claire Smith, Connor Smith, Isaac Sorensen, Samuel Wells and Leeloo Yutuc

Abstract

Citizen science is a research and educational practice that depends on volunteer contributions from the general public. Citizen science projects range from online contributions of nature observations, where the participant rarely meets a scientist face to face, to intensive in person expeditions in which researchers and volunteers work side by side. Citizen science and its participants come in many different forms but, regardless of the form it takes, citizen science has enhanced biodiversity research for many decades and is poised to address biodiversity questions at a global scale. Pollinators are one important aspect of biodiversity that is directly linked to both human and ecosystem health. Pollinators provide the ecosystem services necessary for the majority of all plant reproduction, the biodiversity of which underpins the web of life. Human health relies on the ecosystem services provided by pollinators, which result in agricultural crops (food), fibers for textiles, and other plant-based products essential to human existence. Although many vertebrates and invertebrates are pollinators, bees play an outsized role in the process. Many bee species around the world are thought to be on the decline, which has generated unprecedented interest in their conservation given the role of bees in sustaining ecosystem services. In this chapter we present a case study from the United States of America to describe a citizen science program that is focused on bee conservation, and to highlight the need for more citizen science programs that explicitly link EcoHealth measures to their research goals.

20.1 Introduction

Citizen science is a research practice in which data are generated by engaged volunteers who are sometimes motivated by politically relevant issues (Vohland et al., 2021). However, because citizen science encompasses such a broad array of perspectives, fields of study, and contexts, it is difficult to link a single, sweeping definition to the term. For example, Vohland et al.

Susan C. Cork and Douglas P. Whiteside (eds) **Case Studies in EcoHealth**
DOI: 10.52517/9781789183368.020, © 5m Books Ltd, 2024

(2021) compiled a list of 34 different definitions of the term citizen science! Eitzel et al. (2017) described how the term citizen science holds different meanings depending on the country or region in which it takes place, and how terminology related to terms describing the scientists and citizens who work together comes with many caveats. Regardless, citizen science has a longstanding history in the natural sciences (Brossard et al., 2005; Frigerio et al., 2018; Irwin, 1995), where it has been used to perform research across many disciplines (Kullenberg and Kasperowski, 2016). Examples include the discovery of new diatom species (Bahls, 2014), documentation of bird distribution (Sullivan et al., 2014), and description of galaxies in space (Raddick et al., 2019). Citizen science is particularly effective in the realm of biodiversity studies, with an estimated 1.3 million volunteers annually participating in citizen science projects around the world (Theobald et al., 2015). The focus of our chapter is the application of citizen science to biodiversity studies. We will not address the debate over the term 'citizen science'.

As one might expect, citizen science as related to vertebrates is overrepresented while citizen science related to invertebrates is underrepresented (Theobald et al., 2015). Within the invertebrates, projects have been biased towards butterflies and shellfish (Theobald et al., 2015). Although butterflies are important pollinators, they are not as efficient at pollination as bees (Barrios et al., 2016). Because they are the most efficient of all pollinators, bees consist of an important aspect of biodiversity that is directly linked to both human and ecosystem health. Bees and other insects provide the ecosystem service of pollination that is necessary for the majority of all plant reproduction (Ollerton et al., 2011). A functioning ecosystem relies on healthy plant populations because of their role as primary producers. Human health relies on the ecosystem services

provided by pollinators, which result in agricultural crops (food), fibers for textiles, and other plant-based products essential to human existence (Hristov et al., 2020; Vanbergen and the Insect Pollinators Initiative, 2013). In turn, pollinator health relies on a healthy environment with low levels of parasites, pesticides, diseases, and plant monocultures (Goulson et al., 2015).

The One Health approach can be defined as an approach that recognizes the close relationship between human health and the health of animals and the environment (Centers for Disease Control and Prevention, 2018), and also one that purposefully designs and implements programs, policies, legislation, and research with input from a variety of partners to achieve better public health outcomes (World Health Organization, 2017). From a One Health perspective, pollinators are directly connected to human health outcomes, environmental quality, and the health of other animals because of the ecosystem services pollinators provide (López-Uribe et al., 2020), which means that humans should be focused on measuring pollinator health. However, despite over a decade of known pollinator declines (Potts et al., 2010), standardized measures of pollinator health are only just now being established, especially for wild bees (López-Uribe et al., 2020). A great potential exists for incorporating measures of pollinator health into citizen science projects, but the connection between pollinator health and citizen science programs is in the nascent stages, as evidenced by the low number of published studies in which the two are connected (Figure 20.1).

One reason for the low number of citizen science studies focused on pollinator health is the similarly low number of citizen science projects related to bees (Koffler et al., 2021). Very few bee-focused citizen science data were published between 1992 and 2010, but that number now appears to be on the rise (Koffler

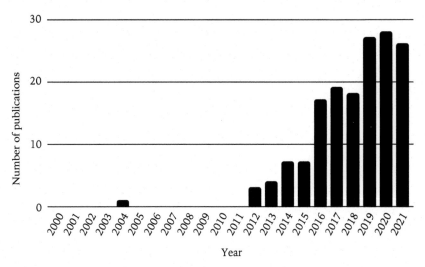

Figure 20.1 An informal survey of published papers on Google Scholar as revealed by the search terms 'pollinator health' AND 'citizen science'. The search was performed by Jacqualine Grant on January 18, 2021, and is a proxy for measuring the annual rate of peer-reviewed articles that explicitly link pollinator health with citizen science projects (currently <30 per year).

et al., 2021), in part because of national policies such as the United States' Presidential Memorandum on pollinators (Obama, 2014). One program that supports citizen science pollinator research is Earthwatch (Earthwatch, 2021). As described on their website, 'Earthwatch is an international nonprofit organization that connects citizens with scientists to improve the health and sustainability of the planet.' Over its 50-year history, Earthwatch has sponsored over 1,430 citizen science projects in over 130 countries. The results of these citizen science projects have informed 1,200 policies and increased the knowledge base of at least 10,000 species. A typical Earthwatch expedition brings scientists and paying volunteers together in the field to collect data for a peer-reviewed research project (Chandler et al., 2012). In this chapter we describe one of the pollinator citizen science projects in the Earthwatch portfolio.

20.2   Case study: citizen scientists study pollinators in southern Utah

Our programs take place in southern Utah (Figure 20.2) because it is a botanically diverse area near the boundary of three major centers of plant speciation: the Colorado Plateau, the Mojave Desert, and the Utah High Plateaus. Elevations in the region are as low as 600 m but soar to over 3300 m. This elevational breadth captures an assemblage of diverse vegetation types, which include desert shrublands, montane coniferous forests, alpine meadows, and riparian woodlands (Fertig, 2005). As might be expected, the region is home to thousands of plant species. For instance, over 1000 named plant taxa have been found in one of the region's national monuments, the Grand Staircase-Escalante National Monument (Fertig, 2005), and another, Cedar Breaks National Monument, is home to over 340 plant species (Fertig, 2009). Because of the close association between flowering plants and pollinators, the region is expected to contain a similarly high amount of pollinator diversity. Indeed, systematic surveys for pollinators in the Grand Staircase-Escalante National Monument between 2000 and 2003 revealed an incredible amount of bee diversity – over 650 species (Carril et al., 2018)! This number does not include wasps, butterflies, flies, or beetles, which are also known to pollinate flowering plants. Regional threats to pollinators include degazetting of federally protected areas (Wilson et al., 2018), industrial honey bee farming (Cane and Tepedino, 2017), and habitat fragmentation (Howell et al., 2017). The combination of high levels of pollinator diversity in an understudied area, a lack of basic natural history information on many species, and

looming threats to pollinator diversity make southern Utah an outstanding area for citizen science-based pollinator research.

A common motivating interest in many citizen science projects is the documentation of local and regional diversity (Silvertown, 2009). This interest can be facilitated and driven by available field guides and online resources. Projects that capture the most interest tend to focus on taxonomic groups that are fairly well known and from which species identifications are possible by non-specialists. Surprisingly, our research on the pollinators of southern Utah does not easily involve these groups. Our primary targets are bees and the local flora they visit. Bee species alone number in the hundreds. One study in the Grand Staircase-Escalante National Monument

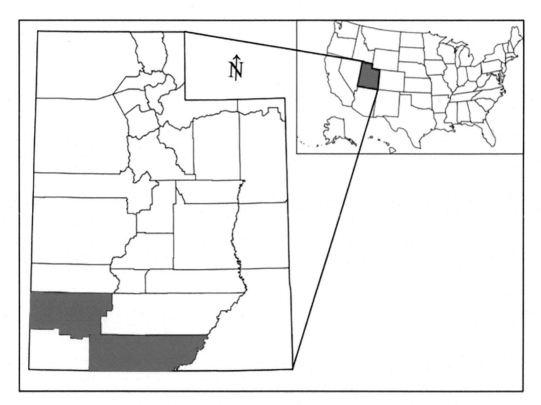

Figure 20.2 Field work locations in southern Utah were primarily located in Iron County in 2021, and are planned for Kane County in 2022. Both counties are located in southwestern Utah and are shaded on the map. Inset map shows the position of the state of Utah within the USA.

(just north of the Arizona border) yielded 660 species of bees in 55 genera (Carril et al., 2018), which is almost as many bee species as can be found east of the Mississippi River in the United States of America! The outstanding diversity of native bees in Utah is due in part to the wide range of elevation, temperature, and geologic gradients throughout the state, which lead to a diverse flora of over 3,000 plant species (Welsh et al., 1987).

The local array of plant species can be similarly intimidating to non-specialists. Three of the most common plant genera in the area are *Penstemon*, *Erigeron*, and *Astragalus* (beardtongues, fleabanes, and milkvetches, respectively). Together these three genera contain over 355 species within the borders of the state of Utah (Welsh et al., 1987). The identification keys for these species challenge even experienced taxonomists. Voucher specimens of plants and insects that might be used for comparison of species were often housed in institutions geographically distant from our location in Cedar City, Utah. The combination of immense biodiversity and lack of regional specialists and voucher specimens could be a barrier to the establishment of a robust citizen science program. Therefore, we considered each of these challenges in turn before heading into the field with citizen scientists.

We handled these hurdles in essentially three different ways. First, we focused on a restricted number of geographies and habitats, which allowed us to significantly reduce the number of species that we might encounter. Second, we recruited regional taxonomists to help identify the key taxa in our program areas. Third, we created data collection protocols that were tailored to the brief training window allowed by our program's structure. For example, we focused taxonomy on well-known groups such as bumble bees, and developed simplified protocols that did not require volunteers to be able to distinguish invertebrate pollinators beyond the taxonomic level of order (bees and wasps, flies, beetles, and butterflies). We spent the year prior to the onset of our citizen science field program developing our own reference collection at Southern Utah University, and created a plan to use pinned specimens and online images to familiarize the volunteers with all of the species that might be encountered.

To ensure that our data collection was meaningful to both volunteers and land managers, we worked with government agencies for several years in advance of our citizen science program. Iron County in southern Utah is home to 503,024 hectares of federally managed land. The majority of this land is managed by three agencies, the Bureau of Land Management (BLM), the United States Forest Service (USFS), and the National Park Service (NPS). These agencies are charged with the responsible management of public lands throughout the United States. Public lands and the pollinators that inhabit them are affected by global change, federal policies, and public demand for access, which makes them excellent candidates for studies supported by citizen science volunteers.

Global change in Utah and much of the United States' Intermountain West has come in the form of damaged habitat due to overgrazing (Longmore and Forrest, 2016), off-road vehicle recreation (Switalski, 2018; Switalski and Jones, 2012), introduction of non-native and invasive plants (Pilliod et al., 2017), and extractive industries (Haugen, 2018). All of the aforementioned activities affect native pollinators, and pressure to place managed hives of non-native honey bees (*Apis mellifera*) on public lands is on the rise (Cane and Tepedino, 2017). Therefore, we focused our research on questions that would have direct conservation applications and contribute to regional pollinator management planning. With these pollinator conservation issues in mind, we were able to apply to work with volunteers from the Earthwatch Institute.

Our initial program consisted of four major research objectives: (1) measure bee diversity in unexplored areas of southern Utah; (2) measure the attractiveness of individual plant species to pollinators from four taxonomic orders: bees and wasps, flies, beetles, and butterflies; (3) collect tissue samples to construct a library of genetic markers; and (4) collect preliminary acoustic recordings from bumble bees to construct a library of bee sounds. The first two objectives were well suited for citizen scientists, but the second two require further development on our part before they will be truly successful in a citizen science setting.

In spite of – or perhaps because of – the pandemic conditions that prevailed in 2021, our Earthwatch program was enrolled to full capacity. International travel was below average in 2021, which meant that many Earthwatch expeditions were unavailable to regular participants. However, because our program was based in the United States of America, it was still open to domestic travelers and high school students. We ran two sessions of two weeks in length and two sessions of one week in length for a total of 60 participants, accompanied by four undergraduate assistants, and four faculty researchers who took turns directing field activities. The age range of the participants was from 16 to 79 years old, which meant that some activities required modifications to keep everyone

engaged. We estimated a total of 2,700 hours of volunteer labor during the course of the field season. These hours do not translate one to one in terms of faculty or student hours, but they are still incredibly valuable.

Beyond its scientific goals, our citizen science project had two major objectives: first, to empower undergraduate students as educators and scientists in their own right; and second, to meet as many of the European Citizen Science Association's Ten Principles of Citizen Science (Robinson et al., 2018) as were within our reach. In some ways, preparing undergraduate students to work with citizen scientists can be similar to teaching a child to swim by throwing them into the water. We read books, we identified bees, we learned about plants and habitats, yet there were still so many questions that we could not answer when the volunteers arrived. It can be easy for imposter syndrome to take hold during a citizen science program but, by encouraging joy in not knowing all the answers, we maintained our composure and student engagement in our team.

We gathered data on many native bee species (Table 20.1) and the bumblebee species that can be found on the Markagunt Plateau just east of Cedar City, Utah. These species included *Bombus huntii, B. morrissoni, B. nevadensis, B. centralis, B. bifarius, B. insularis,* and *B. apositus* (Table 20.2). We also included one species of conservation concern, *B. occidentalis,* in our studies.

Table 20.1 Total number of bees collected in each taxonomic family between 2020 and 2021.

Bee taxonomic family name	Common name	No.
Andrenidae	Miner and fairy bees	162
Apidae	Cuckoo, carpenter, digger, bumble, and honey bees	1486
Colletidae	Plasterer, masked, or yellow-faced bees	136
Halictidae	Sweat bees	1437
Megachilidae	Leafcutter bees	1187
Melittidae	Melittid bees	0
	Total	4,408

Table 20.2 Number of bumble bee species in the genus *Bombus* collected in 2020 and 2021.

Scientific name	Common name	Regional status	No. collected in 2020	No. collected in 2021
Bombus apositus	White-shouldered bumble bee	Common	16	15
B. bifarius	Two-form bumble bee	Very common	36	11
B. centralis	Central bumble bee	Common	117	60
B. fervidus	Golden northern bumble bee	Common	3	0
B. flavifrons	Yellow-head bumble bee	Common	0	0
B. huntii	Hunt bumble bee	Common	139	134
B. insularis	Indiscriminate cuckoo bumble bee	Common	26	1
B. melanopygus	Black tail bumble bee	Common	1	0
B. mixtus	Fuzzy-horned bumble bee	Common	0	0
B. morrisoni	Morrison bumble bee	Common	37	5
B. nevadensis	Nevada bumble bee	Common	12	2
B. occidentalis	Western bumble bee	Rare	2	2
B. rufocinctus	Red-belted bumble bee	Common	0	0
B. vandykei	Van Dyke bumble bee	Uncommon	0	0
B. vosnesenskii	Vosnesensky bumble bee	Very common	0	0
		Total	389	230

This group of bumble bee species is a very manageable list that allowed our undergraduate assistants and volunteers to quickly learn how to identify them by sight in the field. Bumble bees quickly became the focus of much of our research questions because of their relative ease of identification and clear connections to ecosystem health. For example, we observed bumble bees on the first day of emergence because adults immediately fly to and pollinate the flowers of native plants with a high degree of fidelity. Other bee species were more likely to move back and forth between different species of native plants, or spend time on non-native plant species. Bumble bees at our study site established nests in old rodent burrows and copses of conifers, which demonstrated unexpected connections between bumble bees, mammals, and plant community structure. We collected fewer bumble bees (Table 20.2) in 2021 (with

volunteers) than we did in 2020 (without volunteers), but our research questions were more intentional in 2021 than in 2020. Working with citizen scientists really made us think about why we were collecting bees, and whether or not non-lethal approaches could answer some of our questions.

By concentrating most of the research effort on a small number of species, volunteers were able to take ownership in the studies in a way that focused taxonomic attention on real ecological issues. After approximately two weeks in the field, many volunteers sounded like professional biologists when discussing our pollinator project. They knew scientific binomial names, but they also recognized the individual contributions of often overlooked species and unique biological components of local habitats. Undergraduate participants also grew in confidence and ability throughout the field season. Common themes among the undergraduate student leaders (the 'bee crew') were their growth in confidence and surprise at how much they learned from the Earthwatch volunteers.

In addition to the collection of bee biodiversity data, we found that it was quite useful to collect information on the experience of the participants themselves (Figure 20.3). One of the principles of citizen science is to evaluate the participant experience (Robinson et al., 2018). Research on the impact of people's experiences in nature has gained momentum in recent years. Actively noticing nature increases people's connection to nature and prompts them to hone their values and ethics regarding nature (Nisbet, et al., 2019; Richardson et al., 2021). To assess whether participation in the Earthwatch program increased participants' feelings of connectedness to the natural world, we used the *Inclusion of Nature in Self Scale* (Schultz, 2001). The scale is a series of seven Venn diagrams depicting degrees of relationship between self and nature. Adjacent circles indicate a feeling of disconnect with nature (score

of 1), while greater feelings of connection are indicated by increasingly overlapping circles (2–6). The image of a single circle indicates the feeling that oneself is not at all separate from the natural world (7). Participants were administered the scale before Earthwatch field activities commenced and at the end of the program. Statistical analysis of the pre–post scores indicated a significant increase in participants' interrelationship with nature (pre-mean, 4.59 and post-mean, 5.62). Furthermore, in an open-ended question prompting reflection on the nature immersion experience, participants mentioned their desire for future conservation activities. This is evidence of the success of this citizen science program.

20.3 Recommendations for future citizen science programming

As expected, our first field season with Earthwatch revealed intriguing and promising areas for future development and many areas

Figure 20.3 Earthwatch volunteers participating in research about their own perspectives on nature with Dr. Julie Pynn from the Department of Psychology at Southern Utah University. Photo courtesy of Mike Mao, Earthwatch Volunteer (2021).

for improvement. These areas include expanding the scope of the questions addressed by the project, integrating further with federal land management partners, simplifying our data collection protocols, adding variety to volunteer activities, and extending the outreach component of the project to bring in more partners from the local community.

Our first field season took place in 2021, which meant that we worked under a cloud of COVID-19 restrictions that limited our ability to interact in the ways normally expected from field programs. For example, all volunteers needed to have their own rooms instead of sharing a room, face coverings were expected at all times (indoors or out, Figure 20.4), cleaning and disinfecting took up time during the day, we were limited in our ability to interact with community partners, and we could not eat together indoors or provide typical buffet-style field meals. The pandemic conditions resulted in increased anxiety when participants became ill, even if the illness was a common cold or allergies. At the time, home testing kits were not readily available, which meant that when

Figure 20.4 Face coverings were required during the 2021 field season because of uncertainties surrounding the pandemic. Volunteers found the combination of masks and the tools of taxonomy to be a novel experience. Photo courtesy of Mike Mao, Earthwatch Volunteer (2021).

a person became ill they needed to be transported to a testing center far from the field site. Furthermore, because our program was host to teenage volunteers, we ran into difficulties at test sites because of the lack of parental signatories. Participants who came down with an illness during the program needed to be isolated at a separate location, which became challenging when large groups of people fled to our rural area to escape crowded urban conditions or to work remotely. All of these issues should be considered during the organization of future citizen science programs.

One of the biggest realizations that we came to during our first season was that most of our study organisms lacked even the most basic natural history information. Therefore, much of the data we collected in the 2021 field season was focused on basic characterization of the biotic community simply because so little is known about bees and their plant mutualists in our area. We were able to use this insight to develop new research protocols for 2022 that focus more on natural history observations and the collection of basic, but critical, information for managing pollinators and native plants on millions of acres. We also became aware of how much we, as faculty at a primarily undergraduate university with a heavy emphasis on teaching, appreciated and needed to have the motivation of a citizen scientist team to get us out into the field (Figure 20.5).

Another important realization was that the volunteers were in alignment with the bee conservation community, specifically in relation to decreasing the amount of lethal bee trapping that typically takes place during bee monitoring programs (Figure 20.6). Some scientists in the bee conservation community have been calling for a rethinking of native bee research programs since 2020 (Portman, et al., 2020; Portman and Tepedino, 2021). The recommendations of Portman et al. (2020) and Portman and Tepedino (2021) are to reduce the amount of

Figure 20.5 Earthwatch volunteers contributing to the Utah Pollinator Pursuit bumble bee database with Dr. Samuel Wells from the Department of Biology at Southern Utah University. Photo courtesy of Mike Mao, Earthwatch Volunteer (2021).

Figure 20.6 Traditional insect survey programs rely on pinned specimens; however, many volunteers find the pinning of insects irreconcilable with conservation efforts. This volunteer sentiment aligned with a growing movement in the bee conservation community and informed the revision of our protocols for the next season. Photo courtesy of Mike Mao, Earthwatch Volunteer (2021).

lethal trapping events and focus on more effective methods that can improve management and monitoring of native bee populations. We were able to integrate volunteer feedback with published recommendations from the scientific community to revise our protocols in a way that reduced lethal sampling but maintained and even improved our field research.

Moving forward, we plan to strengthen connections with tribal and federal land management agencies in our area, primarily the Paiute Indian Tribe of Utah, Forest Service, and BLM. Our work with the Paiute Indian Tribe of Utah focuses on native plants and their use, but

until now pollinators had not been a part of that conversation. The Forest Service and BLM manage the majority of public land in the area, and pollinator management has become a federal concern (Obama, 2014). Their permitting systems are easier to navigate than some other federal land management agencies, which is relevant when planning to bring citizen scientists onto public lands. Leveraging one of the main strengths of citizen science work, namely that it supplements person-hours that can be devoted to survey work, relieves to some extent the need of these agencies to triage their personnel among multiple competing management priorities. Management research areas that we had not considered prior to our first field season, but that were identified as important by our partners, include learning more about the site characteristics of subterranean bumble bee nests, creating a library of pollen types, and identifying nocturnal pollinators. This natural history information is essentially unknown, but its collection is important and interesting to citizen scientists, which is a critical component of any citizen science program.

Our future research will build on the toolbox of field techniques we developed for our initial questions. However, we are also keenly interested in learning how best to train citizen scientists to address more hypothesis-driven questions within the time constraints offered by a week-long program that must also include time for acclimation to a new environment. In addition to our interest in questions of management importance, we are interested in looking more closely at trait-driven interactions between plants and pollinators. For example, what is the role of flower and stamen structure in driving buzz pollination behaviors in bees? Do certain bee species favor buzz-pollinated flowers, and how might these interactions affect plant and insect community structure both spatially and temporally? Which bees play more specialist or more generalist pollination roles, and how does the prevalence of these species influence plant prevalence? The answers to these questions can help facilitate an EcoHealth approach to the management of pollinators in southern Utah and elsewhere.

20.4 Conclusion

Citizen science leverages the passion of volunteers to facilitate biodiversity data collection in diverse, but understudied, regions. In this chapter we described our partnership with Earthwatch volunteers and how working with citizen scientists drove the development of our pollinator research program. Pollinator decline is a worldwide problem that requires a One Health or EcoHealth approach because of the interrelationship between pollinators, people, and the environment. Through our citizen science program, we learned how to integrate the needs of managers, researchers, and citizen scientists in a way that resulted in useful data and a quality experience for volunteers. Pollinator conservation is dependent upon a

foundational knowledge of pollinator natural history, as well as upon an understanding of how people interact with pollinators in their natural environment.

References

Bahls, L. L. (2014). New diatoms from the American West—A tribute to citizen science. *Proceedings of the Academy of Natural Sciences of Philadelphia, 163*(1), 61–84. https://doi.org/10.1635/053.163.0109

Barrios, B., Pena, S. R., Salas, A., & Koptur, S. (2016). Butterflies visit more frequently, but bees are better pollinators: The importance of mouthpart dimensions in effective pollen removal and deposition. *AoB Plants, 8*. https://doi.org/10.1093/aobpla/plw001

Brossard, D., Lewenstein, B., & Bonney, R. (2005). Scientific knowledge and attitude change: The impact of a citizen science project. *International Journal of Science Education, 27*(9), 1099–1121. https://doi.org/10.1080/09500690500069483

Cane, J. H. (2003). Exotic nonsocial bees (Hymenoptera: Apoidea) in North America: Ecological implications. In K. Strickler & J. H. Cane (Eds.). *For non-native crops, whence pollinators of the future?* (pp. 113–126). Thomas Say Publications in Entomology.

Cane, J. H., & Tepedino, V. J. (2017). Gauging the effect of honey bee pollen collection on native bee communities. *Conservation Letters, 10*(2), 205–210. https://doi.org/10.1111/conl.12263

Carril, O. M., Griswold, T., Haefner, J., & Wilson, J. S. (2018). Wild bees of Grand Staircase-Escalante National Monument: Richness, abundance, and spatio-temporal beta-diversity. *PeerJ, 6*, e5867. https://doi.org/10.7717/peerj.5867

Centers for Disease Control and Prevention. (2018). One health basics. https://www.cdc.gov/onehealth/basics/index.html

Chandler, M., Bebber, D. P., Castro, S., Lowman, M. D., Muoria, P., Oguge, N., & Rubenstein, D. I. (2012). International citizen science: Making the local global. *Frontiers in Ecology and the Environment, 10*(6), 328–331. https://doi.org/10.1890/110283

Donkersley, P., Elsner-Adams, E., & Maderson, S. (2020). A One-Health model for reversing honeybee (*Apis mellifera* L.) decline. *Veterinary Sciences, 7*(3), 119. https://doi.org/10.3390/vetsci7030119

Earthwatch. (2021). Our mission and values. https://earthwatch.org/about/overview

Eitzel, M. V., Cappadonna, J. L., Santos-Lang, C., Duerr, R. E., Virapongse, A., West, S. E., Kyba, C. C. M., Bowser, A., Cooper, C. B., Sforzi, A., Metcalfe, A. N., Harris, E. S., Thiel, M., Haklay, M., Ponciano, L., Roche, J., Ceccaroni, L., Shilling, F. M., Dörler, D., . . . Jiang, Q. (2017). Citizen science terminology matters: Exploring key terms. *Citizen Science: Theory and Practice, 2*(1), 1. https://doi.org/10.5334/cstp.96

Fertig, W. (2005). *Overview of the vegetation of Grand Staircase-Escalante National Monument.* Moenave Botanical Consulting.

Fertig, W. (2009). *Annotated checklist of vascular flora: Cedar Breaks National Monument. Natural resource technical report NPS/NCPN/NRTR—2009/173.* National Park Service.

Frigerio, D., Pipek, P., Kimmig, S., Winter, S., Melzheimer, J., Diblíková, L., Wachter, B., & Richter, A. (2018). Citizen science and wildlife biology: Synergies and challenges. *Ethology, 124*(6), 365–377. https://onlinelibrary.wiley.com/doi/10.1111/eth.12746. https://doi.org/10.1111/eth.12746

Goulson, D., Nicholls, E., Botías, C., & Rotheray, E. L. (2015). Bee declines driven by combined stress from parasites, pesticides, and lack of flowers. *Science, 347*(6229), 1255957. https://doi.org/10.1126/science.1255957

Haugen, G. M. (2018). Rage, grief, ceremony, and praise in the Grand Staircase-Escalante National Monument. *Ecopsychology, 10*(2), 73–76. https://doi.org/10.1089/eco.2018.0005

Howell, A. D., Alarcón, R., & Minckley, R. L. (2017). Effects of habitat fragmentation on the nesting dynamics of desert bees. *Annals of the Entomological Society of America, 110,* 233–243. https://doi.org/10.1093/aesa/saw081

Hristov, P., Neov, B., Shumkova, R., & Palova, N. (2020). Significance of Apoidea as main pollinators. ecological and economic impact and implications for human nutrition. *Diversity, 12*(7), 280. https://doi.org/10.3390/d12070280

Irwin, A. (1995). *Citizen science: A study of people, expertise and sustainable development.* Routledge.

Koffler, S., Barbiéri, C., Ghilardi-Lopes, N. P., Leocadio, J. N., Albertini, B., Francoy, T. M., & Saraiva, A. M. (2021). A buzz for sustainability and conservation: The growing potential of citizen science studies on bees. *Sustainability, 13*(2), 959. https://doi.org/10.3390/su13020959

Kullenberg, C., & Kasperowski, D. (2016). What is Citizen Science? – A scientometric meta-analysis. *PLOS ONE, 11*(1), e0147152. https://doi.org/10.1371/journal.pone.0147152

Longmore, A. T., & Forrest, T. (2016). The history and overview of Utah's grazing improvement program. *Rangelands, 38*(5), 250–255. https://doi.org/10.1016/j.rala.2016.08.007

López-Uribe, M. M., Ricigliano, V. A., & Simone-Finstrom, M. (2020). Defining pollinator health: A holistic approach based on ecological, genetic, and physiological factors. *Annual Review of Animal Biosciences, 8,* 269–294. https://doi.org/10.1146/annurev-animal-020518-115045

Nisbet, E., Zelenski, J. M., & Grandpierre, Z. (2019). Mindfulness in nature enhances connectedness and mood. *Ecological Psychology, 11,* 1–8. https://doi.org/10.1089/eco.2018.0061

Obama, B. (2014). Presidential memorandum – Creating a federal strategy to promote the health of honey bees and other pollinators. https://obamawhitehouse.archives.gov/the-press-office/2014/06/20/presidential-memorandum-creating-federal-strategy-promote-health-honey-b

Ollerton, J., Winfree, R., & Tarrant, S. (2011). How many flowering plants are pollinated by animals? *Oikos, 120*(3), 321–326. https://doi.org/10.1111/j.1600-0706.2010.18644.x

Pilliod, D. S., Welty, J. L., & Toevs, G. R. (2017). Seventy-five years of vegetation treatments on public rangelands in the Great Basin of North America. *Rangelands, 39*(1), 1–9. https://doi.org/10.1016/j.rala.2016.12.001

Portman, Z. M., Bruninga-Socolar, B., & Cariveau, D. P. (2020). The state of bee monitoring in the United States: A call to refocus away from bowl traps and towards more effective methods. *Annals of the Entomological Society of America, 113*(5), 337–342. https://doi.org/10.1093/aesa/saaa010

Portman, Z. M., & Tepedino, V. J. (2021). Successful bee monitoring programs require sustained support of taxonomists and taxonomic research. *Biological Conservation, 256,* 109080. https://doi.org/10.1016/j.biocon.2021.109080

Potts, S. G., Biesmeijer, J. C., Kremen, C., Neumann, P., Schweiger, O., & Kunin, W. E. (2010). Global pollinator declines: Trends, impacts and drivers. *Trends in Ecology and Evolution, 25*(6), 345–353. https://doi.org/10.1016/j.tree.2010.01.007

Raddick, M. J., Prather, E. E., & Wallace, C. S. (2019). Galaxy zoo: Science content knowledge of citizen scientists. *Public Understanding of Science,* *28*(6), 636–651. https://doi.org/10.1177/096366 2519840222

Richardson, M., Hamlin, I., Butler, C. W., Thomas, R., & Hunt, A. (2022). Actively noticing nature (not just time in nature) helps promote nature connectedness. *Ecopsychology.* https://www.liebertpub.com/doi/epub/10.1089/eco.2021.0023, 14(1), 8–16. https://doi.org/10.1089/eco.2021.0023

Robinson, L. D., Cawthray, J. L., West, S. E., Bonn, A., & Ansine, J. (2018). Ten principles of citizen science. In *Citizen science: Innovation in open science, society and policy* (pp. 27–40). UCL Press.

Silvertown, J. (2009). A new dawn for citizen science. *Trends in Ecology and Evolution, 24*(9), 467–471. https://doi.org/10.1016/j.tree.2009.03.017

Sullivan, B. L., Aycrigg, J. L., Barry, J. H., Bonney, R. E., Bruns, N., Cooper, C. B., Damoulas, T., Dhondt, A. A., Dietterich, T., Farnsworth, A., Fink, D., Fitzpatrick, J. W., Fredericks, T., Gerbracht, J., Gomes, C., Hochachka, W. M., Iliff, M. J., Lagoze, C., La Sorte, F. A., . . . Kelling, S. (2014). The eBird enterprise: An integrated approach to development and application of citizen science. *Biological Conservation, 169*, 31–40. https://doi.org/10.1016/j.biocon.2013.11.003

Switalski, A., & Jones, A. (2012). Off-road vehicle best management practices for forestlands: A review of scientific literature and guidance for managers. *Journal of Conservation Planning, 8*, 12–24.

Switalski, A. (2018). Off-highway vehicle recreation in drylands: A literature review and recommendations for best management practices. *Journal of Outdoor Recreation and Tourism, 21*, 87–96. https://doi.org/10.1016/j.jort.2018.01.001

Theobald, E. J., Ettinger, A. K., Burgess, H. K., DeBey, L. B., Schmidt, N. R., Froehlich, H. E., Wagner, C., HilleRisLambers, J., Tewksbury, J., Harsch, M. A., & Parrish, J. K. (2015). Global change and local solutions: Tapping the unrealized potential of citizen science for biodiversity research. *Biological Conservation, 181*, 236–244. https://doi.org/10.1016/j.biocon.2014.10.021

Vanbergen, A. J. & the Insect Pollinators Initiative (2013) Threats to an ecosystem service: Pressures on pollinators. *Frontiers in Ecology and the Environment, 11*(5), 251–259. https://doi.org/10.1890/120126

Vohland, K., Land-Zandstra, A., Ceccaroni, L., Lemmens, R., Perelló, J., Ponti, M., Samson, R., & Wagenknecht, K. (2021). What is citizen science? The challenges of definition. In K. Vohland et al. (Eds.), *The science of citizen science.* Springer.

Welsh, S. L., Atwood, N. D., Goodrich, S., & Higgins, L. C. (1987). A Utah flora. *Great Basin Naturalist Memoirs, 9.*

Wesley Schultz, P. (2001). The structure of environmental concern: concern for self, other people, and the biosphere. *Journal of Environmental Psychology, 21*(4), 327–339. https://doi.org/10.1006/jevp.2001.0227.

Wilson, J. S., Kelly, M., & Carril, O. M. (2018). Reducing protected lands in a hotspot of bee biodiversity: Bees of Grand Staircase-Escalante National Monument. *PeerJ, 6*, e6057. https://doi.org/10.7717/peerj.6057

World Health Organization. (2017). *One Health.* https://www.who.int/news-room/questions-and-answers/item/one-health

Chapter 21

Participatory research and community engagement in the Arctic

Matilde Tomaselli, Ian Hogg and Sylvia Checkley

Abstract

This chapter provides an overview of participatory research and community engagement principles and presents a case study in which a participatory approach was applied to wildlife health monitoring in the Arctic. We explore general strategies and considerations for implementing participatory projects and for engaging with Indigenous communities, including outreach and capacity building.

21.1 Introduction: principles of participatory research and community engagement

Participatory approaches to research have arisen from Participatory Action Research (PAR). This was developed in the 1960s in low-income countries and in the field of adult education (i.e. Paulo Freire's work), following the recognition that (1) conventional 'top-down' research approaches were inadequate for enabling meaningful actions for change, and (2) local people held a depth and breadth of knowledge locally acquired relevant to improved understanding of many processes and, therefore, better enabling change (Baum et al., 2006; Green and Thorogood, 2009a). From this PAR foundation, participatory approaches have grown as research methods, enabling researchers to work in partnership with local communities supporting decision making for effective and sustainable interventions.

Since the late 1970s, participatory research methods have been increasingly used in many disciplines – including agriculture and farming, community and sustainable development, environmental management, ecology and human and animal health (e.g. Calheiros et al., 2000; Farrington and Martin, 1988; Kuhnlein et al., 2009; Kwiatkowski, 2011; Mariner et al., 2012; Pretty, 1995; Trimble and Berkes, 2013; Wiber et al., 2004) – and in a range of both developing and developed countries (Cornwall and Jewkes, 1995; Reed, 2008). 'Putting the last first', empowerment, equity, trust and two-way learning (between participants and between researchers and participants) are the guiding principles of participatory research

Susan C. Cork and Douglas P. Whiteside (eds) **Case Studies in EcoHealth**
DOI: 10.52517/9781789183368.021, © 5m Books Ltd, 2024

(Reed, 2008). Given these characteristics and principles, participatory research becomes particularly relevant when working with Indigenous peoples that have been historically colonized. In these cases, participatory approaches assume a connotation that goes beyond research, becoming enabling tools for the reconciliation process between Western (dominant) institutions and Indigenous groups (Kutz and Tomaselli, 2019).

As participatory research becomes more widely used and even expected in many contexts, practitioners and scholars have raised concerns on the misuse of the term 'participatory' (Castleden et al., 2012; Catley et al., 2012; Mosurska and Ford, 2020). Fundamental critiques are the tokenistic use of participation; this means that participation becomes a mere box-ticking exercise (e.g. for accessing dedicated funds) or a way to obtain data from marginalized groups (Catley et al., 2012; Mosurska and Ford, 2020). Indeed, local participation by itself is neither sufficient nor adequate to claim a research project as 'participatory'. While there are multiple participatory approaches that seek local participation, the notion of 'active' versus

'passive' participation – or 'true participation' versus 'non-participation' – is what qualifies the research as truly participatory or not (Pretty et al., 1995; Reed, 2008). This is an important aspect to consider in the field of EcoHealth since the participatory and action-research framework is a key characteristic (Charron, 2012).

Whether participation is conceptualized as an incremental process or 'ladder of participation', or defined by typology within a 'wheel of participation' (Figure 21.1), scholars converge in recognizing that participatory research practitioners should reflect from the start on the nature and degree of participation their project is set to achieve, and aim for the highest possible level of participation (Catley et al., 2012; Mariner and Paskin, 2000; Pretty et al., 1995; Reed, 2008). While research objectives can also influence the participatory process and determine what level of participation can be adequate (Reed, 2008), local participants must be considered partners in the research rather than 'research subjects'. Inclusion and balanced power between contributors should be the conceptual framework within which participation is realized.

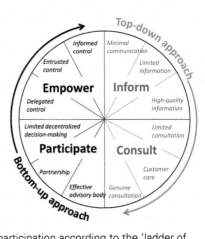

Figure 21.1 Schematic representations of types and levels of participation according to the 'ladder of participation' or 'wheel of participation' conceptualizations (schematics adapted from Pretty et al., 1995 and Reed, 2008, respectively).

With the progressive increase of decision-making models that move from centralized governments to more decentralized community-based forms, participatory processes and engagement with local stakeholders have increasingly become a legislated requirement. For example, in jurisdictions such as the Canadian Arctic, co-management is essential for decision making regarding natural resources (Tomaselli and Barry, 2022). In these contexts, implementation of participatory research can benefit from the existing frameworks for local engagement and significantly improve local participation in the co-management process, enhancing outcomes (Tomaselli and Barry, 2022).

The following sections examine the application of participatory research and community engagement in the Arctic. We first provide context to better understand relevant socio-ecological and governance aspects in the Arctic. We then present a specific example of a participatory research project applied to wildlife heath in the Canadian Arctic, highlighting project design, local participation/engagement and long-term project sustainability. Finally, we summarize our experience and provide recommendations to assist EcoHealth practitioners in the implementation of participatory research in and beyond the Arctic context.

21.2 The Arctic: socio-ecological and governance context

The Arctic is the north polar region of the world. It is a vast and sparsely populated geographical area located above the tree line, spanning from North America to northern Eurasia. It is a region with ecosystems characterized by permafrost (permanently frozen ground) covered by tundra (low vegetation adapted to extreme cold conditions). While there are seasonal variations, for most of the year Arctic temperatures remain below zero and much of the landscape is covered by ice and snow. A number of wildlife species inhabit this area both on land and in the waters. Some species remain in the Arctic year round (e.g. caribou, muskox, Arctic wolves, ringed seals) while many are seasonally migratory (e.g. shore birds, geese) and visit only during the short Arctic summer.

The Arctic has been inhabited for thousands of years by Indigenous peoples who have adapted to, and thrived, in this extreme environment. Lifestyle, culture and traditions of Arctic Indigenous groups have been shaped over millennia through the deep connection of Indigenous peoples to the land they have inhabited (Kuhnlein et al., 2009; Myers et al., 2005). While culture and language can differ across groups and locations, traditional livelihoods associated with harvesting renewable resources (e.g. fishing, hunting, reindeer herding) are a common characteristic that continue to this day (Kuhnlein and Humphries, 2017).

Historic colonization of the Arctic has profoundly changed the life of Indigenous peoples, socially, culturally and economically (Angohiatok et al., 2023). Arctic governance arrangements have evolved since initial colonization with the rise of Indigenous political organization and the increasing recognition of rights of Indigenous peoples. This has led to negotiation of land claim agreements or similar legislative processes, devolution of government responsibilities and, more generally, an inclusive approach to decision making (e.g. the wildlife co-management regime) (Tomaselli and Barry, 2022). Today, rights to land and natural resources are an integral component of Arctic governance systems and key to the survival of Arctic Indigenous people and their culture.

The EcoHealth paradigm recognizes the inseparable linkages between the health of all species, including humans, and their environments and is obvious in the Arctic. Hunting wildlife is part of the way of life, culture and traditions of northern peoples, and also contributes significantly to northern food security (Kuhnlein and Humphries, 2017; Meakin and Kurvits, 2009). In the Arctic, healthy ecosystems including healthy wildlife populations are key to sustaining healthy communities. Today, environmental change driven by rapid global warming[1] is challenging the resilience of the northern socio-ecological system (Meakin and Kurvits, 2009). Warming temperatures are triggering many processes that pose a direct threat to the health of Arctic wildlife[2], which can further exacerbate food insecurity for northern peoples.

This rapidly changing context necessitates an equally rapid response to support the resilience of the socio-ecological system. Wildlife health surveillance and monitoring programmes are enabling tools for the early detection of changes around wildlife health (comprising wildlife population trends and disease status, including zoonoses), and the consequent implementation of timely actions finalized to improve wildlife management and public health protection. However, there are many logistical and technical limitations that challenge the implementation of wildlife health surveillance and monitoring initiatives, especially in the remote and resource-scarce Arctic (Tomaselli, 2018). The application of participatory methods to wildlife surveillance has been demonstrated to overcome many of these limitations while also improving surveillance outcomes in terms of the reliability and timeliness of information (Tomaselli, 2018; Tomaselli et al., 2018). In the Arctic, participatory research applied to wildlife health also strengthens wildlife co-management by enabling the direct inclusion of Indigenous and local voices in the decision-making process (Tomaselli and Barry, 2022). Below we provide a specific example illustrating the value of the participatory approach as it applies to wildlife health in the Arctic.

21.3 Monitoring muskox health in the Canadian Arctic: moving from a university-led research project to a community-led programme

In 2013, following community concerns of local decline of muskoxen harvested to support local food security and economy, a graduate research project started in the community of Cambridge Bay in the Kitikmeot region of Nunavut in Canada's Arctic. Its objectives were to explore the magnitude of, and reasons for, that decline while developing a participatory surveillance system for muskox health characterized by the combined use of Indigenous, local and scientific knowledge (Figure 21.2) (Tomaselli, 2018).

Initially this project was university led although its development responded to community concerns. Efforts were made to build partnerships with local organizations and to engage with local community members meaningfully and respectfully. This process led to building a solid foundation of trust between parties that eventually culminated in the development of a successful participatory research programme that serves as a model for the broader implementation of participatory approaches for wildlife health surveillance and monitoring (Kutz and Tomaselli, 2019; Tomaselli, 2018, 2022).

Active participation of local community members was key to the development of the programme and determined much of its success. Multiple project components were designed in partnership with local hunters or benefitted from their contribution. For example, a hunter-based sample collection programme was developed to obtain samples for laboratory analyses. Logistics around sample collection – including discussions about types of sample that could be collected under Arctic field conditions and how to collect them – benefitted from the experience and input of local hunters (Figure 21.3; Tomaselli and Curry, 2019). Hunters also shared their

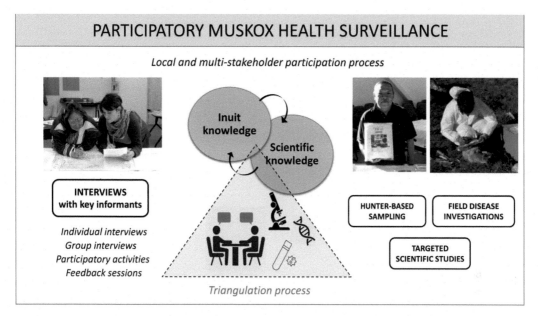

Figure 21.2 Schematic representation of the participatory muskox health surveillance project developed in Cambridge Bay, Nunavut, Canada, including its surveillance components that accessed different knowledge systems. Final surveillance output was generated by corroborating and cross-checking information through the implementation of multiple surveillance components and methods (triangulation process). Local participation was key to obtaining missing epidemiologic data on muskox health, as well as designing and implementing the overall surveillance system.

knowledge on muskox health, providing a new understanding on the status of the local muskox population and enabling interpretation of scientific knowledge derived from conventional veterinary diagnostics in context (e.g. Tomaselli et al., 2016, 2018, 2019). Participatory epidemiology methods and techniques were key to documenting muskox health information from experiential-based knowledge of Inuit hunters (Figure 21.2). Through the use of participatory epidemiology techniques and methods, Inuit knowledge was mobilized in a format that could be more effectively combined with Western scientific knowledge and more successfully utilized in co-management than that previously achieved (Tomaselli et al., 2018).

This 5-year project ended in 2018 with the completion of the graduate (PhD) research programme. However, the trust built, the strong

partnerships established and the sense of local ownership that this project generated through the years were key elements that allowed the programme to evolve from university led to community led. The level of participation achieved over the first 5 years of the project promoted self-mobilization and empowerment at the community level, which are the highest forms of participation (Figure 21.1).

In 2019 the Ekaluktutiak Hunters and Trappers Organization (EHTO), an organization that actively participates in wildlife co-management in Nunavut and one of the initial project partners, became the programme lead. The EHTO recruited funding to continue the project in collaboration with its partners from government and academia. It also shaped the programme according to the evolved local priorities and under the direction of the board

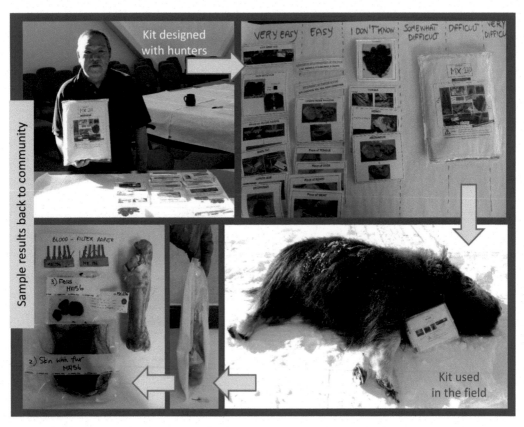

Figure 21.3 Representation from the design to the field use of the sample collection kit used for the hunter-based sampling project component of the participatory muskox health surveillance programme implemented in Cambridge Bay, Nunavut, Canada. Pre-packaged and lightweight kits are designed in concert with hunters, and include samples and information on easy collection under harsh field conditions; once collected, kits remain relatively light and contained making it easy to transport them. After samples are analysed, results are reported back to the community. Pictorial representation from Checkley et al., 2022.

members of the organization, who represent the interest and voices of the hunters of their community. The project objectives broadened to monitor multiple wildlife species including caribou and predators – wolves, grizzly bears and wolverines – in addition to muskox. In addition, monitoring activities were expanded beyond the hunter-based sample collection and interviews of Inuit hunters to include the use of remotely triggered wildlife cameras and GPS tablet devices used by hunters to record and geolocate observations made during hunting

trips (Figure 21.4). Since 2019 project activities, with the exception of laboratory analyses, have been performed in the community through the participation of community members.

Capacity building has been an integral component of the programme and one of the most important objectives. Outreach activities have been integrated within the project by creating opportunities for intergenerational learning and knowledge exchange. However, more structured outreach activities will be prioritized in future as the programme evolves into a long-term

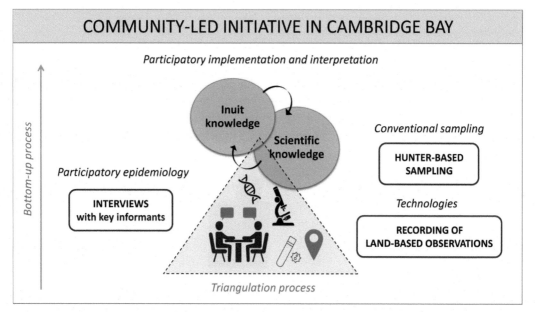

COMMUNITY-LED INITIATIVE IN CAMBRIDGE BAY

Participatory implementation and interpretation

Bottom-up process

Inuit knowledge

Scientific knowledge

Conventional sampling

HUNTER-BASED SAMPLING

Participatory epidemiology

INTERVIEWS with key informants

Technologies

RECORDING OF LAND-BASED OBSERVATIONS

Triangulation process

Figure 21.4 Schematic representation of the community-led initiative in Cambridge Bay, Nunavut, Canada, finalized at monitoring the health of muskoxen, caribou and predators in the Cambridge Bay area. Project components are designed to access different knowledge systems and to cross-check information using multiple methods and sources (triangulation process). This project evolved from the participatory muskox health surveillance project previously implemented in Cambridge Bay (Figure 21.2), and was shaped and implemented by the local hunters and trappers organization with the assistance of the Government of Canada (Polar Knowledge Canada) and academic institutions.

initiative. Engagement with schools, including organization of camps on the land with Inuit Elders and youth from the community, will be considered as opportunities to foster intergenerational connection, build a positive legacy and improve future outcomes.

The global pandemic of SARS-CoV-2 (COVID-19) was a major disruptor to the planned work because some activities (e.g. interviews) had to be cancelled to comply with public health regulations and requirements. However, despite the challenging circumstances the project continued to operate. Many project activities, such as the hunter-based sampling, the use of GPS tablets and placement/retrieval of wildlife cameras, were able to continue as planned. The COVID-19 pandemic highlighted even more the importance of programmes that are locally led and

locally executed for environmental monitoring in the Arctic. Indeed, in 2020 and 2021 many Arctic monitoring and research activities heavily reliant on the efforts of southern researchers and institutions had to be cancelled, creating gaps in longer-term datasets, while projects that relied on local efforts were able to continue operating with only minor adjustments. For example, this project was readily adapted to the new reality and valuable samples, data and observations were continuously collected.

The transition of the project from university led to community led indicates the value and success of the initiative, with a local organization and its members taking ownership of the programme and investing significant efforts to continue it. As the EHTO has taken increasing control over the research, participation has progressively transformed into a process of empowerment for both the organization and its representing members, allowing them to more effectively participate in and influence management decisions. However, the transition of the project from university led to community led has also highlighted challenges to the full realization of the empowerment process. In our context, challenges relate to limited local capacity to operationalize all aspects of the project and to operate within often complex bureaucratic frameworks associated with funding applications and project management (e.g. paperwork for project submission and reporting, coordination with southern-based laboratories, necessary coordination with existing passive wildlife disease surveillance frameworks including multi-stakeholder collaboration/communication). These challenges were mostly overcome through the assistance of project partners. However, it is important to fully acknowledge and evaluate potential challenges to support local organizations in leading wildlife research and monitoring initiatives.

21.4 Consideration for implementation of participatory research within EcoHealth projects

Here we provide general considerations for the development and application of participatory research within EcoHealth projects, including considerations for engaging with local communities and stakeholders and elements on which to reflect to ensure sustainability of participatory initiatives. EcoHealth practitioners should view these as initial points for reflection rather than an exhaustive list.

21.4.1 Understand the working context

Invest time and effort to understand the local context, including the decision-making

framework, as well as opportunities for engagement with different stakeholders as the programme develops. Hiring of local community researchers is essential at the start of the project and, in some cases, may be required. Local community researchers can provide significant guidance in this initial phase, as well as throughout the research process.

21.4.2 Demonstrate respect, cultural sensitivity and research ethic

Local communities may refer to different norms and belief systems to those that external researchers are accustomed to. It is important to be familiar with the local culture, including norms, beliefs and customs. Throughout the project it is essential to adopt a non-judgemental approach, by first understanding our own biases and accepting other viewpoints (Tomaselli and Barry, 2022). When working with Indigenous groups it is also critical to understand and respect human rights, languages and customs to build trust and develop productive relationships. Listen, learn and adhere to high standards of research ethics, including principles regarding data ownership, control, access and possessions that are of particular importance when working with Indigenous peoples (e.g. The First Nations principles of OCAP®).

21.4.3 Change the mindset

Recognize the bottom-up nature of the participatory research process, which is the opposite to the 'conventional' or 'southern' top-down approach. Participatory researchers must be open-minded and able to modify the research process (including research questions) as local participants contribute. Adhere to the five guiding principles of participatory research – 'putting the last first', empowerment, equity, trust and two-way learning – and recognize that participatory research must be action oriented.

21.4.4 Account for time and trust building

Participatory research requires lead time for research preparation (including comprehension of the local context and needs), establishing trust with local partners and designing/implementing an inclusive research process. It is important to acknowledge at the outset the time commitments required to avoid generating unrealistic expectations and frustrations among partners.

21.4.5 Identify and engage with the right stakeholders

An understanding of the local context is key to understanding the appropriate knowledge holders to engage with (e.g. resource users, harvesters, livestock owners). Since participatory research is action oriented, consider opportunities of engagement also with decision makers to communicate research outputs. This will ensure that information generated is successfully utilized for targeted interventions. Knowledge and understanding of the local decision-making framework are essential to design and actualize this component.

21.4.6 Reflect on the nature and type of participation

The types and nature of participation, recognizing 'passive' and 'active' forms of participation need to be considered at the outset (Figure 21.1). Aim to achieve the highest possible level of active participation given the research objectives and local context. The participatory research design and methods must be clear and transparent,

including who participated and in what capacity. Indeed, clarity and transparency of research methods are important criteria for evaluation of qualitative research, including those resulting from the application of participatory approaches (Green and Thorogood, 2009b).

21.4.7 Consider the opportunity for capacity building and outreach

To ensure sustainability and improve future outcomes of participatory initiatives, include a plan for capacity building and opportunities for outreach. Consider that community organizations partnering in participatory initiatives may experience capacity challenges and limited resources that can interfere with their ability to effectively participate in the initiative and, potentially, to take ownership of the programme and continue it as a community-led initiative. Capacity building and outreach activities that ensure intergenerational and multi-stakeholder knowledge exchange can better enable long-term programme sustainability and positive legacy.

21.4.8 Evaluate the programme for future improvement

Monitor and evaluate the performance of the programme by soliciting direct feedback from all participants. Identifying what works and what could be improved enables ongoing positive development of the participatory research and the ability to respond to evolving needs.

21.4.9 Learn from other participatory programmes

Finally, it is important to learn from other participatory programmes from the same community and/or geographical area of study. Locally specific insights from previous participatory initiatives can provide key elements for consideration for the implementation of a successful programme that is locally relevant and customized.

21.5 Conclusion

Participatory and action-oriented research is a key characteristic of EcoHealth projects. This chapter has provided EcoHealth practitioners with an overview of participatory research principles and applications. Genuine participation requires genuine collaboration. Therefore, defining the nature and degree of participation is an essential step in the participatory research process. This includes both technical and social skills and competencies to work collaboratively with local communities and stakeholders. While participatory approaches generally require more time to create fruitful partnerships and local collaborations, they have the potential for long-lasting positive outcomes. Continually learning lessons from the broad participatory research field of study and acquisition of skills to effectively work in teams and across cultures are key assets for EcoHealth practitioners.

Acknowledgements

We thank the EHTO of the community of Cambridge Bay (Nunavut, Canada) for their continued support of, and leadership in, this project. A sincere thank you to the many community members and Inuit knowledge holders of Cambridge Bay who actively participated in the project; their generous collaboration and intellectual contribution that made this project possible and successful. We would also like to thank Dr Alain Leclair, Director, Polar Knowledge Canada, for his support of this work.

Endnotes

1 The Arctic is warming more than twice as fast as the global average, a phenomenon known as Arctic amplification (Cohen et al., 2014).
2 For example, wildlife species once restricted to southern regions are now moving north, bringing with them new parasites against which Arctic wildlife may not have natural immunity (Burek et al., 2008; Kutz et al., 2005).

References

Angohiatok, G. B., Vandenbrink, B., Hogg, I., & McIlwraith, T. (2023). History of the Canadian Arctic: An Inuit perspective. In *History of the polar regions*. Cambridge Press.

Baum, F., MacDougall, C., & Smith, D. (2006). Participatory action research. *Journal of Epidemiology and Community Health*, *60*(10), 854–857. https://doi.org/10.1136/jech.2004.028662

Burek, K. A., Gulland, F. M., & O'Hara, T. M. (2008). Effects of climate change on Arctic marine mammal health. *Ecological Applications*, *18*(2) Suppl., S126–S134. https://doi.org/10.1890/06-0553.1

Calheiros, D. F., Seidl, A. F., & Ferreira, C. J. A. (2000). Participatory research methods in environmental science: Local and scientific knowledge of a limnological phenomenon in the Pantanal wetland of Brazil. *Journal of Applied Ecology*, *37*(4), 684–696. https://doi.org/10.1046/j.1365-2664.2000.00524.x

Castleden, H., Morgan, V. S., & Lamb, C. (2012). 'I spent the first year drinking tea': Exploring Canadian university researchers' perspectives on community-based participatory research involving Indigenous peoples. *Canadian Geographer / Le Géographe Canadien*, *56*(2), 160–179. https://doi.org/10.1111/j.1541-0064.2012.00432.x

Catley, A., Alders, R. G., & Wood, J. L. (2012). Participatory epidemiology: Approaches, methods, experiences. *Veterinary Journal*, *191*(2), *151–160*. *https://doi.org/10.1016/j.tvjl.2011.03.010*

Charron, D. F. (2012). EcoHealth research in practice. In D. Charron (Ed.), *EcoHealth research in practice. Insight and innovation in international development*, *1*. Springer. *https://doi.org/10.1007/978-1-4614-0517-7_22*

Checkley, S. L., Tomaselli, M., & Caulkett, N. (2022). Wildlife health surveillance in the Arctic. In

M. Tryland (Ed.), *Arctic one health*. Springer. https://doi.org/10.1007/978-3-030-87853-5_23

Cohen, J., Screen, J. A., Furtado, J. C., Barlow, M., Whittleston, D., Coumou, D., Francis, J., Dethloff, K., Entekhabi, D., Overland, J., & Jones, J. (2014). Recent Arctic amplification and extreme mid-latitude weather. *Nature Geoscience, 7*(9), 627–637. https://doi.org/10.1038/ngeo2234

Cornwall, A., & Jewkes, R. (1995). What is participatory research? *Social Science and Medicine, 41*(12), 1667–1676. https://doi.org/10.1016/0277-9536(95)00127-s

Farrington, J., & Martin, A. M. (1988) Farmer participatory research: A review of concepts and recent fieldwork. *Agricultural Administration and Extension, 29*(4), 247–264. https://doi.org/10.1016/0269-7475(88)90107-9

Green, J., & Thorogood, N. (2009a). Qualitative methodology and health research: participatory research. In J. Seamen (Ed.), *Qualitative methods for Health Research* (2nd ed). SAGE.

Green, J., & Thorogood, N. (2009b). Developing qualitative research design and generating and analysing data. In J. Seamen (Ed.), *Qualitative methods for Health Research* (2nd ed). SAGE.

Kuhnlein, H. V., Erasmus, B., & Spigelski, D. (2009). Using traditional food and knowledge to promote a healthy future among Inuit. In H. V. Kuhnlein, B. Erasmus & D. Spigelski (Eds.). *Back to the Future, Indigenous Peoples' food systems: The many dimensions of culture, diversity and environment for nutrition and health*. Food and Agriculture Organization of the United Nations (FAO).

Kuhnlein, H. V., & Humphries, M. M. (2017). *Traditional Animal Foods of Indigenous Peoples of Northern North America*. Centre for Indigenous Peoples' Nutrition and Environment, McGill University Montreal. http://traditionalanimalfoods.org/

Kutz, S. J., Hoberg, E. P., Polley, L., & Jenkins, E. J. (2005). Global warming is changing the dynamics of Arctic host-parasite systems. *Proceedings. Biological Sciences, 272*(1581), 2571–2576. https://doi.org/10.1098/rspb.2005.3285

Kutz, S., & Tomaselli, M. (2019). 'Two-eyed seeing' supports wildlife health. *Science, 364*(6446), 1135–1137. https://doi.org/10.1126/science.aau6170

Kwiatkowski, R. E. (2011). Indigenous community based participatory research and health impact assessment: A Canadian example. *Environmental Impact Assessment Review, 31*(4), 445–450. https://doi.org/10.1016/j.eiar.2010.02.003

Mariner, J. C., & Paskin, R. (2000). *Manual on participatory epidemiology: Methods for the collection of action-oriented epidemiological intelligence*. Food and Agriculture Organization of the United Nations (FAO) Animal health Manual, Rome.

Mariner, J. C., House, J. A., Mebus, C. A., Sollod, A. E., Chibeu, D., Jones, B. A., Roeder, P. L., Admassu, B., & van 't Klooster, G. G. (2012). Rinderpest eradication: Appropriate technology and social innovations. *Science, 337*(6100), 1309–1312. https://doi.org/10.1126/science.1223805

Meakin, S., & Kurvits, T. (2009). Assessing the impacts of climate change on food security in the Canadian Arctic. Affairs, IaN (Ed). GRID-Arendal, Ottawa, Canada.

Mosurska, A., & Ford, J. D. (2020). Unpacking community participation in research: A Systematic Literature Review of Community-based and Participatory Research in Alaska. *Arctic, 73*(3), 347–367. https://doi.org/10.14430/arctic71080

Myers, H., Fast, H., Berkes, M. K., & Berkes, F. (2005). Feeding the family in times of change. In F. Berkes, R. Huebert, H. Fast, M. Manseau & A. Diduck (Eds.), *Breaking ice: Renewable resource and ocean management in the Canadian North*. University of Calgary Press.

Pretty, J. N. (1995). Participatory learning for sustainable agriculture. *World Development, 23*(8), 1247–1263. https://doi.org/10.1016/0305-750X(95)00046-F

Pretty, J. N., Guit, I., Thompson, J., & Scoones, I. (1995). *A Trainer's guide for participatory learning and action*. IIED Participatory Methodology Series.

Reed, M. S. (2008). Stakeholder participation for environmental management: A literature review. *Biological Conservation, 141*(10), 2417–2431. https://doi.org/10.1016/j.biocon.2008.07.014

The First Nations principles of OCAP®. https://fnigc.ca/ocap-training

Tomaselli, M. (2018). *Improved wildlife health and disease surveillance through the combined use of local knowledge and scientific knowledge*. Doctoral thesis. University of Calgary. http://hdl.handle.net/1880/107597

Tomaselli, M., & Barry, R. (2022). Stakeholder engagement for collaborative wildlife health management. In C. Stephen (Ed.), *Wildlife population health*. Springer. https://doi.org/10.1007/978-3-030-90510-1_21

Tomaselli, M., & Curry, P. (2019). Wildlife health and disease surveillance. In S. Cork & R. Halliwell (Eds.), *The Veterinary Laboratory, and Manual* (3rd ed). 5m Books.

Tomaselli, M., Dalton, C., Duignan, P. J., Kutz, S. J., van der Meer, F., Kafle, P., Surujballi, O., Turcotte, C., & Checkley, S. (2016). Contagious ecthyma, rangiferine brucellosis, and lungworm infection in a muskox (*Ovibos moschatus*) from the Canadian Arctic, 2014. *Journal of Wildlife Diseases, 52*(3), 719–724. https://doi.org/10.7589/2015-12-327

Tomaselli, M., Kutz, S. J., Gerlach, C., & Checkley, S. (2018). Local knowledge to enhance wildlife population health surveillance: Conserving muskoxen and caribou in the Canadian Arctic. *Biological Conservation, 217*, 337–348. https://doi.org/10.1016/j.biocon.2017.11.010

Tomaselli, M., Elkin, B., Kutz, S., Harms, N. J., Nymo, H. I., Davison, T., Leclerc, L. M., Branigan, M., DuMond, M., Tryland, M., & Checkley, S. (2019). A Transdisciplinary Approach to *Brucella* in Muskoxen of the Western Canadian Arctic 1989–2016. *EcoHealth, 16*(3), 488–501. https://doi.org/10.1007/s10393-019-01433-3

Tomaselli, M. (2022). Participatory epidemiology and surveillance for wildlife health. In C. Stephen (Ed.), *Wildlife population health*. Springer. https://doi.org/10.1007/978-3-030-90510-15

Trimble, M., & Berkes, F. (2013). Participatory research towards co-management: Lessons from artisanal fisheries in coastal Uruguay. *Journal of Environmental Management, 128*, 768–778. https://doi.org/10.1016/j.jenvman.2013.06.032

Wiber, M., Berkes, F., Charles, A., & Kearney, J. (2004). Participatory research supporting community-based fishery management. *Marine Policy, 28*(6), 459–468. https://doi.org/10.1016/j.marpol.2003.10.020

Chapter 22

Striving for success with community engagement in African conservation practice and research

Donna J. Sheppard, Joy Sammy and Karen Landman

Abstract

To effectively stem the ongoing, accelerated loss of the Earth's biodiversity, particularly in the species-rich tropics, conservation organizations have increasingly explored diverse collaborations. Community-based conservation, with its dual goals of conservation and poverty alleviation, is an approach that recognizes as essential the inclusion of local citizens in conservation programmes.

Over the past decade, however, community-based approaches to conservation have been criticized for failing to deliver tangible, sustainable benefits to either biodiversity or humans (Brooks et al., 2013; Lammers et al., 2017). One of the reasons for this is because, in practice, pitfalls are encountered when bringing together disparate scientific and non-scientific perspectives, nature values and worldviews.

Before embarking on a career in conservation, young scientists need to familiarize themselves with concepts of community and positionality in order to meet with success, particularly when they find themselves working in unfamiliar cultural situations. In this chapter, we provide guidelines for respectful and effective engagement with ecosystem-adjacent citizens and buttress these strategies with personal reflections gathered across long careers in African conservation.

The discussion includes personal considerations in preparation for field work, the use of locally engaged staff, tools for handling challenges that arise while in the field, and broader considerations when selecting one's preferred organizational approach to programme implementation. By entering conservation work with an awareness of what to expect when engaging communities in research, young scientists will be more able to collaborate in resonant, locally relevant solutions to better achieve conservation goals.

Susan C. Cork and Douglas P. Whiteside (eds) **Case Studies in EcoHealth**
DOI: 10.52517/9781789183368.022, © 5m Books Ltd, 2024

22.1   Introduction

In this chapter, we will be exploring the process of engaging community participants meaningfully in nature through conservation research and conservation practice. We will be drawing on lessons learned from field conservation experiences in a number of African countries over many years. Using the power of narrative, we bring these experiences to life for the reader. Each of the text boxes below describes an actual experience from a life spent interacting with community participants in African contexts. The location of each of the personal narratives is captured in Figure 22.1.

In these pages, we offer sound guidelines for respectful engagement with ecosystem-adjacent citizens, offering a point of reference for conservationists who find themselves in unfamiliar cultural situations. These perspectives will be useful to young scientists who are commencing careers in the environmental sector in general, and to those embarking on careers in international ecology, biology, development or conservation in particular.

22.1.1   What is 'community'?

Before discussing methods for engaging communities in research, it is important to first establish what is meant by the term 'community'. Although this word appears simple and familiar, in reality 'community' has a multitude

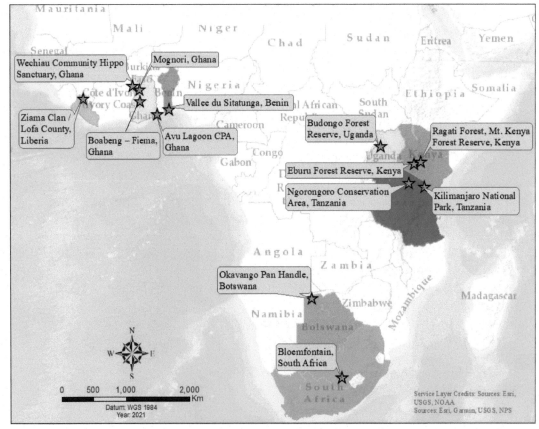

Figure 22.1 Personal reflections – storytelling locations. Source: Brichieri-Colombi, 2021; adapted from Esri, DeLorme, USGS, NPS and NOAA.

of meanings. Community can be a spatial unit, a physical entity or a place that can be drawn on a map. Community can be a social structure that shares personal values, cultural values, business goals, attitudes or a worldview – for example, 'the Christian community'. Community can be a group of people possessing a set of similar characteristics or shared norms – for example, the LGBTQIAS2+ community or the Black community (Agrawal and Gibson, 2001). In its social understanding, community can be applied to define structures, interactions, relationships and philosophical/value statements (Smith, 1996). A useful definition for community comes from Etzioni (1996, 2015), who defines community as:

> A group of people who share affective bonds and a culture, and are defined by two characteristics: first, communities cannot exist without a web of affect-laden relations among a group of individuals (rather than merely one-on-one, or a chain of individual interactions). These relations often crisscross and reinforce one another. Second, being a community entails having a measure of commitment to a set of shared values, norms and meanings.

Community, as a term, has universal usage and appeal; it speaks of togetherness and suggests that everyone is working harmoniously to manage their shared natural resources. When it is used to discuss rural African settlements with strong ties to the environment, it is often seen as a unifying entity, whereas in reality, a community often contains heterogeneous constituents and internal dynamics that should be acknowledged (de Beer, 2013). The 'rural community', for example, comprises individuals with varying degrees of social capital: young and old, men and women, rich and poor, and powerful and powerless individuals. In defining a community as harmonious, it is useful to be aware of how the group is normatively represented. Minorities, women

and the poor tend to be ignored, and dominant cultural norms are expected to be applied to the entire entity using mechanisms of social control that belie differences within the group (Kumar, 2005; Smith, 1996; United Nations Department of Economic and Social Affairs, 2016). It is important to envision communities beyond stereotypes and to focus instead on the multiple interests and actors within them, on roles within the decision-making power dynamics and on internal and external institutions shaping decision-making processes.

22.1.2 How do we engage communities in research?

As a visiting scientist, it is important to understand that there will be an exchange of ideas and knowledge among visitors and hosts, rather than making the assumption that the Western-educated person has much to offer local community members, who, in return, have nothing to contribute to the process.

In most African nations, where people have historically relied on the abundant tropical wild resources surrounding them to exist and prosper, longstanding and multifaceted relationships with nature have developed. A deep appreciation for nature has linked the environment to culture through the use of oral traditions that incorporate these relationships into folklore, metaphors, worldviews and belief systems. These perspectives will be explored further through a discussion of nature values, traditional ecological knowledge (TEK) and cultural dimensions of engagement.

22.1.2.1 Nature values

Nature, in the broadest sense, is the physical universe all around us, with its plants, animals and natural processes. And, although humankind is a small portion of the planet's living biomass, our relative influence is significant. Most often, nature is interpreted by humans in anthropocentric ways, which may include – but are not limited to – market-driven monetary perspectives that commodify nature or perspectives that recognize the health values derived from the conservation of nature. These interpretations lead to decisions about values and usage benefitting humanity.

Stemming from this perspective is the concept of ecosystem services. The term defines the benefits gained by humans from nature and was popularized by the United Nation's Millennium Ecosystem Assessment, conducted from 2001 to 2005 (World Resources Institute, 2005). Four broad categories of services have been identified: provisioning, regulating, cultural and supporting services (World Resources Institute, 2005). While provisioning services, such as drinking water, timber, wood fuel, herbal medicine and bush meat, are commonly linked to community relationships with nature in Africa, there are other, equally valued, ecosystem services identified as important by community members. These might include heritage and spiritual values, historical or traditional sites and the role that ecosystems provide to a sense of identity (Barrow, 2019; Burger et al., 2008).

Examples of some of these less well-known values ascribed to nature include the sacred mountains in the eastern Himalayas for Tibetan Buddhists (Anderson et al., 2005; Salick et al., 2007; Shen et al., 2012); the relationship to the land for many First Nations, Métis People and Inuit of North America, who view the position of humanity as being within nature (Cunfer and Waizer, 2016; Jocks and Sullivan, 2020; Wemigwans and Anderson, 2012); and the extensive knowledge and use of medicinal plants by the Yanomami, Matses and other rainforest tribes of Brazil (Hance, 2015; Milliken and Albert, 1996). These values are culturally based and reflect the wider cultural dimensions of the surrounding society.

22.1.2.2 Traditional ecological knowledge

People within rural African communities are often experts when it comes to knowledge of their local natural environment. This intimate relationship is borne out of years, often generations, of being directly dependent on the natural world for their survival. The intimate nature of this connection has led to an ecological knowledge that is determined through experiential learning (Berkes et al., 2000). Even though formal education may be limited, community members' understanding of nature, natural processes – such as locally relevant climate change and the behavioural ecology of wild species can be rich and detailed, and can span over an extended period of time. This knowledge is often gender-based, with men having one body of knowledge and women having another (Aswani et al., 2018; da Costa et al., 2021) and is passed down from generation to generation through oral history (Eisenberg, 2019).

Ecological knowledge infused with nature values, as previously mentioned, gives shape to traditional knowledge, or TEK, which is often at the heart of socio-ecological structures that create land and resource management systems, social institutions, and worldviews (Pierotti and Wildcat, 2000; Shackeroff and Campbell, 2007). A knowledge–practice–belief complex is suggested by Berkes et al. (2000) within which traditional empirical observations lead to knowledge about land, plants and animals. This knowledge, in turn, results in the establishment of a set of practices, tools and techniques employed by the community in their interactions with nature. These systems of management require social institutions, rules and codes of interpersonal relationships, eventually leading to a fixed ecological practice (West et al., 2006). Over time, as practices become entrenched within the society, beliefs and worldviews are shaped by those who are interacting with nature and ecosystems on a daily basis.

22.1.2.3 Cultural dimensions of engagement

Cultural links to the environment, displayed by different groups of people and societies the world over, are a product of the wider cultural setting in which they exist. Different cultures possess different social rules of engagement. Visiting scientists, existing as guests within a new culture, may struggle at first to grasp local social norms and morays. This is especially true in cultures that differ greatly from one's own. Expected behaviour is not something written down; it is not learned in books. It is, rather, something that tends to be learned through keen observation and a growing familiarity. See Box 22.1 for an example of a lesson learned by a visiting conservationist about the sharing of food.

Interpreting and correctly responding to the behaviours of communities and locally engaged staff members will help collaborations to be successful. As well, internalizing locally important social expectations will ensure that, as a visiting scientist or conservation practitioner, you are able to interact in a way that builds trust and respect. One should expect to be required to consider numerous cultural dimensions in any given circumstance.

In the 1970s, Geert Hofstede generated a cultural dimensions model based on a study of IBM employees in over 50 countries. He defined culture as: 'the collective programming of the mind that distinguishes the members of one group or category of people from others' (Hofstede, 2011). He identified five dimensions or 'problem areas' representing differences among national cultures (Hofstede, 1980): power distance, uncertainty avoidance, individualism/collectivism, masculinity/femininity and long-term orientation; Hofstede et al. (2010) then added a sixth dimension: indulgence versus self-restraint. This chapter explores individualism versus collectivism, hierarchy versus

> ### Box 22.1 Personal reflections: the sharing of food
>
> In my Western approach to fieldwork, I felt it was valuable to be efficient with regard to both time and money. To decrease travel and fuel costs when getting supplies for local community workshops in the Northern Okavango Panhandle of Botswana, I decided to purchase all necessities, such as rice and oil, in bulk all at one time. On the first night of a week-long field assignment, members of the community cooked and ate all the food that was supposed to last several communities for over a week!
>
> I could not believe it and was angry; it made no sense to me. The next day I realized that I had made a mistake by losing my temper and chastising people, in particular people senior to me, for eating all the food. I was embarrassed and had to apologize. I realized after some discussion with my team-mates that the culture of the San people had always been feast or famine. If food was available, it was meant to be eaten!
>
> I had not understood the cultural norms around food and had to adjust my approach for future workshops. This adjustment used more time and more funding, but was also more culturally appropriate and respectful.

egalitarianism, as well as fatalism versus self-efficacy, in more detail as the most relevant to our discussion.

Individualism versus collectivism: a belief in the power and value of the collective, versus a belief in the power and value of the individual, plays a role in individual decision making. In collective societies, people are less interested in personal gain but are motivated by things that advance the group to which they belong. If the family needs one of its members to perform a certain role, then that duty is more likely to be fulfilled. The collective needs of the family are placed before the needs of the individual. Ethnographic studies in West Africa, for example, show that the practice of sending children away to be raised by relatives and non-relatives is widespread among many ethnic groups (Akresh, 2005; Coe, 2017; Isiugo-Abanihe, 1985). Children can be relocated or exchanged among extended family networks as a solution to economic challenges and gender-based labour needs.

While a more individualistic culture leads to more innovation and growth, since such societies attach higher social status to innovators (Mickiewicz and Shepotylo, 2020), societies that value individual freedoms may be worse at responding to issues that require collective action, such as climate change or pandemic response.

Hierarchy versus egalitarianism: Hofstede (2011) describes the degree to which less powerful members of organizations or families expect and accept the fact that power is unequally distributed as the 'power distance index'. A higher degree within the Index indicates that hierarchy is strongly established, without doubt or reason; a lower degree signifies that people distribute power and question authority (Hofstede, 2011).

Many traditional African societies hold a great deal of respect for their elders, who are considered custodians of communal wisdom and accepted leadership in the affairs of the people (Ezenweke and Nwadialor, 2013). Conversely, the youth – a term used in West Africa to refer to individuals up to the age of 35 years (African Youth Charter of the African Union Commission, 2006) – are often not welcomed to provide their opinions nor to take initiative (D.J. Sheppard, Upper West Region, Ghana, 2004–2014). With conservation employment, such as ecotourism, it is the youth who tend to take the lead. Through these activities, they grow their capacity in natural resource

management and build confidence to interact with diverse global cultures. Meanwhile, their elders do not enjoy the same exposure and training and may fail to adopt the 'new' concepts and perspectives. Ultimately, this built capacity is not always capitalized on because the elders, who are most influential in decision making, are not ready to listen to the opinions of the youth.

Fatalism versus self-efficacy: Self-efficacy refers to an individual's belief in their capacity to execute the behaviours necessary to produce a specific outcome (Bandura, 1989). However, the belief that one can determine their own course of action is not a universally accepted concept. In some regions of the world, a more common belief is that control and decision making are left to fate and are in the realm of higher spiritual beings.

In strongly religious societies, such as those found in many parts of Africa, there is an assumption that a supreme being (e.g. God or Allah) is in absolute control. The sentiment, for example, that 'God is in control' is regularly stated by West Africans when trying to understand life, the future and what will happen next. There is a sense of submission to God, and a belief that people are unable to influence this master plan: 'Man proposes, God disposes' is something that people in Ghana are frequently heard to say (Sheppard, 2021). In the conservation context, the deep-seated belief that human intervention is impotent in determining the future can lead people to fail to recognize situations in which they are able to influence environmental outcomes (Sheppard, 2021).

When outside actors attempt to engage with rural, locally based citizens, it is valuable to understand the motivations of your host community as well as uncover your own bias and positionality within the work. Hofstede's cultural dimensions model (1980) offers a tool for shaping your understanding of community and may act as a guide for critical introspection as you become familiar with local TEK systems and their underpinning nature values. A deeper understanding of diverse locally understood and valued ecosystem services helps create a rich and more nuanced understanding of local context. This context may support, as well as hinder, conservation efforts by shaping conservation perspectives. Figure 22.2 illustrates how these different components come together to create conservation perspectives.

Conservation perspectives are formed in the interaction between these elements. Through this holistic interpretation, conservation scientists are better able to understand the basis through which local people will engage and participate in conservation programmes. These concepts will be explored further in the next section.

22.2 Considering different conservation perspectives

In this section we will discuss different conservation perspectives and how positionality knowingly or unknowingly shapes perspectives

and assumptions. By entering the field aware of one's own inherent bias, the conservationist can strive towards best practices within the context of their work. Considerations regarding positionality include insider versus outsider perspectives, formal versus informal education, observer empathy, historical hangovers, language challenges, divergent worldviews and conservation values. Each of these elements will be discussed further below.

22.2.1 Insider versus outsider perspectives

The role of the scientist or conservation practitioner as an outsider has both advantages and disadvantages. One major advantage is that an outsider is considered impartial, because they do not have familial or societal ties within the local communities. However, investigators are not just bystanders and they can be influential actors as well. One disadvantage is that outsiders often represent access to financial benefits for individuals (conservation-related employment

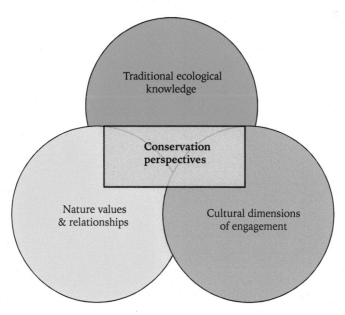

Figure 22.2 Relationship between TEK, nature values and cultural dimensions of engagement.

for insiders), and investment in subjectively selected environmental programmes or sectors within ecosystem-adjacent communities. The desire to influence decisions regarding where and how investments are made can create challenges between outsiders and insiders.

22.2.2 Formal versus informal education

Western scientists engaged in conservation practice and research bring a specific approach to the work stemming from their Western-educated perspective. For them, conservation may be a passion but it is also a form of employment – even a career. By contrast, community constituents of conservation initiatives rarely have formal training and include all of society, from newborn babies to elderly grandmothers and everyone in between. For rural people, historically living within nature, there is a tendency to see a multiplicity of connections infusing their worldviews. Nature conservation is not something people do merely for employment – it is integral to survival and contains spiritual and cultural meaning. During one part of the day, a community member may go into a protected area to illegally collect bushmeat or building materials and within that same day be found using their deep ecological knowledge to guide a Western scientist aiming to determine the conservation status of an endangered wildlife species. The two behaviours may seem contradictory to an outsider but for the community member there are multiple value systems operating simultaneously. See Box 22.2 for an example of the way that relationships with nature can vary around the world.

Recognizing conservation as a profession versus more holistic human relationships with nature begs the question: 'What happens when these diverse worldviews come in contact with each other?' By considering our own cultural

> **Box 22.2 Personal reflections: how far do you travel to be in nature?**
>
> In one community forestry programme in Lofa County, northwestern Liberia, communities of people were living in small settlements completely encircled by old growth forest. They were intertwined with nature. From their doorsteps they could look out at their sacred forests, long protected through traditional means. They would walk past these culturally protected areas to get to their farms and their crops each day (Sheppard, 2021).
>
> For many North Americans, being in nature often means putting on expensive, specialized clothing and driving out of the city and into a designated protected area for an experience or an outdoor adventure. Nature is a place to visit. Nature is seen as something separate.

perspectives and personal biases and by being aware, for example, of having the perspective of a Western-educated scientist, it is possible to open the door to discovery beyond one's own lived experience.

22.2.3 Observer empathy

For international scientists and conservation practitioners, relationship building for improved communication and deeper understanding ensures increased success with programme implementation. However, relationship building can also lead to outside observers empathizing with the situation experienced by inside participants.

For example, major environmental issues facing the planet, such as climate change, require the measurement of very real and long-term impacts of humanity's global actions. However, getting intimately involved with the realities of people living in poverty and in need of immediate solutions to basic survival challenges puts these two issues at odds. How can one make conservation recommendations relevant at the global level as well as at the local level? This dilemma of scale, in which the scientist can empathize with conflicting positions experienced on different levels, poses a very real challenge.

22.2.4 Historical hangovers

As a result of slavery, colonialism and religious conversion, stereotyped relationships have long been established in Africa regarding race and ethnicity. European colonizers established themselves as political elite and landowners, pushing Indigenous peoples into marginalized ecosystems. Traditional African religions were sometimes 'demonized' by missionaries (Bailey and Peoples, 2014; Chitakure, 2017; Eneji et al., 2012), a scenario that has shaped the narrative of current African cultural practices (Ellis, 2006), sometimes driving societies inward. In Liberia, the original 'Spirit of the Forest' became known instead as the 'Bush Devil', a name inherited from early Christian missionaries (Taryor, 1989). To build trust with community members who are believers of African religions, it is worthwhile to set aside past religious stereotypes and talk about all religions in neutral, or even favourable, language.

Grappling with the legacy of these 'historical hangovers' can be a daunting task. By informing oneself of the history of a given place, an unfamiliar scientist or conservation practitioner is able to build a more self-aware approach to engagement, including the realization of the potential and various nuanced perspectives of their host communities.

22.2.5 Language challenges

Intercultural work often requires interaction with different linguistic groups. This leads to language and translation challenges. Difficulties arise when translating complex concepts, since language barriers often prevent a true exchange of ideas and insights across cultures (Sharma, 2010). Previous studies have shown that intercultural research in which translation is required tend to experience problems that fall into two general areas: (1) differences in language content (i.e. word equivalents across translations, accuracy and reliability of translations and language knowledge of translation staff); and (2) differences in acceptable emotional expression (Sharma, 2010).

22.2.6 Divergent worldviews

Engaging in cross-cultural research often involves communication about broad concepts and ideas in which the researcher and informants have divergent worldviews. This makes the clarification of those concepts challenging. Each side is unable to comprehend foundational assumptions that the other needs to know in order to progress with their understanding. See Box 22.3 for an example of the way that worldviews shape the way we interpret our surroundings.

22.2.7 Conservation values: going beyond economic perspectives of nature

As discussed in the introduction, the economic values of nature and natural resources are well known. The extractive industries of logging, hunting and fishing have exploited forests and wildlife in the interest of human survival the world over. However, other perspectives on the values of nature also exist, including for

---

Box 22.3   Personal reflections: my bi-racial cultural worldview

I am mixed race, half Canadian of European descent and half Trinidadian of Indian descent. I grew up consciously and unconsciously navigating two cultures with very differing sets of norms and values. I learned to adapt almost seamlessly between the two and I took this ability with me when I began working overseas. If you are of the dominant cultural group (e.g. white European descent in Canada) then you have most likely not had to think about adapting to others as much as others have adapted to you. It is important to start thinking about your own cultural norms, biases and expectations before you begin work in another culture. It is also important to understand the impact of colonialism around the world and the culture you may represent when you work overseas.

From the moment I set foot in Ghana, I began adjusting to the dominant culture around me. I watched what people wore and how they acted, and I tried to do the same. I noticed that other foreigners did not adjust in the same way. It was common to see tourists wearing items that would be appropriate for summer fun in North America – baggy white t-shirts, tank tops, shorts and flip-flops. This wardrobe did not translate well in the endless heat (30 to 40°C) and dust of Ghana, nor the need for constant laundering. Travellers were frequently dressed in dirty and stretched-out clothes and/or clothes that were considered inappropriate, such as shorts for women (who should not show their knees) or flip-flops (only acceptable footwear for home use).

Because I was a foreigner, local Ghanaians would ask me to explain why my fellow foreigners were dressed so poorly, and it was only through these interactions that I realized that I had adapted – at least with respect to dress code – in a way deemed acceptable to the local people. At the same time, I often wondered at perspectives and experiences of those other foreigners, and how their worldviews were – or were not – being shifted through these diverse interactions.

---

example the links between science and religion. What is the role that socio-spirituality can and does play in conservation? Gaps in the literature concerning spiritual dimensions of conservation persist. While the International Union for Conservation of Nature (IUCN) documented the need to protect sacred natural sites of Indigenous and traditional people (Oviedo et al., 2005), and Papayannis and Mallarach (2009) argued the sacred dimensions of protected areas through their IUCN Delos Initiative, recent meta-analyses of community-based natural resource management programmes, both within Africa (Galvin et al., 2018) and worldwide (Brooks et al., 2013), continue to overlook spiritual dimensions

within their reviews. See Box. 22.4 for a story that further illustrates this point.

In the following, key considerations were reviewed with respect to conservation perspectives and personal positionality. In section 22.3 we move from the personal to the institutional level and consider how different perspectives impact the ability to achieve success in international conservation programmes and citizen science.

22.3   International field conservation dilemmas

Many donor organizations involved in international conservation are based in wealthier

---

Box 22.4 Personal reflections: cultural conservation values at the Wechiau Community Hippo Sanctuary (WCHS), Ghana

The Wechegee people of the WCHS have a number of plant and animal species that are considered sacred, meaning that it is a taboo to hunt or kill them. These taboos have been passed on through storytelling over the generations. There are several creation stories and other survival stories that link the Wechegee people to the hippos. One such story was shared with me by Gurungu-Naa Banda-naa, in which he explained that:

> During the times of our ancestors, wars were fought between different clans and then also, during the slave trade, our people were trying to escape capture and were running away. When we arrived at the Black Volta River we did not know how to cross. All of a sudden, we saw the hippos. The hippos let us sit on their backs and led us safely across the river. It is for this reason that our people refuse to eat hippo meat. (Banda-naa Cheilinah, Upper West Region, Ghana, 2014, personal communication).

The Birifor people, also of WCHS, have beliefs linked to conservation. At puberty, the youth receive a new name given to them by the spirit that lives in the Black Volta River. The hippopotamus is considered to be a child of the river spirit and so it is said that, at puberty, children go to the river and get their 'hippo name'. This cultural practice encourages respect and protection of both the hippopotamus and the river.

Western countries, far from ecosystems and species that they are striving to protect. Often arrangements and fund transfers are agreed from a distance, with relatively few inspection/supervision visits to the country or countries that are receiving the investment. This disconnect leads to a lack of familiarity with work being conducted by local partners in the name of conservation on the ground. There is a danger that foreign donors and conservation organizations miss the mark with their efforts, achieving less than what they claim or, at worst, perpetuating global misunderstandings. See Box 22.5 for an illustration of how misunderstandings and misinterpretations occur when donors and recipients function from a distance.

An alternative concept – and one with which the authors of this chapter have engaged for more than 20 years – involves people who, in one way or another, are permanently based in the country of programming implementation. Advantages to this approach include:

- a deep knowledge of the cultural milieu in general, and of the target collaboration communities in particular
- extensive time investment enabling familiarity with complexities of the topic or focus subject, particularly valuable if the study/work area differs substantially from the outsider's original point of reference
- ongoing relationships established before and continuing after the research, thereby providing motivation to conduct research ethically, and with a high standard of integrity
- opportunities to build on the foundations of your original research or applied conservation activities, offering increasingly higher quality work understandings
- opportunities to be seen and identified as an individual person, moving beyond established stereotypes regarding foreigners
- outsiders are considered impartial, and their actions are seen as objective even

---

Box 22.5   Personal reflections: keeping your distance

I witnessed a team meeting for a Canadian conservation organization, in which the death of one local boat driver from a partnership in far-away Ghana was announced. The tragic news was taken seriously, and so commenced an internal review of the policy regarding the handling and response to staff deaths and injuries of African collaboration partners in the line of duty.

The only problem was that the news was not entirely accurate. The individual in question was indeed a boat driver, but not a boat driver employed by the project. The report had in fact been made by a clever and opportunistic individual, hoping to receive financial support for the upcoming funeral of an elderly community member.

Ghana's rich history and cultural heritage includes vibrant festivals and elaborate traditional funerals. Typically held after 40 days of mourning, funerals are often a lively commemoration to the departed. The announcement of these funerals is something that is communicated far and wide. This is done because every death triggers the flow of money, because money and death are inextricably connected and the funeral business flourishes (de Witte, 2003).

If the Canadian donor had been more aware of this feature of Ghanaian culture, there would have been a deeper understanding of the circumstances. In this case, the cultural unfamiliarity and physical distance from the event meant that assumptions and erroneous interpretations were made.

---

when embedded within the country you are able to remain as an outsider, and even as connections with local citizens grow and intensify

- most importantly, you can develop the ability to hold two or more worldviews as true at the same time. In being able to see things from different perspectives, Western science no longer provides the only benchmark of validation.

There is no set formula for this type of 'on the ground' engagement. Some organizations establish permanent regional and local offices. In 2021, Conservation International (CI) reported on having global offices in more than 30 countries, with large staff teams of both international and local members (Conservation International, 2023). Active in over 100 and 40 countries, respectively, the World Wildlife Fund for Nature (WWF) and the World Conservation Society (WCS) are two other well-known conservation organizations with a large international presence.

Other organizations, such as the Frankfurt Zoological Society (FZS), have fewer international programmes, preferring to concentrate on one or two regions of the world. The longevity of these programmes, however, indicates a serious commitment to nature conservation issues within particular countries and ecosystems. For example, since 1959 FZS has partnered with the government agencies tasked with wildlife protection in the Serengeti, Tanzania (Frankfurt Zoological Society, 2021).

Other organizations set up smaller operations, preferring to limit resources needed to establish an in-country or regional office, and embed a staff member(s) within a qualified, reputable domestic conservation organization instead. This has been the approach of the Wilder Institute/Calgary Zoo since 1999. For relatively minimal cost, tangible benefits of an embedded staff member within an in-country partner organization may include:

- use of a qualified technician/scientist/biologist/ field manager for 'free' to the beneficiary organization
- salary, residential and programming costs covered by the donor organization
- comes with a desirable educational and scientific background and financial accountability
- embedded staff member able to build capacity in citizen understudies
- ability to amplify impact through international connections to media and funding opportunities.

The bigger the organization, the larger the available budgets but also the larger the overhead required for administration and, typically, the deeper the bureaucracy (Mote et al., 2016); global programmes have their own unique organizational challenges (Aarseth et al., 2013). Over the years, the big conservation organizations have been accused of misplaced values, projects that seemed to fail as often as they succeeded, taking credit for the hard work of outside conservationists, and bungled local relationships (Hance, 2016). They have also been criticized for being uncomfortably close to large nature-destroying businesses and corporations (Vidal, 2014).

So, what is the solution? If overseas donor partners lack the local knowledge required to make lasting positive impacts but locally engaged implementation schemes have criticisms of their own, what is the answer? While long-term, in-country commitments are preferable, they are not without their challenges and there is no single engagement plan that can be considered a panacea for complex conservation challenges. Furthermore, the majority of groups and individuals with an interest in supporting international conservation are without the financial clout to run international field offices around the world. Instead, most donate funds to granting agencies in the name of conservation. Before joining or investing in a conservation organization, what

are the key features you ought to be seeking out? Track record, reputation and reports from beneficiaries and collaborators are likely sources of information. Typical dilemmas into which international field-based conservation programmes may fall are reviewed below.

22.3.1 Time crunch and exit strategy

Avoid participation in programmes that have a short implementation window, a rigid end date or enforce an 'exit strategy'. The day after the conservation project 'ends', the local collaborators will be back on the landscape as usual, farming and fishing. This scenario – with the external actor thinking in terms of months or years and the local citizen seeing time through more of a lifelong lens – creates the foundation for misunderstanding. Better options include partnerships and organizations willing to recognize that meaningful growth and change in environmental practice requires long-term commitment and investment. Ecological processes move at a measured pace, as do cultural shifts. A mindset in which short-term fixes are conceivable is not a useful approach when it comes to conservation practice. Grants with fixed start and end dates can be secured for specific components within a larger, long-term partnership, never losing sight of the strength of the overarching collaboration being the structure everything hangs on.

Box 22.6 provides a story about how African communities become used to the short-term comings and goings of international NGO conservation staff and international research teams.

22.3.2 Self-sufficiency

Programmes that require 100% external funding to operate should be avoided. While external funding sources for pilot projects and other

---

Box 22.6   Personal reflections: our borrowed blue chairs

When we first moved into our house in Wechiau in the Upper West Region of Ghana, we had to figure everything out locally – what to eat, how to cook it, how to sleep and how to manage a house in a culturally acceptable way to our neighbours who were attentively watching our every move.

One of the prominent families in the area, with a divisional chief as their head of household, brought four very sturdy blue chairs for us to use. I was confused by the gesture. 'Aren't you using them?' I asked. 'No problem, you will give them back to us when you leave,' came the reply.

Up to that point, the experience with foreign visitors was that they came to stay for just a short time – maybe a few months or maybe a few years, but never more. The community was already getting ready for us to leave.

The longer we stayed, the more layers were revealed to us – a deeper and richer understanding of the local way of life, the society and its complex links to the natural world. Finally, we did return those chairs – but it was 11 years later and the chairs had faded and cracked under the weight of so much use. The owners of the chairs had long ago become like family to us, and we were able to reminisce and laugh at all that we had learned about one another during the time of the borrowed blue chairs.

---

start-up programmes can be vital resources, programmes should include self-sustaining and/or revenue-generating features to reduce their dependency on external actors.

22.3.3   Keeping donors happy

Donors and granting bodies often require that specific measures and indicators be met in order to secure funding. These demands may align well with the situation on the ground, or they may lead activities away from the goals of community beneficiaries. A situation may develop whereby operational plans, that have stopped being relevant to local circumstances and stakeholders, are taking up substantial amounts of time in order to please the donor. As a conservation practitioner, it is important to regularly assess whether your programmes are continuing to provide tangible, measurable benefits to the ecosystems and communities you are striving to support.

22.3.4   Charity versus give and take

When conservation organizations striving to safeguard a particular ecosystem or protected area engage with communities surrounding that ecosystem, budgets tend to include programmes that are aimed not only at biodiversity goals but also at improving the quality of human life. Beneficiaries may be collective (for example, construction of wood lots, tree nurseries or tourist structures) or they may be individual (for example, mobile phones, personal field clothing, solar lights, beehives or bicycles). For individual beneficiaries, whenever possible, a fee should be attached to receipt of the item. By charging a nominal fee, recipients end up with 'skin in the game' and become more invested as a result (Forbes Business Council; Akhrin, 2020). Funds accrued from the 'sale' of individual items can be placed into a maintenance trust fund to help repair damaged items, or to be used towards training fees and other project-specific needs.

22.3.5 Financial integrity and transparency

To engage communities in conservation, money will have to change hands. How can rural community members manage conservation finances for successful programme implementation? An environment of accurate spending and reporting should be enabled through the creation of simple, true-to-life budget templates. Start with small weekly or monthly budgets, providing funds for only what is required. Build capacity gradually, knowing that success breeds success. Make itemized spending agreements and set clear expectations that the funds will be spent on only the agreed terms.

The issues discussed above have reflected on some common programming challenges:

- successfully navigating conservation programmes where funding requirements force fixed start and end dates and require that you determine your 'exit strategy'
- building self-reliance and sustainability over the longer term so that your conservation programme is not perpetually in search of funding
- keeping donors happy while continuing to hold true to best-case scenarios for community partners and critical natural resources
- giving monetary value to community improvement programmes whenever possible
- ensuring financial integrity and transparency.

22.4 Community engagement toolkit for conservation practitioners

Following the broad guidelines set out in Section 22.3, you have found an organization whose values and core tenets match your own. Now, in this section, we will go deeper into the world of international conservation and provide a toolkit for use while in the field. The issues discussed below may not comprise an exhaustive list nor apply to all circumstances; however, they do represent common features of community interactions from East, Southern and West African contexts. Each of the concepts presented below will be explained further through narration and storytelling using the authors' own experiences.

22.4.1 Relationship building

The importance of taking time to build relationships cannot be overstated. It is important to adhere to local communication norms and take time to connect with people at a personal level in addition to a professional level. Trust, accountability and respect all lead to increased community investment and 'buy-in' for projects. Often community members are buying into you as a person more than the project itself, at least initially. When projects slow, or expectations are not met, it will be your relationship with the community that sustains it. In the field, there are fewer boundaries between personal and professional relationships. While continuing to remain professional, you must demonstrate a commitment to the well-being of members of the community.

22.4.2 Setting realistic expectations

When projects and programmes first begin, expectations of benefits, quick results and change are very high in most communities. It can take a very long time, however, for there to be any tangible results. It is important to set realistic expectations and for participants to be aware of how and when they may begin to see changes. Avoid making promises about things that are not certain, or of things that might or might not happen in future.

22.4.3 Power dynamics and differentials

One of the general assumptions of community-based conservation is that, by working with the local community you are creating a decentralized structure for the management of natural resources as compared with previous top-down approaches to conservation (Galvin et al., 2018). Instead, researchers and practitioners are often faced with a continued monopoly of power. This control is typically exerted by the ruling elites or by particular families. If external actors fail to recognize the pre-existing hierarchies and configurations of power that favour certain families, clans and/or ruling elites, their foreign-born, imposed democratic institutions will probably not meet with success (Campbell and Hatcher, 2019; Grayson, 2010).

Where pre-existing community mechanisms have already been established, the creation of new structures should be avoided. Community-based organizations are typically well connected, aware of local needs and able to respond effectively; they may even be able to mitigate the dominance of the local elite (Opare, 2007; Sithole, 2004). If your introduced collaborative management structures can merge with traditional power structures, there is a better chance of representation of all interests within a local community (Sithole, 2004). In fact, a useful approach is to invite communities to use their own processes to select suitable representatives to take up the various positions offered by external conservation organizations and other external actors. Although this does not guarantee an equitable distribution of power, it does provide the opportunity to build on and integrate conservation into local structures. Box 22.7 provides an example of making use of local mechanisms and societal structures to implement a conservation education programme.

Box 22.7 Personal reflections: I think we can teach those sessions ourselves

In 2005, while based at the WCHS in northern Ghana, a series of locally relevant environmental education sessions were designed and funded. Each session had a colourful, locally illustrated booklet that went with the lesson plan. For many families this would be the first book they had ever owned. There were 10,000 books printed for each of the three titles in the series, and the books were to be distributed in 17 widely spaced settlements by a team of two trained education officers who bicycled from place to place.

Although the two traditional divisional chiefs had been centrally involved with the planning of the programme throughout, when the time came to implement the plan, these elders suggested that they would be the ones to disseminate the information rather than the youth education officers. Since the chiefs were not able to ride bicycles, nor would the budget or terrain support driving around in vehicles, the whole thing had to be deferred until a solution was found. The booklets were stacked on shelves in my home and the lesson plans put away.

Some months later, when the issue had lost its intensity, a compromise was reached. The chiefs were provided a stipend to oversee the programme launch in the first few settlements found in readily reachable locations. The trained youth continued from there, covering the material beautifully and building their capacity by leading the environmental education sessions. The chiefs were respected and felt recognized by those first few audiences – and by me and my team for the pioneering roles they played in establishing the community venture in the first place.

22.4.5 Transparency/petty jealousies

Envy and petty jealousy towards locally engaged staff, participants and/or volunteers by community members who expect to be in central, influential roles are issues that practitioners need to be ready to face (DeGeorges and Reilly, 2009). Petty jealousies related to real or perceived perks of working in conservation have the power to infuse a community and cause great disharmony. Jealousy is closely related to power dynamics and relationships and, as an outsider, it can be easy to feed into jealousies without realizing. The success and material gain of locally engaged citizens can amplify another's misfortune. Where you purchase supplies and food, secure housing and even park your car can contribute to misunderstandings in communities. Transparency and efforts towards the equitable distribution of benefits are two ways to combat jealousy. See Box 22.8 for an example of a perceived act of favouritism performed accidentally by a visiting international conservationist.

> **Box 22.8 Personal reflections: a shady place to park**
>
> After several months of parking under a large tree in a village in the Okavango Delta, northern Botswana, I was asked by a community member why I parked there. I answered that it was for the shade and because it was close to the home of the interpreter who was my point of contact in the village. I was gently but firmly informed that it was a major 'faux pas' and disrespectful to the head of the community.
>
> There was a bit of status associated with the vehicle and, more importantly, with the order in which people in the community ought to be greeted by outsiders. I had inadvertently insulted the head of the community. I apologized and changed my approach immediately, parking in front of the Chief's compound from that point forward. It is important to be humble and to understand local protocol.

22.4.6 Time management

Different cultures have different relationships with time. Understanding the parameters of those relationships enables an individual to manage their time to maximum productivity and efficiency. If one lives in a place where it is possible to predict the time required to drive from A to B with certainty, or to be assured that the electricity will remain on throughout the day, this enables time to be well scheduled and used. In places with less time assurance and increased unpredictability – where driving may include any number of delays or difficulties or where power outages are a familiar occurrence – the ability to adapt and alter plans, goals and activities is essential. Realistic programme deliverables and time frames take into consideration disruptions within the wider society outside of the conservation programme's control. See Box 22.9 for an example of the ways that time management can vary across cultures and countries.

22.4.7 Setting smaller goals and getting good at multi-tasking

Goal setting is an important part of any career. However, setting achievable goals, especially for multi-year initiatives, is a difficult thing to accomplish successfully. Project management typically includes activities linked to higher-level objectives and goals, measured by pre-set outcomes and outputs. Project management

---

Box 22.9   Personal reflections: sorry, am I late?

I remember my first years in Ghana, arriving to a scheduled meeting about 15 minutes late and apologizing to the host. She laughed and said: 'You are *Obruni* late' (Obruni being a local term used to mean 'foreigner' or 'white person'; Holsey, 2008). In other words, in her cultural expectation I was not considered late at all.

Later I learned how to plan my arrival to coincide with the arrival time of my fellow meeting attendees. In rural areas, start times can be loosely set, or not set at all. Because transport challenges are endless, a morning meeting simply starts when the majority of people have found their seats.

As I have matured in my career I aim to be the last one to arrive at a meeting so that I do not use valuable time waiting for everyone else to find their way. Of course, the challenges are real; however, some take advantage of the lax start time expectations. If challenged on a late arrival time, you will often hear: 'ahhh, traffic'!

---

ought to also include ample time and space for delays, and active participation and input from local partners and beneficiaries.

Things always take longer than expected, and the temptation to push ahead of the understanding of local stakeholders is tangible because failure to reach set targets may put you at odds with your donor or employer. Leaving the community members behind, however, will guarantee that after your departure everything that you implemented is more than likely to collapse. Programmes must move at a pace that allows locally engaged participants to grow

their capacity to ensure they can fulfil their roles effectively.

To combat long delays that can, and do, occur with some frequency, it is useful to get many balls up in the air at the same time. When one activity meets with a roadblock, time and energy can be re-focused on the various other elements that continue to progress. Naturally, to be successful with this approach, organizational skills are essential.

22.4.8   Secure a strong cultural foothold: stay for a generation

At the start of your career you can expect to be involved in a number of short-term opportunities. Eventually, however, the combination of education and practical experience should lead to securing a more stable, long-term position. If it is possible within your employment arrangement to remain in a single location or, in addressing a particular species, it is advisable to settle down. In other words, once you find your conservation site, species or issue, stick with it. Rather than spending a year or two in a revolving multitude of destinations and/or ecosystems in which you never secure a strong cultural foothold, invest as much time as possible into a single ecosystem or location. See Box 22.10 for an illustration of different time perspectives and scales.

You and your conservation organization's specific handling of the above-mentioned considerations will impact programme success, in-country stakeholder communities and locally engaged staff members. In the next section we will look deeper into the hiring of local community members in conservation staffing roles, reviewing the advantages and disadvantages of local staff as well as recommendations for success.

Box 22.10   Personal reflections: I am building my house

When I was new in Ghana, I remember visiting people at their homes for the first time and finding a building under construction. The family would be dwelling in the original house, but it was clear that improvements were under way. However, when I went back to see that person again months later, the construction work would not have changed at all. This was confusing. And it was not just something that one person was doing, but more of a trend. It seemed that many people were reporting that they were constructing a house, when actually they were not!

Gradually – after years of living in Ghana – I began to see these houses being completed and then I understood. When someone tells you that they are building a house in Ghana, they *are* building a house! It is just a matter of time. It is probably going to take them many years to secure the resources to complete their endeavours, but they will keep building whenever they get a small windfall of cash.

I had taken my understanding of what it meant to build a house in Canada and assumed it meant the same thing in Ghana. In the same way, Africa-based community conservation programmes will progress; it is just a matter of time.

22.5   Hiring local community members in conservation staffing roles

Hiring paid local staff is a way of engaging the community more deeply, beyond community environmental education and awareness programmes, and providing an economic incentive for conservation initiatives. With paid locally engaged staff, skills and capacity are built leading to an expansion in the area's ability to handle more technical and scientific roles, and the commitment to conservation becomes more permanent. Typical paid roles include wildlife trackers, tourist guides, trail and road maintenance units, firefighting brigades, caretakers, cooks and camera trapping and other technical research teams.

Local staff members can lead their communities into new and diverse relationships with nature and ensure stronger community advocacy for conservation issues. Local staff members can also bring conflict and disharmony between neighbours and families when members belong to different and opposing natural resource user groups. Below, the advantages and disadvantages of hiring local staff – as well as recommendations for success with local staff – are presented.

22.5.1   Advantages to employing local staff

Knowledge of the area is gained over a lifetime and is often intergenerational. As discussed earlier in this chapter, TEK is often a vital ingredient to understanding local geography, climate, behavioural ecology and other nuances within ecosystems. Change-over-time scenarios can be understood better through the eyes of the elders and other citizens who have lived long in that place and the surrounding environments.

By establishing good relationships with local staff, it is possible to create a conduit through which to broaden relationships of trust and respect with the local citizens.

Local staff members can become leaders and role models for community members at large to see what is possible to achieve. People can relate to them and feel proud of them.

A higher degree of conservation buy-in and pride within the local community can be created, both in their place in the world of conservation and in the world in general.

Local financial benefits from conservation go directly to community members, ensuring the support of those community members for at least as long as funds are available.

It is possible to access knowledge bases that are not shared with outsiders. This is particularly important for the interpretation of more private topics or issues that are seen as controversial.

22.5.2   Disadvantages to employing local staff

Local staff may take advantage of their position by looking the other way when laws are broken, especially when it concerns friends or family. For example, illegal logging or hunting may occur and community rangers may be conflicted in reporting the activity due to their relationship with the people breaking the rules.

Dismissals for failed community staffing roles can be difficult and there can be repercussions within the community. In some cases, it may be almost impossible to 'retire' local staff members. If you enter community situations already knowing this, then you will appreciate the importance of not rushing into firm commitments with people.

Often local staff are selected from the younger society members. The youth tend to have higher levels of literacy, speak English and have been exposed to environmental and conservation concepts through their formal education. Although youth work well with the more global NGO organizations, they may lack credibility and respect within their home communities. It might be important to think about who

you hire and how to include traditional authorities or senior members of the community in addition to youth.

Local staff frequently act as translators. This is a powerful position when others are unable to counter-check the interpretation. If the translator disagrees with the message, they can change the meaning of the sentiment expressed or include their own opinions in the translation.

Paid employees of conservation are at an economic advantage. International research and conservation teams often provide high-quality/-paying opportunities, and these conditions can impact community cohesion. Resentment builds in those excluded from the opportunity, and petty jealousy can cause community members to turn on each other.

Local community members have intimate knowledge about how to do a 'work around' in their home societies. For example, a community member may knowingly collect their transport allowance for attendance at a meeting but get a free ride home with a family member.

22.5.3   Recommendations for success with local staff

Check the local labour laws in the country of employment. Many positions will fit the definition of casual labour but, in cases of long-term employment, all staff should be able to access standard health benefits and other insurance schemes available in country.

Select local staff as slowly as possible. First, become familiar with the site and the people. Focus groups or a household survey are helpful tools for becoming familiar with the community. After observing for many months, only then commence with interviews and a hiring process.

When hiring local staff in paid positions, first use monthly probationary contracts for 6 months to a year. These short-term contracts do not necessarily need to be labelled 'probationary', but they will allow for a natural termination point if the arrangement does not work out. It also enables the person to maintain their pride in front of their neighbours and family.

Be as transparent and open as possible about the hiring process with the community leaders and elders. Be inclusive, ensuring that all subzones within the study or focus area have been informed of the opportunity and invited to nominate a candidate.

Be sure to explain the time frame – especially if it is a short-term research contract or other employment contract that will have a definitive end date; be clear about the salary and the other conditions of employment, such as access to a vehicle or a transport allowance, uniform, daily food stipend and so forth. This will help to avoid unrealistic expectations of both the staff and the wider community.

For casual labour or short contract positions, per diem agreements work well. If funding becomes limited, the days of work can more easily be reduced or terminated.

When you are dismissing staff, make sure all uniforms and other branded materials are returned to the employer to avoid confusion in the future. It is important to share with community elders and administrative leaders these staffing developments and, where necessary, the reasons for the change.

22.6 Conclusion

The sections in this chapter read as a loose checklist of concepts, issues and challenges to be borne in mind when engaging communities in conservation field research and practice. Sections 22.1–22.3 discuss the complexities of understanding communities and your position as a researcher within those communities; they give you things to think about in preparation for fieldwork. The ways in which conservation perspectives vary based on education, access to

nature and the multifaceted – and sometimes contradictory – relationships that communities and cultures have with nature are better understood by embedded conservation organizations with a greater understanding of the local context.

In sections 22.4 and 22.5 we have given you some insight into common field conservation pitfalls impacting your ability to work well with local communities, and we have reviewed the hiring of locally engaged staff. These sections give you things to think about while doing fieldwork. The issues presented reflect key junctures of misunderstanding across cultures. Expectations and standards may differ and, when problems arise and it is difficult to understand the root cause, reflection on our own personal perspectives and assumptions about these key concepts brings us ever-closer to finding a shared understanding.

What does success look like in community-based conservation? Success within communities includes tangible and measurable environmental programmes that involve local citizens in every possible way – long-lasting programmes across the lifetimes of community members that strengthen their advocacy for the health of their natural environment. The creation of stronger networks and cohesion within communities will increase the resilience required to face not only conservation issues, but other challenges that impact the community as well.

Success is also a process. Do not rush things. Go more slowly than you are comfortable with, and more slowly than you are accustomed to working. Make sure that everyone is on board and possesses the same vision before moving on to the next step. Observe the impacts of small actions before moving to larger, more complicated ones. Notice who benefits and who does not; notice who holds power and who does not; notice which internal structures are operating within communities that either support or undermine conservation work. The setbacks along the road are opportunities for learning for both the practitioner and community members.

Success is also about working respectfully with communities, which includes learning and observing local customs and traditions as well as being transparent about what programmes and initiatives you can and cannot bring to a community. The phrase 'relationships move at the speed of trust' should always be considered. By taking your time, you will have the opportunity to develop trusting relationships that will strengthen initiatives and ensure that you are able to make a lasting contribution.

For the authors, the benefits of working with communities in conservation far outweigh the challenges presented by the numerous twists and turns. Although more time may be required to establish baselines, build relationships and capacity and create a shared vision, we have come to understand that complex environmental challenges require patient, long-term solutions involving all participants if they are going to succeed.

Acknowledgements

We recognize and appreciate the many people we have learned from across our conservation careers in Africa. Thanks as well to Dr Typhenn Brichieri-Colombi for graphics support.

References

Aarseth, W., Rolstadås, A., & Andersen, B. (2013). Managing organizational challenges in global projects. *International Journal of Managing Projects in Business, 7*(1), 103–132. https://doi.org/10.1108/IJMPB-02-2011-0008

African Union. (2006). African Youth Charter. Commission. In *Executive Council of the African Union Commission 6th Ordinary Session*. Banjul, The Gambia, June 2006.

Agrawal, A., & Gibson, C. C. (2001). The role of community in natural resource conservation. In A. Agrawal & C. C. Gibson (Eds.), *Communities and the environment: Ethnicity, gender, and the state in community-based conservation* (pp. 1–31). Rutgers University Press.

Akhrin, D. (2020). Why you shouldn't give away your product for free. Forbes Business Council. https://www.forbes.com/sites/forbesbusinesscouncil/2020/10/20/why-you-shouldnt-give-away-your-product-for-free/?sh=4f02fafb3604

Akresh, R. (2005). Risk, Network Quality, and Family Structure: Child fostering decisions in Burkina Faso. Center Discussion Paper no. 902. *SSRN Electronic Journal*. Economic Growth Center, Yale University, 1–42. https://doi.org/10.2139/ssrn.643163

Anderson, D. M., Salick, J., Moseley, R. K., & Xiaokun, O. (2005). Conserving the sacred medicine mountains: A vegetation analysis of Tibetan sacred sites in Northwest Yunnan. *Biodiversity and Conservation, 14*(13), 3065–3091. https://doi.org/10.1007/s10531-004-0316-9

Aswani, S., Lemahieu, A., & Sauer, W. H. H. (2018). Global trends of local ecological knowledge and future implications. *PLOS ONE, 13*(4), e0195440. https://doi.org/10.1371/journal.pone.0195440

Bailey, G., & Peoples, J. (2014). *Essentials of cultural anthropology* (3rd ed). Cengage Learning.

Bandura, A. (1989). Human agency in social cognitive theory. *American Psychologist, 44*(9), 1175–1184. https://doi.org/10.1037/0003-066x.44.9.1175

Barrow, E. (2019). *Our Future in Nature: Trees, spirituality and ecology*. Balboa Press.

Berkes, F., Colding, J., & Folke, C. (2000). Rediscovery of traditional ecological knowledge as adaptive management. *Ecological Applications, 10*(5), 1251–1262. https://doi.org/10.1890/1051-0761(2000)010[1251:ROTEKA]2.0.CO;2

Brichieri-Colombi, T. (2021). *Personal Reflection Storytelling Locations*. Adapted from Esri, DeLorme, USGS, NPS and NOAA.

Brooks, J., Waylen, K. A., & Mulder, M. B. (2013). Assessing community-based conservation projects: A systematic review and multilevel analysis of attitudinal, behavioural, ecological, and economic outcomes. *Environmental Evidence, 2*(1). https://doi.org/10.1186/2047-2382-2-2

Burger, J., Gochfeld, M., Pletnikoff, K., Snigaroff, R., Snigaroff, D., & Stamm, T. (2008). Ecocultural Attitudes: Evaluating ecological degradation in terms of ecological goods and services versus subsistence and tribal values. *Risk Analysis, 28*(5), 1261–1272. https://doi.org/10.1111/j.1539-6924.2008.01093.x

Campbell, B., & Hatcher, P. (2019). Neoliberal reform, contestation and relations of power in mining: Observations from Guinea and Mongolia. *Extractive Industries and Society, 6*(3), 642–653. https://doi.org/10.1016/j.exis.2019.06.010

Centers for Disease Control and Prevention. (2021). Science Brief: Community use of cloth masks to control the spread of SARS-CoV-2. https://www.cdc.gov/coronavirus/2019-ncov/science/science-briefs/masking-science-sars-cov2.html?CDC_AA_refVal=https%3A%2F%2Fwww.cdc.gov%2Fcoronavirus%2F2019-ncov%2Fmore%2Fmasking-science-sars-cov2.html

Chitakure, J. (2017). *African Traditional Religion encounters Christianity: The resilience of a demonized religion.* Pickwick Publications.

Coe, C. (2017). Child circulation and West African migrations. In C. N. Laoire, A. White & T. Skelton (Eds.), *Movement, mobilities and journeys* (pp. 389–407). Springer.

Conservation International (2023). *Global Offices.* Retrieved September 17, 2023, from https://www.conservation.org/about/global-office

Cunfer, G., & Waizer, B. (2016). *Bison and people on the North American Great Plains: A deep environmental history.* Texas A & M University Press.

da Costa, F. V., Guimarães, M. F. M., & Messias, M. C. T. B. (2021). Gender differences in traditional knowledge of useful plants in a Brazilian community. *PLOS ONE, 16*(7), e0253820. https://doi.org/10.1371/journal.pone.0253820

De Beer, F. (2013). Community-based natural resource management: Living with Alice in Wonderland? *Community Development Journal.* http://cdjoxfordjournals.org Retrieved December 26, 2018, 48(4), 555–570. https://doi.org/10.1093/cdj/bss058

DeGeorges, P. A., & Reilly, B. K. (2009). The realities of community based natural resource management and biodiversity conservation in sub-Saharan Africa. *Sustainability, 1*(3), 734–788. https://doi.org/10.3390/su1030734

de Witte, M. (2003). Money and death: Funeral business in Asante, Ghana. *Africa, 73*(4), 531–559. https://doi.org/10.3366/afr.2003.73.4.531

Eisenberg, C. (2019). Defining and integrating traditional ecological knowledge to create a more sustainable earth. *Bulletin of the Ecological Society of America, 100*(4), e01585. https://doi.org/10.1002/bes2.1585

Ellis, S. (2006). *The Mask of Anarchy: The destruction of Liberia and the religious dimension of an African Civil War.* New York University Press.

Eneji, C. V. O., Ntamu, G. U., Unwanade, C. C., Godwin, A. B., Bassey, J. E., Willaims, J. J., & Ignatius, J. (2012). Traditional African Religion in natural resources conservation and management in Cross River State, Nigeria. *Environment and Natural Resources Research, 2*(4). https://doi.org/10.5539/enrr.v2n4p45

Etzioni, A. (1996). Positive aspects of community and dangers of fragmentation. *Development and Change, 27*(2), 301–314. https://doi.org/10.1111/j.1467-7660.1996.tb00591.x

Etzioni, A. (2015). *The New Normal. Finding a balance between individual rights and the common good.* Transaction Publishing.

Ezenweke, E. O., & Nwadialor, L. K. (2013). Understanding human relations in the African traditional religious context in the face of globalization: Nigerian perspectives. *American International Journal of Contemporary Research, 3*(2), 61–70.

Frankfurt Zoological Society. (2021). Country offices. https://fzs.org/en/about-us/organization/country-offices/

Galvin, K. A., Beeton, T. A., & Luizza, M. W. (2018). African community-based conservation: A systematic review of social and ecological outcomes. *Ecology and Society, 23*(3), 39. https://doi.org/10.5751/ES-10217-230339

Grayson, K. (2010). Human security, neoliberalism and corporate social responsibility. *International Politics, 47*(5), 497–522. https://doi.org/10.1057/ip.2010.20

Ham, A., Duthie, S., & Kaminski, A. (2018). *Kenya Travel Guide.* Lonely Planet.

Hance, J. (2015). Amazon tribe creates 500-page traditional medicine encyclopedia. Mongabay. http://news.mongabay.com/2015/06/amazon-tribe-creates-500-page-traditional-medicine-encyclopedia/

Hance, J. (2016). Has big conservation gone astray? Mongabay. https://news.mongabay.com/2016/04/big-conservation-gone-astray/

Hofstede, G. (1980). *Culture's consequences: International differences in work-related values.* SAGE.

Hofstede, G. (2011). Dimensionalizing cultures: The Hofstede model in context. Online. *Readings in Psychology and Culture, 2*(1). https://doi.org/10.9707/2307-0919.1014 Accessed October 17, 2021.

Hofstede, G., Hofstede, G. J., & Minkov, M. (2010). *Cultures and organizations: Software of the mind* (3rd ed). McGraw-Hill.

Holsey, B. (2008). *Routes of remembrance: Refashioning the slave trade in Ghana*. University of Chicago Press.

Isiugo-Abanihe, U. C. (1985). Child fostering in West Africa. *Population and Development Review, 11*(1), 53–73. https://doi.org/10.2307/1973378

Jocks, C., & Sullivan, L. E. (2020). Native American religions. *Encyclopaedia Britannica*. https://www.britannica.com/topic/Native-American-religion

Kumar, C. (2005). Revisiting "community" in community-based natural resource management. *Community Development Journal, 40*(3), 275–285. https://doi.org/10.1093/cdj/bsi036

Lammers, P. L., Richter, T., Lux, M., Ratsimbazafy, J. H., & Mantilla-Contreras, J. M. (2017). The challenges of community-based conservation in developing countries – A case study from Lake Alaotra, Madagascar. *Journal for Nature Conservation, 40*(11), 100–112. http://doi.org/10.1016/j.jnc.2017.08.003

Lu, J. G., Jin, P., & English, A. S. (2021). Collectivism predicts mask use during COVID-19. *Proceedings of the National Academy of Sciences of the United States of America, 118*(23). https://doi.org/10.1073/pnas.2021793118

Du Mickiewicz, T., J., & Shepotylo, O. (2020). Are individualistic societies worse at responding to pandemics? *The conversation*. https://theconversation.com/are-individualistic-societies-worse-at-responding-to-pandemics-147386

Milliken, W., & Albert, B. M. (1996). The use of medicinal plants by the Yanomami Indians of Brazil. *Economic Botany, 50*(1), 10–25. https://doi.org/10.1007/BF02862108

Mote, J., Jordan, G., Hage, J., Hadden, W., & Clark, A. (2016). Too big to innovate? Exploring organizational size and innovation processes in scientific research. Science and Public Policy, 43(3), 332–337. https://doi.org/10.1093/scipol/scv045

Murphree, M. W. (1994). The role of institutions in community-based conservation. In D. Western & M. Wright (Eds.), *Natural connections: Perspectives in community-based conservation*. Island Press.

Murphree, M. W. (2000). Community-based conservation: Old ways, new myths and enduring challenges. In Conference on African Wildlife Management in the New Millennium. College of African Wildlife Management.

Mwema, F. M., & Nyika, J. M. (2020). Challenges in facemasks use and potential solutions: The case study of Kenya. *Scientific African, 10*, e00563. https://doi.org/10.1016/j.sciaf.2020.e00563

Opare, S. (2007). Strengthening community-based organizations for the challenges of rural development. *Community Development Journal, 42*(2), 251–264. https://doi.org/10.1093/cdj/bsl002

Oviedo, G., Jeanrenaud, S., & Otegui, M. (2005). *Protecting Sacred Natural Sites of Indigenous and Traditional Peoples: An IUCN perspective*. International Union for Conservation of Nature and Natural Resources.

Papayannis, T., & Mallarach, J. M. (Eds.). (2009). The sacred dimension of protected areas. In *Proceedings of the Second Workshop of the Delos Initiative—Ouranoupolis 2007*. IUCN and Med-INA

Pew Research Institute. (2018). Young adults around the world are less religious by several measures. https://www.pewforum.org/2018/06/13/young-adults-around-the-world-are-less-religious-by-several-measures/

Pierotti, R., & Wildcat, D. (2000). Traditional ecological knowledge: The third alternative (commentary). *Ecological Applications, 10*(5), 1333–1340. https://doi.org/10.1890/1051-0761(2000)010[1333:TEKTTA]2.0.CO;2

Quinlan, M. B., Quinlan, R. J., Council, S. K., & Roulette, J. W. (2016). Children's acquisition of ethnobotanical knowledge in a Caribbean horticultural village. *Journal of Ethnobiology, 36*(2), 433–456. https://doi.org/10.2993/0278-0771-36.2.433

Salick, J., Amend, A., Anderson, D., Hoffmeister, K., Gunn, B., & Zhendong, F. (2007). Tibetan sacred sites conserve old growth trees and cover in the Eastern Himalayas. *Biodiversity and Conservation, 16*(3), 693–706. http://doi.org/10.1007/s10531-005-4381-5

Shackeroff, J. M., & Campbell, L. M. (2007). Traditional ecological knowledge in conservation research: Problems and prospects for their constructive engagement. *Conservation and Society, 5*(3), 343–360.

Sharma, B. (2010). Language and cultural barriers. In A. J. Mills, G. Durepos & E. Wiebe (Eds.), *Encyclopedia of case study research* (pp. 519–521). SAGE.

Shen, X., Lu, Z., Li, S., & Chen, N. (2012). Tibetan sacred sites: Understanding the traditional management system and its role in modern conservation. *Ecology and Society*, *17*(2), 693–706. http://doi.org/10.5751/ES-04785-170213

Sheppard, D. J. (2021). *Improving community-based natural resource management practice in West Africa: Mapping changing spiritual values within belief-based conservation networks in Ghana and Liberia*. PhD thesis. University of Guelph.

Sithole, B. (2004). New configurations of power around Mafungautsi State Forest in Zimbabwe. In C. Fabricius & E. Koch (Eds.), *Rights, Resources and Rural Development: Community-based natural resource management in Southern Africa*. Earthscan Publications.

Smith, G. (1996). *'Community-arianism' Community and communitarianism; concepts and contexts*. UK Communities Online. http://www.communities.org.uk

Somers, M. (2021). How cultural psychology influences mask-wearing. *Ideas made to matter*. MIT Press Sloan School of Management. https://mitsloan.mit.edu/ideas-made-to-matter/how-cultural-psychology-influences-mask-wearing

Taryor, N. K. (Ed.). (1989). Religions of Liberia. *Liberia-forum*, *5*(8), 3–17.

United Nations Department of Economic and Social Affairs. (2016). Leaving no one behind: The imperative of inclusive development. *Report on the world social situation 2016*. United Nations.

Vidal, J. (2014). WWF International accused of "selling its soul" to corporations. https://www.theguardian.com/environment/2014/oct/04/wwf-international-selling-its-soul-corporations

Wemigwans, J., & Anderson, D. (2012). *Forest Directions Teachings*. National Indigenous Literacy Association and Invert Media, Inc. http://www.fourdirectionsteachings.com/

West, P., Igoe, J., & Brockington, D. (2006). Parks and peoples: The social impact of protected areas. *Annual Review of Anthropology*, *35*(1), 251–277. https://doi.org/10.1146/annurev.anthro.35.081705.123308

World Conservation Society. (2021). *About us*. https://www.wcs.org/about-us/offices

World Resources Institute. (2005). *Millennium Ecosystem Assessment, 2005. Ecosystems and human well-being: Synthesis*. Island Press.

World Wildlife Fund for Nature. (2021). *Who we are*. https://wwf.panda.org/discover/about_wwf/

Chapter 23

Citizen science photography in scientific literature: An efficient tool for assessing EcoHealth

Oana Catalina Moldoveanu, Emiliano Mori, Luca Petroni and Alessandro Massolo

Abstract

Advanced technological means are progressively available at low (and decreasing) cost and increasing choice in the Information Age. For example, small electronic devices with high-resolution cameras, often with integrated smartphones, are now widely owned. Concurrently we are also experiencing an era of remarkable environmental alterations and ecosystem deterioration. Increasing EcoHealth field actions require greater, faster and more efficient data collection and analysis; therefore, photographic reports from citizen science may provide information that is efficient, easy to acquire and analyse across the planet. For example, wildlife photography has become more and more popular during the last two decades, with implementation of citizen science platforms and websites with the specific purpose of assessing health and conservation of animal and plant species. This chapter reports the results of a scoping review of indexed publications to assess means and methods, as well as benefits and drawbacks, of citizen science photography in scientific literature. Data collected by citizens, opportunistically accessed, can be helpful for assessing natural resources; however, these data need to be even more carefully analysed and used. Supporting citizen science photography can enhance the efficiency of EcoHealth research and conservation policies, as well as educating and bringing communities closer to nature.

23.1 Introduction and objectives

Global loss of biodiversity, climate change, pollution and land use are only some of the threats leading to ecosystem degradation, along with the loss of fundamental ecosystem functions for human and animal well-being (Cardinale et al., 2012; Parsons et al., 2018). Those concerns must be solved through conservation actions supported by scientific data, enacted by policies and endorsed by the public (Forrester et al., 2017).

Several studies have emphasized the role of citizen science (also defined as Public

Susan C. Cork and Douglas P. Whiteside (eds) **Case Studies in EcoHealth**
DOI: 10.52517/9781789183368.023, © 5m Books Ltd, 2024

Participation in Scientific Research: the engagement of volunteers and professionals in collaborative research to generate new science-based knowledge; Bonney et al., 2009) as a means to collect biological data at large scales and in almost real time (Bonney et al., 2009; Tulloch et al., 2013). Furthermore, others regard it as an important way of raising public awareness through indirect (active-learning) education about the natural world (Cooper et al., 2007; Dickinson et al., 2012).

Because well-being of humans and environmental health are both subject to social and political conditions and ecological changes, there is an increasing need to tackle health problems according to their many perspectives (Cunha et al., 2017). Citizen Science projects are very well suited to the use of technology and various communications, generating large amounts of data at low cost. Although involvement of citizens in gathering data for scientific research is not novel, new technology and devices have increased participation and impact. Citizen Science has reached more and more stakeholders, especially in studies on nature, as noted by Bonney et al., who asserted that '[i]n the past two decades, Cornell Laboratory of Ornithology projects have engaged thousands of individuals in collecting and submitting data on bird observations' (Bonney et al., 2009; Kullenberg and Kasperowski, 2016). In those types of studies, citizens can contribute in many ways, including submitting observations and/or analysing data that will subsequently be available for experts and researchers (Kullenberg and Kasperowski, 2016).

The concepts and definitions of citizen science itself are still debated. According to Irwin (1995), citizen science is a kind of science performed by citizens to answer and address their necessities and worries. Currently, one of those concerns is nature conservation along with ecosystems health assessment. Citizen Science has dramatically changed both the space – time scale of ecological studies and relationships between scientists and non-scientists (Dickinson et al,. 2010). By engaging in scientific research (Kruger and Shannon, 2000), e.g. environmental monitoring (Donnelly et al., 2014), people can establish a new connection with nature and also support biodiversity monitoring. The level of participation of citizens in those studies can be classified as recommended (Bonney et al., 2009): (1) contributory level, with scientists organizing study design and citizens collecting data; (2) collaborative level, with citizens accumulating data and helping with project organization, analysis and diffusion of results and discoveries; and (3) co-created level, with experts from the scientific community involving general people in almost every step of the study.

In the last two decades there has been an increase in camera trap technology use by both scientists and natural photography professionals and amateurs. Thus, a major challenge for scientists is storage and analysis of huge numbers of images (Green et al., 2020). However, this can be solved by involving citizen scientists in classifying data, and promoting user-friendly technology as platforms, applications and websites (Green et al., 2020). Photographs from citizen science projects can be an effective way to monitor and evaluate wildlife and human health, because these can reach virtually every country on the planet, every habitat, any animal species and be immediately available for researchers.

Because of the variety of applications and technologies, a scoping review was deemed necessary for evaluating the kinds and ranges of existent and ongoing searches in this context, and to provide an overview (Grant and Booth, 2009). Thus, this chapter aimed to review indexed scientific literature for assessing both the pros and cons of the use of citizen science photography for assessment of the health of terrestrial and aquatic wildlife and plants. Trends in space and time regarding the use of photographic data in citizen science projects for purposes that could be related to EcoHealth assessment are described, along with the most popular tools and platforms. Some notable examples of studies conducted in terrestrial and aquatic ecosystems are then reported, followed by some important examples of studies whose purpose was explicitly linked to education and community outreach.

23.2 Trends, spatial distribution and platforms for the use of photography in citizen science projects for EcoHealth research

The scoping review was conducted through a literature search focused on three main concepts: 'ecosystem health', 'citizen science', 'photo*' and their synonyms. We considered only peer-reviewed papers in any language, using the same keyword combinations in several databases of indexed scientific literature (Web of Science, Scopus and JStor). Papers were screened and analysed based on time and space distribution, on topics and types of and citizens' engagement and on methods and tools used. The screening phases followed the PRISMA Statement (Page et al., 2021) (Figure 23.1).

More than 3,000 papers were identified during the bibliographic search but, after removing duplicates, and screening titles and abstracts, 156 papers were left for full-text eligibility. Ultimately, 82 papers dealt comprehensively with our theme, with the majority assessing terrestrial ecosystems (Appendix 23.1).

Remarkable differences in the number and types of projects, levels of public engagement, data collection techniques and data analysis emerged from papers selected for our review. Since 2000, the number of research studies using citizen science photography in ecosystem health assessment grew exponentially, and in 2020 reached >20 papers published annually (Figure 23.2).

Most papers report studies carried out in the USA, Canada and northern Europe and,

less frequently, in underdeveloped countries (Figure 23.3). Differences across countries were probably determined by economy, technology and education. As this type of research increases concurrently with technological progress, richer countries have many more possibilities for citizen science.

A great number of papers using this technology in the context of EcoHealth and citizen science had as their main topics conservation and monitoring purposes (Figure 23.4), whereas others were research studies focused on invasive alien species, both terrestrial and aquatic, or on native species that may become invasive, and on their phenology, mainly regarding birds. A smaller proportion of papers focused on climate change, habitat suitability and land use (ecological papers in Figure 23.4). There were also surveillance studies that regarded mainly insects as disease vectors for humans, animals and plants. Another topic was education, with some papers providing guidelines on how classes could participate with citizen science programmes and educate children on nature conservation.

Crowdsourcing and advanced technology have made collecting data on nature faster and cheaper (Arts et al., 2015; Koh and Wich, 2012; van Tamelen, 2004; Willi et al., 2019) and information volumes and resolution have rapidly increased, enabling assessment of entire populations or ecosystems (Arts et al., 2015).

In addition, many specific platforms were developed to facilitate storage and analysis of large amounts of data; otherwise, much information would remain unanalysed (Table 23.1). Among those most widely used, Zooniverse (www.zooniverse.org) is mostly known and used in North America and Africa, whereas the Global Biodiversity Information Facility (GBIF; www.gbif.org) and iNaturalist (www.inaturalist.org) are better known in Europe. For projects involving mainly camera traps, eMammal (www.emammal.si.edu) is preferred in North America. This platform was developed by the Smithsonian Institution and

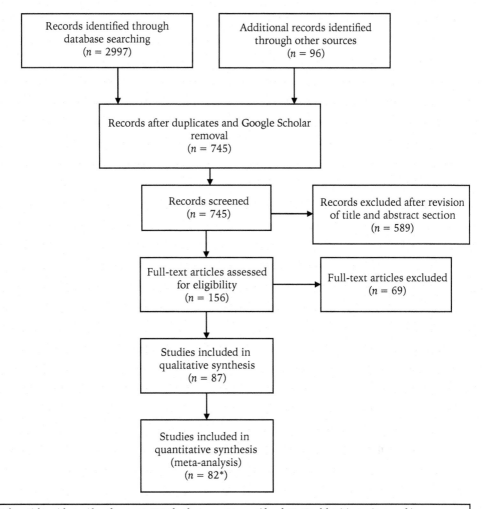

Only articles with specific reference at ALL the three concepts considered: EcoHealth, citizen science and images

Figure 23.1 PRISMA 2009 flow diagram citizen science photography and EcoHealth.

the North Carolina Museum of Natural Sciences, with all data stored in the Smithsonian Data Repository (McShea et al., 2016). Its main features include online tools for training and guiding volunteers, means for data upload and download and interactive interfaces for identifying wildlife photography (McShea et al., 2016).

European researchers have also adopted iNaturalist for camera trap data storage and management. Surprisingly, none of the papers in our selection discussed specific camera trap

databases such as Wildlife Insights (www.wild lifeinsights.org).

Zooniverse may be the world's most popular platform for 'people-powered research'. Millions of people around the world, with a wide range of age and skills, are currently volunteering on Zooniverse to help professional researchers by classifying images (Raddick et al., 2009). They can classify either images containing animals or empty images – that is, images containing no animals, perhaps triggered by the wind (Egna

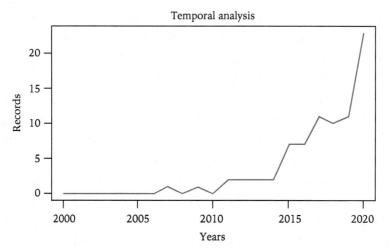

Figure 23.2 Trends in published papers. During the last 20 years and, particularly in the last decade, papers involving citizens in natural science research have increased rapidly. Compared with 2010, papers in 2020 have almost tripled.

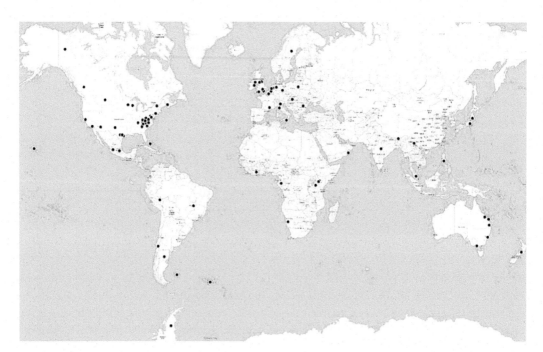

Figure 23.3 Main areas of interest of selected papers. There is a consistent density in the USA and in northern Europe, occasionally in strategic points for EcoHealth such as coral reefs and tropical habitats.

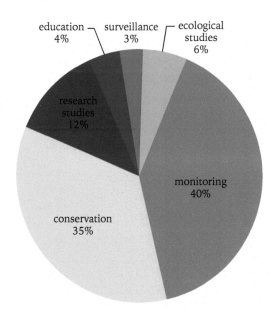

Figure 23.4 Percentages of the main goals of selected papers. Almost half were monitoring studies and a great portion were conservation studies. Research studies included searches on invasive alien species both terrestrial and aquatic and studies on phenology, mainly for bird species. Ecological studies considered wide searches on habitat suitability and climate change. Most of the surveillance studies focused on insects such as mosquitoes and crickets, or important plant parasites.

et al., 2020). One of the first Zooniverse projects and, currently one of the most important camera traps projects, is Snapshot Serengeti. It hosts camera trap images collected from stations located in the Serengeti National Park, Northern Tanzania (~1.5 million hectares) and aims at studying spatial–temporal interspecies dynamics (Swanson et al., 2015).

iNaturalist is another important citizen science platform that hosts a huge community of scientists and citizens. It is supported internationally by several organizations and academies. Photos or records on iNaturalist must reach a consensus from a specific number of identifiers and experts to be classified, minimizing

accuracy concerns (Wilson et al., 2020); furthermore, observations of any user can be included in specific projects by other members of the community, allowing collaborative participation among contributors. The iNaturalist App and many other platforms aim to build and strengthen communities among people so they can upload records and observations of nature, learn, discuss and discover and help experts through collection of data for important ecological searches (Arts et al., 2015). Some of those platforms are Open Air Laboratories (OPAL; www.imperial.ac.uk/opal/), eBird (www.ebird.org), the Atlas of Living Australia (www.ala.org.au), WikiAves (www.wikiaves.com) and CoralWatch (www.coralwatch.org).

Social media have found increasing use for scientific disclosure in the past few years, offering the opportunity of collecting images as well as reaching new information and communicating ideas between experts and the general public (Barratclough et al., 2019; Bombaci et al., 2016). Social media such as Flickr (www.flickr.com), Instagram (www.instagram.com) and Facebook (www.facebook.com) permit opportunistic collection of large amounts of photos. Records can be filtered through hashtags, but these often lack geographic and date information. Nevertheless, social media can be efficiently used for recruiting citizens in science projects by word-of-mouth, advertising and sharing links to other platforms.

23.3 Terrestrial ecosystems

The complexity of terrestrial ecosystems, both natural and anthropogenic, requires a huge amount of data for appropriate status assessment and management. Furthermore, compared with aquatic ecosystems, they are more directly under human threats and are undergoing continuous and rapid changes. Undoubtedly, more and more conservation and monitoring studies will be engaging citizen scientists at various levels. Whereas several plant, bird and insect species are among the best-studied taxa thanks to standardized surveys across landscapes (McShea et al., 2016), similar data on distributions of mammals are missing (McShea et al., 2016; O'Connell et al., 2006). Most terrestrial mammals are nocturnal and elusive, prefer to avoid humans and live in hostile-to-human habitats (McShea et al., 2016; O'Connell et al., 2006).

23.4 The use of motion-activated cameras (camera traps)

Despite many studies appreciating the power of citizen science monitoring, many others are concerned about low-quality data from sightings or when observers incorrectly identify subject species (Suzuki-Ohno et al., 2017). From the early 2000s, technological advancements in camera traps led to a rapid increase in their use to monitor terrestrial species (Rovero et al., 2013) because they collect robust data on species occurrence (i.e. photos or videos). Camera traps, field grids and transects are spreading

across terrestrial ecosystems (Forrester et al., 2017; McShea et al., 2016), because cameras are relatively cheap, easy to set up and supervise, enable collection of a great number of photos and videos, are activated by simple motions and can monitor biodiversity of almost all species. Widespread deployment of accessible and easy-to-use instruments is well suited for integration with contributive citizen science – for example, by training citizen volunteers on how to install motion-activated cameras, as well as classify and analyse the resulting photographic collections (Figures 23.5 and 23.6) (Karlin and De La Paz, 2015; Suzuki-Ohno et al., 2017).

The disadvantages of using camera traps for biodiversity monitoring are several, but most are solvable through study designs. When camera-trapping projects are extensive in space and time, such as those involving many countries contemporarily and for months or years, exhaustive resources may be needed to classify images and revise classifications (Jones et al., 2018; Norouzzadeh et al., 2018); for example, Fennell et al. estimated a mean rate of 500 images per hour classified manually (range, 300–1000, based on expertise) (Fennell et al., 2022). Furthermore, many studies still focus

Figure 23.5 An early-morning view of a zebra from Serengeti Season 15, project activated on the platform Zooniverse (Snapshot Safari, University of Minnesota Lion Center & Wake Forest University).

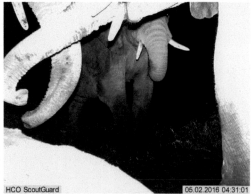

Figure 23.6 A group of elephants from Snapshot Serengeti, project activated on the platform Zooniverse (Snapshot Safari & Snapshot Serengeti).

on single species of interest and many 'bycatch' images are even left unanalysed (Green et al., 2020). Other problems include the quality of images and the potential of misclassification of species with very similar morphological traits (McShea et al., 2016; Swanson et al., 2015; Yu, et al., 2013); many citizens may recognize only one of those species but not the other, causing an under- or over-estimation of species biodiversity for that site. Even camera setups, especially regarding sensor sensitivity and the use of single versus multiple shots, and habitat of camera placement, were identified as factors potentially affecting the accuracy of volunteers in identifying images (Egna et al., 2020). However, regular training and attendance of volunteers can reduce bias and enable high-quality data collection (Danielsen et al., 2014), similar to that obtained by trained scientists. Furthermore, even just-in-time training is enough to obtain high accuracy rates in image classification, even if done by volunteers with no background in biology (Katrak-Adefowora et al., 2020). Basing studies on standardized and clear protocols, enhancing resources for volunteer training (Kosmala et al., 2016), engaging a sufficient number of classifiers per image (Hsing et al., 2018) and using multiple consensus methods and a hierarchical workflow in image classification (Gadsden et al., 2021) are among the solutions available to improve handling of large datasets.

Conservation of natural resources and ecosystems through camera traps surveys relies mainly on data on populations – for example, abundance, distribution and habitat suitability – but also inter- and intraspecific relationships, reproductive success and health status (Agha et al., 2018; Carricondo-Sanchez et al., 2017; Fisher et al., 2014; McShea et al., 2016; Moo et al., 2018; O'Connor et al., 2019).

Since the use of camera traps surveys has increased, many studies focus on how collected images can be analysed for estimating animal density for both marked (Karanth and Nichols, 2011) and unmarked species (Palencia et al., 2021). For example, Cusack et al. (2015) used a random encounter model (Rowcliffe and Carbone, 2008) to determine lion density from camera trapping in Serengeti National Park in Tanzania. Images were collected and classified through Snapshot Serengeti Project from Zooniverse whereas, to estimate lion density, camera trap encounter rates were estimated considering the number of independent photographic events relative to total camera effort, as well as the radius and angle of the camera trap detection zone and the speeds of movement (Cusack et al., 2015).

Other studies adopted occupancy models (MacKenzie et al., 2002) to evaluate species occurrence, accounting for imperfect detection – and the variables possibly affecting it – even of anthropogenic origin (Erb et al., 2012; Parsons et al., 2019; Wevers et al., 2021). Extensions of the occupancy framework are powerful tools to monitor various aspects of natural populations. For example, multi-state models (Mackenzie et al., 2009) were coupled with camera trapping to investigate disease dynamics and reproductive success in wildlife populations (Fisher et al., 2014; Murray et al., 2021) whereas more complex co-occurrence models (Mackenzie et al., 2004) were developed to investigate occupancy probabilities accounting for interspecific interaction (which may include host–pathogen interactions) (MacKenzie et al., 2018). Co-occurrence models are built on detection and non-detection rates collected concurrently for two or more species, and are used to study spatial associations (Parsons et al., 2019; Rota et al., 2016), species relationship dynamics and positive or negative interactions (Anderson et al., 2016). This type of information is crucial for conservation management because communities are composed of various species that share habitats and other natural resources, can be competitors or have complementary roles in ecosystems (Rota et al., 2016).

Most of these studies used imagery data from large platforms such as Zooniverse – e.g. the Snapshot Serengeti Project, WildWatch Kenya or AmazonCam Tambopata. In this way, citizen scientists are involved only in contributory projects by identifying animals in photographs stored on these platforms. On another level of citizen engagement we have eMammal (emammal.si.edu) and MammalWeb (mammalweb.org), collaborative projects where experts point out locations and sites for volunteers to run cameras, with citizens involved in placing cameras and in data collection and classification (Forrester et al., 2017). For instance, recruiting and training volunteers such as citizens or students to deploy camera traps in protected areas across forests in the USA enabled assessment of the impacts of hunting and hiking on wildlife communities (Kays et al., 2017). Similarly, Snapshot Wisconsin (Townsend et al., 2021) has volunteer community scientists trained by the Wisconsin Department of Natural Resources that deploy camera traps on their property or on public land. Finally, Schaus et al. (2020) engaged the public by placing camera traps in urban gardens to evaluate hedgehog density.

The association of camera trapping and citizen science also benefited ecological studies aiming to assess habitat suitability or abundance of non-native species. For example, the 'Elephant Expedition' Project was created in collaboration with Zooniverse.org to better understand the role of forest edges for elephants in the forest-savanna mosaic habitat of Gabon (Cardoso et al., 2020). Camera trap surveys documented a two-way relationship: the local alternation of forests and savannas may be sustained by elephant communities, whereas elephants are provided with water and fruits available only during specific seasons (Cardoso et al., 2020).

eMammal software has been used in several ecological studies – e.g. to assess effects of feral species (e.g. domestic cats) on native wildlife, and how enhancing natural predators (like coyotes) could reduce impacts on native species (Kays et al., 2015). eMammal was also used to assess anthropogenic landscape alteration on competitive species (Statham et al., 2012) and how these may conflict with humans for food and territory (Gosselink et al., 2003; Kellner et al., 2020).

Phenology is another important element in terrestrial species conservation that can be easily investigated through strategic placement of camera traps (Jachowski et al., 2015). Many species of mammals, birds and also invertebrates manifest shifts in phenology due to natural seasonal responses or to climate change at geographical and temporal scales (Nowak et al., 2020). For species like bison (*Bison bison* Linnaeus, 1758) and muskoxen (*Ovibos moschatus* Zimmermann, 1780), data on phenology are relatively easy to achieve (Wilkinson, 1974) but difficult for most species characterized by seasonal activity across a wide latitudinal and altitudinal range (Beltran et al., 2018). Image datasets from crowdsourcing enable follow-up of species phenological activities across broad geographical range thanks to the spread of citizen science camera trapping and, at large temporal scales, due to already available historical data (Cooper et al., 2014; MacPhail and Colla, 2020; Taylor et al., 2019).

Because motion-activated cameras can be placed and left in the field for long intervals (depending only on volunteers periodically recharging batteries or downloading data) reducing human presence in the areas of interest, preventing weather condition damage, and also avoiding invasive traditional techniques that involve handling and stressing animals, they are currently preferred in many kinds of landscape studies (Karlin and De La Paz, 2015; McCallum, 2013).

However, analysis of large numbers of audio files and images is both computationally and labour intensive. Algorithms to provide automatic animal identification (AAI) are being developed to reduce classification times. Whereas AAI was initially based on time-consuming machine-learning algorithms that required highly specific manual processing for matching species-specific marks in images, the use of convolutional neural networks (CNNs) is becoming more common (Willi et al., 2019). Furthermore, AAI is improving and, although reported accuracies for CNNs are generally lower or comparable to human accuracies (Norouzzadeh et al., 2018), such models do not need manual training even if they rely on complex and substantial parameters and model instructions. CNN models can markedly reduce human efforts and, by processing smaller clusters of datasets instead of a large amount of imagery, accurate outputs may be obtained (Willi et al., 2019). Many CNN models are being developed, and some researchers are trying to make them smaller and lighter for potential implementation in mobile devices (e.g. camera traps) and capable of being processed with common-use PCs (Choiński et al., 2021; Yang et al., 2021).

However, even the most recent models suffer from non-generalizability to new sites, environments and communities (Beery et al., 2018; Schneider et al., 2020) and, since developing AAI models is complex, integration of citizen science and artificial intelligence in camera trap image processing is straightforward (Figure 23.7). In fact, the inclusion of object detectors that distinguish among blank images and images containing wildlife, humans and even vehicles in the identification workflow may speed up the process of data handling. For example, Fennell et al. (2022) used MegaDetector (Beery et al., 2019) to filter images, resulting in an 840% increase in classification speed. Furthermore, using detectors to filter images in citizen science projects, and subsequently proposing only wildlife pictures to volunteers, may encourage them to keep identifying images (Adam et al., 2021). Therefore, it is not surprising that many web-based platforms such as Wildlife Insights (www.wildlifeinsights.org) and Agouti (https://www.

agouti.eu/) are implementing and developing AAI algorithms to facilitate the image annotation process for their users.

Balantic and Donovan developed AMMonitor as 'an open-source R package dedicated to collecting, storing, organizing and analyzing AMU' (autonomous monitoring units; Balantic and Donovan, 2020). It is an R library based on SQLite database engine, a public domain that allows storage and analysis of a large amount of data and various information, using statistical tools such as species occurrence models (Balantic and Donovan, 2020). Tabak et al. (2019) proposed an R package (Machine Learning for Wildlife Image Classification, MLWIC) that allows even researchers unfamiliar with programming to run a detector model and even species identification models based on training pictures of wildlife from the North American Camera Trap Images dataset.

23.4.1 The use of citizen photograph archives

Human population growth leads to increasing urbanization and, consequently, to changes in environmental variables such as temperatures (higher in cities), light conditions (light pollution can affect birds and insects phenology) and food and space availability (human interactions affect natural behaviours of wildlife), and thus animals need to adapt to new conditions (de Jong et al., 2018). Bird species are one of the taxa most affected by urbanization, but long-distance migratory patterns and small-bodied features are problematic for constant monitoring (Jachowski et al., 2015). The National Audubon Society's annual Christmas bird count is considered the oldest example of a citizen science project on birds (Jachowski et al., 2015; Karlin and De La Paz, 2015), claiming a consistent dataset usable by avian ecologists. Simple

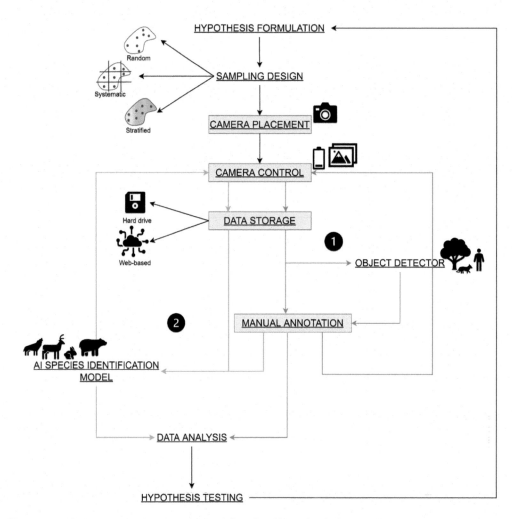

Figure 23.7 A diagram representing the workflow from hypothesis formulation to testing in camera trap studies. Blue-shaded components are those in which volunteers and citizen science might play a major role, and numbers 1 and 2 represent two alternative cycles for camera trap data handling and annotation: cycle 1 represents the classical manual image annotation, which may be speeded up by the use of object detectors; cycle 2 detaches from the former and involves the training and use of AAI models to reduce human work in subsequent cycles.

photographs from citizen science projects or opportunistic Internet-collected images can be considered rather than bird camera trapping that may be not as successful as for large mammals, especially for small-bodied species. Bird photography is useful not only for assessing population status and migratory behaviour but also to understand diets and health problems due to human settlement adaptation – for example, studying variation in the distribution of colour aberrations (Zbyryt et al., 2021).

The same approach can be used for insects and other invertebrates and plants for which camera trap monitoring would not be profitable. Platforms such as iNaturalist and GBIF provide great quality and a large number of photos employed in invasive alien insect species assessment (Ciceoi et al., 2017), or in conservation of specific target insects taxa such as Lepidoptera (Linnaeus, 1758) and Coleoptera (Linnaeus, 1758) (Campanaro et al., 2017) or Hymenoptera (Linnaeus, 1758) (Suzuki-Ohno et al., 2017).

Many similar platforms exist specifically for plants such as Invaders of Texas (texasinvasives. org), The Invasive Plant Atlas of New England (invasive.org) and PlantNet (plantnet.org). Using citizen science image data from Invaders of Texas, Gallo and Waitt (2011) determined that *Arundo donax* (Linnaeus, 1753) is close to become an invasive species, despite not previously being seen as a conservation concern by scientists and policymakers. A similar application can result as an effective tool for citizen science to assess invasiveness of target species, the presence of rare plant species or forest health status (Gallo and Waitt, 2011; Hallett and Hallett, 2018). Hallett and Hallett evaluated the spread of emerald ash borer (*Agrilus planipennis* Fairmaire, 1888) through community-based monitoring photography, this being an important means to manage urban forests containing ash trees that may represent a risk for people (Hallett and Hallett, 2018).

23.4.2 Social media as sources of photographic evidence

Social media are another way of collecting natural data, especially for umbrella species. People can be motivated by the desire to share travels or experiences on their social media, whereas their materials such as posts or audio can be seen by experts as a source of citizen science (Mohd Rameli et al., 2020). For example, photographic reports from citizen scientists were useful in determining the presence of small ape species in habitats from which no presence information was available (Mohd Rameli et al., 2020). Also, the New Zealand Garden

Bird Survey saw nearly 75,000 interactions from posts to comments, likes and shares on Facebook in only one year (Liberatore et al., 2018). A Facebook group called 'Amphibians and Reptiles of Bhutan Search Group', in less than 5 years, collected >1000 images from mobile phones and cameras resulting in 48 new records of species previously unrecorded in Bhutan (Wangyal et al., 2020).

23.4.3 The use of smartphone applications

Lastly, surveillance studies also employed citizen science photography mostly through specific designed smartphone applications. Among the myriad of threats such as ecosystem degradation and the concomitance of climate change, globalization, increasing urbanization and habitat fragmentation, emergence and transport of disease vectors or reservoirs is increasing concerns worldwide (Pataki et al., 2021). Mosquito-borne diseases rapidly extended their boundaries; therefore, surveillance applications such as Mosquito Alert have been expanding. Users can upload mosquito photographs from all over the globe, they can identify species through expert entomologist classification or by using CNN interfaces and then data are shared with public health agencies for control and further supervision (Pataki et al., 2021). Disease control applications involving citizen science are also used for diseases affecting common animal species such as squirrels, which people can easily observe and photograph (Stephenson et al., 2013). Submission of those images, along with location and timing, supported studies on severe epidemics, such as notoedric mange in western grey squirrel populations (Lurz et al., 2015) or mange in urban foxes (Scott et al., 2020).

According to Parsons et al. (2017), 'Animal density is often difficult to estimate, and several other metrics are used instead as measures of relative abundance.' Accuracy in estimating species is crucial for wildlife management, because even tiny oversights may lead to over- or underestimation of species richness (Costa et al., 2015; Miller et al., 2011; Molinari-Jobin et al., 2012; Royle and Link, 2006) and nullify conservation efforts (McKelvey et al., 2008).

For the previously described characteristics, citizen science camera trap is turning to an ever-more popular tool for studying animals' distribution and behaviour, reducing time, costs and efforts for experts and increasing sensitivity and knowledge in the general public (McShea et al., 2016). However, classifying images can be risky if they are blurred, shaded or animals are partially covered (Meek et al., 2013); consequently, teams of experts are needed to remove incorrect classifications. As camera trap use has increased among non-scientists, protocols for users were set and improved over time. These protocols point out the technical features of camera traps, the correct ways of positioning them, trigger and exposure time and image analysis techniques. Missing from those concerns are also legal and privacy problems in using photography from citizen science. Any camera trap project must remove sensitive images and not publish them on websites or platforms, to protect the privacy of people who may have been caught on camera and also to protect rare species from poachers when their explicit location is made public (Willi et al., 2019).

23.5 Freshwater and marine ecosystems

Whereas terrestrial ecosystems initially seemed to benefit more from larger numbers of citizen science projects focused on photography than freshwater and marine ecosystems, in recent years that tendency seems to have changed (Bolt et al., 2022). Marine habitats are threatened by multiple anthropogenic sources such as ocean acidification and sea-level rise (van Hooidonk et al., 2014), eutrophication (Jessen et al., 2013), microplastics and other xenobiotic pollutants, invasive alien species and overfishing (Loh et al., 2015). As a result, there is a need for large-scale and accurate ecosystem assessment; however, monitoring marine areas is often more time consuming and expensive than for terrestrial ecosystems. Whereas detailed methodologies have been developed for land habitats (Belay et al., 2015), surveying aquatic habitats is often hindered by the lack of adequate underwater equipment, trained volunteers and difficulties in coming across wide submerged regions (Leonardsson et al., 2016). Consequently, many studies on subaquatic life are based mainly on predictive models (Rengstorf et al., 2013).

Figure 23.8 An example of frames (extracted from videos by free-diving fishermen) that can be used to obtain estimates of diversity and abundance of fish species.

Global climate change, however, has forced researchers to build up citizen science projects focusing on basic methodologies for assessing underwater species' health (Figure 23.8). The Australian Great Barrier Reef is currently one of the most popular reefs and tourist destinations, but is threatened by massive coral bleaching and species loss (Marshall et al., 2012). In 2012, Marshall et al. launched CoralWatch Project as 'a global citizen-science program that integrates education and global reef monitoring to examine coral bleaching' (Marshall et al., 2012). The 'Coral Health Chart' can be easily used at various depths unlike making underwater activities, and it does not require previous knowledge or skills in corals identification – it is enough to take a photo or an annotation of the matching between the colours on the Coral Chart and the coral itself (Marshall et al., 2012). However, such simplified methodologies are often questioned by experts because they contemplate many fewer variables, thereby depressing accuracy, resolution and standardization (Raoult et al., 2016). Over the past few decades, new marine cameras and GoPro™ use have spread among divers and surfers, mostly for recreational purposes. Consequently, new ways of marine assessment have become available by involving both citizens and tourists.

One of the major concerns regarding marine ecosystems is invasive alien species spreading through new areas that are difficult to detect and monitor. Plants, invertebrates and fish species introduced by humans or accidentally through ballast waters or escape from aquariums can cause declines in native species via competition, parasitism or predation (Blum et al., 2007; Carlton, 2001). Seawatchers (observadoresdelmar.es) is an online citizen science website that launched the project 'Invasive Fishes' to assess the distribution of non-native species in the Mediterranean region through community-collected data (Azzurro et al., 2013). Citizens are allowed to upload pictures of targeted species as

well as information on the date, depth, geographical coordinates and even marine water temperature and habitat type characteristics, enabling researchers to periodically review records submitted by users (Azzurro et al., 2013).

Scepticism regarding taxonomic identification of specimens from photographs is common among scientists, because underwater species are often difficult to classify without accurate physical examination (Newcomer et al., 2019) and many studies have documented how photographic methods may bypass a consistent number of animal species or result in low-quality and incorrect species information (Bicknell et al., 2016; Borgman et al., 2007; Newcomer et al., 2019). Combining capture–recapture models with presence information from citizen science can be an effective way of estimating population size and species abundance, thereby increasing study accuracy (Parham et al., 2017; Pollicelli et al., 2020). However, this requires frequent field campaigns involving experts, resulting in the same difficulties as for landscape photographic surveys – i.e. a large number of images to process (Pollicelli et al., 2020).

A less invasive technique than capture–recapture models has been used in many studies on large-bodied marine mammals such as dolphins (Pollicelli et al., 2020) and grey seals (Sayer et al., 2019), involving PID (Marshall and Bennett, 2010) and RoI extraction (Mitroi et al., 2020; Pollicelli et al., 2020). The PID method allows studies over a lifespan of target species, providing insights on reproductive behaviour and geographic sites of foraging (Lawson et al., 2015), population structure and dynamics and possible threats or sources of disturbance (Hayes, 2011). The RoI method was proposed as a valid approach in cleaning imagery data, saving unnecessary computational effort and reducing bias from superfluous or confounding photographs – for example, when the shape of the dorsal or caudal fin is marked by singular scars it is possible to identify single individuals from two or more images (Pollicelli et al., 2020). Recently, Structure from motion (SFM) photogrammetry has become available for several fields including palaeontology (Falkingham et al., 2014), forestry (Bohlin et al., 2012) and fisheries biology (Rohner et al., 2011). SFM is a new way of processing still images or videos and building three-dimensional models and textures, due to the availability of high-resolution cameras and powerful computers (Westoby et al., 2012).

A study by Raoult and colleagues compared the efficiency of the traditional snorkel survey technique with the use of citizen science imagery from GoPro™ coupled with off-site SFM analysis for sampling a structurally complex coral reef flat habitat (Raoult et al., 2016). Whereas a traditional survey requires time and financial support for human resources and training, the second approach needs only adequate technology and basic instructions. This method had both strengths and weaknesses, but the use of GoPro™ photography proved significantly lower benthic flora and fauna species estimation density, one motivation being the low resolution of images (Raoult et al., 2016).

Whereas the past years have seen an increase in projects such as those mentioned above regarding marine ecosystems, freshwater ecosystems have received very scant attention. Most citizen science projects are about monitoring proliferation of potentially toxic cyanobacteria (Mitroi et al., 2020). They are often available for developed countries but not for poorer countries where they might be even more needed, because underdeveloped countries are more likely to have bad freshwater management rendering them susceptible to eutrophication compared with others (Cunha et al., 2017). For example, the CyanoTRACKER project, launched in the USA, is based on volunteer observations of algal blooms in urban ponds and lakes (Mishra et al., 2020), with similar projects conducted in Canada (Mitroi et al., 2020), plus observations

submitted through social media such as Facebook and Twitter for monitoring water bodies (Mishra et al., 2020). Furthermore, in 2020 some African countries and France managed a collaboration to involve local people in a participatory project on monitoring algal blooms, also aimed at increasing awareness and building up management frameworks for protection of freshwater bodies from pollution and at establishing the best way to obtain data on water conditions, fishing and water vegetation (Mitroi et al., 2020). Nevertheless, none of these projects rely on photography because volunteers make direct in situ measurements of water quality parameters because understanding the dynamics of algal blooms in urban water may be difficult due to the many variables involved, and also photography of the water surface might not be helpful in taxonomic identification of phytoplankton species (Ram.rez et al., 2009). Improvements in equipment for underwater photography will lead to increasingly more citizen science projects regarding marine life assessment and conservation, because many studies have proved that citizen scientists working as a team can provide powerful environmental information on several natural elements such as animal and plant biodiversity, water, soil and air quality (Cunha et al., 2017).

23.6   Outreach, education and research combined

Climate change, biodiversity loss and ecosystems degeneration are enhanced, reinforcing themselves and being worsened by perpetual incorrect management decisions. Despite this, many people are apparently incapable of perceiving incoming dangers for nature and human health whereas many seem not to care (Schuttler et al., 2019).

Citizen science projects promoted by conservationists have proved capable of increasing

Figure 23.9 Two images obtained from the same camera trap in Apuan Alps Regional Park, Central Italy. On the left, a young Italian wolf with the typical phenotype; on the right, an Italian wolf with cutaneous lesions consistent with mange.

awareness on animal and plant species, on good and bad behaviours and relationships with natural resources, concurrently collecting reliable information on human commitment threats (Forrester et al., 2017; Schuttler et al., 2019). By understanding the value of informal education and long-term monitoring, both powered by citizen science projects, societies might discover ways to increase population knowledge about how nature health affects human well-being and how they can help restore it (Figure 23.9) (Parsons et al., 2018). For example, the CoralWatch Project (Marshall et al., 2012) was introduced with an educational purpose in regard to the coral bleaching phenomenon and other threats affecting marine habitats, offering people the opportunity to learn about them and to engage in the conservation of such a priceless ecosystem (Marshall et al., 2012). In a similar perspective, citizen science camera traps projects are an effective tool in monitoring wildlife populations and assessing biodiversity richness, but also stimulate children, students and potentially

any type of citizen to conduct science and help experts (Karlin and De La Paz, 2015).

School-age children are very receptive and at an age when they are forming and deciding on their values and interests; consequently, education about nature conservation can be extremely effective (Nugent, 2017). In addition, citizen science projects simultaneously educate about nature conservation and human relationships. Many can cross large territorial and language barriers. Schools in wealthy countries can collaborate with those in poorer countries, even sending materials and resources to provide opportunities for communities that otherwise could not afford to participate (Schuttler et al., 2019).

Youth is the most easily accessible sector of society by citizen science new technology, but also the most challenging in regard to engaging in serious and important conservation projects. As Schuttler et al. (2019) stated: 'By incorporating citizen science in classrooms, scientists not only have an effective means to monitor biodiversity across a large scale but also a means for youth to experience nature and perhaps to offer them a chance to become stewards for nature.'

23.7   Conclusion and perspectives

Developing management strategies and policies requires data and information about animal density, population and community structures and dynamics, and the influence of biotic and abiotic factors and resources on them, including anthropogenic presence (Schaus et al., 2020). Contextually, ecological long-term and wide-range monitoring is important for validating conservation efforts and decisions (Lindenmayer and Likens, 2010; Parsons et al., 2018).

The increasing popularity of citizen science natural data and new advanced technologies, such as high-resolution cameras, upgraded camera traps for landscape and underwater, GIS software and improved connectivity (Zhang, 2019), have overcome many challenges in collecting and validating information on wildlife (Schaus et al., 2020).

Moreover, data from community-based monitoring are less time absorbing, cheaper and more able to reach every place on Earth than classical biological censuses and surveys involving only experts (Dickinson et al., 2012). Data credibility, positional and identification accuracy and data storage are the central problems regarding citizen science; however, various methods have been developed for managing the integrity of mass-produced data as training programmes, standardized protocols along with specific software and coordinated work between citizens and experts are powerful solutions (Zhang, 2019).

Undoubtedly photography is one of the most efficient, low-cost and suitable-for-all means of citizen science. Among all techniques, camera trap imagery fits perfectly for mammals and sometimes for birds, but less so for invertebrates and fishes. Regardless, remote cameras have highlighted new ways of making science

less invasive, less demanding and more habitat suitable (O'Connell et al., 2011). Photography, either from camera trapping or other tools, is a captivating hobby for much of the population, can arouse interest in people of all ages and thus it can be a readily accessible means for citizen science and a solid contribution to ecosystems health management and education on nature and its threatening circumstances.

References

Adam, M., Tomášek, P., Lehejček, J., Trojan, J., & Jůnek, T. (2021). The role of citizen science and deep learning in camera trapping. *Sustainability*, 13(18), 10287. https://doi.org/10.3390/su131810287

Adriaens, Tim & Sutton-Croft, Michael & Owen, Katy & Brosens, Dimitri & van Valkenburg, Johan & Kilbey, Dave & Groom, Quentin & Ehmig, Carolin & Thürkow, Florian & Hende, Peter & Schneider, Katrin. (2015). MBI 2015 Adriaens etal Supplement.

Agha, M., Batter, T., Bolas, E. C., Collins, A. C., Gomes da Rocha, D., Monteza-Moreno, C. M., Preckler-Quisquater, S., & Sollmann, R. (2018). A review of wildlife camera trapping trends across Africa. *African Journal of Ecology*, 56(4), 694–701. https://doi.org/10.1111/aje.12565

Anderson, T. M., White, S., Davis, B., Erhardt, R., Palmer, M., Swanson, A., Kosmala, M., & Packer, C. (2016). The spatial distribution of African savannah herbivores: Species associations and habitat occupancy in a landscape context. *Philosophical Transactions of the Royal Society of London. Series B, Biological Sciences*, 371(1703), 20150314. https://doi.org/10.1098/rstb.2015.0314

Antoń, S., Denisow, B. (2018). Pollination biology and breeding system in five nocturnal species of *Oenothera* (Onagraceae): reproductive assurance and opportunities for outcrossing. Plant Syst Evol 304, 1231–1243. https://doi.org/10.1007/s00606-018-1543-y

Arts, K., van der Wal, R., & Adams, W. M. (2015). Digital technology and the conservation of nature. *Ambio*, 44(4) Suppl. 4, 661–673. https://doi.org/10.1007/s13280-015-0705-1

Azzurro, E., Broglio, E., Maynou, F., & Bariche, M. (2013). Citizen science detects the undetected: The case of Abudefduf saxatilis from the Mediterranean Sea. *Management of Biological Invasions*, *4*(2), 167–170. https://doi.org/10.3391/mbi.2013.4.2.10

Bagnolini, G., Da Costa, G., Gerino, M., Roth, M., & Tran, C. (2017). Multidisciplinarity for biodiversity management on campus through citizen sciences. 2017 IEEE SmartWorld, Ubiquitous Intelligence & Computing, Advanced & Trusted Computed, Scalable Computing & Communications, Cloud & Big Data Computing, Internet of People and Smart City Innovation (SmartWorld/SCALCOM/UIC/ATC/CBDCom/IOP/SCI), 1–6.

Bagwyn, R., K. Bao, Z. Burivalova, and D.S. Wilcove. (2020). Using citizen-science data to identify declining or recently extinct populations of Bahamian birds. Journal of Caribbean Ornithology 33:104–110.

Bahr, K.D., Severino, S.J.L., Tsang, A.O. et al. (2020). The Hawaiian Koʻa Card: coral health and bleaching assessment color reference card for Hawaiian corals. SN Appl. Sci. 2, 1706 https://doi.org/10.1007/s42452-020-03487-3

Balantic, C., & Donovan, T. (2020). AMMonitor: Remote monitoring of biodiversity in an adaptive framework with r. *Methods in Ecology and Evolution*, *11*(7), 869–877. https://doi.org/10.1111/2041-210X.13397

Barratclough, A., Wells, R. S., Schwacke, L. H., Rowles, T. K., Gomez, F. M., Fauquier, D. A., Sweeney, J. C., Townsend, F. I., Hansen, L. J., Zolman, E. S., Balmer, B. C., & Smith, C. R. (2019). Health assessments of common bottlenose dolphins (Tursiops truncatus): Past, present, and potential conservation applications. *Frontiers in Veterinary Science*, *6*, 444. https://doi.org/10.3389/fvets.2019.00444

Beery, S., Morris, D., & Yang, S. (2019). Efficient pipeline for camera trap image review. *Arxiv* [Abs.]/1907.06772.

Beery, S., Van Horn, G., & Perona, P. (2018). *Recognition in terra incognita. Computer vision. ECCV.* Springer International Publishing.

Belay, K. T., Van Rompaey, A., Poesen, J., Van Bruyssel, S., Deckers, J., & Amare, K. (2015). Spatial analysis of land cover changes in eastern Tigray (Ethiopia) from 1965 to 2007: Are there signs of a forest transition? *Land Degradation and Development*, *26*(7), 680–689. https://doi.org/10.1002/ldr.2275

Beltran, R. S., Burns, J. M., & Breed, G. A. (2018). Convergence of biannual moulting strategies across birds and mammals. *Proceedings. Biological Sciences*, *285*(1878), 20180318. https://doi.org/10.1098/rspb.2018.0318

Bicknell, A. W. J., Godley, B. J., Sheehan, E. V., Votier, S. C., & Witt, M. J. (2016). Camera technology for monitoring marine biodiversity and human impact. *Frontiers in Ecology and the Environment*, *14*(8), 424–432. https://doi.org/10.1002/fee.1322

Blum, J. C., Chang, A. L., Liljesthröm, M., Schenk, M. E., Steinberg, M. K., & Ruiz, G. M. (2007). The non-native solitary ascidian Ciona intestinalis (L.) depresses species richness. *Journal of Experimental Marine Biology and Ecology*, *342*(1), 5–14. https://doi.org/10.1016/j.jembe.2006.10.010

Bohlin, J., Wallerman, J., & Fransson, J. E. S. (2012). Forest variable estimation using photogrammetric matching of digital aerial images in combination with a high-resolution DEM. *Scandinavian Journal of Forest Research*, *27*(7), 692–699. https://doi.org/10.1080/02827581.2012.686625

Bolt, M. H., Callaghan, C. T., Poore, A. G. B., Vergés, A., & Roberts, C. J. (2022). Using the background of fish photographs to quantify habitat composition in marine ecosystems. *Marine Ecology Progress Series*, *688*, 167–172. https://doi.org/10.3354/meps14027

Bombaci, S. P., Farr, C. M., Gallo, H. T., Mangan, A. M., Stinson, L. T., Kaushik, M., & Pejchar, L. (2016). Using Twitter to communicate conservation science from a professional conference. Conservation Biology, 30(1), 216–225. https://doi.org/10.1111/cobi.12570.

Bonney, R., Cooper, C. B., Dickinson, J., Kelling, S., Phillips, T., Rosenberg, K. V., & Shirk, J. (2009). Citizen science: A developing tool for expanding science knowledge and scientific literacy. *BioScience*, *59*(11), 977–984. https://doi.org/10.1525/bio.2009.59.11.9

Bonter, D. N., Gauthreaux, S. A. Jr, Donovan, T. M. (2009). Characteristics of important stopover locations for migrating birds: remote sensing with radar in the Great Lakes basin. Conserv Biol. Apr;23(2): 440–8. doi: 10.1111/j.1523-1739.2008.01085.x. Epub 2008 Oct 20. PMID: 18983598.

Borgman, C. L., Wallis, J. C., & Enyedy, N. (2007). Little science confronts the data deluge: Habitat ecology, embedded sensor networks, and digital libraries. *International Journal on Digital Libraries*, *7*(1–2), 17–30. https://doi.org/10.1007/s00799-007-0022-9

Campanaro, A., Hardersen, S., Redolfi De Zan, L. R., Antonini, G., Bardiani, M., Maura, M.,

Maurizi, E., Mosconi, F., Zauli, A., Bologna, M. A., Roversi, P. F., Sabbatini Peverieri, G., & Mason, F. (2017). Analyses of occurrence data of protected insect species collected by citizens in Italy. *Nature Conservation*, 20, 265–297. https://doi.org/10.3897/natureconservation.20.12704

Caravaggi, A., Gatta, M., Vallely, M., Hogg, K., Freeman, M., Fadaei, E., Dick, J.T.A., Montgomery, W.I., Rei,d N., Tosh, D.G. (2018). Seasonal and predator-prey effects on circadian activity of free-ranging mammals revealed by camera traps. PeerJ 6:e5827 https://doi.org/10.7717/peerj.5827

Cardinale, B. J., Duffy, J. E., Gonzalez, A., Hooper, D. U., Perrings, C., Venail, P., Narwani, A., Mace, G. M., Tilman, D., Wardle, D. A., Kinzig, A. P., Daily, G. C., Loreau, M., Grace, J. B., Larigauderie, A., Srivastava, D. S., & Naeem, S. (2012). Biodiversity loss and its impact on humanity. *Nature*, *486*(7401), 59–67. https://doi.org/10.1038/nature11148

Cardoso, A. W., Malhi, Y., Oliveras, I., Lehmann, D., Ndong, J. E., Dimoto, E., Bush, E., Jeffery, K., Labriere, N., Lewis, S. L., White, L. T. J., Bond, W., & Abernethy, K. (2020). The role of forest elephants in shaping tropical forest-savanna coexistence. *Ecosystems*, *23*(3), 602–616. https://doi.org/10.1007/s10021-019-00424-3

Carlton, J. (2001). *Introduced species in US coastal waters: Environmental impacts and management priorities.*

Carricondo-Sanchez, D., Odden, M., Linnell, J. D. C., & Odden, J. (2017). The range of the mange: Spatiotemporal patterns of sarcoptic mange in red foxes (Vulpes vulpes) as revealed by camera trapping. *PLOS ONE*, *12*(4), e0176200. https://doi.org/10.1371/journal.pone.0176200

Choiński, M., Rogowski, M., Tynecki, P., Kuijper, D. P. J., Churski, M., & Bubnicki, J. W. (2021). *A first step towards automated species recognition from camera trap images of mammals using AI in a European temperate forest. Computer information systems and industrial management* (pp. 299–310). Springer International Publishing.

Ciceoi, R., Badulescu, L., Minodora, G., Mardare, E., & Pomohaci, C.-M. (2017). Citizen-generated data on invasive alien species in Romania: Trends and challenges. *Acta Zoologica Bulgarica Suppl.*, *9*, 255–260.

Connors, J. P., Lei, S., & Kelly, M. (2012). Citizen science in the age of neogeography: Utilizing volunteered geographic information for environmental monitoring. Annals of the Association of American Geographers, 102(6), 1267–1289.

Cooper, C., Dickinson, J., & Phillips, T. R. J. E. Bonney and society. (2007). *Citizen science as a tool for conservation in residential ecosystems 12.*

Cooper, C. B., Shirk, J., & Zuckerberg, B. (2014). The invisible prevalence of citizen science in global research: Migratory birds and climate change. *PLOS ONE*, *9*(9), e106508. https://doi.org/10.1371/journal.pone.0106508

Costa, H., Foody, G. M., Jiménez, S., & Silva, L. (2015). Impacts of species misidentification on species distribution modeling with presence-only data. *ISPRS International Journal of Geo-Information*, *4*(4), 2496–2518. https://doi.org/10.3390/ijgi4042496

Cunha, D. G. F., Casali, S. P., de Falco, P. B., Thornhill, I., & Loiselle, S. A. (2017). The contribution of volunteer-based monitoring data to the assessment of harmful phytoplankton blooms in Brazilian urban streams. *Science of the Total Environment*, *584–585*, 586–594. https://doi.org/10.1016/j.scitotenv.2017.01.080

Cunha, D. G. F., Marques, J. F., Resende, J. C., Falco, P. B., Souza, C. M., & Loiselle, S. A. (2017). 'Citizen science participation in research in the environmental sciences: key factors related to projects' success and longevity. *Anais da Academia Brasileira de Ciências*, *89*(3) Suppl., 2229–2245. https://doi.org/10.1590/0001-3765201720160548

Cusack, J. J., Swanson, A., Coulson, T., Packer, C., Carbone, C., Dickman, A. J., Kosmala, M., Lintott, C., & Rowcliffe, J. M. (2015). Applying a random encounter model to estimate lion density from camera traps in Serengeti National Park, Tanzania. *Journal of Wildlife Management, 79*(6), 1014–1021. *https://doi.org/10.1002/jwmg.902*

Danielsen, F., Jensen, P. M., Burgess, N. D., Coronado, I., Holt, S., Poulsen, M. K., Rueda, R. M., Skielboe, T., Enghoff, M., Hemmingsen, L. H., Sørensen, M. and Pirhofer-Walzl, K. (2014). Testing focus groups as a tool for connecting indigenous and local knowledge on abundance of natural resources with science-based land management systems. *Conservation Letters*, *7*, 380–389. https://doi.org/10.1111/conl.12100

de Jong, M., van den Eertwegh, L., Beskers, R. E., de Vries, P. Pd., Spoelstra, K., & Visser, M. E. (2018). Timing of Avian Breeding in an Urbanised World. *Ardea*, *106*(1), 31–38. https://doi.org/10.5253/arde.v106i1.a4

Dickinson, J. L., Zuckerberg, B., & Bonter, D. N. (2010). Citizen science as an ecological research tool: Challenges and benefits. *Annual Review of*

Ecology, Evolution, and Systematics, 41(1), 149–172. https://doi.org/10.1146/annurev-ecolsys-1022 09-144636

Dickinson, J. L., Shirk, J., Bonter, D., Bonney, R., Crain, R. L., Martin, J., Phillips, T., & Purcell, K. (2012). The current state of citizen science as a tool for ecological research and public engagement. *Frontiers in Ecology and the Environment, 10*(6), 291–297. https://doi.org/10.1890/110236

Donnelly, A., Crowe, O., Regan, E., Begley, S., & Caffarra, A. (2014). The role of citizen science in monitoring biodiversity in Ireland. *International Journal of Biometeorology, 58*(6), 1237–1249. https://doi.org/10.1007/s00484-013-0717-0

Egna, N., O'Connor, D., Stacy-Dawes, J., Tobler, M. W., Pilfold, N., Neilson, K., Simmons, B., Davis, E. O., Bowler, M., Fennessy, J., Glikman, J. A., Larpei, L., Lekalgitele, J., Lekupanai, R., Lekushan, J., Lemingani, L., Lemirgishan, J., Lenaipa, D., Lenyakopiro, J., . . . Owen, M. (2020). Camera settings and biome influence the accuracy of citizen science approaches to camera trap image classification. *Ecology and Evolution, 10*(21), 11954–11965. https://doi.org/10.1002/ece3.6722

Epps MJ, Menninger HL, LaSala N, Dunn RR. (2014). Too big to be noticed: cryptic invasion of Asian camel crickets in North American houses. PeerJ 2:e523 https://doi.org/10.7717/peerj.523

Erb, P. L., Mcshea, W. J., & Guralnick, R. P. (2012). Anthropogenic influences on macro-level mammal occupancy in the Appalachian trail corridor. *PLOS ONE, 7*(8), e42574. https://doi.org/10.1371/jour nal.pone.0042574

Falkingham, P. L., Bates, K. T., & Farlow, J. O. (2014). Historical Photogrammetry: Bird's Paluxy River Dinosaur Chase Sequence Digitally Reconstructed as It Was prior to excavation 70 years Ago. *PLOS ONE, 9*(4), e93247. https://doi.org/10.1371/jour nal.pone.0093247

Fennell, M., Beirne, C., & Burton, A. C. (2022). Use of object detection in camera trap image identification: Assessing a method to rapidly and accurately classify human and animal detections for research and application in recreation ecology. *Global Ecology and Conservation, 35*, e02104. https://doi. org/10.1016/j.gecco.2022.e02104

Fisher, J. T., Wheatley, M., & Mackenzie, D. (2014). Spatial patterns of breeding success of grizzly bears derived from hierarchical multistate models. *Conservation Biology, 28*(5), 1249–1259. https:// doi.org/10.1111/cobi.12302

Forrester, T. D., Baker, M., Costello, R., Kays, R., Parsons, A. W., & McShea, W. J. (2017). Creating advocates for mammal conservation through citizen science. *Biological Conservation, 208*, 98–105. https://doi.org/10.1016/j.biocon.2016.06.025

Gadsden, G. I., Malhotra, R., Schell, J., Carey, T., & Harris, N. C. (2021). Michigan ZoomIN: Validating crowd-sourcing to identify mammals from camera surveys. *Wildlife Society Bulletin, 45*(2), 221–229. https://doi.org/10.1002/wsb.1175

Gallo, T., & Waitt, D. (2011). Creating a successful citizen science model to detect and report invasive species. *BioScience, 61*(6), 459–465. https://doi. org/10.1525/bio.2011.61.6.8

Gibson, K.J., Streich, M.K., Topping, T.S., Stunz, G.W. (2019) Utility of citizen science data: A case study in land-based shark fishing. PLOS ONE 14(12): e0226782. https://doi.org/10.1371/journal.pone.0226782

Gooliaff, T. J., & Hodges, K. E. (2018). Measuring agreement among experts in classifying camera images of similar species. Ecology and Evolution, 8, 11009–11102. 10.1002/ece3.4567

Gosselink, T. E., Deelen, T. R. V., Warner, R. E., & Joselyn, M. G. (2003). Temporal habitat partitioning and spatial use of coyotes and red foxes in east-central Illinois. *Journal of Wildlife Management, 67*(1), 90. *https://doi.org/10.2307/3803065*

Grant, M. J., & Booth, A. (2009). A typology of reviews: An analysis of 14 review types and associated methodologies. *Health Information and Libraries Journal, 26*(2), 91–108. https://doi.org/ 10.1111/j.1471-1842.2009.00848.x

Green, S. E., Rees, J. P., Stephens, P. A., Hill, R. A., & Giordano, A. J. (2020). Innovations in camera trapping technology and approaches: The integration of citizen science and artificial intelligence. *Animals: An Open Access Journal from MDPI, 10*(1), 132. https://doi.org/10.3390/ ani10010132

Grol Monique G. G., Vercelloni Julie, Kenyon Tania M., Bayraktarov Elisa, van den Berg Cedric P., Harris Daniel, Loder Jennifer A., Mihaljević Morana, Rowland Phebe I., Roelfsema Chris M. (2021) Conservation value of a subtropical reef in southeastern Queensland, Australia, highlighted by citizen-science efforts. Marine and Freshwater Research 72, 1–13. https://doi.org/10.1071/MF19170

Hallett, R., & Hallett, T. (2018). Citizen science and tree health assessment: How useful are the data? *Arboriculture and Urban Forestry, 44*(6), 236–247. https://doi.org/10.48044/jauf.2018.021

Hayes, R. A. (2011). Analyzing Animal Societies: Quantitative Methods for Vertebrate Social Analysis. *Austral Ecology, 36*(5), e23–e23. https://doi.org/10.1111/j.1442-9993.2010.02208.x

Hepler SA, Erhardt R, Anderson TM. (2018). Identifying drivers of spatial variation in occupancy with limited replication camera trap data. Ecology. 2018;99(10):2152–2158. https://doi.org/10.1002/ecy.2396

Hidalgo-Hermoso, E., Cabello, J., Vega, C., Kroeger-Gómez, H., Moreira-Arce, D., Napolitano, C., Navarro, C., Sacristán, I., Cevidanes, A., Di Cataldo, S., Dubovi, E. J., Mathieu-Benson, C., & Millán, J. (2020). An eight-year survey for canine distemper virus indicates lack of exposure in the endangered Darwin's fox (*Lycalopex fulvipes*). *Journal of Wildlife Diseases, 56*(2), 482–485. https://doi.org/10.7589/2019-08-195

Hsing, P.-Y., Bradley, S., Kent, V. T., Hill, R. A., Smith, G. C., Whittingham, M. J., Cokill, J., Crawley, D., & Stephens, P. A. (2018) Economical crowdsourcing for camera trap image classification. *Remote Sensing in Ecology and Conservation, 4*(4), 361–374. https://doi.org/10.1002/rse2.84

Irwin, A. (1995). *Citizen science: A study of people, expertise and sustainable development.* Routledge.

Jachowski, D. S., Katzner, T., Rodrigue, J. L., & Ford, W. M. (2015). Monitoring landscape-level distribution and migration Phenology of Raptors using a volunteer camera-trap network. *Wildlife Society Bulletin, 39*(3), 553–563. https://doi.org/10.1002/wsb.571

Jessen, C., Roder, C., Villa Lizcano, J. F., Voolstra, C. R., & Wild, C. (2013). In-situ effects of simulated overfishing and eutrophication on benthic coral reef algae growth, succession, and composition in the central Red Sea. *PLOS ONE, 8*(6), e66992. https://doi.org/10.1371/journal.pone.0066992

Johnson Brian Alan, André Derek Mader, Rajarshi Dasgupta, Pankaj Kumar. (2020). Citizen science and invasive alien species: An analysis of citizen science initiatives using information and communications technology (ICT) to collect invasive alien species observations, Global Ecology and Conservation, Volume 21,e00812. https://doi.org/10.1016/j.gecco.2019.e00812.

Joly, A., Goëau, H., Champ, J., Dufour-Kowalski, S., Müller, H., & Bonnet, P. (2016, October). Crowdsourcing biodiversity monitoring: how sharing your photo stream can sustain our planet. In

Proceedings of the 24th ACM international conference on Multimedia (pp. 958–967).

Jones, F. M., Allen, C., Arteta, C., Arthur, J., Black, C., Emmerson, L. M., Freeman, R., Hines, G., Lintott, C. J., Macháčková, Z., Miller, G., Simpson, R., Southwell, C., Torsey, H. R., Zisserman, A., & Hart, T. (2018). Time-lapse imagery and volunteer classifications from the Zooniverse penguin Watch project. *Scientific Data, 5*(1), 180124. https://doi.org/10.1038/sdata.2018.124

Jong, A., Kleven, O., Østnes, J., Kroglund, R., Vahlström, I., Nilsson, J., & Spong, G. (2019). Birds of different feather flock together - genetic structure of Taiga Bean Goose in Central Scandinavia. Bird Conservation International, 29(2), 249–262. doi:10.1017/S0959270918000205

Karanth, K. U., & Nichols, J. D. (2011). Estimating tiger abundance from camera trap data: Field surveys and analytical issues. In A. F. O'Connell, J. D. Nichols and K. U. Karanth (Eds.) Camera Ttraps in Aanimal Eecology: Methods and Aanalyses (pp. 97–117). Springer.

Karlin, M., & De La Paz, G. (2015). Using camera-trap technology to improve undergraduate education and citizen-science contributions in wildlife research. *Southwestern Naturalist, 60*(2–3), 171–179.

Katrak-Adefowora, R., Blickley, J. L., & Zellmer, A. J. (2020). Just-in-time training improves accuracy of citizen scientist wildlife identifications from camera trap photos. *Citizen Science: Theory and Practice, 5*(1), 8. https://doi.org/10.5334/cstp.219

Kays, R., Costello, R., Forrester, T., Baker, M. C., Parsons, A. W., Kalies, E. L., Hess, G., Millspaugh, J. J., & McShea, W. (2015). Cats are rare where coyotes roam. *Journal of Mammalogy, 96*(5), 981–987. https://doi.org/10.1093/jmammal/gyv100

Kays, R., Parsons, A. W., Baker, M. C., Kalies, E. L., Forrester, T., Costello, R., Rota, C. T., Millspaugh, J. J., & McShea, W. J. (2017). Does hunting or hiking affect wildlife communities in protected areas? *Journal of Applied Ecology, 54*(1), 242–252. https://doi.org/10.1111/1365-2664.12700

Kays, R., Dunn, R.R., Parsons, A.W., Mcdonald, B., Perkins, T., Powers, S.A., Shell, L., McDonald, J.L., Cole, H., Kikillus, H., Woods, L., Tindle, H. and Roetman, P. (2020), The small home ranges and large local ecological impacts of pet cats. Anim Conserv, 23: 516–523. https://doi.org/10.1111/acv.12563

Kellner, K. F., Hill, J. E., Gantchoff, M. G., Kramer, D. W., Bailey, A. M., & Belant, J. L. (2020).

Responses of sympatric canids to human development revealed through citizen science. *Ecology and Evolution, 10*(16), 8705–8714. https://doi.org/10.1002/ece3.6567

Kelly, R., Fleming, A., Pecl, G. T., Richter, A., & Bonn, A. (2019). Social license through citizen science: a tool for marine conservation. Ecology and Society, 24(1). https://www.jstor.org/stable/26796907

Koh, L. P., & Wich, S. A. (2012). Dawn of drone ecology: Low-cost autonomous aerial vehicles for conservation. *Tropical Conservation Science, 5*(2), 121–132. https://doi.org/10.1177/194008291200500202

Kosmala, M., Wiggins, A., Swanson, A., & Simmons, B. (2016). Assessing data quality in citizen science. *Frontiers in Ecology and the Environment, 14*(10), 551–560. https://doi.org/10.1002/fee.1436

Kruger, L., & Shannon, M. (2000). Getting to know ourselves and our places through participation in civic social assessment. *Society and Natural Resources, 13*.

Kullenberg, C., & Kasperowski, D. (2016). What is citizen science? – A scientometric meta-analysis. *PLOS ONE, 11*(1), e0147152. https://doi.org/10.1371/journal.pone.0147152

Lavariega, M.C., Ríos-Solís, J.A., Flores-Martínez, J.J., et al. Community-Based Monitoring of Jaguar (Panthera onca) in the Chinantla Region, Mexico. Tropical Conservation Science. 2020;13. doi:10.1177/1940082920917825

Lawson, B., Petrovan, S. O., & Cunningham, A. A. (2015). Citizen science and wildlife disease surveillance. *EcoHealth, 12*(4), 693–702. https://doi.org/10.1007/s10393-015-1054-z

Leonardsson, K., Blomqvist, M., & Rosenberg, R. (2016). Reducing spatial variation in environmental assessment of marine benthic fauna. *Marine Pollution Bulletin, 104*(1–2), 129–138. https://doi.org/10.1016/j.marpolbul.2016.01.050

Liberatore, A., Bowkett, E., MacLeod, C. J., Spurr, E., & Longnecker, N. (2018). Social media as a platform for a citizen science community of practice. *Citizen Science: Theory and Practice, 3*(1), 3. https://doi.org/10.5334/cstp.108

Lindenmayer, D. B., & Likens, G. E. (2010). The science and application of ecological monitoring. *Biological Conservation, 143*(6), 1317–1328. https://doi.org/10.1016/j.biocon.2010.02.013

Loh, T. L., McMurray, S. E., Henkel, T. P., Vicente, J., & Pawlik, J. R. (2015). Indirect effects of overfishing on Caribbean reefs: Sponges overgrow reef-building corals. *PeerJ, 3*, e901. https://doi.org/10.7717/peerj.901

Lurz, P., Bertolino, S., Koprowski, J., Willis, P., Tonkin, M. E. L., & Gurnell, J. (2015). Squirrel monitoring: Snapshots of population presence and trends to inform management. *Ecology, Conservation and Management in Europe, 17,* 281–300.

McCarthy, M. S., Lester, J. D., Cibot, M., Vigilant, L., & McLennan, M. R. (2020). Atypically High Reproductive Skew in a Small Wild Chimpanzee Community in a Human-Dominated Landscape. Folia Primatologica, 91(6), 688–696. https://doi.org/10.1159/000508609

Mackenzie, D. I., Bailey, L. L., & Nichols, J. D. (2004). Investigating species co-occurrence patterns when species are detected imperfectly. *Journal of Animal Ecology, 73*(3), 546–555. https://doi.org/10.1111/j.0021-8790.2004.00828.x

MacKenzie, D. I., Nichols, J. D., Lachman, G. B., Droege, S., Andrew Royle, J. A., & Langtimm, C. A. (2002). Estimating site occupancy rates when detection probabilities are less than one. *Ecology, 83*(8), 2248–2255. https://doi.org/10.1890/0012-9658(2002)083[2248:ESORWD]2.0.CO;2

MacKenzie, D. I., Nichols, J. D., Royle, J. A., Pollock, K. H., Bailey, L. L., & Hines, J. E. (2018). *Occupancy estimation and modeling: Inferring patterns and dynamics of species occurrence, academic press.*

Mackenzie, D. I., Nichols, J. D., Seamans, M. E., & Gutiérrez, R. J. (2009). Modeling species occurrence dynamics with multiple states and imperfect detection. *Ecology, 90*(3), 823–835. https://doi.org/10.1890/08-0141.1

MacIntosh, A. (2019). A Message from Down Under: Calling for Marsupial Toxicology. Environ Toxicol Chem, 38: 2353–2354. https://doi.org/10.1002/etc.4575

MacPhail, V. J., & Colla, S. R. (2020). Power of the people: A review of citizen science programs for conservation. *Biological Conservation, 249,* 108739. https://doi.org/10.1016/j.biocon.2020.108739

Marshall, A. D., & Bennett, M. B. (2010). Reproductive ecology of the reef manta ray Manta alfredi in southern Mozambique. *Journal of Fish Biology, 77*(1), 169–190. https://doi.org/10.1111/j.1095-8649.2010.02669.x

Marshall, N. J., Kleine, D. A., & Dean, A. J. (2012). CoralWatch: Education, monitoring, and sustainability through citizen science. *Frontiers in Ecology and the Environment, 10*(6), 332–334. https://doi.org/10.1890/110266

Mazzolli, M., Haag, T. A. I. A. N. A., Lippert, B. G., Eizirik, E., Hammer, M. L., & Al Hikmani, K. (2017). Multiple methods increase detection of large and medium-sized mammals: working with volunteers in south-eastern Oman. Oryx, 51(2), 290–297.

McCallum, J. (2013). Changing use of camera traps in mammalian field research: Habitats, taxa and study types. Mammal Review, 43(3), 196–206. https://doi.org/10.1111/j.1365-2907.2012.00216.x

McKelvey, K. S., Aubry, K. B., & Schwartz, M. K. (2008). Using anecdotal occurrence data for rare or elusive species: The illusion of reality and a call for evidentiary standards. BioScience, 58(6), 549–555. https://doi.org/10.1641/B580611

McShea, W. J., Forrester, T., Costello, R., He, Z., & Kays, R. (2016). Volunteer-run cameras as distributed sensors for macrosystem mammal research. Landscape Ecology, 31(1), 55–66. https://doi.org/10.1007/s10980-015-0262-9

Meek, P. D., Vernes, K., & Falzon, G. (2013). On the reliability of expert identification of small-medium sized mammals from camera trap photos. Wildlife Biology in Practice, 9(2), 1–19. https://doi.org/10.2461/wbp.2013.9.4

Miller, D. A., Nichols, J. D., McClintock, B. T., Grant, E. H. C., Bailey, L. L., & Weir, L. A. (2011). Improving occupancy estimation when two types of observational error occur: Non-detection and species misidentification. Ecology, 92(7), 1422–1428. https://doi.org/10.1890/10-1396.1

Mishra, D. R., Kumar, A., Ramaswamy, L., Boddula, V. K., Das, M. C., Page, B. P., & Weber, S. J. (2020). CyanoTRACKER: A cloud-based integrated multi-platform architecture for global observation of cyanobacterial harmful algal blooms. Harmful Algae, 96, 101828. https://doi.org/10.1016/j.hal.2020.101828

Mitroi, V., Ahi, K. C., Bulot, P. Y., Tra, F., Deroubaix, J. F., Ahoutou, M. K., Quiblier, C., Koné, M., Coulibaly Kalpy, J. C., & Humbert, J. F. (2020). Can participatory approaches strengthen the monitoring of cyanobacterial blooms in developing countries? Results from a pilot study conducted in the Lagoon Aghien (Ivory Coast). PLOS ONE, 15(9), e0238832. https://doi.org/10.1371/journal.pone.0238832

Mohd Rameli, N. I. A., Lappan, S., Bartlett, T. Q., Ahmad, S. K., & Ruppert, N. (2020). Are social media reports useful for assessing small ape occurrence? A pilot study from Peninsular Malaysia.

American Journal of Primatology, 82(3), e23112. https://doi.org/10.1002/ajp.23112

Molinari-Jobin, A., Kéry, M., Marboutin, E., Molinari, P., Koren, I., Fuxjäger, C., Breitenmoser-Würsten, C., Wölfl, S., Fasel, M., Kos, I., Wölfl, M., & Breitenmoser, U. (2012). Monitoring in the presence of species misidentification: The case of the Eurasian lynx in the Alps. Animal Conservation, 15(3), 266–273. https://doi.org/10.1111/j.1469-1795.2011.00511.x

Moo, S. S. B., Froese, G. Z. L., & Gray, T. N. E. (2018). First structured camera-trap surveys in Karen State, Myanmar, reveal high diversity of globally threatened mammals. Oryx, 52(3), 537–543. https://doi.org/10.1017/S0030605316001113

Murray, M. H., Fidino, M., Lehrer, E. W., Simonis, J. L., & Magle, S. B. (2021). A multi-state occupancy model to non-invasively monitor visible signs of wildlife health with camera traps that accounts for image quality. Journal of Animal Ecology, 90(8), 1973–1984. https://doi.org/10.1111/1365-2656.13515

Newcomer, K., Tracy, B. M., Chang, A. L., & Ruiz, G. M. (2019). Evaluating performance of photographs for marine citizen science applications. Frontiers in Marine Science, 6, 336. https://doi.org/10.3389/fmars.2019.00336

Nordstrom, B., James, M. C., Martin, K., & Worm, B. (2019). Tracking jellyfish and leatherback sea turtle seasonality through citizen science observers. Marine Ecology Progress Series, 620, 15–32.

Norouzzadeh, M. S., Nguyen, A., Kosmala, M., Swanson, A., Palmer, M. S., Packer, C., & Clune, J. (2018). Automatically identifying, counting, and describing wild animals in camera-trap images with deep learning. Proceedings of the National Academy of Sciences of the United States of America, 115(25), E5716–E5725. https://doi.org/10.1073/pnas.1719367115

Nowak, K., Berger, J., Panikowski, A., Reid, D. G., Jacob, A. L., Newman, G., Young, N. E., Beckmann, J. P., & Richards, S. A. (2020). Using community photography to investigate phenology: A case study of coat molt in the mountain goat (Oreamnos americanus) with missing data. Ecology and Evolution, 10(23), 13488–13499. https://doi.org/10.1002/ece3.6954

Nugent, J. C. F. (2017). 'Where the wildlife are: See wildlife and do science with eMammal.'. Science Scope, 41(3), 10–13.

O'Connell, A., Nichols, J. D., & Karanth, K. U. (2011). *Camera traps in animal ecology: Methods and analyses.* Springer.

O'Connell, A. F., Talancy, N. W., Bailey, L. L., Sauer, J. R., Cook, R., & Gilbert, A. T. (2006). Estimating site occupancy and detection probability parameters for meso- and large mammals in a coastal eosystem. *Journal of Wildlife Management, 70*(6), 1625–1633. https://doi.org/10.2193/0022-541X (2006)70[1625:ESOADP]2.0.CO;2

O'Connor, D., Stacy-Dawes, J., Muneza, A., Fennessy, J., Gobush, K., Chase, M. J., Brown, M. B., Bracis, C., Elkan, P., Zaberirou, A. R. M., Rabeil, T., Rubenstein, D., Becker, M. S., Phillips, S., Stabach, J. A., Leimgruber, P., Glikman, J. A., Ruppert, K., Masiaine, S., & Mueller, T. (2019). Updated geographic range maps for giraffe, Giraffa spp., throughout sub-Saharan Africa, and implications of changing distributions for conservation. *Mammal Review, 49*(4), 285–299. https://doi.org/10.1111/mam.12165

Ohno, Y., Yokoyama, J., Nakashizuka, T., & Kawata, M. (2017). Utilization of photographs taken by citizens for estimating bumblebee distributions. Scientific reports, 7(1), 11215.

Page, M. J., McKenzie, J. E., Bossuyt, P. M., Boutron, I., Hoffmann, T. C., Mulrow, C. D., Shamseer, L., Tetzlaff, J. M., Akl, E. A., Brennan, S. E., Chou, R., Glanville, J., Grimshaw, J. M., Hróbjartsson, A., Lalu, M. M., Li, T., Loder, E. W., Mayo-Wilson, E., McDonald, S., . . . Moher, D. (2021). The PRISMA 2020 statement: An updated guideline for reporting systematic reviews. *BMJ, 372,* n71. https://doi.org/10.1136/bmj.n71

Palencia, P., Rowcliffe, J. M., Vicente, J., & Acevedo, P. (2021). Assessing the camera trap methodologies used to estimate density of unmarked populations. *Journal of Applied Ecology, 58*(8), 1583–1592. https://doi.org/10.1111/1365-2664.13913

Parham, J., Crall, J., Stewart, C., Berger-Wolf, T., & Rubenstein, D. (2017). Animal population censusing at scale with citizen science and photographic identification. AAAI Spring Symposium—Technical Report.

Parsons, A. W., Forrester, T., McShea, W. J., Baker-Whatton, M. C., Millspaugh, J. J., & Kays, R. (2017). Do occupancy or detection rates from camera traps reflect deer density? *Journal of Mammalogy, 98*(6), 1547–1557. https://doi.org/10.1093/jmammal/gyx128

Parsons, A. W., Goforth, C., Costello, R., & Kays, R. (2018). The value of citizen science for ecological

monitoring of mammals. *PeerJ, 6,* e4536. https://doi.org/10.7717/peerj.4536

Parsons, A. W., Rota, C. T., Forrester, T., Baker-Whatton, M. C., Mcshea, W. J., Schuttler, S. G., Millspaugh, J. J., & Kays, R. (2019). Urbanization focuses carnivore activity in remaining natural habitats, increasing species interactions. *Journal of Applied Ecology, 56*(8), 1894–1904. https://doi.org/10.1111/1365-2664.13385

Pataki, B. A., Garriga, J., Eritja, R., Palmer, J. R. B., Bartumeus, F., & Csabai, I. (2021). Deep learning identification for citizen science surveillance of tiger mosquitoes. *Scientific Reports, 11*(1), 4718. https://doi.org/10.1038/s41598-021-83657-4

Pfeffer, S. E., Spitzer, R., Allen, A. M., Hofmeester, T. R., Ericsson, G., Widemo, F., ... & Cromsigt, J. P. (2018). Pictures or pellets? Comparing camera trapping and dung counts as methods for estimating population densities of ungulates. Remote Sensing in Ecology and Conservation, 4(2), 173–183.

Podgórski, T Niedziałkowska, M., Tarnowska, E., Ligmanowska, J., Jędrzejewska, B., Radziszewska, A., ... & Woźniak, M. (2021). Clear phylogeographic pattern and genetic structure of wild boar Sus scrofa population in Central and Eastern Europe. Scientific Reports, 11(1), 9680.

Pollicelli, D., Coscarella, M., & Delrieux, C. (2020). RoI detection and segmentation algorithms for marine mammals photo-identification. *Ecological Informatics, 56*(1), 101038. https://doi.org/10.1016/j.ecoinf.2019.101038

Raddick, M. J., Bracey, G., Gay, P. L., Lintott, C. J., Murray, P., Schawinski, K., Szalay, A. S., & Vandenberg, J. (2009). Galaxy zoo: Exploring the motivations of citizen science volunteers. *Astronomy Education Review, 9*(1), 10. https://doi.org/10.3847/AER2009036

Rameli, N. I., Lappan, S., Bartlett, T. Q., Ahmad, S. K., & Ruppert, N. (2020). Are social media reports useful for assessing small ape occurrence? A pilot study from Peninsular Malaysia. American journal of primatology, 82(3), e23112.

Ramírez, A., De Jesús-Crespo, R., Martinó-Cardona, D. M., Martínez-Rivera, N., & Burgos-Caraballo, S. (2009). Urban streams in Puerto Rico: What can we learn from the tropics? *Journal of the North American Benthological Society, 28*(4), 1070–1079. https://doi.org/10.1899/08-165.1

Raoult, V., David, P. A., Dupont, S. F., Mathewson, C. P., O'Neill, S. J., Powell, N. N., & Williamson, J. E. (2016). GoPros(TM) as an underwater photogrammetry tool for citizen science. *PeerJ, 4,* e1960. https://doi.org/10.7717/peerj.1960

Rengstorf, A. M., Yesson, C., Brown, C., & Grehan, A. J. (2013). High-resolution habitat suitability modelling can improve conservation of vulnerable marine ecosystems in the deep sea. *Journal of Biogeography, 40*(9), 1702–1714. https://doi.org/10.1111/jbi.12123

Rohner, C. A., Richardson, A. J., Marshall, A. D., Weeks, S. J., & Pierce, S. J. (2011). How large is the world's largest fish? Measuring whale sharks Rhincodon typus with laser photogrammetry. *Journal of Fish Biology, 78*(1), 378–385. https://doi.org/10.1111/j.1095-8649.2010.02861.x

Rota, C. T., Wikle, C. K., Kays, R. W., Forrester, T. D., McShea, W. J., Parsons, A. W., & Millspaugh, J. J. (2016). A two-species occupancy model accommodating simultaneous spatial and interspecific dependence. *Ecology, 97*(1), 48–53. https://doi.org/10.1890/15-1193.1

Rovero, F., Zimmermann, F., Berzi, D., & Meek, P. (2013). 'Which camera trap type and how many do I need?' A review of camera features and study designs for a range of wildlife research applications. *Hystrix, the Italian Journal of Mammalogy, 24*(2), 148–156.

Rowcliffe, J. M., & Carbone, C. (2008). Surveys using camera traps: Are we looking to a brighter future? *Animal Conservation, 11*(3), 185–186. https://doi.org/10.1111/j.1469-1795.2008.00180.x

Rowley, J.J.L., Stuart, B. L., (2020). A new Leptobrachella (Anura: Megophryidae) from the Cardamom Mountains of Cambodia. Zootaxa 4834: 556–572 (https://doi.org/10.11646/zootaxa.4834.4.4)

Royle, J. A., & Link, W. A. (2006). Generalized site occupancy models allowing for false positive and false negative errors. *Ecology, 87*(4), 835–841. https://doi.org/10.1890/0012-9658(2006)87[835:gsomaf]2.0.co;2

Rudd, R. J., & Davis, A. D. (2016). Rabies Virus. Clinical Virology Manual, 473–491.

Santangeli, A., Chen, Y., Kluen, E., Chirumamilla, R., Tiainen, J., Loehr, J. Integrating drone-borne thermal imaging with artificial intelligence to locate bird nests on agricultural land. Sci Rep. 2020 Jul 14;10(1):10993. doi: 10.1038/s41598-020-67898-3. PMID: 32665596; PMCID: PMC7360548.

Sayer, S., Allen, R., Hawkes, L. A., Hockley, K., Jarvis, D., & Witt, M. J. (2019). Pinnipeds, people and photo identification: The implications of grey seal movements for effective management of the species. *Journal of the Marine Biological Association of the United Kingdom, 99*(5), 1221–1230. https://doi.org/10.1017/S0025315418001170

Schaus, J., Uzal, A., Gentle, L. K., Baker, P. J., Bearman-Brown, L., Bullion, S., Gazzard, A., Lockwood, H., North, A., Reader, T., Scott, D. M., Sutherland, C. S., & Yarnell, R. W. (2020). Application of the Random Encounter Model in citizen science projects to monitor animal densities. *Remote Sensing in Ecology and Conservation, 6*(4), 514–528. https://doi.org/10.1002/rse2.153

Schneider, S., Greenberg, S., Taylor, G. W., & Kremer, S. C. (2020). Three critical factors affecting automated image species recognition performance for camera traps. *Ecology and Evolution, 10*(7), 3503–3517. https://doi.org/10.1002/ece3.6147

Schüttler, E., Klenke, R., Galuppo, S., Castro, R. A., Bonacic, C., Laker, J., & Henle, K. (2017). Habitat use and sensitivity to fragmentation in Americas smallest wildcat. Mammalian Biology, 86, 1–8.

Schuttler, S. G., Sears, R. S., Orendain, I., Khot, R., Rubenstein, D., Rubenstein, N., Dunn, R. R., Baird, E., Kandros, K., O'Brien, T., & Kays, R. (2019). Citizen science in schools: Students collect valuable mammal data for science, conservation, and community engagement. *BioScience, 69*(1), 69–79. https://doi.org/10.1093/biosci/biy141

Scott, D. M., Baker, R., Tomlinson, A., Berg, M. J., Charman, N., & Tolhurst, B. A. (2020). Spatial distribution of sarcoptic mange (Sarcoptes scabiei) in urban foxes (Vulpes vulpes) in Great Britain as determined by citizen science. *Urban Ecosystems, 23*(5), 1127–1140. https://doi.org/10.1007/s11252-020-00985-5

Statham, M. J., Sacks, B. N., Aubry, K. B., Perrine, J. D., & Wisely, S. M. (2012). The origin of recently established red fox populations in the United States: Translocations or natural range expansions? *Journal of Mammalogy, 93*(1), 52–65. https://doi.org/10.1644/11-MAMM-A-033.1

Stephenson, N., Swift, P., Villepique, J. T., Clifford, D. L., Nyaoke, A., De la Mora, A., Moore, J., & Foley, J. (2013). Pathologic findings in Western gray squirrels (Sciurus griseus) from a notoedric mange epidemic in the San Bernardino Mountains, California. *International Journal for Parasitology. Parasites and Wildlife, 2,* 266–270. https://doi.org/10.1016/j.ijppaw.2013.09.004

Suzuki-Ohno, Y., Yokoyama, J., Nakashizuka, T., & Kawata, M. (2017). Utilization of photographs taken by citizens for estimating bumblebee distributions. *Scientific Reports, 7*(1), 11215. https://doi.org/10.1038/s41598-017-10581-x

Swanson, A., Kosmala, M., Lintott, C., Simpson, R., Smith, A., & Packer, C. (2015). Snapshot Serengeti, high-frequency annotated camera trap images of 40 mammalian species in an African savanna. *Scientific Data, 2*(1), 150026. https://doi.org/10.1038/sdata.2015.26

Tabak, M. A., Norouzzadeh, M. S., Wolfson, D. W., Sweeney, S. J., Vercauteren, K. C., Snow, N. P., Halseth, J. M., Di Salvo, P. A., Lewis, J. S., White, M. D., Teton, B., Beasley, J. C., Schlichting, P. E., Boughton, R. K., Wight, B., Newkirk, E. S., Ivan, J. S., Odell, E. A., Brook, R. K., . . . Miller, R. S. (2019). Machine learning to classify animal species in camera trap images: Applications in ecology. *Methods in Ecology and Evolution, 10*(4), 585–590. https://doi.org/10.1111/2041-210X.13120

Taylor, S. D., Meiners, J. M., Riemer, K., Orr, M. C., & White, E. P. (2019). Comparison of large-scale citizen science data and long-term study data for phenology modeling. *Ecology, 100*(2), e02568. https://doi.org/10.1002/ecy.2568

Tiralongo, F., Messina, G., Lombardo, B. M., Longhitano, L., Li Volti, G., & Tibullo, D. (2020). Skin mucus of marine fish as a source for the development of antimicrobial agents. Frontiers in Marine Science, 760.

Townsend, P. A., Clare, J. D. J., Liu, N., Stenglein, J. L., Anhalt-Depies, C., Van Deelen, T. R., Gilbert, N. A., Singh, A., Martin, K. J., & Zuckerberg, B. (2021). Snapshot Wisconsin: Networking community scientists and remote sensing to improve ecological monitoring and management. *Ecological Applications, 31*(8), e02436. https://doi.org/10.1002/eap.2436

Tulloch, A. I. T., Possingham, H. P., Josph, L. N., Szabo, J., Martin, T. G. (2013). Realising the full potential of citizen science monitoring programs. *Biological Conservation, 165*, 128–138. https://doi.org/10.1016/j.biocon.2013.05.025

Turner, A. D., Stubbs, B., Coates, L., Dhanji-Rapkova, M., Hatfield, R. G., Lewis, A. M., ... & Algoet, M. (2014). Variability of paralytic shellfish toxin occurrence and profiles in bivalve molluscs from Great Britain from official control monitoring as determined by pre-column oxidation liquid chromatography and implications for applying immunochemical tests. Harmful Algae, 31, 87–99.

van Hooidonk, R., Maynard, J. A., Manzello, D., & Planes, S. (2014). Opposite latitudinal gradients in projected ocean acidification and bleaching impacts on coral reefs. *Global Change Biology, 20*(1), 103–112. https://doi.org/10.1111/gcb.12394

van Tamelen, P. G. (2004). A comparison of obtaining field data using electronic and written methods. Fisheries Research, 69(1), 123–130. https://doi.org/10.1016/j.fishres.2004.01.006

Wangyal, J. T., Phuntsho, S., Tshewang, S., Koirala, B., Wangdi, D., Gyeltshen, G., Das, I., Tenzin, J., & Ghalley, B. (2020). New herpetofaunal records from the Kingdom of Bhutan obtained through citizen science. *Herpetological Review, 51*, 790–798.

Westoby, M. J., Brasington, J., Glasser, N. F., Hambrey, M. J., & Reynolds, J. M. (2012). 'Structure-from-Motion' photogrammetry: A low-cost, effective tool for geoscience applications. *Geomorphology, 179*, 300–314. https://doi.org/10.1016/j.geomorph.2012.08.021

Wevers, J., Beenaerts, N., Casaer, J., Zimmermann, F., Artois, T., & Fattebert, J. (2021). Modelling species distribution from camera trap by-catch using a scale-optimized occupancy approach. *Remote Sensing in Ecology and Conservation, 7*(3), 534–549. https://doi.org/10.1002/rse2.207

Wilkinson, P. F. (1974). Wool shedding in musk oxen. *Biological Journal of the Linnean Society, 6*(2), 127–141. https://doi.org/10.1111/j.1095-8312.1974.tb00718.x

Willi, M., Pitman, R. T., Cardoso, A. W., Locke, C., Swanson, A., Boyer, A., Veldthuis, M., & Fortson, L. (2019). Identifying animal species in camera trap images using deep learning and citizen science. *Methods in Ecology and Evolution, 10*(1), 80–91. https://doi.org/10.1111/2041-210X.13099

Wilson, J. S., Pan, A. D., General, D. E. M., & Koch, J. B. (2020). More eyes on the prize: An observation of a very rare, threatened species of Philippine Bumblebee, Bombus irisanensis, on iNaturalist and the importance of citizen science in conservation biology. *Journal of Insect Conservation, 24*(4), 727–729. https://doi.org/10.1007/s10841-020-00233-3

Wood, S. H., Hindle, M. M., Mizoro, Y., Cheng, Y., Saer, B. R. C., Miedzinska, K., ... & Loudon, A. S. (2020). Circadian clock mechanism driving mammalian photoperiodism. Nature Communications, 11(1), 4291.

Yang, D. Q., Ren, G. P., Tan, K., Huang, Z. P., Li, D. P., Li, X. W., Wang, J. M., Chen, B. H., & Xiao, W. (2021). An adaptive automatic approach to filtering empty images from camera traps using a deep learning model. *Wildlife Society Bulletin, 45*(2), 230–236. https://doi.org/10.1002/wsb.1176

Yu, X., Wang, J., Kays, R., Jansen, P. A., Wang, T., & Huang, T. (2013). Automated identification of animal species in camera trap images. *EURASIP Journal on Image and Video Processing, 2013*(1), 52. https://doi.org/10.1186/1687-5281-2013-52

Zbyryt, A., Mikula, P., Ciach, M., Morelli, F., & Tryjanowski, P. (2021). A large-scale survey of bird plumage colour aberrations reveals a collection bias in Internet-mined photographs. *Ibis, 163*(2), 566–578. https://doi.org/10.1111/ibi.12872

Zhang, G. (2019). Integrating citizen science and GIS for wildlife habitat assessment. *Wildlife Population Monitoring*, 1–19.

Appendix 23.1 Sample papers following screening.

Reference	Year	Taxa		Country	Main scope	Secondary scopes	Public participation in scientific research — Contributory Projects, Collaborative Projects, Co-created Projects	Data type	Statistical approach	Platform
		terrestrial	marine (FW/MW)							
Anderson et al.	2016	mammal		Kenya, Tanzania	ecological studies	habitat suitability	collaborative projects	photogram (CT)	Bayesian modeling	Zooniverse
Anton et al.	2018	mammal		New Zealand	monitoring	monitoring	collaborative projects	photogram (CT)	GLM	Identifyanimals.co.nz
Arts et al.	2015	all		UK	conservation	conservation	multiple	multiple		multiple
Azzurro et al.	2013		fish (MW)	Italy, Spain	research studies	invasive species	contributory project	photogram		Seawatchers
Bagnolini et al.	2017	all		France	conservation	conservation	contributory project	multiple		BiodiverCity
Bagwyn et al.	2020	bird		Bahamas	conservation	conservation	contributory project	photogram	presence/absence	eBrd
Bahr et al.	2020		corals (MW)	Hawaii	monitoring	education	contributory project	photogram	regression functions	project specific platform
Balantic et al.	2020	all		USA	monitoring	monitoring	contributory project	photogram	occurence based monitoring	project specific platform
Bicknell et al.	2016		mammals, invertebrates, fish (MW)	UK	monitoring	monitoring	contributory project	photogram(CT)		project specific platform
Bonter et al.	2009	bird		USA	research studies	invasive species	collaborative projects	checklists		Project FeederWatch
Campanaro et al.	2017	invertebrate		Italy, Spain	conservation	conservation	contributory project	photogram	presence/absence	lifemipp.eu

Appendix 23.1 (continued)

Reference	Year	Taxa terrestrial	Taxa marine (FW/MW)	Country	Main scope	Secondary scopes	Public participation in scientific research: Contributory Projects, Collaborative Projects, Co-created Projects	Data type	Statistical approach	Platform
Caravaggi et al.	2018	mammal		Northern Ireland	conservation	invasive species	contributory project	photogram(CT)	regression cross-correlation function/ frequence	project specific platform
Cardoso et al.	2020	mammal		Gabon	ecological studies	habitat suitability	contributory project	photogram(CT)	random encounter model, chi-square tests	Zooniverse
Ciceoi et al.	2017	invertebrate		Romania	research studies	invasive species	contributory project	checklists		GBIF, iNaturalist, IAS, social media
Cunha et al.	2017		phytoplancton (FW)	Brazil	monitoring	monitoring	co-created projects	checklists	ROC analyses	FreshWater Watch
Cusack et al.	2015	mammal		Tanzania	monitoring	monitoring	collaborative projects	photogram(CT)	REM	Snapshot Serengeti
Danielsen et al.	2007	all		Philippine	conservation	conservation	co-created projects	photogram		project specific platform
Donnelly et al.	2013	all		Ireland	ecological studies	climate change	collaborative projects	multiple		multiple
Egna et al.	2020	mammal		Kenya, Tanzania, Perú	monitoring	monitoring	contributory project	photogram(CT)		Snapshot Serengeti, Wildwatch Kenya, and AmazonCam Tambopata
Epps et al.	2014	invertebrates		USA	research studies	invasive species	contributory project	checklists		yourwildlife.org
Forrester et al.	2017	mammal		USA	conservation	conservation	collaborative projects	photogram(CT)	ANOVA	eMammal

Author	Year	Taxa	Country			Project type	Data type	Statistics	Platform
Gibson et al.	2019	fish (MW)	USA	conservation	conservation	contributory project	photogram	confusion matrices	texassharkrodeo.com
Gooliaff et al.	2018	mammal	Canada	monitoring	monitoring	contributory project	photogram(CT)		fluidsurveys.com
Green et al.	2020	all	UK	conservation	conservation	collaborative projects	photogram(CT)		multiple
Grol et al.	2021	invertebrates, corals, fish (MW)	Australia	conservation	conservation	contributory project	checklists	PERMANOVA, nMDS	CoralWatch
Hallett et al.	2018	plants	USA	monitoring	monitoring	collaborative projects	checklists		project specific platform
Hepler et al.	2018	mammal	Kenya, Tanzania	monitoring	monitoring	contributory project	photogram(CT)	occupancy models	Snapshot Serengeti
Hermoso et al.	2020	mammals, invertebrates,fish (MW)	Chile	conservation	conservation	collaborative projects	multiple		project specific platform
Hsing et al.	2018	mammal	UK	monitoring	monitoring	contributory project	photogram(CT)		MammalWeb
Jachowski et al.	2015	bird	Usa, Canada	monitoring	monitoring	collaborative projects	photogram(CT)	mixed model logistic regression	AEMP
Johnson et al.	2020	all	Japan	research studies	invasive species	multiple	multiple		multiple
Jones et al.	2018	bird	Falkland Islands, South Georgia, South Sandwich Islands, Antartica	monitoring	monitoring	collaborative projects	photogram(CT)		Zooniverse Penguin Watch project
Jong et al.	2019	bird	The Netherlands	research studies	phenology studies	collaborative projects	checklists	LMMs	NESTKAST
Karlin et al.	2015	mammal	USA	education	education	contributory project	photogram(CT)	abundance based species richness estimators	project specific platform
Kays et al.	2020	mammal	USA	conservation	conservation	collaborative projects	photogram(CT)	ecological modelling	eMammal

Appendix 23.1 (continued)

Reference	Year	Taxa terrestrial	Taxa marine (FW/MW)	Country	Main scope	Secondary scopes	Public participation in scientific research (Contributory Projects, Collaborative Projects, Co-created Projects)	Data type	Statistical approach	Platform
Kays et al.	2017	mammal		USA	conservation	conservation	contributory project	photogram(CT)	occupancy models	eMammal
Kays et al.	2015	mammal		USA	research studies	invasive species	contributory project	photogram(CT)	occupancy models	eMammal
Kellner et al.	2020	mammal		USA	ecological studies	land use changes	contributory project	checklists	generalized linear mixed models	project specific platform
Kelly et al.	2019		mammals, invertebrates, fish (MW)	Germany	conservation	conservation	contributory project	checklists		project specific platform
Lavariega et al.	2020	mammal		Mexico	conservation	conservation	collaborative projects	photogram(CT)	Bayesian methods, Spatially explicit capture–recapture models	project specific platform
Liberatore et al.	2019	bird		New Zealand	monitoring	monitoring	collaborative projects	photogram		social media/facebook groups
Lurz et al.	2015	mammal		Uk	monitoring	monitoring	contributory project	multiple		project specific platform
Macintosh et al.	2019		mammal (MW)	Australia	monitoring	monitoring	contributory project	photogram		project specific platform
Marshall et al.	2012	invertebrate		Australia	monitoring	education	contributory project	photogram		CoralWatch
Mazzolli et al.	2017	mammal		Oman	monitoring	monitoring	collaborative projects	photogram(CT)	presence/absence	project specific platform

Author	Year	Taxon	Taxon (detail)	Country			Project type	Data collection	Analysis	Platform
McCarthy et al.	2020	mammal		Ivory Coast	conservation	conservation	contributory project	photogram(CT)	multiple regression quadratic assignment procedure (MROAP)	Chimp&See Zooniverse
McShea et al.	2016	mammal		USA	monitoring	monitoring	multiple	photogram(CT)		eMammal
Mitrol et al.	2020		phytoplancton (FW)	Ivory Coast	monitoring	monitoring	co-created projects	checklists		WaSAF Project
Newcomer et al.	2019		mammals, invertebrates, fish (MW)	USA	conservation	invasive species	contributory project	photogram	diversity indices	project specific platform
Nordstrom et al.	2019	invertebrate, reptile		Canada	conservation	conservation	contributory project	checklists	GLM	project specific platform
Nowak et al.	2020	mammal		Canada	research studies	phenology studies	contributory project	photogram	logistic equation?	iNaturalist, CitSci
Nugent	2017	mammal		USA	monitoring	monitoring	multiple	photogram(CT)		eMammal
Ohno et al.	2017	invertebrate		Japan	conservation	conservation	contributory project	photogram	species distribution	project specific platform
Parsons et al.	2017	mammal		USA	monitoring	monitoring	collaborative projects	photogram(CT)	occupancy models	eMammal
Parsons et al.	2018	mammal		USA	monitoring	monitoring	contributory project	photogram(CT)	occupancy models	eMammal
Parsons et al.	2018	mammal		USA	monitoring	monitoring	contributory project	photogram(CT)	occupancy models	eMammal
Pataki et al.	2011	invertebrates		global	surveillance	surveillance	contributory project	photogram	ROC AUC score	Mosquito Alert
Pfeffer et al.	2018	mammal		Sweden	monitoring	monitoring	contributory project	photogram(CT)	randoma encounter model	project specific platform
Podgorski et al.	2021	mammal		global	monitoring	monitoring	contributory project	photogram(CT)		ENETWILD
Pollicelli et al.	2020		mammals (MW)	Argentina	conservation	conservation	contributory project	photogram		project specific platform
Rameli et al.	2020	mammal		Malaysia	conservation	conservation	contributory project	photogram	occupancy models	social media
Raoult et al.	2016		marine mammal, invertebrates, fish (MW)	Australia	conservation	conservation	contributory project	photogram		CoralWatch & other

Appendix 23.1 (continued)

Reference	Year	Taxa terrestrial	Taxa marine (FW/MW)	Country	Main scope	Secondary scopes	Public participation in scientific research (Contributory Projects, Collaborative Projects, Co-created Projects)	Data type	Statistical approach	Platform
Rota et al.	2016	mammal		USA	monitoring	monitoring	contributory project	photogram(CT)	two-species occupancy models	eMammal
Rowley et al.	2020	amphibian		Australia	conservation	conservation	contributory project	acoustic data		FrogID
Rudd et al.	2016	mammal		USA	surveillance	surveillance	contributory project	photogram	logistic regression	project specific platform
Santangeli et al.	2020	bird		Namibia	monitoring	monitoring	contributory project	photogram(CT)	multi event CMR model	project specific platform
Sayer et al.	2019		mammal (MW)	UK	monitoring	monitoring	contributory project	photogram		project specific platform
Schaus et al.	2020	mammal		UK	monitoring	monitoring	contributory project	photogram(CT)	spatial capture-recapture, REM	project specific platform
Schuttler et al.	2017	mammal		USA	education	education	contributory project	photogram(CT)		eMammal
Schuttler et al.	2019	mammal		USA, Kenya, India, Mexico	education	monitoring	contributory project	photogram(CT)	frequency	eMammal
Tiralongo et al.	2020		fish (MW)	Italy, Spain	conservation	invasive species	contributory project	photogram		social media
Turner	2014	all		USA	monitoring	monitoring	collaborative projects	photogram(CT)		multiple
Wangyal et al.	2020	amphibian, reptile		Australia	conservation	conservation	contributory project	photogram		social media

Study	Year	Taxon	Location	Aim	Aim	Project type	Method	Platform
Willi et al.	2019	mammal	Serengeti, USA	conservation	conservation	collaborative projects	photogram(CT)	Zooniverse
Wilson et al.	2020	invertebrate	Philippine	conservation	conservation	contributory project	photogram	iNaturalist
Wood et al.	2021	mammal	USA	conservation	conservation	contributory project	photogram(CT)	Zooniverse
Zbyryt et al.	2020	bird	Poland	conservation	conservation	contributory project	photogram	Internet
							multivariate and univariate phylogenetic generalized least squares PGLS	
Zhang	2019	mammal	China	ecological studies	habitat suitability	co-created projects	checklists	3dMapper
							density etimation-based method	
Adriaens et al.	2015	plants	UK, France, Germany, Belgium, The Netherlands	research studies	invasive species	contributory project	photogram	Rinse/Korina
Connors et al.	2012	plants	USA	monitoring	monitoring	contributory project	GIS	OakMapper
Gallo et al.	2011	plants	USA	research studies	invasive species	co-created projects	multiple	The Invaders of Texas
Joly et al.	2016	plants	global	monitoring	monitoring	contributory project	photogram	PlantNet

Index